COMPARATIVE VERTEBRATE MORPHOLOGY

COMPARATIVE VERTEBRATE MORPHOLOGY

Douglas Webster and Molly Webster

New York University

ACADEMIC PRESS
New York and London

A Subsidiary of
Harcourt Brace Jovanovich, Publishers

ORIGINAL ART BY HOWARD S. FRIEDMAN

COPYRIGHT © 1974, BY ACADEMIC PRESS, INC.
ALL RIGHTS RESERVED.
NO PART OF THIS PUBLICATION MAY BE REPRODUCED OR TRANSMITTED IN ANY FORM OR BY ANY MEANS, ELECTRONIC OR MECHANICAL, INCLUDING PHOTOCOPY, RECORDING, OR ANY INFORMATION STORAGE AND RETRIEVAL SYSTEM, WITHOUT PERMISSION IN WRITING FROM THE PUBLISHER.

ACADEMIC PRESS, INC.
111 Fifth Avenue, New York, New York 10003

United Kingdom Edition published by
ACADEMIC PRESS, INC. (LONDON) LTD.
24/28 Oval Road, London NW1

CAMROSE LUTHERAN COLLEGE
LIBRARY
QL
805
W38
14,931

Library of Congress Cataloging in Publication Data

Webster, Douglas, DATE
Comparative vertebrate morphology.

Includes bibliographies.
1. Vertebrates–Anatomy. 2. Morphology (Animals)
3. Anatomy, Comparative. I. Webster, Molly, joint author. II. Title.
QL805.W38 596'.04 72-12185
ISBN 0–12–740850–9

PRINTED IN THE UNITED STATES OF AMERICA

CONTENTS

PREFACE

Understanding vertebrate morphology requires an appreciation of both the diversity and the uniformity of structure found among living vertebrates. Such an appreciation in turn requires knowing something of the evolutionary adaptations to environmental situations that have occurred within closely related groups.

By definition, whether a structure is adaptive or maladaptive depends on the uses of that structure in the animal's environment. This is another way of saying that evolution is related to structure and function. The structure-function concept at the level of organs and organ systems is thus fundamental to an understanding of comparative evolutionary morphology. It is upon these three interrelated aspects—structure, function, and evolution—that we have organized and presented the diversity of vertebrate organization of each organ system. We have put more than the usual amount of emphasis on the nervous system because of the recent dramatic advances in this area and because students have a great inherent interest in this most complicated organ system.

This book is designed to meet the needs of a one-semester course for students who have already had an introductory course in biology. It is assumed that the lectures will be supplemented by a laboratory with its own laboratory manual. The organization of the text allows the instructor to coordinate the laboratory and lecture portions of the course.

The illustrations are almost all original. All artwork was executed, under our direction, by Mr. Howard S. Friedman, whose skill we greatly appreciate. All the microscopic photography was done by DW and the gross photography by MW. The American Museum of Natural History generously lent or gave us many specimens for dissection, photography, or both. The human skull bones in Figure 4-17 were lent by Dr. F. Baker-Cohen of the Anatomy Department, Albert Einstein Medical School. Several persons have kindly reviewed parts or all of the manuscript; we particularly thank Drs. C.B.G. Campbell, P.F.A. Maderson, J. Moulton, and H. Rosenberg. Their comments have been most helpful but only we are responsible for errors which, despite vigilance, may be found herein. We also appreciate the efforts of the staff of Academic Press, who did their best to help us through some of the vicissitudes involved in producing a book.

This book was written at University College of New York University, during the last two years of the College's life. We appreciate the assistance given by the college administration to this project. Now University College is effectively dead, the University Heights Campus has been sold, and—like most of the students who will use this book—we have moved on to medical school (Department of Otolaryngology, Louisiana State University, Medical Center, New Orleans, Louisiana). We hope that this book will help the undergraduate student—whether pre-med or not—to share the intellectual excitement of comparative morphology by emphasizing some of the *concepts* of a subject that is often known for its facts alone. Finally, the students who for 11 years labored at comparative morphology at the Heights deserve a word of appreciation for their part in the development of this book and its approach.

Douglas B. Webster
Molly Webster

1 HOW WE GOT HERE: INTELLECTUAL BACKGROUND

1

The history of Western civilization has been greatly influenced by attitudes toward the nature of human beings, how they came to be, and how they relate to the rest of the world. Plato pictured man as the most perfect form on earth, the Ideal. All else, as he described in his "Timaeus," was a degraded form of this Ideal and could be arranged in a descending order of less and less perfect forms: woman, other mammals, nonmammals, and so on down to rocks, minerals, and soil.

Plato's student, Aristotle, was almost modern in the way he compared humans and other organisms, for he based these comparisons not on philosophic contemplation but on actual observations of internal and external structure and behavior. Thus Aristotle distinguished the Sanguinae (= vertebrates), or animals with red blood, from the Exsanguinae (= invertebrates), or animals without red blood. Within the Sanguinae he distinguished various subgroupings, separating man, the viviparous biped, from all other mammals, the viviparous quadrupeds. He placed whales and porpoises in a separate group, not realizing that they are mammals but noting that they are viviparous and thus not like fishes.

Animate beings, as he described in his "Natural History," had one or more souls. All possessed a "vegetal soul," permitting nutrition, growth, and reproduction. Only animals had in addition an "animal soul" necessary for sensation and movement. A "rational soul" was possessed only by man, and allowed intellectual activity.

Aristotle believed, much as had Plato, that all things could be arranged into one logical, ladderlike series of gradations of form and function. This series included both the inanimate and the animate, and ranged from the simplest to the most complex; at the top of this series, or ladder, was man. This idea was to become prominent many centuries later.

Meanwhile, the Judaeo-Christian tradition held that man was made in God's image on the last day of Creation, was a little lower than the angels, and was meant to have dominion over the earth and all its inhabitants. When this tradition became blended with Aristotle's idea of the ladderlike series, increased complexity of form was taken to mean increased perfection; the further up the ladder any being stood, the "better" it was. Man was regarded as the achieved goal of creation, the purpose for which other earthly beings existed, and this teleology in one guise or another has remained a locked-in value judgment of the Western mind ever since.

But by the third century AD and throughout the Dark Ages the emphasis in the natural sciences had shifted away from the things that had intrigued Aristotle; anatomy and physiology became parts of medicine and were concerned with man and his ills rather than with diversity of form. The physician, Galen, wrote 62 treatises on the structure and function of the human body, proving, for example, that nerves control muscles. His writings, approved by the Christian Church, were accepted as true and were not questioned during the following twelve stagnant centuries. The errors that coexisted with his many acute observations, therefore, were doubly regrettable.

For instance, Galen outlined an anatomical framework for Aristotle's concept of the three souls. According to Galen, the "vegetal soul" entered the blood from the liver. From there the blood flowed to the heart, where it acquired the "animal soul"—marked by the change in blood color at this point. The blood then passed through invisible pores from the right to the left ventricle (an anatomical impossibility); it flowed up to the head where, in man only, the "human soul" was added to it. Here "vital spirits" were distilled from the blood and then passed to all parts of the body. This is one of the earliest, most dramatic examples of what can happen when one ascribes a function to a structure without experimentation!

Galen's writings were based not only on observations (mostly of cattle, swine, and Barbary apes) but also on theology, for, although he was not a Christian, his beliefs were influenced by Christianity. Thus he could declare that man has one less rib on one side than on the other, because it was from one of Adam's ribs that God created Eve. If much of Galen's work had been regarded as hypothesis rather than dogma it would have stimulated active investigation, and the errors would not have been so critical. As it was, however, his work greatly impeded understanding in many areas of biology. That of course was not Galen's fault as much as it was a result of the intellectual climate.

The freeing of man's mind to observe and experiment, rather than merely to accept the word of authority, came later in biology than it did in other intellectual fields. In the sixteenth century, at the same time that actual comparative studies were beginning (Fig. 1-1), biology received a needed jolt from Andreas Vesalius. He was a medical student at the University of Padua who so excelled in his ability to demonstrate and dissect parts of the human body that on the day he passed his examinations he was made Professor of Surgery (which at that time included all anatomical studies). Vesalius somewhat timidly but accurately voiced his objections to Galen's work when he found it wrong; for example, he pointed out that human beings have in fact an equal number of ribs on right and left sides and that there is no observable anatomical way for blood to pass from one side of the heart to the other.

He and his followers tried to look objectively at vertebrate structure. Their understanding of function, however, was clouded by mysterious vitalism. They thought the blood flowed out from the heart, irrigating the rest of the body with the "animal forces" necessary for life. They realized that Galen was correct in stating that nerves cause muscles to contract, but their understanding of how this happened was no clearer than Galen's—they spoke of some "vital spirit." In short, they "explained" what they could not understand in mysterious, almost superstitious terminology, and called it physiology.

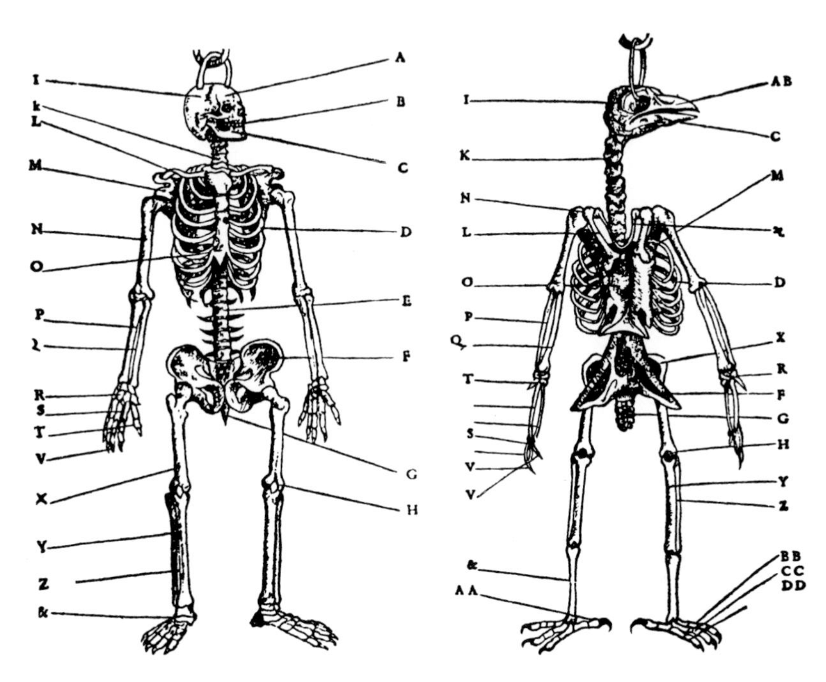

Fig. 1-1. This sixteenth century illustration by Pierre Belon is the first known to identify the same bones in widely different animals.

With the dawn of the seventeenth century this attitude was to change. The vertebrate body began to be viewed as a machine that was governed by physical laws and was understandable in physical terms rather than as some never-to-be-understood lodging for souls. The first dramatic advance in this mechanistic exploration of physiological phenomena came in 1628.

William Harvey had studied for five years at the University of Padua; he returned to England in 1602, and then, experimenting on some 40 species (including himself), determined the basic circulation of the blood. He observed the beating heart of a freshly killed animal and found that when grasped it could be felt to harden while contracting; therefore he concluded that the heart is a hollow muscle. He also noticed that when the heart contracted, the arteries near it expanded. He noted that if an artery was cut, blood would spurt from the end nearer the heart; if a vein was cut blood would not spurt, but would merely flow from the cut edge that was further from the heart. Within the veins he noticed valves, allowing blood to flow only toward the heart; within the heart he also saw valves, ensuring one-way flow.

Through this series of careful morphological observations and physiological experiments, followed by rational argument, Harvey discovered the structure of the vertebrate circulatory system and its function: to move the blood continuously away from the heart in arteries and back toward the heart in veins. In 1645 Malpighi discovered and described the capillaries, adding the final confirmation to the conclusions reported by Harvey in 1628 by explaining how the blood passes from the arteries to the veins. But Harvey's greatest contribution may have been not the discovery itself, but the mechanistic, functional way in which he approached the problem.

The French philosopher, Descartes, gave intellectual respectability to this approach by describing animals as machines. During the same period came the development and early improvements of the light microscope. With both a favorable intellectual climate and the means thus present, seventeenth century biologists studied diverse forms of animals in detail, hypothesizing about the functions of structures they described, and often experimenting on them.

During the eighteenth century biology split into various specialties whose proponents enjoyed only limited interchange. Animal classification was one "splinter science," fathered by the Swedish scientist, Linnaeus. He established the binomial system of classification which is still used today. In this system, the formal name for each specific type of animal or plant consists of two Latin words, the first being the genus and the second the species. According to the rules of nomenclature there can be no two animal or plant genera of the same name (although an animal genus may have the same name as a plant genus); within a genus there can be no two species of the same name. (These names, i.e., genus and species, are always set in italic type or underlined; they are the only biological names so treated. Generic names are capitalized, species names are not.)

Larger taxonomic ranks were also developed by Linnaeus, approximately although not exactly as they are today. Today the largest taxonomic group is the kingdom—Animal, Plant, or Protist. Each kingdom is divided into several phyla (singular, phylum), each of which may contain several classes; classes are subdivided into orders which are further subdivided into families, containing genera; genera in turn contain species; between these ranks there are often intermediate ranks, such as suborders and superfamilies, and thus the groups increase in number to the infinite delight and despair of taxonomists.

This system of classification not only provided an organization for animals and plants, but it also affected the way in which scientists thought about them. It required that the individual from which a species was first described be treated in a rigorous, formal manner and regarded as representing the entire species. This individual was forevermore to be preserved in a museum and known as the "type specimen" for that species. In this approach to diversity, called typology, a species was regarded as a real and fixed entity, all of whose members were essentially the same; individual variation was considered abnormal. Although remnants of this idea still persist, modern biologists understand that variation among the members of a species is not only normal but an important attribute of the species contributing to its viability.

While Linnaeus and his co-workers were organizing a usable scheme of plant and animal diversity based only on external structures, anatomists of the eighteenth century were developing a comparative anatomy based on internal structure.

The preeminent English anatomist of the time was John Hunter. He studied the gross anatomy of living and fossil forms and developed an educational museum of over 500 species, systematically arranged to

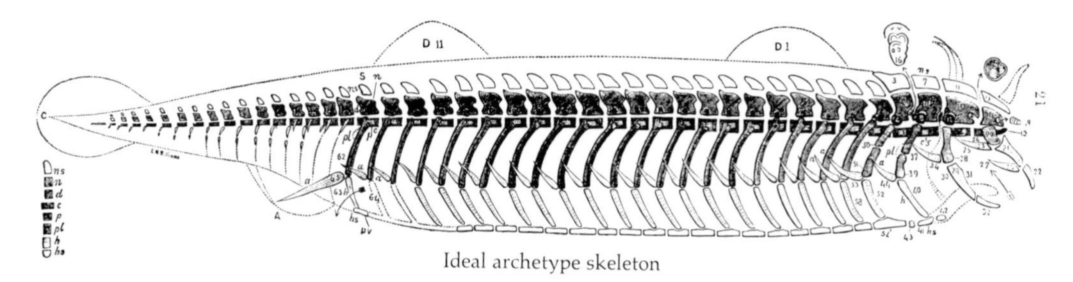

Fig. 1-2. Richard Owen presented this drawing as representing the idealized archetype of all vertebrate skeletons. Note its repeating segmental structure from head to tail.

illustrate the morphological bases of plant and animal actions. His anatomy was accurate and his physiological experiments clever, and his museum eventually formed the nucleus of the British Museum of Natural History. Hunter's interests, however, were not widely shared by eighteenth century anatomists.

The dominant movement in anatomy was in the European School, which resurrected the Aristotelian idea of a scale and called it (in Latin) *scala naturae* or (French) *echelle d'être.* Elaborate lists of animals were compiled, arranged from simple to complex, or imperfect to perfect, or low to high. Man was always at the apex of these lists. Continuity in the Creator's plan and purpose in the development of diverse life forms were thus demonstrated: the purpose toward which all else was directed was Man.

There was also great interest in another idea from antiquity—the Platonic Ideal. The diverse vertebrate forms were thought to be variants of some ideal archetype. The "natural philosophers" of Germany, such as Lorenz Oken and the poet–scientist Johann Wolfgang von Goethe, in the late eighteenth and early nineteenth centuries put much effort into trying to discover what the original, basic form—the archetype—had been.

Most prominent among their theories was that of a segmental archetype (Fig. 1-2). Each segment of the skeletal system was believed to be based upon one vertebra, which often had a wide lateral expansion or appendage. Ribs, jaws, and paired limbs were all considered to be modified from such lateral appendages. The skull was believed to consist of a group of vertebrae which had been highly modified and fused together.

It was considered unimportant whether this archetype had ever actually lived, or had merely existed as an idea or "blueprint" in the mind of the Creator; what was important was that the fundamental plan had been modified in various ways to produce modern animals. Although this theory was finally discredited (by T. H. Huxley in 1858), it and the intellectual climate out of which it grew played an important role in the development of a *comparative* anatomical science. The archetype could best be understood by finding those aspects of a structure that were common to many animals. (In later chapters we will do something very similar, but for somewhat different reasons.)

At the same time there lived the French zoologist, Georges Cuvier, who was much more concerned with the real than with the ideal. His method was almost diametrically opposed to that of the natural philosophers: he observed first and theorized afterward, making the theory fit the observable facts rather than the other way around.

Early in his studies he developed two major principles. The first he called the "subordination of parts." In simplest terms, this stated that some parts of organisms are so important to the life of the organism that they can vary only a little if at all within a major group, and are therefore the ones to be used in any classificatory system since they demonstrate the large natural groupings of animals. The nervous system, the organs of locomotion, and the organs of digestion are the three most important, in that order.

Cuvier's second general principle was that of the "correlation of parts," which states that since each organism is a functioning whole, the structure of any one part dictates the structure of other parts. In this way, Cuvier believed one could determine the entire structure of an animal from a very small number of its parts: if an animal has feathers it will also have a forelimb which is greatly modified to form a wing; the structure of this wing in turn will be correlated with the structure of the beak, the structure of the pelvic girdle and hindlimb, and the structure of the lung. Cuvier could explain many correlations by the requirements of function; others he could not explain but simply observed to be true, such as that horns or antlers are correlated with cloven hooves. Cuvier was a rationalist; he believed that biology could be studied with the same rigor as physics and that these two principles—the subordination of parts and the correlation of parts—were as fundamental to biology as Newton's laws of motion were to physics.

He divided all animals into four large groups, called *embranchements,* on the basis of the structure of their nervous systems. The Vertebrata were animals with a dorsal spinal cord and anterior brain. The Mollusca's nervous system consisted of discontinuous masses. The Articulata (including what we now call arthropods) had a nervous system consisting of two long cords running along the ventral side of the body. The fourth group, Radiata, which would not today be regarded as a natural group, was composed of all forms with a basic radial symmetry and included both coelenterates and echinoderms.

Cuvier did more complete paleontological studies than any of his predecessors. Drawing on his immense knowledge of the structure of vertebrates and applying his principle of the correlation of parts, he described in great detail the structures of whole organisms of which he had only two or three fossil bones. Such an exercise is all the more impressive because in many instances entire fossils were later discovered which closely matched Cuvier's descriptions.

By the early nineteenth century, therefore, vertebrate zoology had three vigorous branches. In Scandinavia were the taxonomists, with the binomial system of classification and well-established procedural rules. In England were the physiological morphologists, heirs of a tradition begun by Harvey and continued by Hunter. On the Continent were what we might call the internal comparative anatomists—Cuvier and his followers in France and the natural philosophers in Germany.

A man greatly influenced by all three of these schools and uniquely able to use the best from each was Richard Owen. He was the son-in-law of John Hunter and was in charge of the museum that Hunter had founded. He had studied with Cuvier shortly before his death and was greatly influenced by Cuvier's rational, empirical approach. He also knew and had been considerably impressed by the work of the natural philosophers.

Richard Owen did many original studies but had few truly original ideas. He is remembered for his contributions toward understanding and synthesizing what had come before. In addition to his rich intellectual heritage, he had at his disposal the vast Hunterian collection, which was constantly augmented by specimens carried by English ships returning from all parts of the world. His monumental books on vertebrate anatomy and physiology were touchstones for generations and are still the standard reference sources for anatomical facts about rare animals.

Firmly convinced of the value of making comparisons between animals, Owen introduced terminology for the purpose. He made use of what the natural philosophers had learned and of Cuvier's principles and formalized the ideas of homology and analogy much as they are still used today.

A *homologue* is the same organ or part of an organ in different animals, under every variety of form and function. For instance, the bones in a whale's flipper are homologous to those in the forelimb of a cat, although their structure and function are quite different. In order to determine homologies one must examine both the embryology of the structure in question and its structural relationships to other parts of the body. In the above example, the bones of the flipper and forelimb develop from embryonic tissue in the limb bud and form a single proximal bone, two parallel bones just distal to this, and several smaller bones most distally. Only in later development do significant differences in the limbs become apparent. In the adult, the shapes and numbers of bones differ significantly; however, their relationships to major muscles, blood vessels, and nerves remain remarkably similar.

On the other hand, *analogues* are organs which serve the same function in different animals but which may be different in structure, origin, or both. Common examples are the lungs of man and the gills of fish, or the wings of butterflies and the wings of birds; but the student will discover more subtle illustrations.

Owen therefore added some principles of his own to those Cuvier had set forth and helped make comparative morphology more rigorous. By the mid-nineteenth century it was a firmly established and respectable science, and its practitioners were systematically examining the diversity of vertebrate forms. Although for a century or more there had been occasional suggestions that animal types might change over time, these ideas had never been convincingly expressed and had had little impact on the development of comparative morphology. However, the situation was to change dramatically in 1859, when comparative morphology and, indeed, the entire scientific and intellectual community were to be challenged and stimulated by the publication of a book by the English naturalist, Charles Darwin.

Son of a physician, Charles Darwin had been the naturalist on a round-the-world voyage of the "H.M.S. Beagle" from 1831 to 1835. During these years he had made numerous observations on animal variety and geological formations whenever the expedition had touched land. He had seen that species differences from island to island were sometimes very slight, but could be significant, for instance, variations in the beaks of otherwise similar birds. Gradually he had begun to doubt the doctrine of the fixity of species and to feel instead that species are continual with one another and may change. Back home he pondered these ideas and certain others current in Victorian England, for instance, the geological works of Lyell and others which apparently established that the world was very much older than had been believed, and an essay by the economist T. R. Malthus which explained that while the food supply increases arithmetically, populations increase exponentially unless checked by disease, famine, or war.

In 1842, Darwin first wrote up but did not publish his ideas on the theory and mechanism of the origin of new species by natural selection. He based his theory on empirical facts, basically the following:

1. All organisms tend to produce more offspring than can survive. That is to say, if all the progeny of any species were to survive, the species would rapidly outpopulate its physical environment (food, space, etc.).
2. Within a species there is variation; all individuals are not exactly the same. Since they are not the same, some will be more "fit" to survive in a given environmental situation than others. (Note that this implies no value judgment; the individuals not fit for one environment may be fit for another, and their inability to survive is due to chance rather than merit.)
3. At least some of these variations are inheritable. Therefore, the individuals that survive long enough to reproduce will frequently pass their more "fit" traits on to their offspring, and each species will over a long period of time gradually become more adapted—more fit—for a particular environment. In this way nature "selects" which animals will survive and which will not. It is not a random selection, nor is it a selection having anything to do with the relationship of human beings and Divinity; it is simply a selection for adaptation to the environment.
4. Finally, since slow geologic changes cause changes in environments, the direction or force of natural selection will also change; there will be continual, albeit gradual, change in the genetically determined structure of the species in any area in response to its changing environment. Given sufficient time, the species can change enormously.

Darwin did not rush these ideas into print once they were formulated. Fearing to lose the day for lack of sufficient evidence, he continued to compile massive amounts of empirical data and to check the data against the theory. So it was not until 1859 (over 17 years after his first notes were written and nearly a quarter of a century after the voyage that started it all) that he finally put it all into his bestselling "On the Origin of Species by Means of Natural Selection, or the

Preservation of Favoured Races in the Struggle for Life." Like all radical and great ideas, this immediately received both passionate approval and hysterical disapproval. One of the leading comparative morphologists of the day, T. H. Huxley, was instantly converted from his ideas on the fixity of species by Darwin's clear exposition and powerful arguments. An apocryphal rumor claims that when Huxley finished reading "The Origin of Species" (on the very day of its publication), he slapped his thigh and exclaimed, "By Jove, how stupid of *me* not to have thought of that!"

Richard Owen, on the other hand, flatly refused to accept the idea that species could change. To his death he remained one of the most adamant adversaries to Darwin's idea.

He was not alone in this opinion, but he was definitely in the minority. Within a short time the majority of both the lay and scientific communities had accepted Darwin's theory. It had far-reaching implications in many fields, but here we can consider only its impact on the interpretation of morphological similarity and diversity.

An important effect of these new ideas was that the great diversity of living forms could now be viewed as a natural, rather than a supernatural, phenomenon. Variation within a species became at least as important a concept as the uniqueness of the species. The species was not a fixed phenomenon, but a group of interbreeding variants. It would never again be theoretically satisfactory to consider individuals of a species as identical; the "type specimen" for a species became merely the first individual of that species to be described in the scientific literature. For the more than one hundred years since then the typologist's approach has been gradually giving way to a consideration of communities or populations.

The idea of a linear *scala naturae* of living forms also should have died in 1859. However, the idea was so entrenched in the minds of both the scientific community and the general public, and so buttressed by organized Western religion, that the misleading model of an evolutionary ladder and the terminology of "higher" and "lower" animals still persist today. We must therefore devote a few paragraphs to examining these notions.

The term "ladder" implies a linear organization in which each step gives way to the one above it. However we know that all living forms are the result of millions of years of evolution: they are all the contemporary terminal points of the evolutionary process, and none has evolved *from* another presently living animal. By definition, each living species must be well adapted to its environment and able to compete successfully with all other organisms in that environment, or else it will become extinct. All living forms, therefore, should perhaps be considered "higher" than all extinct forms.

But would one say monkeys or perch are "higher?" Since their common ancestor millions of years ago they have had totally different histories and are presently adapted to different environments in vastly different ways. Perch in fact could be considered biologically more successful than monkeys because they have more stable populations and many more individuals. Still, most people would regard monkeys as "higher." In practice, then, "higher" means "more like me," and therefore, by extension, "better"—both biologically and morally. This view of life is not supported by the facts of biology.

However, the tenacity with which even biologists cling to these incorrect images is understandable. Long before the Victorian Age, astronomers had shown that the earth was not the center of the universe but only one small planet in one of many solar systems. In the nineteenth century Darwin seemed to detract from the uniqueness of man and treat him as merely one species among millions and the product, like all the others, of a natural rather than a supernatural process. It was a stunning blow to man's egocentricity, but at the same time a splendid opportunity for a better understanding of himself and the world his species inhabits. But in the century since, with the evidence of evolution impossible for most people to refute, many have preferred to integrate the processes of evolution with religious dogma, and teleology has unfortunately gotten a new lease on life: the natural processes are often thought to be supernaturally directed, and man is still believed by many to be the goal toward which all else has been directed and the one earthly being for whose use and pleasure all the rest of nature exists. (It is interesting, however, that this is a peculiarly Western, specifically Judaeo-Christian, idea: in many other philosophical systems man is considered not above or against nature but a part of it, and the American Indians, for instance, call other creatures "my brothers.")

The terms "higher" and "lower," furthermore, when they are not justified by religious dogma, are too often defended on the ground that man is the only species with such technology and with such a well-developed brain—the only one that thinks about thinking, as it were. While this may be man's most significant attribute (especially from the viewpoint of the writers and readers of books), it is not the only point to consider in a biological assessment.

Darwin's theory sparked an awakened interest in the phylogeny, or evolutionary history, of vertebrates, and particularly of humans, as morphologists attempted to discover inherited relationships among species and to trace their evolution. Paleontology of course was one approach, but there was another available in embryology, and the work of embryologists in the first half of the nineteenth century was examined for its bearing on the problem.

In the early nineteenth century J. F. Meckel and E. R. A. Serres, among others, had suggested that during an individual's embryonic development it passes through stages representing all the species "lower" than it on the *scala naturae*. In the mid-nineteenth century Carl Ernst von Baer had demonstrated that the development of any embryo proceeds from a very generalized condition, to a less generalized, to a more specialized, to an extremely specialized form which is the adult of the species, and had pointed out that early stages in all forms resemble one another more closely than do later stages. After Darwin's publication of "The Origin of Species," Ernst Haeckel put these two ideas together and got The Biogenetic Law: ontogeny recapitulates phylogeny, or, less succintly, the life history of any individual repeats the evolutionary history of its species. According to this view, the human embryo should pass through a fish stage, an amphibian stage, a reptilian stage, and a mammalian stage to reach finally the uniqueness of the human being. These ideas stimulated a great deal of work in embryological morphology, and when by the late nineteenth century it had been adequately demonstrated how wrong Haeckel was, a great deal of embryology had been learned. How different the nineteenth century was from the third century when Galen wrote!

Unlike Haeckel's concept, Von Baer's descriptions and the principles he derived from them were shown to be largely valid. Development does indeed proceed from the generalized, unicellular, fertilized egg through progressively less generalized and more specialized forms until it reaches the most specialized form possible for that particular union of sperm and egg—the adult of the species.

It was as a result of these embryological studies that the phylum Chordata, which contains the subphylum Vertebrata, was finally defined, and the common characteristics of all vertebrates were pointed out. By definition, all vertebrates possess vertebrae or remnants of vertebrae. All species of the phylum Chordata, at some time in their ontogeny, possess a dorsal hollow nerve cord; a notochord, which is a stiff, flexible rod made up of turgid connective tissue cells; and pharyngeal pouches which invaginate from the pharynx and which form the gills in fishes and other structures in tetrapods.

The *mechanisms* of heredity did not begin to be understood until the early twentieth century when Gregor Mendel's great work was rediscovered. The burgeoning of modern genetics followed, with the work of T. H. Morgan on *Drosophila* being replaced not long after by the more modern approach of biochemical molecular genetics.

By the 1930s, however, the concept of natural selection, integrated with what had been learned of genetics and paleontology, formed the modern, synthetic view of evolution by natural selection. Each species is made up of one or more interbreeding populations in which genetic diversity exists. Processes such as gene mutation and recombination tend to increase genetic diversity, while the process of natural selection tends to decrease it by eliminating genetic types that are poorly adapted to the environment. As the environment changes, the force and direction of selective pressures change. Thus the amount of a species' genetic diversity is in flux, balanced between the effects of gene mutation and recombination, on the one hand, and natural selection, on the other.

GENERAL CONCEPTS OF VERTEBRATE EVOLUTION

The evolutionary changes by which a vertebrate that lived 400 million years ago was transformed into a present-day cat, frog, perch, bird, or man were random and, teleologically speaking, inefficient. There were no straight-line developments, but series of "solutions," answering random "problems" with varying degrees of success.

The *rates* of evolutionary change have differed for different times and places, different taxonomic groups, and even different characteristics within the same group. They have depended largely upon the rate at which the "problems" have been posed, that is, the rate and degree of environmental change.

Fig. 1-3. In birds the beak is a general adaptation facilitating food-gathering and object-manipulation. In most birds this general adaptation shows further specializations for diet. After J. C. Welty, 1962. "The Life of Birds." Saunders, Philadelphia, Pennsylvania.

The *types* of evolutionary change which are possible are limited not only by the available genetic diversity, but also by the necessity for any change to "work" with the rest of the animal's morphology. Thus what might be the best solution to a problem, from an engineer's view, may not develop, and instead there may be a solution which is less efficient but still workable.

In other words, an adaptive change may be simply making the best of a bad situation. For instance, it might be best for a vertebrate to have enough eyes to see in all directions at once, but these genetic mutations have not occurred. Instead one pair of eyes combined with increased mobility in the neck region of most groups have allowed rapid scanning of the environment and have eliminated the need to move the entire body in order to survey an area.

Most evolutionary changes have involved adaptations related to specific environmental situations, but some even more significant changes have involved whole new levels of organization; these were adaptive not just to a specific environment, but gave their possessors a general advantage in many types of environments. The evolution of paired eyes is an example. It occurred early, for we know of no vertebrates without paired eyes and visual centers in the brain (although they are underdeveloped in some cave-dwellers). Other general adaptations include a movable jaw, efficient swimming muscles (fish) or limbs capable of sustaining an animal's weight (land animals, or tetrapods), the ability to maintain constant body temperature independent of environmental temperatures (homeotherms, that is, birds and mammals), the ability to fly (birds and bats), and the persistent care of young (birds and mammals).

The development of one of these new levels of organization was usually followed by a (relatively) rapid burst of evolutionary activity in which several species arose from the stem group that had first developed the general adaptation (Fig. 1-3). These species, adapted to diversified habitats in addition to possessing the general adaptations of the group, comprised an *adaptive radiation* into new environmental niches. During the past 70 million years birds, mammals, and teleosts have been undergoing massive adaptive radiations into what one could imagine must be nearly all possible environments and life situations.

COMPARATIVE MORPHOLOGY AND VERTEBRATE EVOLUTION

The comparative morphologist studies the past record of evolution (temporal diversity, or phylogeny) and the present result of evolution (spatial diversity). Comparisons are first made on a structural level, examining, for instance, the diversity of skeletal structure or reproductive mechanisms from group to group. But we must also look at diversity of function. We ask: *What is the adaptive value* of the structural diversity we see? The answer gives us a clue to the evolutionary history of the organism as well as to its continued survival. For as Cuvier knew, structure and function are opposite sides of the same coin and one cannot be fully appreciated without knowledge of the other.

SUGGESTED READING

Coleman, W. (1964). "Georges Cuvier, Zoologist." Harvard University Press, Cambridge, Massachusetts.

Darwin, C. (1964). "On the Origin of Species by Charles Darwin," A facsimile of the first edition. Harvard University Press, Cambridge, Massachusetts.

Ghiselin, M. T. (1969). "The Triumph of the Darwinian Method." Univ. of California Press, Berkeley, California.

Russell, E. S. (1916). "Form and Function." John Murray, London.

Taylor, G. R. (1963). "The Science of Life." McGraw-Hill, New York.

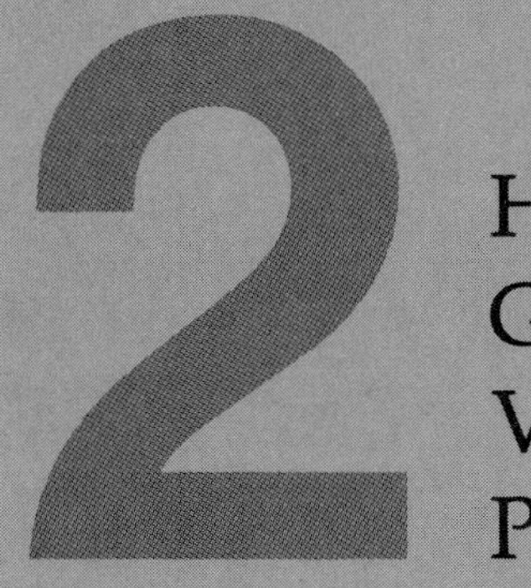

HOW WE GOT HERE: VERTEBRATE PHYLOGENY

2

The fossil record provides an incomplete and biased story of the evolution of vertebrate structure, for it depends upon favorable geologic and climatic conditions, and even then is usually limited to hard parts and often to the teeth alone. Nevertheless it provides solid, direct evidence of much of evolution, and, to those who learn how to read it, can give substantial indirect evidence as well. It is, in short, a starting point from which logical constructs can be made.

The geological time that includes the evolution of vertebrates is divided into three eras; from oldest to most recent these are the Paleozoic, Mesozoic, and Cenozoic. Each era is subdivided into geologic periods as indicated in Fig. 2-1. The Cenozoic era contains both the 70 million year Tertiary period (as indicated on the figure) and the most recent one million years, which comprise the beginning of the Quarternary period and bring us to the present time.

FISHES

EARLIEST FOSSIL VERTEBRATES—AGNATHA

Ostracoderms

The earliest identified vertebrate fossils, from the Ordovician period of the Paleozoic era some 450 million years ago, are so incomplete that they tell us little more than that these vertebrates possessed bone and probably lived in fresh water (suggesting that, unlike most invertebrate groups, vertebrates may have first evolved in fresh water).

From the Silurian and Devonian periods, approximately 400 million years ago, there are various excellent vertebrate fossils classified collectively as the order Ostracodermi of the class Agnatha. Although quite diverse, all lacked a jaw skeleton. A few had lateral extensions from the body, but paired appendages were rare (Fig. 2-2).

In the deep layers of the skin were thick bony scales and plates. (Bone which develops in and is located in the skin's deep layer is called "dermal bone," as opposed to that which develops deep to the skin and is called "endochondral bone.") The dermal bony plates were largest in the head and anterior trunk. In the best-preserved fossils, the entire head was covered by a single plate in which there were only small openings for sense organs, gills, and mouth. More caudally the dermal armor was made up of smaller, but still thick bony scales.

Many ostracoderms had a flattened head with a ventral mouth. The caudal fin projected much further dorsally than ventrally (called a heterocercal tail, whose power stroke sends the fish downward and forward). In modern fish, such characters are correlated with bottom-

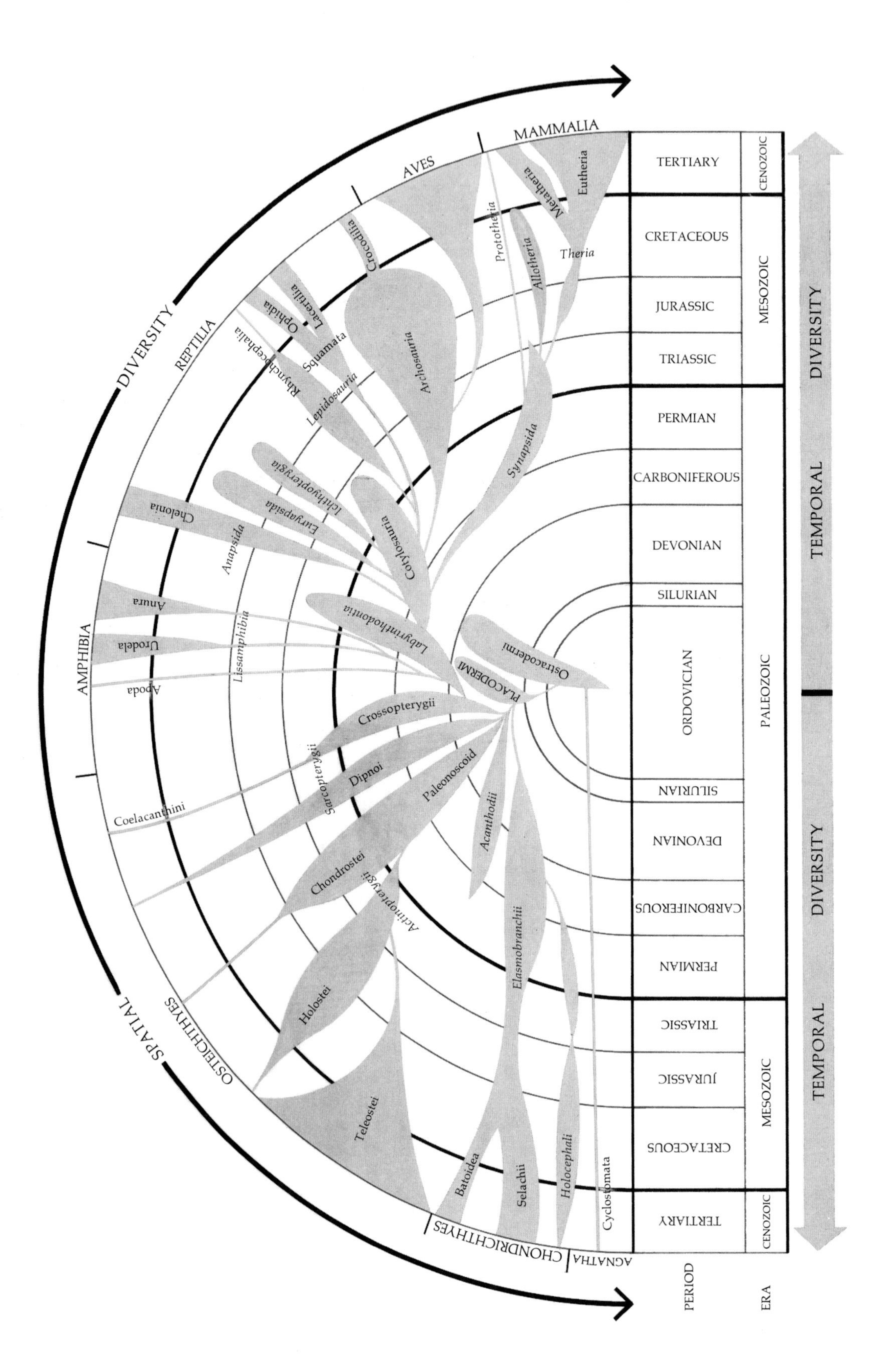

DIVERSITY
SPATIAL
MAMMALIA
AVES
REPTILIA
AMPHIBIA
OSTEICHTHYES
CHONDRICHTHYES
AGNATHA
Eutheria
Metatheria
Theria
Prototheria
Allotheria
Crocodilia
Archosauria
Lacertilia
Ophidia
Squamata
Rhynchocephalia
Lepidosauria
Synapsida
Ichthyopterygia
Euryapsida
Chelonia
Anapsida
Cotylosauria
Anura
Urodela
Apoda
Lissamphibia
Labyrinthodontia
Ostracodermi
PLACODERMI
Crossopterygii
Coelacanthini
Sarcopterygii
Dipnoi
Paleonoscoid
Acanthodii
Chondrostei
Actinopterygii
Holostei
Teleostei
Elasmobranchii
Batoidea
Selachii
Holocephali
Cyclostomata
TERTIARY
CRETACEOUS
JURASSIC
TRIASSIC
PERMIAN
CARBONIFEROUS
DEVONIAN
SILURIAN
ORDOVICIAN
CENOZOIC
MESOZOIC
PALEOZOIC
PERIOD
ERA
TEMPORAL

Fig. 2-1. The adaptive radiation of major vertebrate groups. Earliest vertebrate fossils are from the Ordovician period of the Paleozoic era, some 450 million years ago. Note that the evolutionary rates of different groups vary greatly in time and extent and that there is, in fact, no general evolutionary direction of vertebrates as a group.

living and -feeding environmental niches. Very likely ostracoderms scooped detritus from lake and stream beds with their jawless mouths, filtered it past their gills, and swallowed the solid material.

Other ostracoderms were evidently surface feeders. They had a more rounded head with a terminal mouth, and the ventral lobe of the caudal fin was larger than the dorsal lobe (called a reversed heterocercal tail, whose power stroke sends the fish upward and forward).

Despite their diversity, apparently none of these ancient, jawless fishes was equal to the competition from other groups, and by the end of the Devonian period (345 million years ago) ostracoderms had become extinct.

Cyclostomes

The class Agnatha survives today in only two representatives, the hagfish and the lamprey, both of the order Cyclostomata (Fig. 2-2). Until 1968 this order had no known fossil record, but in that year imprints of a few lamprey-like fossil cyclostomes were reported in a deposit of late Carboniferous shale, at least 300 million years old.

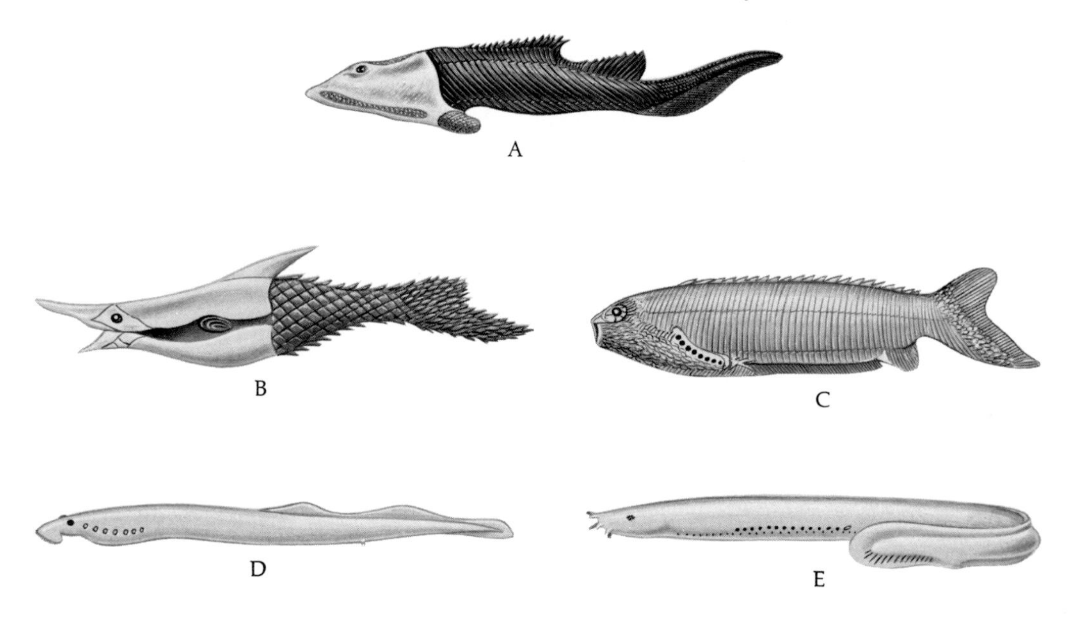

Fig. 2-2. Diversity of agnathan external structure. Reconstructions of ostracoderms are shown in A, B, and C; the living cyclostomes in D (lamprey) and E (hagfish). A, B, and C after L. B. Halstead (1969). "The Pattern of Vertebrate Evolution."

The cyclostomes have no bone; the skeleton is composed of scant cartilages and a prominent notochord. There are no paired appendages, and of course no jaws. They have a single dorsal median nostril, lateral eyes, and a pineal eye. The hagfish is a scavenger of dead or dying fish; the lamprey is a parasite, sucking blood and fluids from other fish. Both have round mouths, as did the fossil cyclostomes, and a large muscular tongue with horny teeth that can rasp away the host's flesh. As is often true of parasites, whose diets are quite simple, their bodies have also become simplified. Undoubtedly many of the internal structures of ancient ostracoderms were lost by fossil as well as modern cyclostomes in their adaptation to parasitic life.

There is still not enough evidence to determine whether modern cyclostomes are direct descendents of the ancient ostracoderms, or whether both groups had a common ancestor dating back to some time earlier than the mid-Ordovician. And as for setting a date for the origin of vertebrates, one can say only that they were present 450 million years ago, and may have been present considerably earlier.

FIRST FISHES WITH TRUE JAWS—PLACODERMI

The first vertebrates that had a jaw apparatus were the class Placodermi; they appeared in the very late Silurian, and most were extinct before the end of the Devonian period. They were primarily freshwater animals and like Ostracodermi were characterized by heavy bony dermal armor. They had in common a skeletal support for the mouth and paired appendages, but beyond that they were of diverse forms.

The jaws were of various types. For instance, in one placoderm, *Dinichthys,* ball and socket joints were present between the pectoral girdle and bones of the skull. The upper jaw and the top of the head and braincase moved upward to the same extent that the lower jaw moved downward.

There were also various types of paired appendages. In some, such as *Bothryolepis,* paired appendages were jointed and superficially resembled those of modern Crustacea; in others, they were little more than bony spines; in still others they evidently had a certain amount of muscle attached internally to bone or cartilage. Most placoderms had two pairs of appendages, a pectoral and a pelvic, but some had as many as seven or eight pairs (Fig. 2-3).

By the end of the Devonian period the only surviving placoderms were in the subclass Acanthodii. Their small fossils (from about 3 to 12 inches long) date from the late Silurian through the early Permian. They are more similar to the living jawed fishes than to any other placoderm group, and authorities disagree as to their proper taxonomic placement. Some regard them as Placodermi, others as Osteichthyes, others as Chondrichthyes, and still others have suggested that they form a class by themselves. In any case, it is almost certainly from the same group that produced Acanthodii that the ancestors of modern fishes arose, and we will regard them as a subclass of Placodermi.

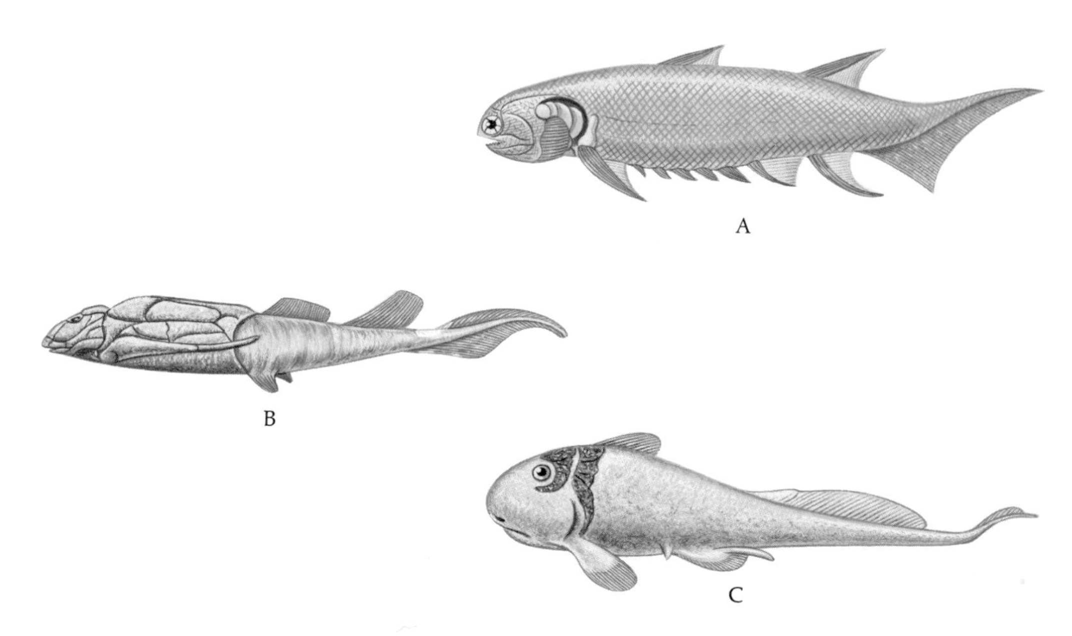

Fig. 2-3. Reconstructions of an acanthodian (A) and two other placoderms (B and C). After L. B. Halstead (1969). "The Pattern of Vertebrate Evolution."

The modern gnathostomes (jawed fishes) are separated into two classes. Chondrichthyes are cartilaginous fishes which lack bone entirely. Osteichthyes are the so-called bony fishes, although a few have predominantly cartilaginous skeletons. These major, extremely successful groups of fishes appear in the Devonian fossil record at approximately the same time. They may have arisen from two different subgroups of acanthodians, although their origins remain uncertain.

CARTILAGINOUS FISHES – CHONDRICHTHYES

During the evolution of the class Chondrichthyes (as well as of cyclostomes) there was a loss of all bone, and the skeleton is now formed completely of cartilage and the notochord. Chondrichthyeans were numerous during the late Devonian and Carboniferous, dwindled dramatically during the Permian and Triassic, and then underwent a second adaptive radiation in the Jurassic and Cretaceous. Unlike ostracoderms and placoderms, their fossils are found in marine deposits, and they remain almost entirely marine in habitat today.

Since the Carboniferous they have been divided into two subclasses, the Holocephali, or rat-fishes, and the Elasmobranchii; since the Cretaceous the Elasmobranchii have been further divided into Selachii, the sharks, and Batoidea, the dorsoventrally flattened skates and rays (Fig. 2-4).

Elasmobranchs, although not remarkably numerous, appear to be on the increase today. They are a highly active, predaceous group, having "conventional" jaws, a pectoral and a pelvic pair of appendages, a het-

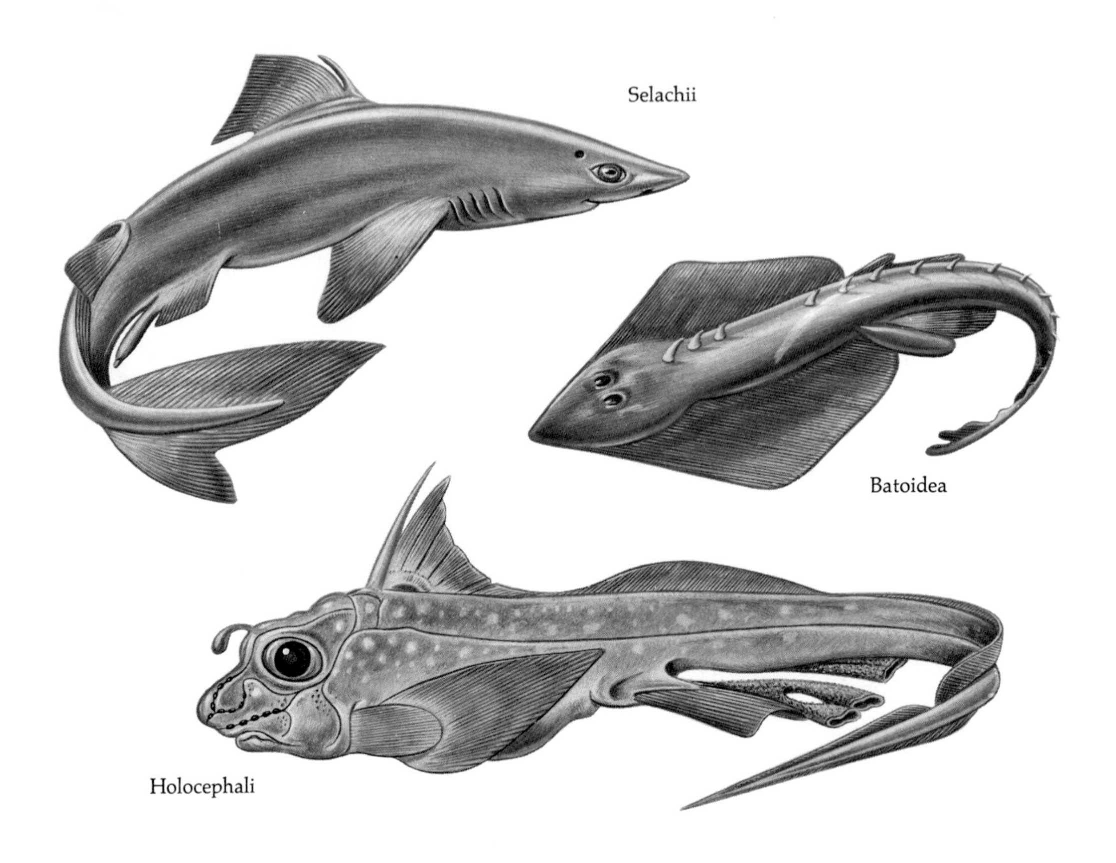

Fig. 2-4. External structural diversity of chondrichthyeans as seen in a selachian, a batoidean, and a holocephalian. From Malcolm Jollie (1962). "Chordate Morphology" © 1962 by Litton Educational Publishing, Inc. Reprinted by permission of Van Nostrand Reinhold Company.

erocercal tail, and gill slits which open independently to the animal's surface.

Embedded in the skin of most chondrichthyeans are tiny dermal denticles, called placoid denticles which are similar in structure to the small tubercles on the outer edges of the dermal bones of placoderms. Not bone themselves, these denticles are nevertheless the closest the Chondrichthyes come to having bone. This adaptive loss of bone decreases bulk and increases the flexibility, and thus the agility, of these active predators.

BONY FISHES—OSTEICHTHYES

Osteichthyes, the bony fishes, possess dermal bone as well as some and usually much endochondral bone. The gill slits, rather than opening independently, are covered by a (usually bony) flap called the operculum, which causes the water from all gill slits to be flushed out of a single posterior opening.

The history of the bony fishes is both better known and more complex than that of Chondrichthyes. The first Osteichthyes, like the first Chondrichthyes, appear in the fossil record in the early Devonian period. Unlike Chondrichthyes, they arose in fresh water; later many migrated to marine environments. From their first appearance, two subclasses of Osteichthyes are apparent, the Actinopterygii and the Sarcopterygii.

Actinopterygii

The paired appendages of actinopterygians are thin, ray-like fins which have little or no muscle and are covered by skin. There are two pairs of nostrils opening to the surface on each side, a subterminal mouth, and usually only a single dorsal fin. The primitive actinopterygians had heavy, thick, bony ganoid scales in which a deep layer of bone is covered by dentine which in turn is covered by an enamel called ganoin.

The complex evolution of the Actinopterygii can be simplified (with some distortions) to three large, overlapping adaptive radiations. From the Devonian period of the Paleozoic through the Triassic period of the Mesozoic the major group of actinopterygians was the superorder Chondrostei. These fish, often called the palaeoniscoid fish, have a subterminal mouth and a strongly heterocercal tail; their internal skeleton is primarily bone, with some cartilage. There is a separate, heavy dermal skeleton of ganoid scales and, in the head region, superficial bony plates. This was the only group of fish to be successful and even increase in numbers during the transition between the Permian and Triassic periods, when the seas of the world were in geologic crisis and most marine forms diminished. Only a few chondrosteans survive today, however; the best known are the bichir (*Polypterus*), the sturgeon (*Acipenser*), and the paddlefish (*Polyodon*). The latter two have secondarily developed a mostly cartilaginous skeleton, whereas *Polypterus* retains most of the generalized characteristics present in palaeoniscoids.

The second major group of Actinopterygii appeared during the early Triassic period. This was the superorder Holostei, characterized by a (frequently) terminal mouth, an abbreviated heterocercal tail whose dorsal part was only slightly larger than the ventral part, a generally more compact, streamlined skeleton and body shape, and thin ganoid scales. Throughout the Triassic and Jurassic the numbers of holosteans increased as the numbers of chondrosteans decreased. Evidently the holosteans possessed more successful general adaptations that allowed them to move into environmental niches previously occupied by chondrosteans and largely to replace them. The holostean fishes dwindled in number throughout the late Mesozoic and Cenozoic; today there are only the bowfin, *Amia*, and the garpike, *Lepisosteus*.

The third group of actinopterygians, the superorder Teleostei, appeared in the Triassic but did not become numerous until the mid-Cretaceous. Teleosts have a terminal mouth, a homocercal tail (externally it is symmetrical dorsoventrally), an almost complete loss of cartilage, a compact skeleton of light bones, and a streamlined shape.

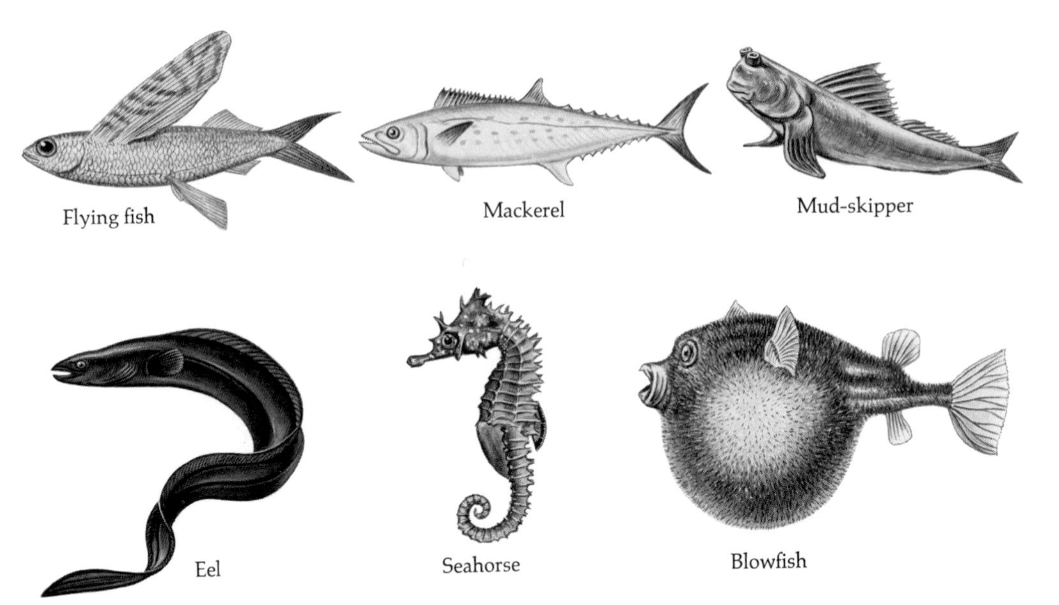

Fig. 2-5. Diverse teleosts.

These characteristics make them agile and rapid swimmers. The scales are thin and bony and are covered by a fine acellular layer of dentine.

Teleosts started their expansion in the Cretaceous and have continued to expand and diversify in an elaborate adaptive radiation. Today there are more species of teleosts than there are of all other vertebrates combined. They have moved into every possible aquatic niche, both freshwater and marine, and some are even amphibious (Fig. 2-5). One expert ichthyologist challenges his lecture audiences to name any function of other vertebrates which is not also performed by at least one teleost, including walking on land, flying, and maintaining a constant body temperature.

In addition, there are remarkable adaptations which belong exclusively to teleosts. For instance, the four-eyed fish, a surface-feeder, swims with half its eye's lens above water and half below; however, because the two halves have different convexities all of the resultant image is in sharp focus. The archer fish swims at the surface and spits water at insects flying above; these fall into the water and are eaten by the fish. The seahorse, which usually swims in a vertical rather than horizontal position, has a prehensile tail which can grasp the seaweed near which it lives. The bottom-dwelling flounder is flattened laterally; both eyes are on the right side of its head, and it swims with the right side up.

Sarcopterygii

Less dramatic than the subclass Actinopterygii is the subclass Sarcopterygii, the lobe-finned fishes, but it is from them that the first land vertebrates evolved. Whereas great numbers of actinopterygians moved into salt water environments, the sarcopterygians remained almost completely freshwater forms. Their lobate fins had an internal skeleton and considerable musculature. They had a terminal mouth and a pair of external nostrils from which paired channels opened into the mouth as the paired internal nostrils. The early forms had two dorsal fins and cosmoid scales in which the deep bony layer of each scale was covered by many tubercles of cosmine, a dentine-like substance, and capped by enamel.

The subclass Sarcopterygii appeared in the fossil record during the Devonian period of the Paleozoic, and from the start was divided into two orders, Dipnoi and Crossopterygii. In fact, these two orders may have evolved independently from separate stocks of Acanthodii.

The Dipnoi, or lungfish, are represented today by three genera (Fig. 2-6) on three continents (Africa, South America, and Australia), all living in lakes or streams which dry up periodically. They survive these dry periods by aestivation. (Aestivation is similar to hibernation but less severe: it involves greatly reduced metabolic activity as a response to metabolic stress, such as that caused by periods of very warm or very dry weather.)

The Crossopterygii retained more generalized characteristics than did the Dipnoi or Actinopterygii, and some of these general characteristics can, from our vantage point, be considered "preadaptations" for land life. They soon divided into two groups, the Rhipidistii and the Coelacanthini. The Rhipidistii gave rise in the very late Devonian to the first amphibians, and thus are the common ancestors of all tetrapods; they became extinct in the late Paleozoic era. The Coelacanthini, after having been quite successful during the Permian and Triassic, were thought to have become extinct during the mid-Mesozoic. However, since 1938 several individuals of an aberrant deep-water marine genus, *Latimeria,* have been found near Madagascar (Fig. 2-6).

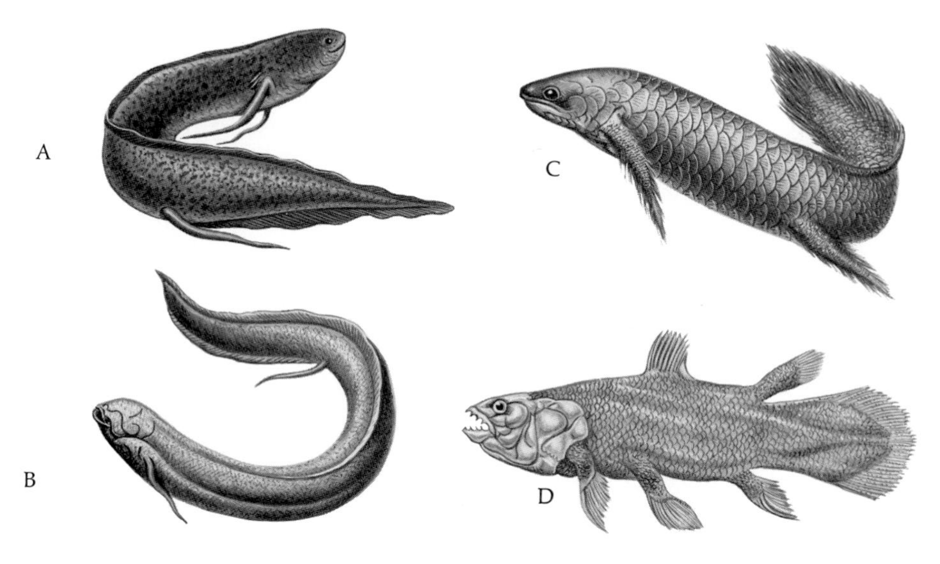

Fig. 2-6. The living Sarcopterygii: A, African lungfish; B, South American lungfish; C, Australian lungfish; and D, the coelacanth, Latimeria.

It is interesting that of the two subclasses of Osteichthyes it is the Sarcopterygii, the fish with the more general adaptations, that gave rise to the terrestrial vertebrates, while the more specialized Actinopterygii (which might have been considered "higher") have exploited the aquatic niches. This illustrates a general rule of evolution, in fact: a group which is extremely specialized to one mode of life is not apt to give rise to forms that can survive or successfully compete in another mode. Actinopterygians and chondrichthyeans underwent extreme adaptive radiations; they conquered and still dominate the aquatic environments of the world. The less prominent and less specialized crossopterygians gave rise to the forms which were to dominate the terrestrial part of the world.

TRANSITION FROM WATER TO LAND

The problems of transforming an aquatically adapted vertebrate into a terrestrially adapted vertebrate are staggering. For instance, a loosely articulated vertebral column and strong trunk and tail musculatures are adaptive in an aquatic environment, for together they make lateral undulations an easy and efficient means of locomotion. When there is no water for support, however, an animal requires a much more firmly articulated vertebral column, and efficient locomotion on land requires strong limb muscles rather than strong trunk and tail muscles.

Sense organs must be adapted differently for aerial than for aquatic environments. The refractive indices of air and water differ, so the eyes must also. Airborne vibrations are considerably different than waterborne vibrations, and hearing necessitates that vibrations be transmitted to fluids (in the inner ear); thus a different auditory system is necessary on land than in water.

Land vertebrates encounter different problems in obtaining, masticating, and swallowing food: there is little buoyancy on land, and the food is usually much harsher. There are new problems encountered in respiration because gills, the respiratory mechanism of most fishes, must be kept wet in order to function. Water conservation is a problem for the first time, and affects both respiration and the elimination of metabolic wastes. Since temperature changes are more extreme and more sudden on land than in water, there is also greater stress on the body's regulatory mechanisms.

Reproduction is still another problem. In an aquatic medium eggs can develop into a swimming larval form without danger of desiccation. For land animals another mechanism is needed to protect the eggs from desiccation, and it is necessary to bypass the swimming larva.

These are some of the problems that needed "solutions" if vertebrates were to utilize terrestrial environmental niches, and although the initial transition from water to land took place between rhipidistian fishes and labyrinthodonts, the first amphibians, it is hardly surprising that the complete transition required some 100 million years and lasted until the emergence of reptiles.

None of these changes took place rapidly, but rather by the gradual

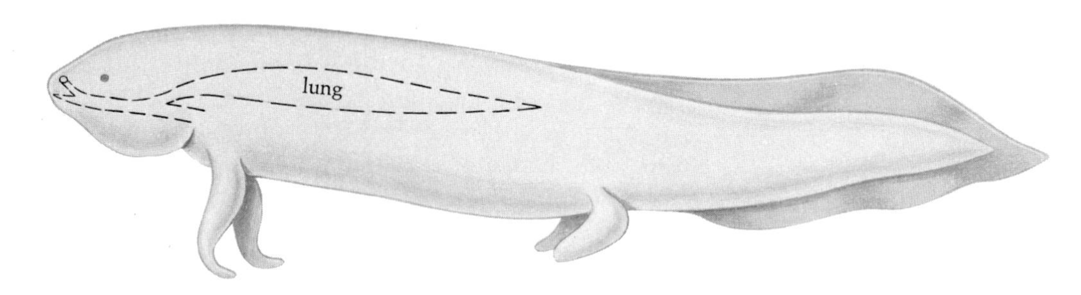

Fig. 2-7. Internal nostrils, lungs, and limbs capable of supporting the body all adapt the lungfish to life in stagnant water. They are also preadaptations to a terrestrial environment.

accumulation of small mutations, each of which was favored by the pressures of natural selection in a changing environment. In the process, structures evolved that were adaptive to this changing environment and were also "preadaptations" to terrestrial life.

For instance, some rhipidistians probably lived in stagnant water and in areas which occasionally dried up completely. A fish in such a tenuous environment could survive if it could wait out a temporary dry period, or if it could somehow get from a drying-up pool to a more favorable one. While ray fins would be useless in such a situation, lobate fins with a strong internal skeleton and musculature could help propel the animal from one pool to another; they could also provide the raw materials for the eventual evolution of terrestrially adapted limbs. Lungs were present in crossopterygian fish as an adaptation to life in oxygen-poor water, and were of course another preadaptation to a nonaquatic mode of life. The development of internal as well as external nostrils was a similar preadaptation.

Thus the term "preadaptation" does not mean the development of something before it has a use, but the development of something which adapts the organism to the present environment *and also* to an environment which has not yet been encountered (Fig. 2-7). Whether or not the latter ever is encountered is purely chance; if it is, the species will be in a better position for survival and perhaps will give rise to a new adaptive radiation.

AMPHIBIA

The first amphibians, and the first tetrapods, were of the subclass Labyrinthodontia. The earliest forms were similar to the rhipidistian fishes from which they evolved, but had a number of adaptations which allowed them to spend more time on land. Even so they were bound to water for at least part of their lives.

Their vertebral columns were stronger and more rigid because of interlocking, articulating processes called zygapophyses, which helped support the animal out of the water. The limb skeleton was not much more efficient than that of rhipidistians; limbs were spread out laterally and the muscles were probably weak. Most of their locomotion was probably accomplished by fishlike lateral undulations across moist areas of land. This is generally less efficient than other tetrapod patterns of locomotion, although as demonstrated by modern salamanders it can be quite effective. But efficient or not it was no doubt adequate, for the early labyrinthodonts had no terrestrial vertebrate competitors or predators.

Terrestrial locomotion was facilitated by two other labyrinthodont developments. Their scales were smaller, which made the animal lighter and more flexible. Further, a sacral vertebra provided a skeletal connection between the vertebral column and the pelvic girdle; thus when the hindlimbs pushed against the substrate the force was transferred directly to the vertebral column.

A comparison of the fossil records of labyrinthodonts and rhipidistians does not indicate much change in feeding mechanisms. The teeth were similar, being conical and bearing the peculiar labyrinthine grooves which give this subclass its name. (Such "labyrinthine" teeth were also found in rhipidistians.) A tongue was probably not developed any further than its rudimentary form in rhipidistians. Most tetrapods, unlike fishes, have a neck which permits independent head and body movements, but this was not so in the labyrinthodonts. We cannot tell whether labyrinthodonts had gills, but probably the adults (at least) did not, and like most living amphibians relied on lungs and skin for respiration. As to water conservation, excretion, reproduction, and temperature control, the fossil record tells us nothing directly. However, the study of diverse living forms allows some logical inferences to be made.

The subclass Labyrinthodontia expanded throughout the Carboniferous and Permian periods of the Paleozoic, decreased in numbers during the Triassic, and became extinct by the end of the Triassic. However, it gave rise to the subclass Lissamphibia, the living amphibians, and to the subclass Cotylosauria of the class Reptilia. It thus played a pivotal role in the evolution of terrestrial vertebrates.

Living amphibians—Lissamphibia

The subclass Lissamphibia includes all the living amphibians, i.e., frogs and toads, salamanders, and caecilians (Fig. 2-8). The fossil currently considered the most likely candidate for their common ancestor comes from the lower Permian and is named *Doleserpeton.* It is the only known amphibian fossil with the modifications of vertebrae and teeth and the flattened, reduced skull which are typical of modern amphibians.

However, the fossil record is sparse, and the stages by which the Lissamphibia evolved to be so different from their ancient amphibian ancestors are still unknown. There are three good reasons for the scarcity of their fossils: they were probably never extremely numerous or large in size; there was a progressive loss of heavy skeletal elements during their phylogeny; and they inhabited damp regions or areas associated with fresh water where good fossil deposits are seldom found.

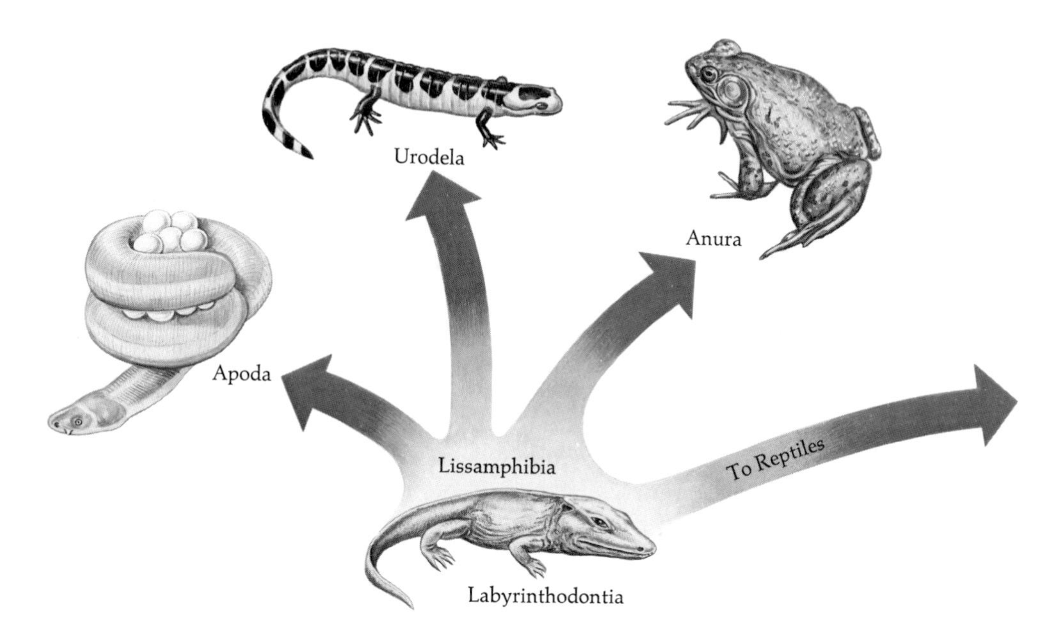

Fig. 2-8. Amphibian evolution.

The smallest of the Lissamphibia are the limbless amphibians of the order Apoda called caecilians. These are secretive, wormlike, tropical, and the only living amphibians which have retained scales. Since they play no major role in today's environment, we will summarily dismiss them from consideration.

Salamanders, the order Urodela, are small and fragile animals compared to labyrinthodonts, with a mostly cartilaginous skeleton. Although they are the most generalized of living amphibians, different members of this order exhibit considerable diversity of structure. Some, such as *Necturus* (often dissected in comparative morphology laboratories), retain larval characteristics such as external gills in the adult stage, that is, they are neotenic: reproductive maturity occurs before somatic maturity and they live and reproduce as larvae.

The most specialized order of Lissamphibia is Anura, the frogs and toads. Toads are somewhat less aquatic than frogs, but both depend upon the aquatic medium for reproduction and maintenance of larvae. They have unusual respiratory systems, large hindlimbs for jumping (saltatorial) locomotion, and a unique auditory system which provides good sensitivity to airborne sounds.

REPTILIA

Reptiles originated in the late Carboniferous period, some 300 million years ago. The stem group was the order Cotylosauria, which, although structurally similar to the Labyrinthodontia, had some further adaptations to terrestrial life. For one, the vertebral column was connected to the pelvic girdle by two sacral vertebrae instead of one. This stronger connection could accommodate larger limbs and more vigorous motion. There was a neck, with several cervical vertebrae; like modern reptiles, but unlike fishes and labyrinthodonts, cotylosaurs could move their heads independently of the rest of their bodies.

All living reptiles, but no amphibians, have a further, very significant adaptation to land life, the cleidoic egg. It is assumed to have evolved in Cotylosauria, although this is not directly evident from the fossil record.

The cleidoic egg has a large amount of yolk and a tough, leathery shell which protects it from desiccation. Extraembryonic membranes form early in development and facilitate respiration and excretion; this in turn makes possible a prolonged embryonic development, and thus eliminates the need for a larval stage. Thus the ancient cotylosaur, like modern reptiles, was not dependent upon an aquatic environment for any part of its life history. Although many dwelt in water, they apparently all had cleidoic eggs which, like the eggs of the modern sea turtle and crocodilians, could develop more safely on land.

The adaptive radiations of reptiles were the most significant events of vertebrate evolution during the Permian period of the Paleozoic era and throughout the Mesozoic era. By comparison, the origins of teleosts, mammals, and birds (which occurred during this same time) seem almost incidental (Fig. 2-9). Cotylosauria, themselves a diversified group, gave rise to several even more diversified subclasses. These became the dominant vertebrate fauna of land, air, and even water for over 200 million years—longer than the class Mammalia has yet existed.

Fossils of the mammal-like reptiles, subclass Synapsida, are first found in the very late Carboniferous period. They rapidly expanded to become the dominant reptiles of the Permian and Triassic periods before giving rise to the class Mammalia and then becoming extinct in the Jurassic. As the Synapsida were dwindling in numbers, the other great subclasses of reptiles became dominant. Two of these subclasses were thoroughly aquatic—the long-necked, paddle-finned plesiosaurs, subclass Euryapsida, and the fish-shaped, short-necked ichthyosaurs, subclass Ichthyopterygia.

The subclass Archosauria, the most diverse of all, included a great variety of terrestrial, arboreal, and aerial forms. In this subclass were the "ruling reptiles" of the latter Mesozoic, such as the bipedal, carnivorous *Tyrannosaurus,* the herbivorous *Brontosaurus,* the tanklike *Ankylosaurus,* the plated *Stegosaurus,* and the three-horned *Triceratops.* In addition there were many small, delicate, bipedal and often arboreal archosaurs and the pterosaurs, the only flying reptiles. But of all the orders that developed in this tremendous adaptive radiation of archosaurs, the only one which survived to modern times is the Crocodilia, which includes alligators and crocodiles.

The final two subclasses of Reptilia also have living representatives. The subclass Lepidosauria includes the present lizards (Lacertilia) and snakes (Ophidia) of the order Squamata, and the one genus, *Sphenodon,*

Fig. 2-9. Adaptive radiation of amniotes, demonstrating the tremendous diversity of reptiles.

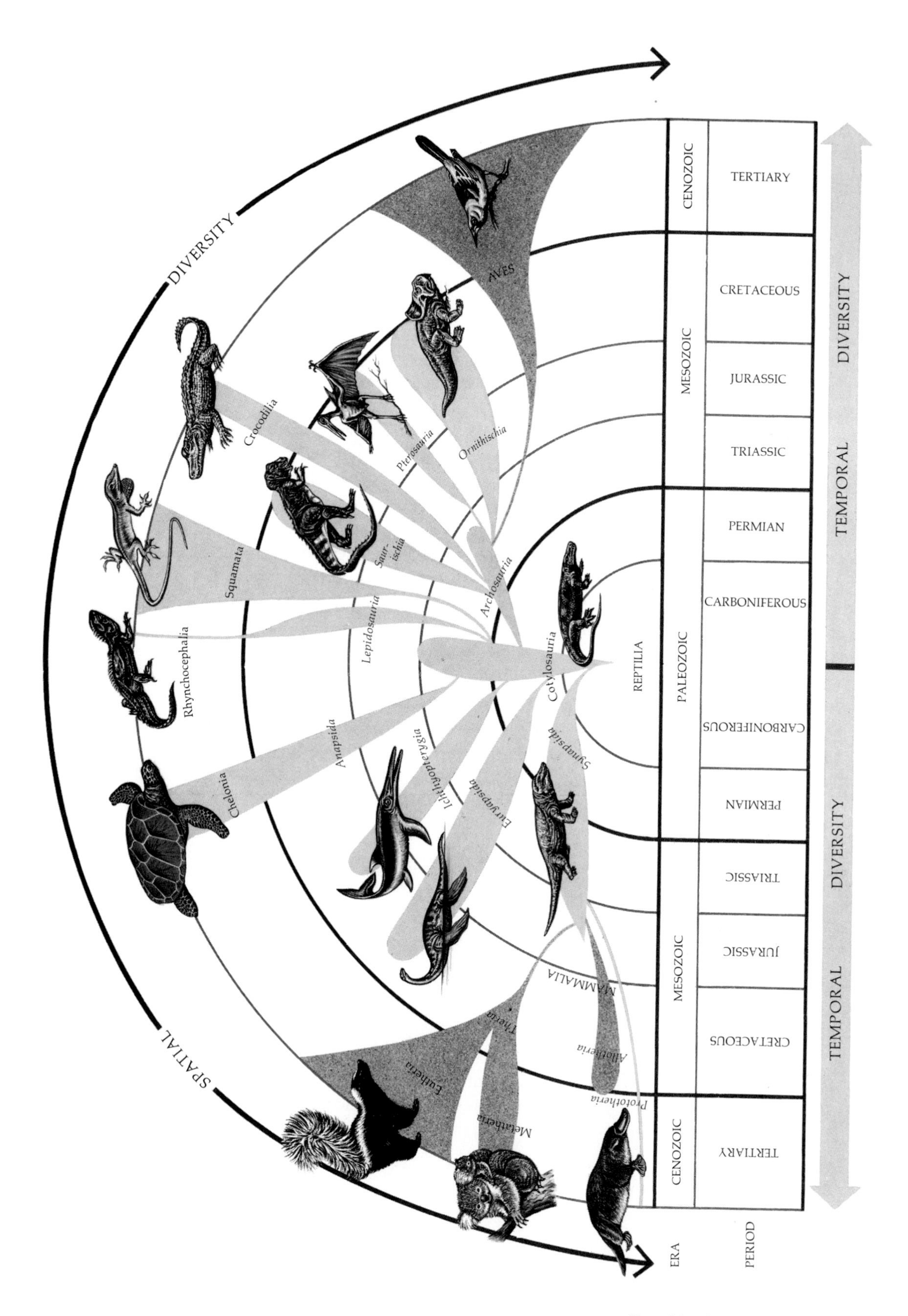
DIVERSITY
SPATIAL
AVES
Crocodilia
Pterosauria
Ornithischia
Saur-ischia
Squamata
Rhynchocephalia
Lepidosauria
Archosauria
Cotylosauria
REPTILIA
Chelonia
Anapsida
Ichthyopterygia
Euryapsida
Synapsida
MAMMALIA
Theria
Eutheria
Metatheria
Allotheria
Prototheria
ERA
PERIOD
CENOZOIC
MESOZOIC
PALEOZOIC
TERTIARY
CRETACEOUS
JURASSIC
TRIASSIC
PERMIAN
CARBONIFEROUS
TEMPORAL
DIVERSITY

of the order Rhynchocephalia. The largely amphibious subclass Anapsida survives today as the order Chelonia, the turtles.

Except for the relatively insignificant members of these living orders (Crocodilia, Squamata, Rhynchocephalia, and Chelonia), no reptiles survived the great geological transformations that marked the Mesozoic–Cenozoic interface.

DEVELOPMENT OF HOMEOTHERMY

Just as the cleidoic egg was all-important to reptilian evolution, so homeothermy facilitated the massive adaptive radiations of birds and mammals that occurred in the Cenozoic era. Homeothermy is the ability to maintain a constant, warm body temperature regardless of ambient temperatures; it completed the vertebrate evolution of homeostatic mechanisms. (Homeostasis is the ability to maintain internal integrity unaffected by external environment. In addition to body temperature, homeostasis includes the regulation of such aspects of the internal milieu as osmotic balance.) Homeothermy requires a constant, high metabolic rate which in turn necessitates efficient digestive, respiratory, circulatory, and excretory systems. There must be adequate heat generation by the musculature, insulation by the integument, and heat dissipation by several systems. Furthermore, all these functions must be intricately controlled by the nervous and endocrine systems. The morphological modifications necessary for homeothermy are almost as complex as those necessary for the transition from water to land. It is remarkable that these evolutionary changes occurred in two independent evolutionary lines—that of reptiles to birds, and that of reptiles to mammals. The fact that such complex events did occur twice is ample evidence of the great adaptive value of homeothermy.

MAMMALIA

The fossil record from Cotylosauria through the Synapsida to the first mammals is one of the most complete. The earliest known fossil mammals were present during the late Triassic period, some 170 million years ago, yet mammals did not become abundant in numbers (i.e., "successful" from an evolutionary point of view) until the end of the Cretaceous, some 75 million years ago. During the intervening 95 million years mammals were present but were small in both size and numbers. Evidently these early mammals were unable to compete successfully with the ruling reptiles, and it was not until the late Cretaceous, when the ruling reptiles became extinct, that mammals flourished and underwent their own multiple adaptive radiations into what must be nearly every possible terrestrial niche as well as many aquatic niches.

Prototheria

Living mammals are divided into two subclasses, Prototheria and Theria. The Prototheria (monotremes) have scarcely any fossil record,

Fig. 2-10. These three animals, from the three basic divisions of the class Mammalia, have parallelly evolved powerful claws and long snouts and tongues for digging out and capturing ants and termites. They are (A) the spiny anteater, a prototherian; (B) the banded anteater, a metatherian; and (C) the giant anteater, a eutherian.

and there are only two living groups, the duck-billed platypus and the spiny anteaters, which are both of particular interest to the comparative morphologist. They possess the diagnostic mammalian characters, e.g., hair, mammary glands, and a "mammalian" jaw joint. They also have some characters which are fundamentally "reptilian," e.g., limb posture, shoulder girdle, and young developing in a cleidoic egg. Possibly monotremes evolved from a different synapsid group than did other living mammals and retained "reptilian" features as well as evolving "mammalian" features.

Theria

The subclass Theria, rich in both fossils and living diversity, includes two infraclasses: Eutheria (placental mammals) and Metatheria (marsupials). These two infraclasses diverged sometime before the mid-Cretaceous. The marsupials underwent two major adaptive radiations—one in South America and the other in Australia (Fig. 2-10).

Placental mammals have had multiple adaptive radiations and are now the dominant mammalian fauna on every continent except Australia. The earliest known placentals were small, semiarboreal, insectivorous forms which are placed in the order Insectivora. They were quite unlike today's specialized Insectivora (moles, shrews, hedgehogs) and all other living mammals, but it was from them that the great mammalian diversity arose. One can consider the early Cenozoic evolution of Eutheria as a huge adaptive radiation which established

Fig. 2-11. Diversity of eutherian mammals.

the major orders of placental mammals: Insectivora, Chiroptera (bats), Cetacea (whales), Rodentia, Edentata (e.g., anteaters), Carnivora, Proboscidia (elephants), Artiodactyla (deer, etc.), Perissodactyla (horses, etc.), Lagomorpha (rabbits and hares), and Primates (Fig. 2-11). Within each order there was continual diversification and smaller adaptive radiations as families, genera, and species came into being, each with its own specializations and environmental limitations. Mammalian evolution, as nonsystematic as the evolution of any other group, has nevertheless almost completely filled all possible niches.

Fig. 2-12. Avian diversity.

AVES

Birds evolved from a group of small, bipedal archosaurs in the early- to mid-Jurassic, about 20 million years after the first mammals. Their earliest fossils (particularly Archeopteryx), although quite reptilian, include feather imprints which by definition makes them birds. Like mammals, their massive adaptive radiations did not begin until the latter part of the Cretaceous, and only in the last 75 million years have birds and mammals dominated the terrestrial scene. The two groups, it will be noted, evolved independently from completely separate lines of reptiles, yet both have developed a high degree of adaptation to terrestrial life.

One of the factors responsible for the success of birds was the evolution of a light, delicate skeleton beautifully adapted to flight. Such a skeleton fossilized only infrequently, and therefore details of avian adaptive radiations are not well known. However, today there are 21 orders containing more than 8600 species—twice as many as the species of living mammals (Fig. 2-12).

SPATIAL AND TEMPORAL DIVERSITY

All vertebrates alive today are the result of some 450 million years of the pressures of natural selection, which they may have survived by retaining generalized characteristics or by developing many and extreme specializations. We can call this present result of vertebrate evolution—that is, those species alive today—the current *spatial diversity of vertebrates.*

How they got this way—how they changed through time—comprises their phylogenies, which we can call the *temporal diversity of vertebrates.* Although the fossil record is limited (usually) to hard parts, it may be possible by various methods to reconstruct the soft parts and their temporal diversity.

One way is by studying the spatial diversity. Let us say, for instance, that we want to know about the temporal diversity of the mammalian heart. If we study the heart in living reptiles, birds, and mammals, we learn that at least one heavy-walled ventricle and thin-walled atrium are found in all these forms. We could thus rather confidently infer that at least this much must have been present in the cotylosaur, which was their common ancestor.

Another way of determining temporal diversity of soft parts is by using Cuvier's principle of the correlation of parts. In the shoulder of all living forms, for instance, the supraspinatus muscle is found only in conjunction with the supraspinous fossa of a shoulder bone, the scapula. Therefore the presence or absence of this muscle in any fossil can be determined by the presence or absence of the supraspinous fossa.

Another useful correlation is the well-understood relationship between structure and habitat. The very rock in which a fossil is found reveals something about the habitat and therefore about the structure of the extinct animal.

Some insight on temporal diversity can be gained from the *ontogeny,* or life history, of the individual. Up to the pharyngula stage (described below), early development is similar in all vertebrates, except that development occurs on a flat, broad plane of yolk in large-yolked eggs (e.g., chondrichthyeans, birds, reptiles), while in small-yolked eggs (e.g., most fish, amphibia, mammals) the entire egg is incorporated in development. As each group develops past the pharyngula stage, it becomes gradually more specialized until it has the precise adult form of its species. The changes which occur during this ontogeny can tell us something about the phylogenetic changes which comprise temporal diversity, but they must be interpreted carefully, since Haeckel was overgeneralizing when he declared that ontogeny recapitulates phylogeny.

EMBRYOLOGICAL DEVELOPMENT

From the egg to the pharyngula stage, however, development is orderly and constant in all vertebrate groups. Initially there is cleavage, in which cells proliferate; the result of this process is the blastula, a mass of morphologically undifferentiated cells. Partly by further cell division but mostly by migration the cells of the blastula become organized into the gastrula, which is composed of three layers: the outer layer, called the ectoderm, will give rise primarily to the outer skin and the nervous system; the innermost layer, the endoderm, will give rise to the digestive system; the intermediate layer, the mesoderm, will give rise to other vertebrate tissues such as muscles and kidneys.

The gastrula starts to elongate and the dorsomedial surface of the ectoderm turns inward to form the long, hollow nervous system or neural tube, and differentiation of body parts begins. This is the pharyngula stage. It has a definite head region and anterior enlargements of the nervous system for the presumptive brain. Between the neural tube and the surface ectoderm lie small isolated bits of ectoderm, the neural crest, which will give rise to several diverse structures. The notochord, which supports the embryo in the axial plane, forms just under the dorsal hollow nerve tube. Ventral to the notochord lies the tube of endoderm which will form the lining of the alimentary canal; anteriorly, pharyngeal clefts (which give it the name pharyngula) are already present (Fig. 2-13).

Deep to the surface ectoderm lies the mesoderm which in the pharyngula has three major components: epimere, mesomere, and hypomere. The *epimere* lies dorsally, between surface ectoderm and nerve cord, and can be further divided into a medial sclerotome giving rise to skeletal structures, a lateral dermatome giving rise to the deep layer of the skin, and an intermediate myotome giving rise to much of the skeletal musculature. The *mesomere,* or intermediate mesoderm, is a constricted area just ventral to the epimere which gives rise to genital and excretory structures. The *hypomere* lies ventrally between endoderm and surface ectoderm; it contains the body cavities, and gives

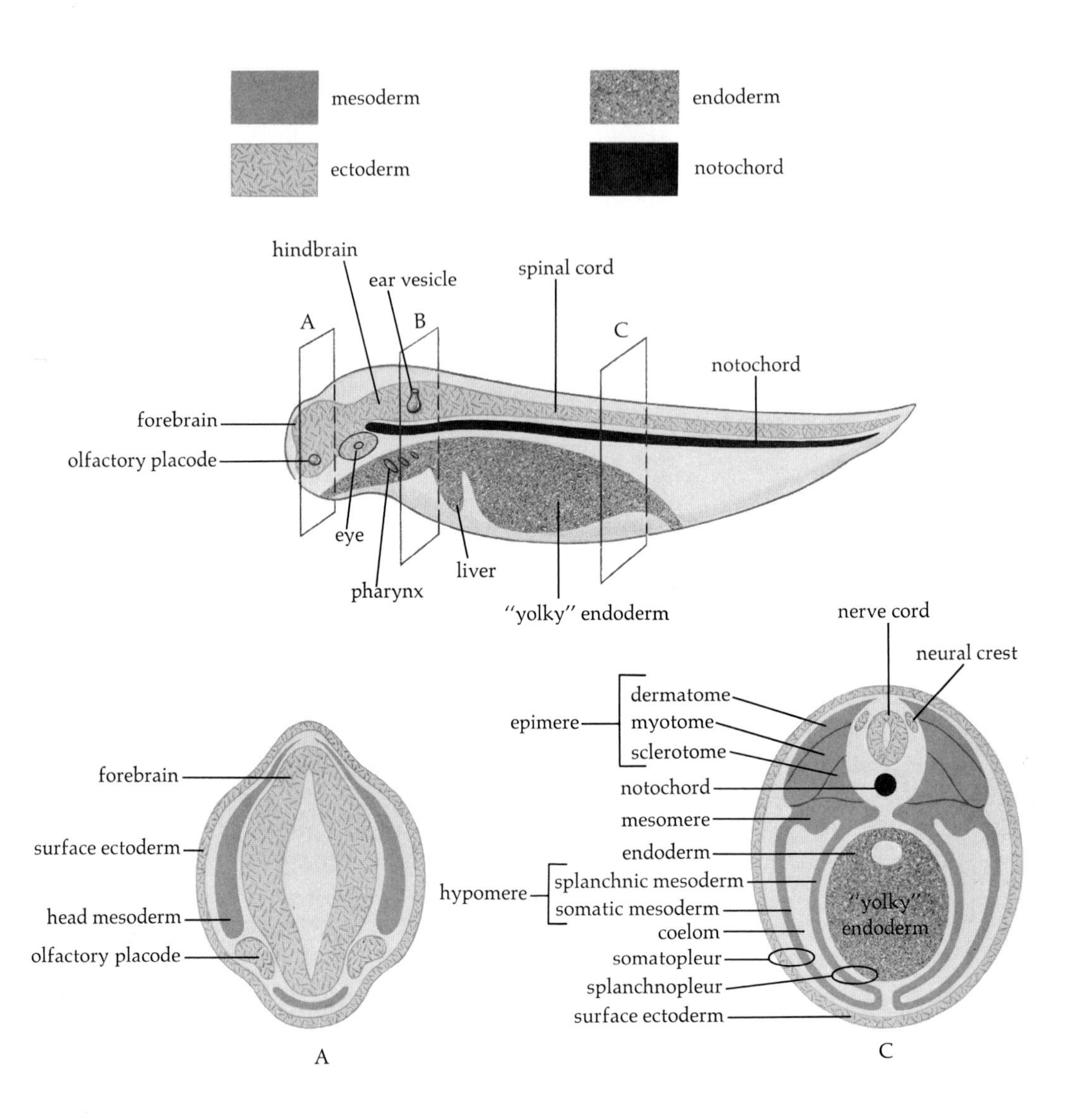
mesoderm
endoderm
ectoderm
notochord
hindbrain
ear vesicle
spinal cord
A
B
C
notochord
forebrain
olfactory placode
eye
pharynx
liver
"yolky" endoderm
nerve cord
neural crest
dermatome
epimere
myotome
sclerotome
notochord
mesomere
endoderm
hypomere
splanchnic mesoderm
somatic mesoderm
coelom
somatopleur
splanchnopleur
surface ectoderm
"yolky" endoderm
forebrain
surface ectoderm
head mesoderm
olfactory placode
A
C

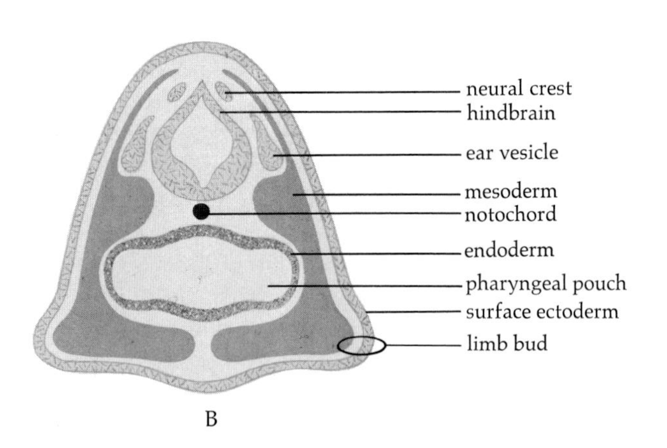
neural crest
hindbrain
ear vesicle
mesoderm
notochord
endoderm
pharyngeal pouch
surface ectoderm
limb bud
B

Fig. 2-13. General morphology of an amphibian pharyngula.

rise medially to smooth muscles and connective tissue and laterally to striated muscles and connective tissue.

This pharyngula stage is of particular significance to the evolutionary morphologist because in it the generalized anlagen (embryonic precursors) of tissues and organs are represented similarly in all vertebrates. Beyond this stage development in any species is accompanied by the emergence of particular characters setting it apart from other species, and the further ontogenetic development proceeds, the more species-specific and individual-specific the animal becomes. By studying embryos which have developed to various stages past the pharyngula it can be shown, for instance, that a jaw cartilage of a shark develops from the same anlage as an ear bone of a mammal. The two are therefore homologous as Owen defined homology in the nineteenth century—the same structure in different animals. The later the ontogenetic stage at which a developing structure becomes different in two animals, the more precisely the homology can be determined. For instance, the forelimb muscles of a perch and a cat are different from a very early stage; we can call them field homologies for they develop from similar, large, undifferentiated cell masses, but there is no basis on which to make closer homologies. On the other hand, the forelimb muscles of a dog and a cat develop very similarly and diverge after individual muscles can be readily distinguished; thus we can point to specific muscles which are homologous in the two species. Even more significant, however, are the facts that these structures have common evolutionary ancestries and that their particular relationships are best determined by comparative embryological studies.

Thus, through comparative anatomy and comparative embryology, correlated with paleontology, the history and diversity of vertebrate life become more meaningful. In the chapters which follow we will look at the origins and phylogenetic development of each organ system, examine the diversity of their forms, and relate this diversity to the habitat and functioning of the individual.

SUGGESTED READING

Ballard, W. W. (1964). "Comparative Anatomy and Embryology." The Ronald Press Co., New York.

Bardack, D., and Zangerl, R. (1968). First fossil lamprey: a record from the Pennsylvanian of Illinois. *Science* **162,** 1265–1267.

Bolt, J. R. (1969). Lissamphibian origins: Possible protolissamphibian from the lower Permian of Oklahoma. *Science* **166,** 888–891.

Colbert, E. H. (1955). "Evolution of the Vertebrates." John Wiley & Sons, New York.

Romer, A. S. (1966). "Vertebrate Paleontology," 3rd ed. Univ. of Chicago Press, Chicago.

Simpson, G. G. (1967). "The Meaning of Evolution," revised ed. Yale Univ. Press, New Haven, Connecticut.

3 SKELETAL COMPONENTS

3

VERTEBRATE SKELETAL TISSUES

Because they are rigid and resist compression, vertebrate skeletal tissues support and protect the soft parts of the body. They also act as levers on which muscles pull for body movements. Biochemically, bone is a storage reservoir for calcium and phosphate ions, and bone marrow is a major site for the manufacture of red blood cells.

Skeletal tissues are of three types: notochord, cartilage, and bone. Each has distinct structural characteristics, but all are composed of living cells in a nonliving matrix which they secrete. New skeletal material can always be laid down and resorbed, and the shape of skeletal elements can change in response to external and internal stresses.

NOTOCHORD

The notochord, a flexible, rodlike structure, is the primary axial skeletal element in the early development of all vertebrates and one of the distinguishing characteristics of the phylum Chordata. By the pharyngula stage, the midline tissue which lies just dorsal to the endoderm and ventral to the nerve cord has differentiated into notochordal tissue. Unlike other skeletal tissue, most of the notochord is cellular; its major acellular portions are the collagen and the elastic fibers of its tough connective tissue outer sheath (Fig. 3-1). Large, turgid, vacuolated cells fill its central core, and together with the connective tissue sheath give the notochord its characteristic stiff flexibility.

In the embryo the notochord extends from the tail through the posterior half of the skull, terminating at the point where, in the adult, the pituitary gland lies. In a few adults the notochord is the primary axial skeleton (Fig. 3-2), but in most adult tetrapods it is replaced to a greater or lesser extent by the vertebral column.

CARTILAGE

Cartilage also develops embryologically in all vertebrates, and in the adult, except Agnatha and Chondrichthyes, is replaced to a greater or lesser extent by bone. It is a milky, clear-to-opaque substance, moderately flexible and elastic. The primary, or parenchymal, cells of cartilage are chondrocytes. They are embedded in a gelatinous matrix of acellular material which the chondrocytes themselves secrete. This acellular matrix is primarily a sulfated mucopolysaccharide, acidic in nature, called chondroitin sulfate.

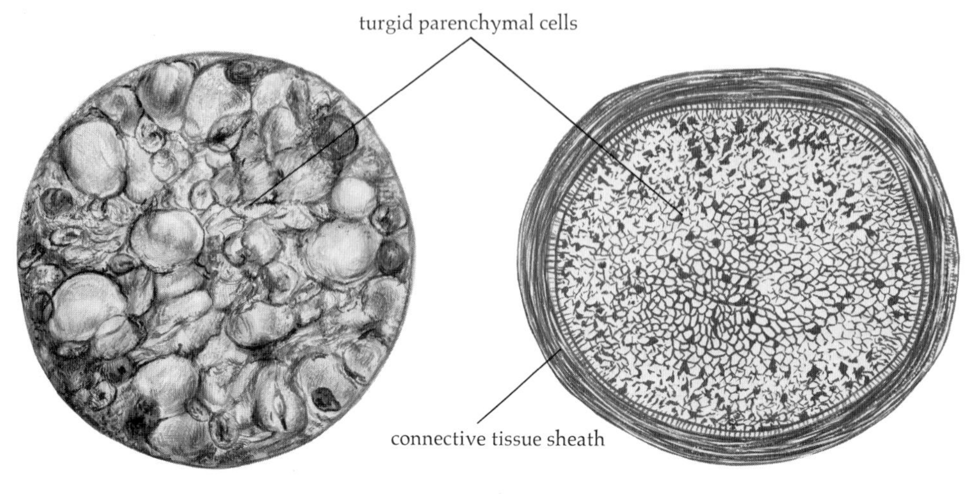

Fig. 3-1. In all vertebrates the embryonic notochord lacks the tough, connective tissue sheath which is present in the adults of some fish such as the lamprey.

Types of cartilage

There are four types of cartilage, distinguished by differences in their matrices (Fig. 3-3). *Hyaline* cartilage has a matrix of almost pure chondroitin sulfate. *Fibrous* cartilage has a dense, tough matrix containing collagen fibers; it is typically found in areas of considerable stress, such as at the attachment points of ligaments. *Elastic* cartilage has elastic fibers rather than collagen fibers in the matrix; it is best known in the mammalian outer ear. *Calcified* cartilage sometimes occurs when cartilages are under great compression forces, as in some of the large chondrichthyean fishes. The cartilaginous matrix becomes infiltrated with calcium salts, making the entire cartilage very hard and brittle. Although it superficially resembles bone, it is technically cartilage because it contains chondrocytes and chondroitin sulfate. Calcified cartilage is also a transitory structure in the process which replaces cartilage with bone, described below.

Unlike most other tissues, cartilage has no direct vascular supply. Blood vessels run in the thin connective tissue covering (perichon-

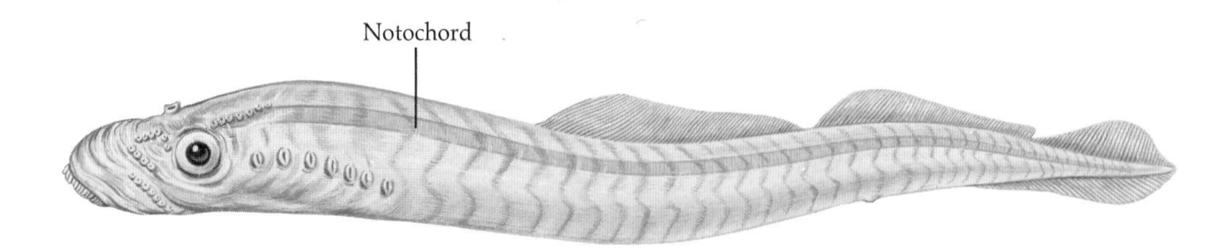

Fig. 3-2. In the lamprey the notochord is the primary axial skeleton.

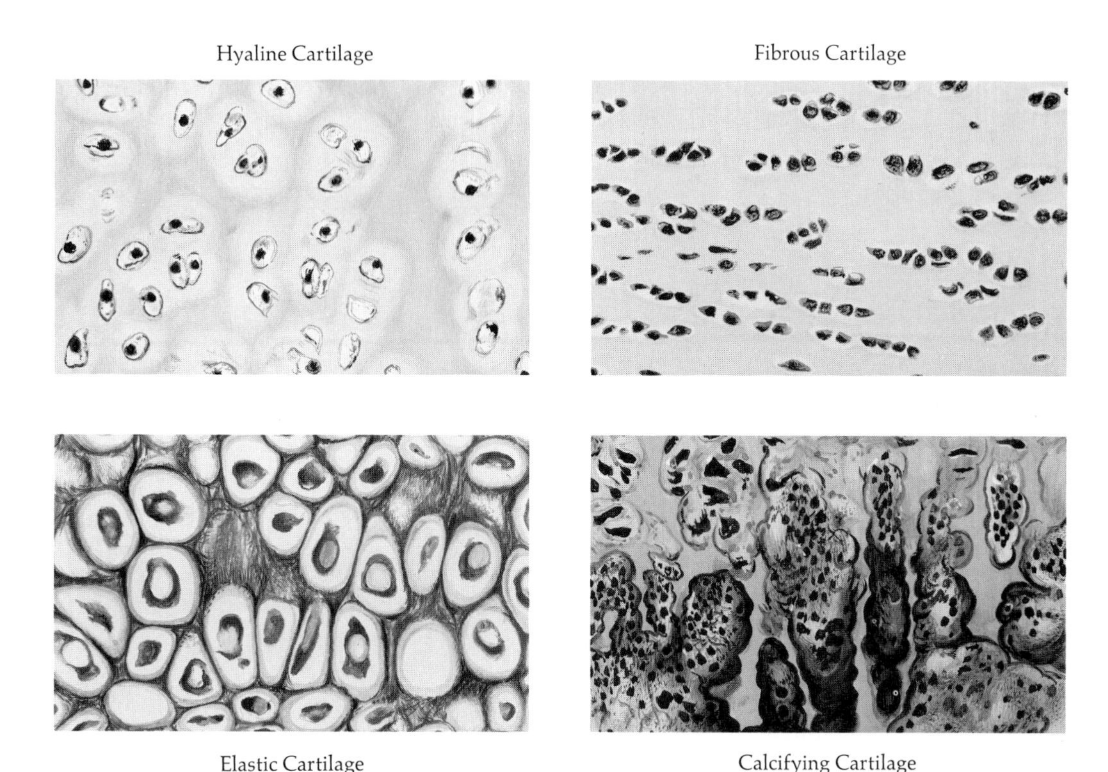

Fig. 3-3. The cells of all cartilage are chondrocytes. It is the structure of the matrix which determines the type of cartilage and its physical properties.

drium), and the metabolic needs of the chondrocytes are satisfied by diffusion through the matrix, often for considerable distances.

Development of cartilage

Embryologically, cartilage develops from two sources. That which forms the visceral skeleton and anterior neurocranium of the head is of neural crest origin and is formed in the following manner. In the pharyngula head region the neural crest element is especially large, and much of it migrates to the anterior splanchnic mesoderm. This tissue, then called mesectoderm, differentiates into the cartilaginous visceral skeleton which supports the gill skeleton and its derivatives.

However, most cartilage is of mesodermal origin (Fig. 3-4). Some differentiates from epimeric mesoderm in the head region: next to the anterior notochord or in the prechordal area anterior to the notochord. Some of these cartilages form protective capsules around the sense organs, and others form the floor, walls, and some of the roof of the braincase. Postcranially, the epimeric mesoderm adjacent to the nerve cord and notochord is called the sclerotome, which gives rise to cartilage for vertebrae. The more lateral mesoderm may also give rise to ribs, and the more ventral to the sternum. Cartilages of the appendages and

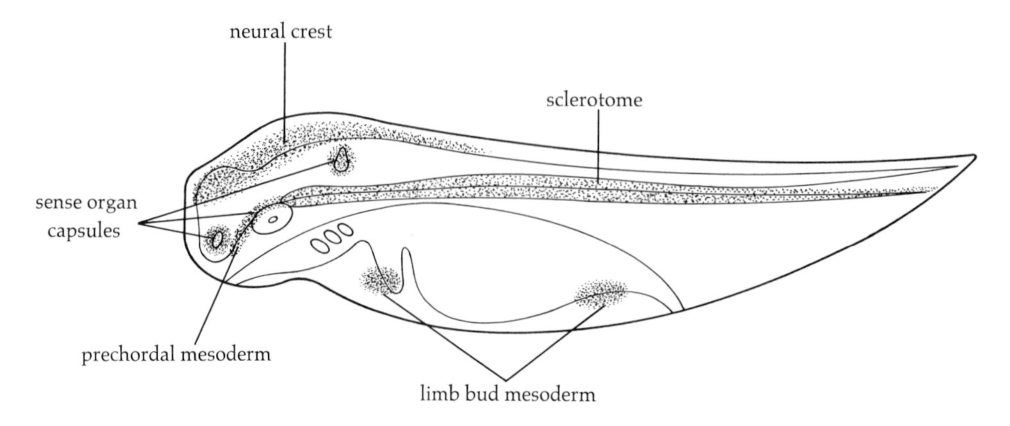

Fig. 3-4. Various embryonic structures of the pharyngula differentiate into cartilage.

their girdle supports form as budlike enlargements of the somatic mesoderm in the presumptive limb regions.

Advantages and disadvantages

Although cartilage develops embryologically as one of the earlier skeletal tissues and is more abundant in fishes than in tetrapods, it should not be regarded as a "primitive" tissue. Indeed, in many situations it is highly adaptive: because it is flexible, it is also resilient and less apt to fracture. On the other hand, it has two potentially major drawbacks.

The greater an animal's size, the more force its muscles exert on their attachments and, as they contract, on nearby soft organs. Normal cartilage, being less resistant to compression than bone, cannot adequately protect the soft parts of animals beyond a certain critical size; calcified cartilage can. However, calcified cartilage is heavier and more

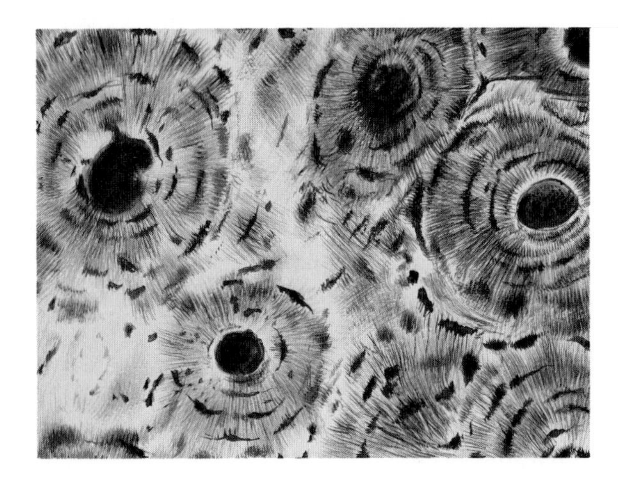

Fig. 3-5. Cross section of Haversian canals.

brittle than bone, and therefore less adaptive. Interestingly, the critical size differs for land- and water-dwelling animals: land animals, which must support and propel themselves without the buoyancy of water, require larger muscles and limbs than aquatic animals of equal size. Not surprisingly, then, it is two classes of fishes which possess entirely cartilaginous skeletons, and calcified cartilage is found only in the largest chondrichthyeans.

The second major drawback to cartilage is that it cannot act as a reservoir for calcium and phosphate. This is not important for marine animals which have an abundant supply of these minerals in their external milieu, but it is important for freshwater and terrestrial forms, and is no doubt another reason why in most classes bone replaces cartilage as the primary skeletal element.

BONE

Structure

Bone is found in all vertebrates except living agnathans and chondrichthyeans. It is a hard, rigid tissue of limited elasticity, which is opaque unless thin. Like cartilage, it is composed of a parenchymal cell, the osteocyte, embedded in a secreted acellular matrix. This matrix is composed of complex inorganic salts (phosphate and carbonate), which combine with calcium to form apatite crystals, and of much organic material which consists primarily of collagen. Within this matrix the cells lie in geometrically organized lacunae, or spaces.

Bone is highly vascularized, both centrally, where the bone marrow frequently manufactures red blood cells, and peripherally, where blood vessels enter the bone in canals called nutrient foramina.

Thicker bones are organized into internal columns called Haversian canal systems. In the center of each column is a Haversian canal in which travel blood vessels and sometimes nerve fibers that enter and leave through transverse canals of Volkmann. The osteocytes within their lacunae remain in nutrient communication with the Haversian canals and with other osteocytes by way of tiny anastomosing canaliculi (Fig. 3-5). This concentration of vascular and neural tissue is responsible for the high metabolic rate and plasticity of bone. (Plasticity, an often misunderstood word, means the capability of being molded or receiving form; it is the opposite of fixity.)

Upon cutting across almost any bone one finds an outer portion called compact or lamellar bone and an inner portion called cancellous bone.

Compact or lamellar bone is dense, hard, and solid, accounting for most of the bone's weight, mass, and rigidity. Externally it is composed of broad layers parallel to the surface, having no Haversian canal systems. There are typically two such layers, at right angles to one another, called the outer and the inner circumferential lamellae. Deep to these layers, the compact bone is organized in long cylindrical Haversian canal systems, or osteons. The osteons are tied together by less regularly shaped columns of compact bone called interstitial lamellae.

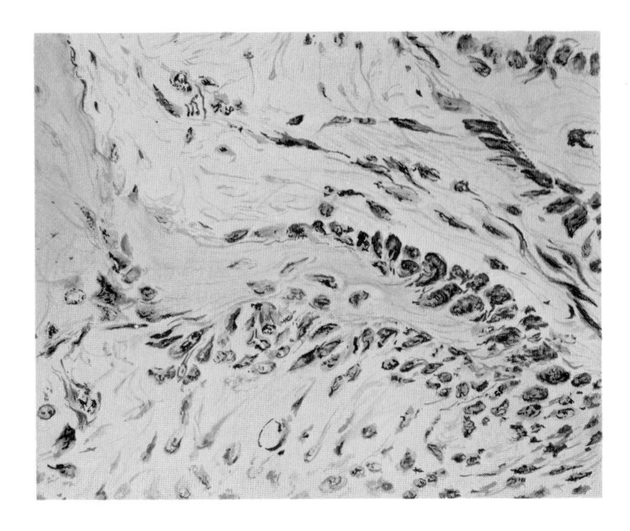

Fig. 3-6. Development of a dermal bone. Note the concentration of osteoblasts laying down the bone matrix.

Cancellous bone is more hollowed out, with trabeculae between which are spaces occupied in the living animal by bone marrow. A careful study of the trabeculae demonstrates that they are oriented much as bridge struts would be positioned by an engineer, giving maximal strength to the bone.

Development of bone

Although all adult bone is structurally similar, it is classified according to its histogenesis as dermal bone, cartilage replacement bone, or heteroplastic bone.

Dermal bones are the clavicle, the flat bones of the skull, and the flat bones found elsewhere such as in turtle and armadillo shells. Dermal bone forms in the presumptive deep layer, or dermis, of the skin by a process called intramembraneous ossification. In this process, embryonic mesodermal cells differentiate into osteoblasts and fibroblasts. The fibroblasts then lay down a network of collagen and reticular fibers, and upon this network the osteoblasts secrete the apatite crystals of calcium, phosphate, and carbonate. It is this secretion of the apatite crystals by the osteoblasts that is called ossification. Each osteoblast continues to secrete apatite crystals until it is completely encased in a matrix; then its active secretion is suspended and it is called an osteocyte (Fig. 3-6).

By the continued activity of fibroblasts and osteoblasts, spicules of bone are formed in broad, flat, platelike structures within the presumptive dermis. Soon there differentiates from the dermal mesenchyme another type of cell, the osteoclast, which is multinuclear in nature and is capable of resorbing or dissolving the matrix of bone.

The osteoblasts lie mostly along the surface of the bone, forming the inner portion of the periosteum which covers the bone. The osteoclasts lie deeper. The bone develops and continues to grow as the osteoblasts lay down new layers of circumferential lamellae, and the osteoclasts resorb the matrix of former outer circumferential and inner circumferential lamellae, reforming these lamellae into osteons, or Haversian canal systems; deeper into the bone, osteoclasts erode the bony matrix of former osteons, forming marrow cavities lined with fine bony trabeculae.

Thus the major growth of membrane, or dermal, bone occurs on the outer surfaces, with resorption and remodeling of bone occurring in the deeper portions. This growth along the surface is called periosteal ossification.

Cartilage replacement, or endochondral, bone comprises parts of the braincase and visceral arch skeleton and, postcranially, most of the

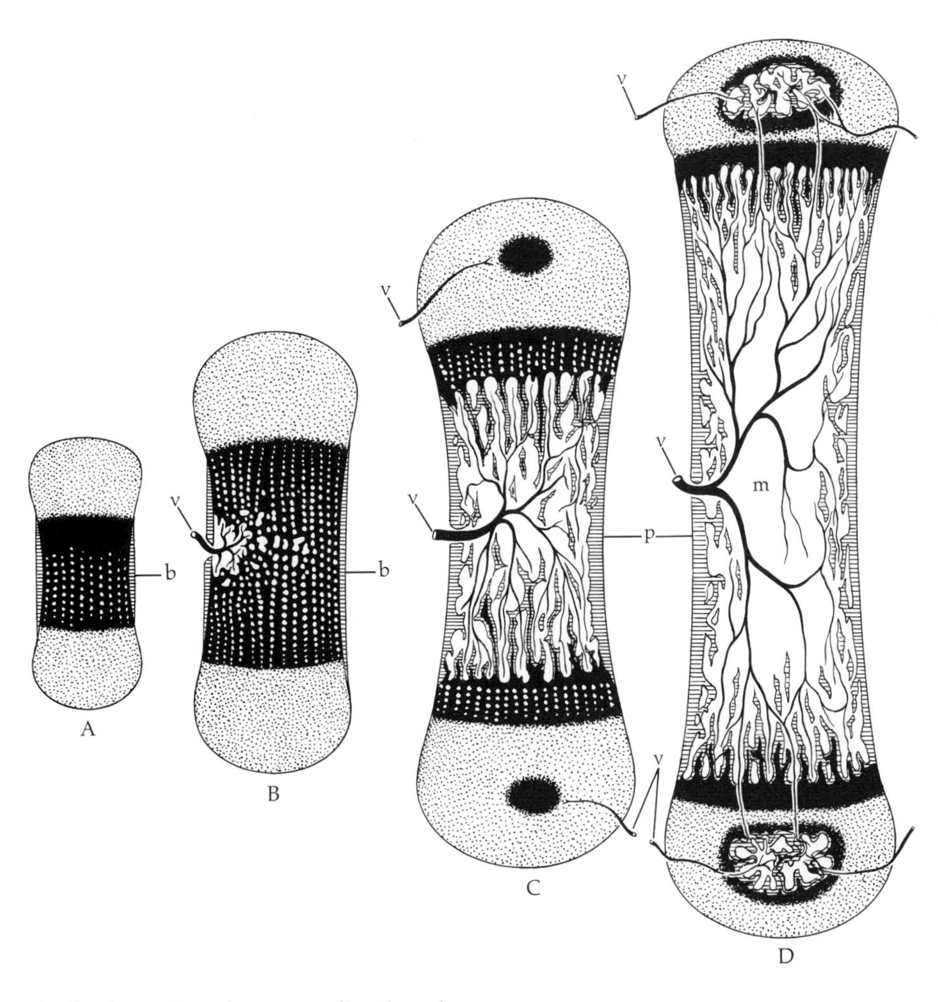

Fig. 3-7. Stages in the formation of a mammalian long bone. Black portions indicate calcified cartilage, and stippled portions indicate hyaline cartilage. Blood vessels (v) enter the marrow cavities (m), which are created as calcified cartilage is eroded during the laying down of bone (b) by osteoblasts. After W. F. Windle (1969). "Textbook of Histology." McGraw-Hill, New York. Used by permission of McGraw-Hill.

vertebral column, ribs, sternum, girdles, and appendage skeletons. Its embryonic development involves a cartilage precursor: a miniature "model" of the bone which is to be, usually including its processes, turns, and shapes. Cartilage cells and cartilage matrix are never transformed into bone; rather they are *replaced* by bone cells and bone matrix of apatite crystals.

In a typical long appendage element such as the humerus, the replacement process begins centrally as the chondrocytes expand and the matrix begins to calcify (Fig. 3-7). When calcification is complete, the chondrocytes die. The area is then invaded by blood vessels and, with them, mesenchymal cells, some of which differentiate into osteoblasts and then start laying down bony spicules; at the same time the calcified cartilage is resorbed by other cells of the invading vascular tissue.

The first product of the cartilage replacement process is spongy, cancellous bone in the center of the shaft. This bone increases in diameter, and in time the outer portion of the shaft becomes compact bone, in the layers already described, and growth in girth then continues by periosteal ossification as it does in membrane bone.

The two ends of a long bone, such as the humerus, remain cartilaginous until much of the shaft is ossified. Then calcified cartilage forms centrally in each end nubbin; this calcified cartilage is invaded by vascular tissue, and the laying down of cancellous bone and replacement of the cartilage occurs. The extreme tips, or articular facets, form the joints; they do not ossify.

The ossifying ends of the bone are called the epiphyses, and the shaft is the diaphysis. Between the diaphysis and each epiphysis, where the epiphyseal center of growth and the diaphyseal center of growth meet, is the metaphyseal plate. This is the only area where the bone can grow in length, which it does by the process of endochondral ossification. In this process, the cartilage along the metaphyseal plate becomes calcified; the cartilage is eroded; osteons of bone are laid down; and new cartilage is formed along the plate. The metaphyseal plate itself remains cartilaginous as long as the bone is capable of growth.

Heteroplastic bone, the third type, forms within other, already differentiated tissues. Some heteroplastic bones are abnormal, apparently being produced by the physiological response of fibrous connective tissues to extreme stress; in fact, under proper stimulation bone can form in almost any tissue including, for instance, heart and kidneys.

Other heteroplastic bones are normal to a species; these are called sesamoid bones: the patella (kneecap) is an example. Splint bones are heteroplastic bones formed in the tendons or ligaments of older animals; in horses and some others they are normal, sesamoid bones, while in human beings they are abnormal, can be quite troublesome, and are sometimes surgically removed.

It must be reemphasized that despite differences in histogenesis, the structure of all adult bone is identical. It is composed of both compact and cancellous bone, each characterized by a matrix of apatite crystals and living cells within lacunae. The few exceptions to this rule occur mainly among teleosts, whose adult bones lose their lacunae and cells

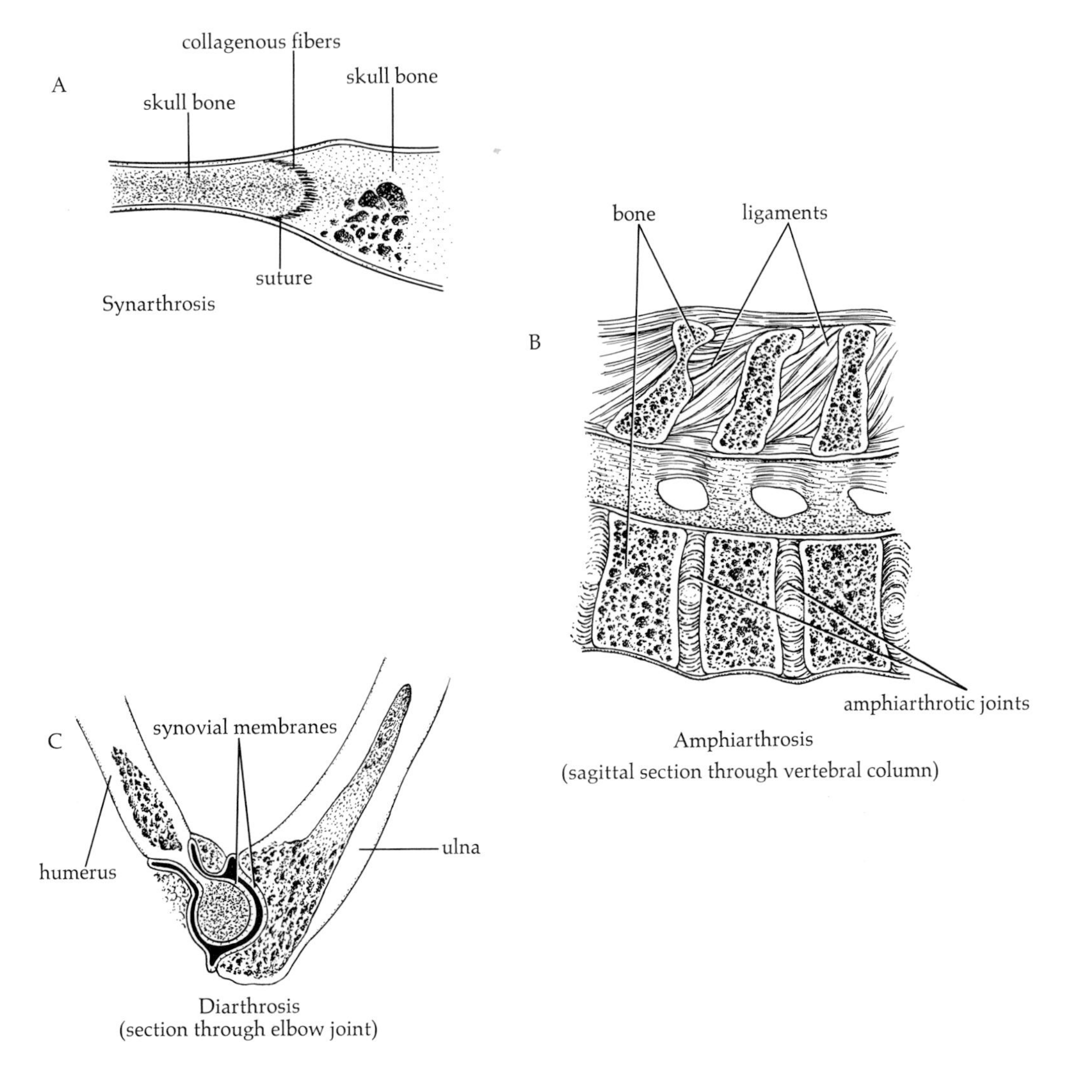

Fig. 3-8. Gross structure of joints.

and become dead, implastic, acellular structures. Otherwise, vertebrate bone is a living, plastic structure, capable of changing shape throughout life.

JOINTS

Where two skeletal elements meet a joint is formed. Sometimes, but not always, they are held together by connective tissue bands known as ligaments (as opposed to tendons, which join a muscle and a skeletal element). There are different types of joints, classified by the degree of movement possible between the elements (Fig. 3-8).

If there is no movement, as is usually the case in the skull, the joint is called a synarthrosis. The two elements are usually connected by heavy, collagenous fibers, or sometimes by fibrous cartilage. However, the two elements may also grow together, i.e., ankylose, without even a suture line to show the junction; this is a bone complex.

An intricate structure is necessary when there is wide movement between the elements, forming a diarthrosis. In this joint the two elements usually meet in cartilage-covered articular facets, whose shape helps to determine the direction and extent of movement; movement is also determined by the structure and the placement of the ligaments. In many diarthrotic joints an epithelial tissue called the synovial membrane lies within the joint cavity; it secretes and encloses the synovial fluids that lubricate the joint. It is not unusual to have a lemniscal fibrous cartilage within the synovial cavity as well, acting as a pad for the joint. (Synovial membranes are not limited to joints. Tendons which must extend for long distances are frequently enclosed in synovial membranes and lubricated by the synovial fluids.)

If there is limited movement between the elements of the joint it is called a symphysis or amphiarthrosis. These are less elaborate than the diarthrotic joints, being composed usually of elastic, fibrous connective tissue.

SUGGESTED READING

Bloom, W. and Fawcett, D. W. (1968). "A Textbook of Histology," 9th ed. W. B. Saunders Co., Philadelphia.

Moss, M. L. (1968). Comparative anatomy of vertebrate dermal bone and teeth. *Acta Anat.* **71,** 178–208.

Patt, D. I. and Patt, G. R. (1969). "Comparative Vertebrate Histology." Harper & Row, New York.

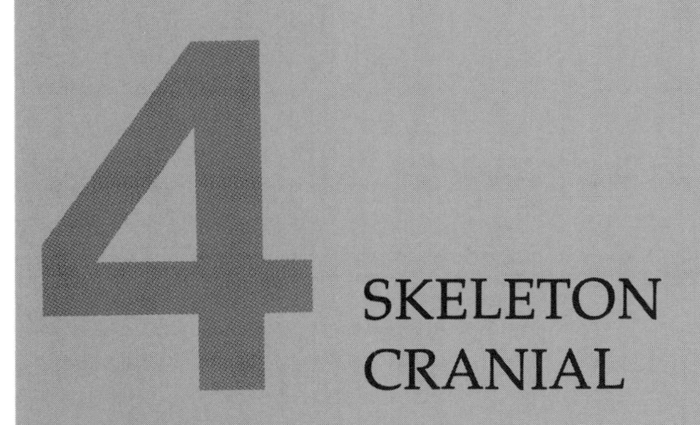

4 SKELETON CRANIAL

4

CEPHALIZATION

A characteristic of vertebrates (and bilaterally symmetrical, highly mobile invertebrates) is cephalization, that is, the concentration of sense organs, brain, food-gathering apparatus, and much of the respiratory system in the head. Thanks to cephalization an animal has immediate information about each new subenvironment it enters, can catch unwarned prey, and can take in water or air not yet contaminated by its own wastes.

The concentration of so many vital organs in one area increases their potential vulnerability and their need for protection. Too much protection, however, would deprive the animal of the very values of cephalization. To be adaptive, the head skeleton must protect the delicate organs; must allow the appropriate stimuli to reach the eyes, ears, and nose; and must facilitate food gathering and oxygen intake. It is not surprising that this is the most complex and intricate portion of the skeleton, exhibiting much structural diversity because natural selection has shaped specific features to specific environments. Simply stated, the head skeleton has three structural and functional components: the neurocranium, which supports the brain and sense organs; the splanchnocranium, which supports the respiratory and food-gathering apparatus; and the dermatocranium, which protects deep-lying delicate tissues.

EMBRYOLOGIC DEVELOPMENT OF THE HEAD SKELETON

THE NEUROCRANIUM (Fig. 4-1)

The primordia of most features of the head are distinguishable at the pharyngula stage, but the only skeleton at this stage of development is the notochord. It lies in the midline just under the hind part of the presumptive brain. The lateral somatic mesoderm flanking its anterior end differentiates into two parachordal cartilages, which enlarge, fuse around the notochord, and form the basal plate of the neurocranium. Paired occipital arch cartilages develop around the hindbrain, unite with the basal plate, and form the posteriormost walls of the neurocranium. Paired otic capsule cartilages develop around the developing inner ears, uniting with the basal plate and the occipital arches. A small cartilage, the synotic tectum, develops above the rear part of the hindbrain and also connects with the occipital arches. Thus the floor, walls, and partial roof of the neurocranium form around the presumptive hindbrain.

At the same time the neurocranium forms anteriorly. From the neural crest develop trabecular cartilages (paired in fishes, single in

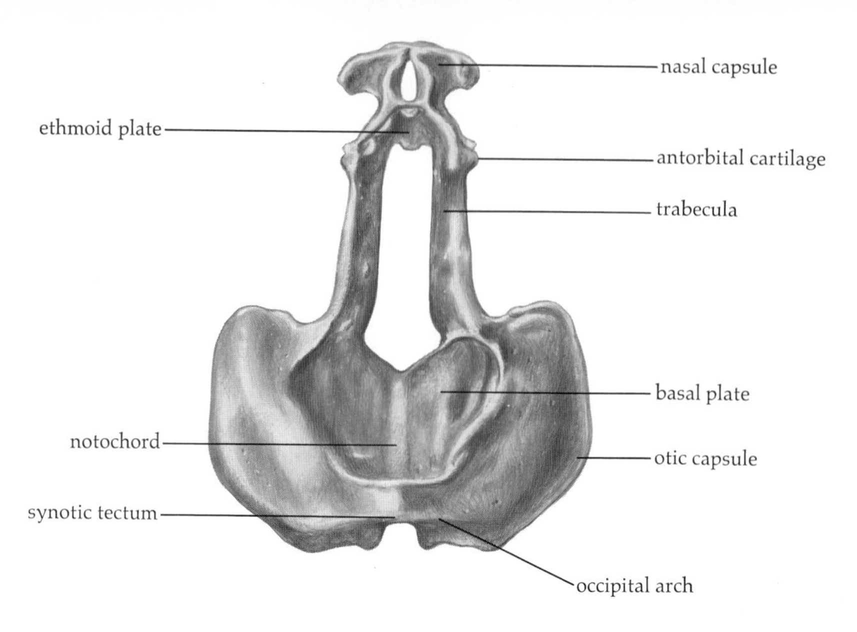

Fig. 4-1. Dorsal view of salamander neurocranium.

amniotes), which lie longitudinally under the presumptive forebrain, and then unite to form the trabecular, or ethmoid, plate. The somatic mesoderm anterior to the notochord, called prechordal mesoderm, gives rise to other elements of the anterior neurocranium: a pair of orbital cartilages forms the walls between eyes and brain; a prominent antorbital process forms the front wall of the orbit; the basitrabecular processes, developing from the posterior trabecular plate, grow ventrally around the presumptive pituitary; thin cartilages encapsulate the nasal sacs.

Finally, the anterior and posterior neurocrania fuse, although in *Latimeria* and extinct crossopterygians a joint persists between them.

At this stage, then, neurocranium forms the floor, partial walls, and posteriormost roof for the brain, and cartilaginous capsules around the major sense organs. Its further development varies in different groups (Fig. 4-1).

THE SPLANCHNOCRANIUM (Fig. 4-2)

Although definitions vary, we will regard the splanchnocranium as all of the skeleton developing in relationship to the pharynx. It arises from neural crest elements which migrate ventrally and differentiate into cartilages associated with pharyngeal pouches. These pouches differentiate into gills in fishes and diverse derivatives in tetrapods, while the visceral mesoderm of this area differentiates into the pharyngeal (branchiomeric) musculature; the splanchnocranium forms their skeletal support.

Of the splanchnocranial cartilages, the most anterior pair is the mandibular arch. In all living vertebrates except cyclostomes the mandibular arch forms the jaw skeleton of at least the early embryo. Its

QL 805 W38 / 14 931 CAMROSE LUTHERAN COLLEGE LIBRARY

primary parts are the palatoquadrate (dorsally) and Meckel's cartilage (ventrally), with the jaw joint lying between them in at least the early embryo.

Behind the mandibular is the hyoid arch, composed of a prominent hyomandibular (dorsally) and smaller elements (ventrally). The ventral end of the hyomandibular abuts against the mandibular arch at the jaw joint. The dorsal end, which normally articulates against the otic capsule, is the only part of the splanchnocranium that, very early in development, articulates with the neurocranium. With some variations, the ventral hyoid elements develop into the supporting skeleton for the tongue.

Behind the hyoid arch develop similar pairs of cartilages, one for each pharyngeal pouch (Fig. 4-2).

THE DERMATOCRANIUM

Dermal bone develops in the dermis of the outer head skin and the dermis underlying the oral cavity epithelium, and is a significant part of the skull (except in chondrichthyeans and living agnathans). However there are two major, distinctly different patterns of dermatocranial elements—one in actinopterygians and the other in crossopterygians and tetrapods—and a third, minor pattern in Dipnoi (which we will discuss only briefly). Evidently the two major patterns evolved separately and so differently that they are not directly homologous. In general, actinopterygians have a larger number of dermal skull bones than do living tetrapods, while in tetrapods there has been greater fusion of dermal with dermal and of dermal with neurocranial or splanchnocranial elements.

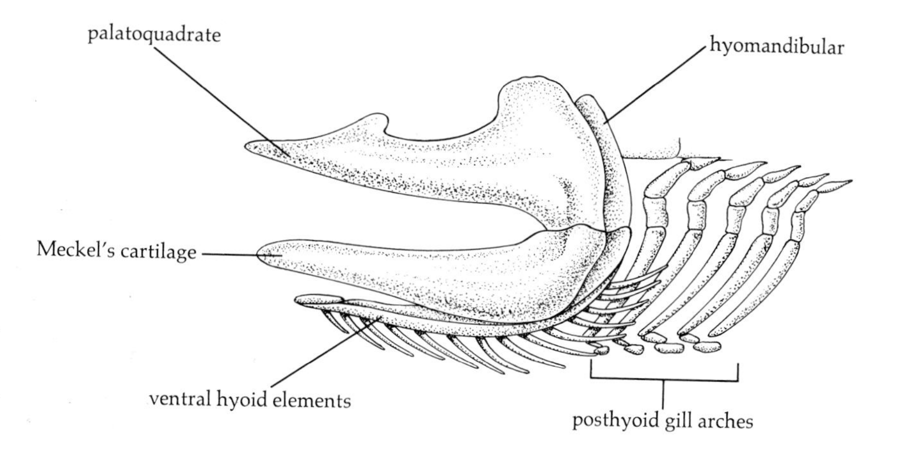

Fig. 4-2. Lateral view of the splanchnocranium of an acanthodian.

A GENERALIZED VERTEBRATE SKULL, *Amia calva*

Because of variations in these three components of the head skeleton, and because of the formation of several bone complexes, the relationships between neurocranium, splanchnocranium, and dermatocranium are often confusing. We will first examine an animal in which each of the three components is mostly, although not entirely, separate from the other two.

We can call this a "generalized vertebrate skull," so long as we are careful to understand the term. It means neither "primitive" nor "typical." It approaches what the natural philosophers wanted: a kind of archetype in which parts and relationships may be clearly seen; a simple statement of a plan often modified into incredibly complex and diverse forms. This animal is the holostean fish, *Amia calva,* familiarly known as the bowfin.

Bowfin neurocranium

More extensive than in most vertebrates, the neurocranium of the bowfin forms the complete roof for the braincase as well as its walls and floor; the only openings are where the nerves, blood vessels, and spinal cord enter and leave the brain. This structure is mostly cartilaginous, but there are several ossifications in the otic capsules and also at the points where the neurocranium fuses with the more superficial dermatocranium (Fig. 4-3).

Bowfin splanchnocranium

Ventral to the neurocranium and quite separate from it is the splanchnocranium. The jaw is based on the mandibular arch, whose dorsal element, the palatoquadrate, forms the main skeleton of the upper jaw. The posteriormost portion of the palatoquadrate is ossified as the quadrate bone, which articulates with the lower jaw. The anterior portion is a cartilaginous plate with three ossifications.

Meckel's cartilage forms the central skeleton of the bowfin's lower jaw, and most of it remains cartilaginous in the adult. However, its caudalmost portion ossifies as a very small articular bone, and anteriorly there are three or four tiny ossifications (Fig. 4-3).

Fig. 4-3. The skull of the bowfin, Amia calva. *The upper picture shows a lateral view of the entire skull. Except for a small area of neurocranium seen in the orbit and small areas of splanchnocranium in the upper jaw and hyomandibular regions, the entire superficial skull is dermatocranium. The middle picture is a lateral view of the isolated neurocranium, and the lower picture a lateral view of the isolated splanchnocranium. After E. P. Allis (1897). "Cranial muscles and cranial and first spinal nerves in* Amia calva." J. Morph. **XII,** *487–808.*

BOWFIN SKULL

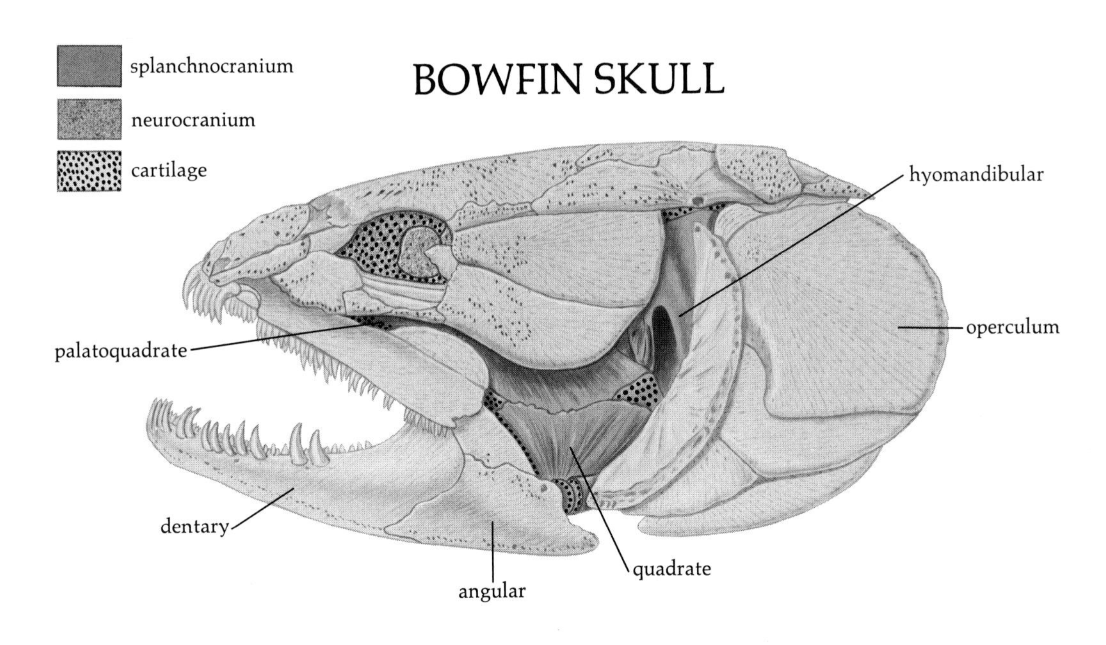

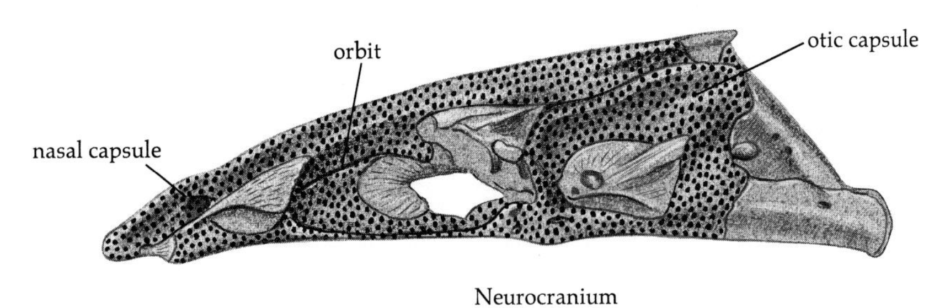

Neurocranium

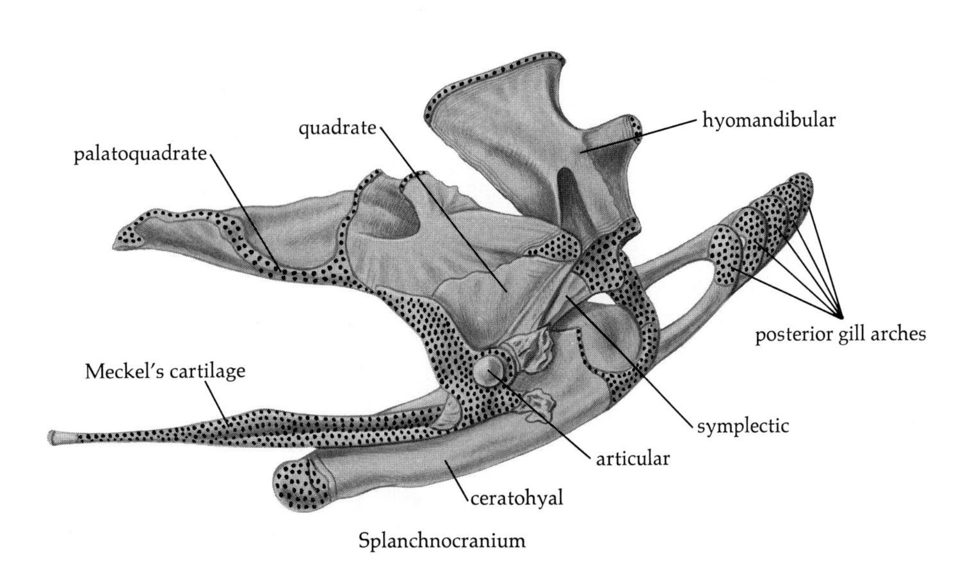

Splanchnocranium

In the bowfin's second visceral arch (the hyoid), the dorsal element is the hyomandibular, which dorsally butts against an ossification of the otic capsule and ventrally articulates with the quadrate by way of a separate, small hyoid ossification, the symplectic. The ventral elements of the hyoid arch are the epihyal, two separate ossifications of the ceratohyal, and finally the hypohyal. This ventral portion forms the primary skeleton of the floor of the oral cavity, and thus supports the rudimentary tongue and the muscles which move the oral cavity when water is pumped past the gills.

Behind the hyoid are five typical, or generalized, gill arch skeletons supporting the gill apparatus and gill slits. Each arch is composed of a dorsal pharyngobranchial element and—moving laterally, ventrally, and then medially—an epibranchial, ceratobranchial, hypobranchial, and usually a midline basibranchial element (Fig. 4-3).

Bowfin dermatocranium

The bowfin's dermatocranium forms an almost complete bony armor superficial to all other head structures except the epidermis (Fig. 4-3). As mentioned, the actinopterygian pattern of dermatocranial bones differs from that in all other living or extinct vertebrates. Although many of these bones are given the same names as bones of other vertebrate skulls, the homologies are dubious at best.

THE BOWFIN SKULL AS A FUNCTIONING UNIT

How well does the bowfin skull fulfill its functional requirements of support and protection, food gathering, and respiration?

The complete neurocranium, mostly cartilaginous but ossified at points of stress, gives the bowfin brain elegant support and protection from shock. Even more protection is provided by the outer, hard, rigid shield of the dermatocranium.

The sense organs are well protected and yet able to receive their proper stimuli. The inner ear is completely embedded in the otic capsule, but since its primary and perhaps only function is as an organ of equilibrium, the stimuli (gravity and acceleration forces) can reach it without difficulty. The eyes are well protected in the orbit, the medial portion of which is ossified as the orbitosphenoid. The opening is kept small and constricted (for protection) by a surrounding series of dermal bones. Similarly, the nasal sacs are protected by dermal bones without and cartilage within. The deep cartilages associated with all these sense organs give additional protection by acting as shock absorbers.

The bowfin's food-gathering machinery is formidable. There are numerous teeth of two kinds: those on the edges of the jaws are large, curved, and sharp, while those on the bones lining the oral cavity and pharynx are blunt. The jaw joint is posteriorly placed, allowing a large gape which is made even larger because the upper jaw can be raised at the same time that the lower jaw is lowered. This extra large gape allows the bowfin to eat larger prey and to grasp rapidly moving prey.

The head skeleton both supports the gills and facilitates respiration by moving water past the gills. Water is brought into the mouth; then the mouth is closed, and its floor raised by contraction of the muscles between the two sides of the lower jaw; this forces water past the gills where gas exchange takes place.

Despite the morphological separation of the bowfin's neurocranium, splanchnocranium, and dermatocranium, therefore, these three components of the head skeleton are intimately connected functionally.

DIVERSITY OF VERTEBRATE SKULL ORGANIZATION

Turning now to the spatial diversity of vertebrate skulls we find two major types of characteristics. First, any vertebrate's skull exhibits the major characteristics of its taxon, inherited over millions of generations. Second, it also exhibits more recently evolved features which are adaptations to its habitat. The pressure for these specific adaptations to evolve is so strong that vertebrates living in similar environmental niches, even if only distantly related, frequently have similar characteristics. For instance, although sharks and teleosts are of quite different phylogenetic stocks, with different jaw morphology and embryologic development, some species in each group have independently evolved powerful, flattened teeth and subsist on a diet of mussels whose hard shells they crush before eating the soft parts (Fig. 4-4).

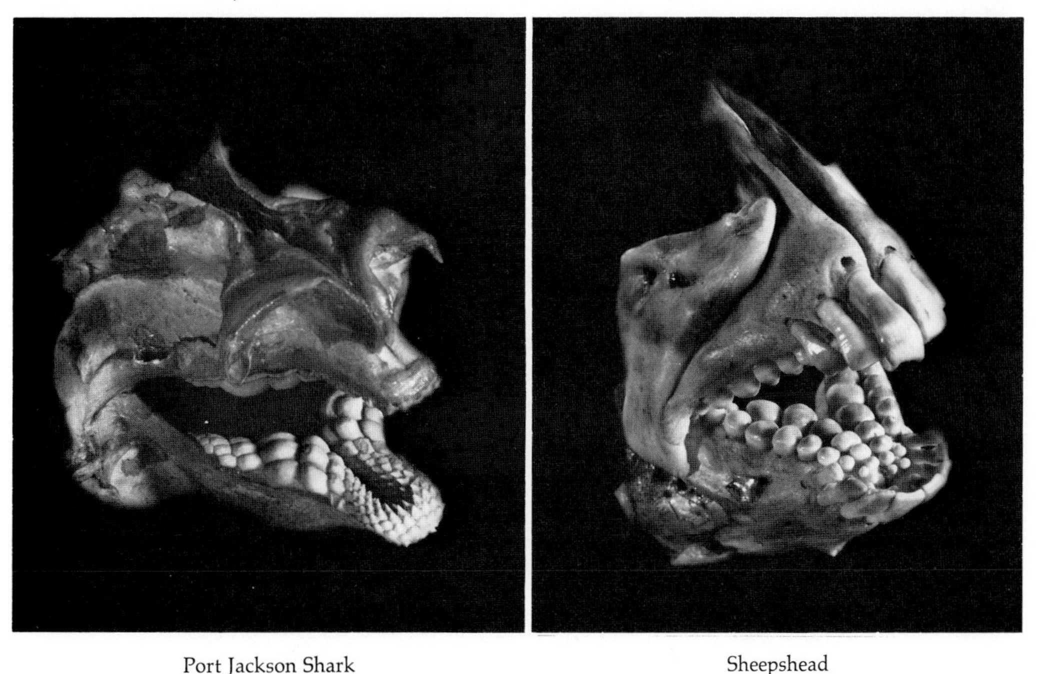

Port Jackson Shark — Sheepshead

Fig. 4-4. The jaws of a chondrichthyean, the Port Jackson shark, and of a teleost, the sheepshead. They have convergently evolved heavy flattened posterior teeth which adapt them to a diet of molluscs.

This is a case of convergent evolution. If similar adaptations arose in more closely related groups, e.g., two teleosts of the same family but different genera, it would be called parallel evolution. These phenomena, convergence and parallelism, can confuse group interrelationships. One must keep in mind that broad evolutionary relationships are indicated by characteristics common to large taxonomic categories, while the more recently evolved, specific adaptations to environment show the greatest diversity within closely related species and also show convergence and parallel evolution (Fig. 4-5, see pages following 177).

AGNATHA

The ancient ostracoderms had a heavy, dermal bone shield with various openings, fused with the neurocranium and splanchnocranium; the head skeleton was so massive and protective that it limited the animal's mobility. In one group, the Osteostraci, bony canals led from the cranial cavity to broad, dorsolateral, crescent-shaped areas and to a medial, dorsal area, both apparently innervated from the brain. It has been suggested that these were sense organs, possibly corresponding to the lateral line organs of modern fishes, or that they were highly modified muscles serving as protective electric organs similar to those in today's electric fish. The latter seems unlikely, for if the Osteostraci had had such protective organs there would seem to have been little or no adaptive value for the massive head skeleton as well; one would have expected instead a great reduction and lightening of the skeleton.

Today's cyclostomes are probably similar to their ancient ostracoderm relatives in the gross structure of their brain and sense organs, but their head skeletons are fundamentally different. Most noticeably, there has been a loss of all bone including dermal bone; thus there is no dermatocranium, not even in the form of scales, in either hagfish or lamprey. The neurocranium is of a soft mucocartilage, and the splanchnocranium is of elastic cartilage.

In addition to supporting its brain and sense organs, the cyclostome's neurocranium has large cartilages known in no other vertebrates extant or extinct. These are "shield cartilages" which protect the dorsum of the head, subocular cartilages which give support ventrolaterally, and complex lingual cartilages which form the tongue skeleton (Fig. 4-6).

The equally bizarre splanchnocranium has fine interwoven elastic cartilages covering each gill slit; these cartilages lie in a position unique among vertebrates, external to the cranial nerves and blood vessels of the pharynx. Other cartilages extend from the gill area caudally to form a pericardial cartilage partially surrounding the heart; this protects the heart and provides a skeletal attachment for the largest of the lingual (tongue) muscles. Still other cartilages extend along the notochord anteriorly into the basal plate of the neurocranium.

Since the agnathan head skeleton lacks jaws, only the cartilages of the tongue and the lamprey's annular cartilage are involved in food gathering. The velar valve separating the oral cavity and the

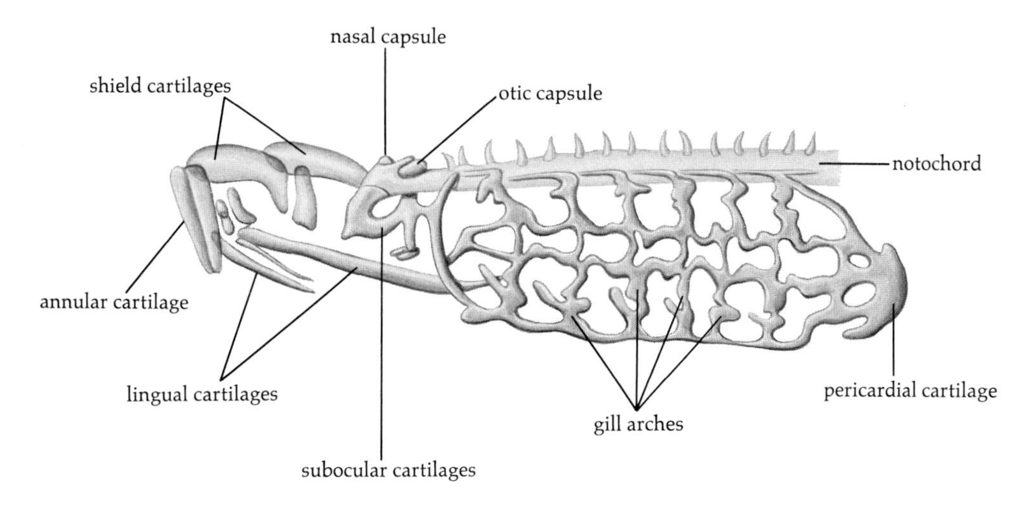

Fig. 4-6. Head and branchial skeleton of the marine lamprey, Petromyzon marinus.

pharyngeal (gill) chamber makes possible the intake of water for gill respiration coincident with feeding. Confronted with a prey, the lamprey pumps water out of its mouth, creating suction which holds it to the side of the (usually) teleost fish. The annular cartilage acts as a washer in this plumbing arrangement. The rasping tongue is then brought back and forth across the host's flesh, allowing the lamprey to take in its body fluids.

The hagfish also has a rasping tongue with horny teeth with which it makes holes in a dead fish before ingesting it. A scavenger, it needs neither holdfast mechanisms nor the ability to respire while feeding, and these adaptations have not evolved in the hagfish.

PLACODERMI

In the diverse placoderm groups there was a neurocranium, often completely roofed, and a splanchnocranium supporting the gill apparatus. Except in Acanthodii the primary jaw skeleton was usually of dermal (rather than splanchnocranial) bone.

The subclass Acanthodii, the oldest known jawed fossil, most likely gave rise to the ancestral stocks of living gnathostomes. Its skull shows more generalized gnathostome characteristics than that of other ancient fishes, including the mandibular arch jaw skeleton. The palatoquadrate, which articulated with the braincase in a movable joint, was ossified in two or three places, while Meckel's cartilage was ossified in one or two; the jaw was partially supported and buttressed by overlying dermal bone. The enlarged hyomandibular arch, ossified in two places, also helped prop the jaws against the braincase. Posterior to the hyoid arch were four more, well-developed skeletal gill arches. Although many other "experiments" in jaw structure evolved after the acanthodian, they were all apparently less advantageous for they remain in no living form and are found only in the fossil record.

In the acanthodian dermatocranium there were many small bony plates which, except for the group surrounding each large eye, had no consistent pattern from one to another species. The neurocranium was usually a complete braincase of cartilage with several centers of ossification.

CHONDRICHTHYES (Fig. 4-7)

Although Chondrichthyes have no dermatocranium, both the teeth and the minute placoid denticles that cover the skin are derivatives of dermal bone. The cartilaginous neurocranium completely encloses the brain. An anterior projection, the rostrum, gives skeletal support to the elongated snout region in sharks; otherwise the chondrichthyean neurocranium is much like that of *Amia*.

The cartilaginous splanchnocranium is prominent. The mandibular arch is loosely articulated to the ventral orbital region of the neurocranium by way of the orbital processes; when the mouth is opened, the upper jaw is slightly raised as the lower jaw is lowered. In most sharks the only other connection of splanchnocranium and neurocranium is by way of the hyomandibular, which runs from the otic capsule to the angle of the jaws where the palatoquadrate and Meckel's cartilage articulate. Muscular attachments allow the hyomandibular, and therefore the jaws, to be protruded when grasping prey. The shark skull thus provides another example of what we saw in the bowfin and will see again: a cranial kinesis in which loose articulation with the braincase permits the upper jaw to move up and both jaws to

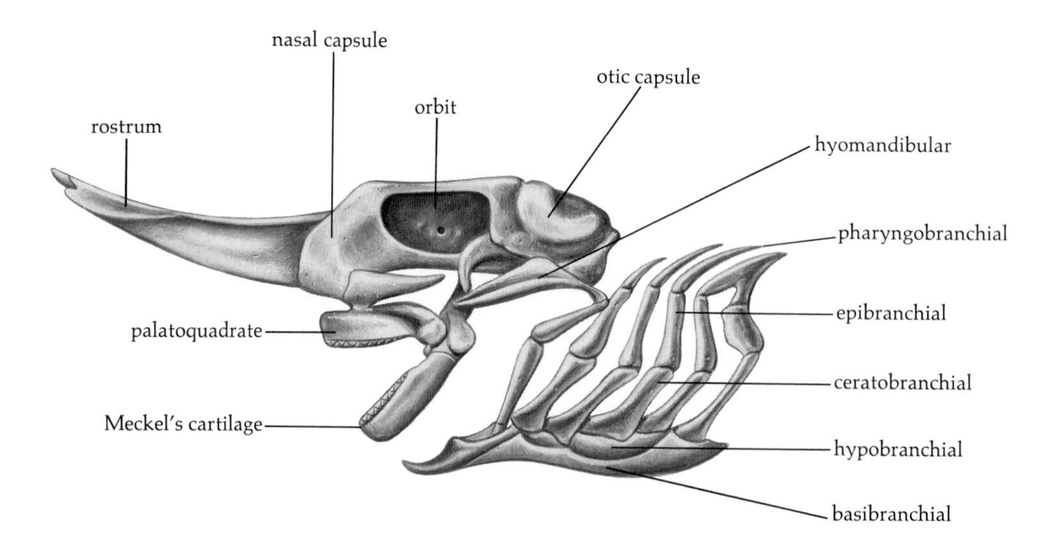

Fig. 4-7. Lateral view of the head skeleton of a skate. Note the loose articulation of the jaws, which permits free movement of both Meckel's cartilage and the palatoquadrate. From J. F. Daniel (1922). "The Elasmobranch Fishes," Univ. of California Press, Berkeley.

move forward, relative to the rest of the skull (Fig. 4-7). Such kinetic skulls evolved independently in several lines when they had an adaptive value for predatory and other feeding habits.

The ventral part of the chondrichthyean hyoid arch is made up of the large ceratohyals and the unpaired midline basihyal. The ceratohyal and basihyal run just behind Meckel's cartilage, forming the skeletal support for the floor of the mouth, particularly for the rudimentary tongue. Behind the hyoid are five typical pairs of gill arches, each comprised—dorsally to ventrally—of paired pharyngobranchial, epibranchial, ceratobranchial, and hypobranchial elements and a single midline basibranchial. In most sharks the posterior arches are incomplete, particularly ventrally in the hypobranchials and basibranchials. In the larger sharks and rays the jaws as well as some other parts of the skeleton become calcified, which makes the cartilage stronger but also more brittle.

OSTEICHTHYES—ACTINOPTERYGII

The head skeleton of the "generalized" actinopterygian, the bowfin, has been described. One diagnostic characteristic of this subclass is a large cavity behind the orbit, located between the base of the neurocranium and the dermatocranial elements of the roof of the mouth and often extending caudally past the level of the pituitary gland. This cavity, the myodome, houses the rectus muscles (four of the six extrinsic ocular muscles). Compared with fishes whose orbit contains many muscles, the actinopterygian orbit and eyeball appear abnormally small.

Another diagnostic characteristic of actinopterygians is the ossification of the hyomandibular into two elements—a large dorsal hyomandibular bone and a smaller ventral symplectic bone. There was a similar double ossification of the hyomandibular in the fossil Acanthodii, suggesting a close relationship between these two groups.

The actinopterygians evolved in three superorders: Chondrostei, Holostei, and Teleostei. Each expanded and radiated, and the first two then subsided in turn as they were largely replaced by the next.

One of the early chondrostean groups, the palaeoniscoid fishes, differed from the holosteans (such as *Amia*) in the neurocranium, which was an ossified, continuous piece similar in shape and extent to the cartilaginous neurocranium of Chondrichthyes. It tightly adhered to, and was often fused with, the more superficial dermatocranium, rather than being mostly separate from the dermatocranium as it is in *Amia.* The dermal bones behind the orbit were of broader extent than they are in *Amia.* The jaws were large, giving a large gape, and their splanchnocranial elements were covered by dermal bony plates. There were several prominent ossifications in the palatoquadrate, a single large articular ossification in Meckel's cartilage, and a fully ossified gill skeleton.

As palaeoniscoid evolution continued, the neurocranial and splanchnocranial ossifications were gradually reduced. Two of today's living chondrosteans, sturgeon and paddlefish, are more cartilaginous

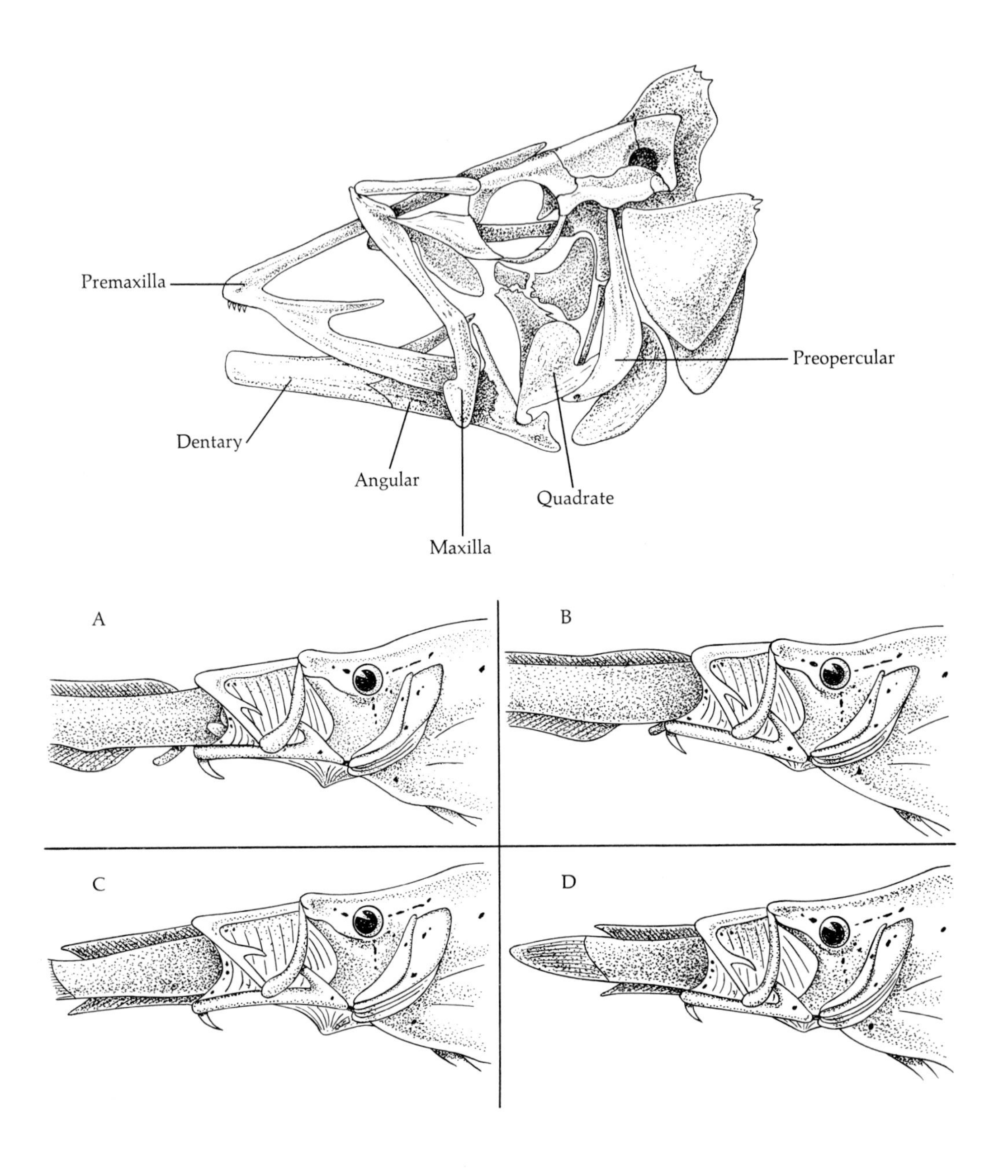

Fig. 4-8. Top: lateral view of a skull of a predatory teleost, Monocirrhus polyacanthus. *Bottom: tracings from motion picture frames of this fish swallowing a smaller teleost. In A, the jaws are protruded and the head of the prey taken in. In B, the jaws are partially retracted, drawing the prey further in. In C, the jaws are protruded again, gaining another hold. In D, retraction draws the prey in still further. This sequence continues until the prey is swallowed whole. After K. F. Liem (1970). "Comparative Functional Anatomy of the Nanidae (Pisces: Teleostei)." Fieldiana: Zoology* **56,** *1–166. The Field Museum of Natural History, Chicago.*

than bony, unlike their chondrostean ancestors; however the bichirs have mostly bone and little cartilage.

Holosteans, represented today by *Amia,* were the most characteristic ray-finned fishes during the middle and latter portions of the Mesozoic era. They differ from chondrosteans mainly by possessing thinned dermal bone, shortened jaws, reduced cheek region and preoperculum, and shortened supratemporal and subocular bones.

Teleosts, the dominant fishes of the very late Mesozoic and the entire Cenozoic, continued the evolutionary trends of the holosteans. Their dermal bones are even thinner and finer than those of the holosteans. The adult skeleton is composed almost entirely of a very light, fine, acellular bone which is an exception to the rule that bone is a living tissue: once teleost bone is fully laid down the cells die. The lower jaw has a reduced number of elements: there are usually only two dermal bones, an anterior dentary and a posterior angular, and further caudally a single splanchnocranial bone, the articular. The dermal maxilla bone of the upper jaw is fully freed from adjoining bones and greatly shortened.

The diverse feeding habits of teleosts are accompanied by interesting modifications for cranial kinesis. The basic movements are diagramed in Fig. 4-8. The hyomandibular and symplectic are brought forward, which forces the angle of the jaws anteriorly, which in turn pushes the entire lower jaw forward by a lever system and protrudes the large premaxillas; these are then buttressed both dorsally and ventrally by the short, separate maxillas. The entire palate is also protruded at the same time and can be moved independently of the maxilla and premaxilla elements. Once the prey is caught it is held by the teeth on the outer parts of the jaws (dentaries and premaxillas) and by the teeth on the inner parts of the oral cavity (including the branchial and palatine elements). Then the premaxillas are retracted, independently of the lower jaw, drawing in the prey and holding it against the palatine and pharyngeal teeth. The premaxillas are then protruded again in order for their teeth to get a new hold and are again retracted. This process can continue for up to an hour, and enables even a relatively small fish to swallow large prey. Such kinesis requires complex musculature, a fine, complexly articulated head skeleton, and a synergy of neurocranial, dermatocranial, and splanchnocranial elements.

OSTEICHTHYES – SARCOPTERYGII

Differing fundamentally from the skulls of the subclass Actinopterygii are those of the subclass Sarcopterygii, orders Dipnoi and Crossopterygii. These were the dominant fishes of the Devonian in both number and variety.

The interesting but aberrant Dipnoi skull is represented today only by the three living species of lungfish.

The order Crossopterygii is divided into the Rhipidistii, which gave rise to the first amphibians and then became extinct, and the Coelacanthini, which were quite successful and then dwindled until today they survive only in one sparse genus, *Latimeria.* Although aberrant and identical with neither its Coelacanthini ancestors nor the Rhipidistii, *Latimeria* nevertheless is more closely related to the latter than is any other living form. Therefore this genus tells us more about what the aquatic ancestor of tetrapods was probably like than does any other source, including the fossil record.

The crossopterygian pattern of dermal bones, comparable to and ancestral to the tetrapod pattern, is so different from the actinopterygian pattern that homologies cannot be reliably made. In the evolutionary lineage from Acanthodii to Crossopterygii, the large number of small dermal skull bones coalesced into a smaller number of larger bony plates with a regular pattern which for convenience sake can be divided into five groups:

1. *Dorsal paired midline series.* From anterior to posterior the dorsal paired midline series includes nasals, frontals, parietals, postparietals, and extrascapulars.
2. *Marginal and palatal upper jaw series.* The upper jaw and palate are formed by the marginal tooth-bearing premaxillas and maxillas, while the roof of the mouth is formed by the palatines, vomers, pterygoids, and a large midline parasphenoid.
3. *Circumorbital series.* The orbit is surrounded by the lacrimal, prefrontal, postfrontal, postorbital, and jugal.
4. *Cheek series.* Caudal to the circumorbital series and extending to the pectoral girdle is a broad cheek region composed of the squamosal, quadratojugal, supratemporal, intertemporal, posttemporal, tabular, preopercular, opercular, and subopercular bones. The last three form a dermal shield over the gill apparatus.
5. *Lower jaw series.* In the lower jaw are many, small dermal elements including the dentary, splenial, angular, surangular, prearticular, and often several coronoid bones.

Thus, except for small openings for nares, eyes, mouth, and gills, the entire head is protected by a layer of bony plates just deep to the epidermis.

A peculiarity in the crossopterygian neurocranium is that the fundamental separation of its anterior portion (based on the trabecular plate) and its posterior portion (based on the basal plate or parachordals) remains throughout life. An intracranial joint between the anterior and posterior parts of the neurocranium divides the braincase into an anterior, ethmosphenoid region and a posterior, otico-occipital region. Since there are loose sutures between parietals and postparietals and between intertemporals and supratemporals, the dermatocranium can also be bent along this same line. Furthermore, as is known from the muscles of *Latimeria* and from muscle scars on fossils, this joint functioned in another kind of cranial kinesis by allowing the upper jaw to move relative to the lower jaw and posterior braincase (Fig. 4-9). The amount of movement possible was greatest for the short-jawed forms and least for the long-jawed forms.

It is thought that most rhipidistians and coelacanths lived in shallow waters and that many were surface feeders, for which it would be efficient to raise the anterior part of the skull above the posterior part.

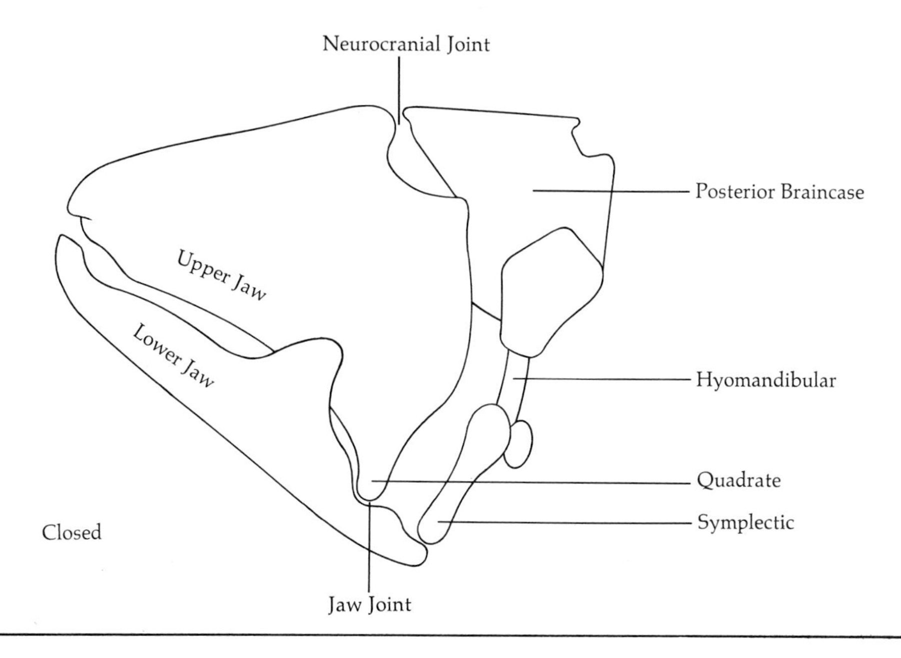

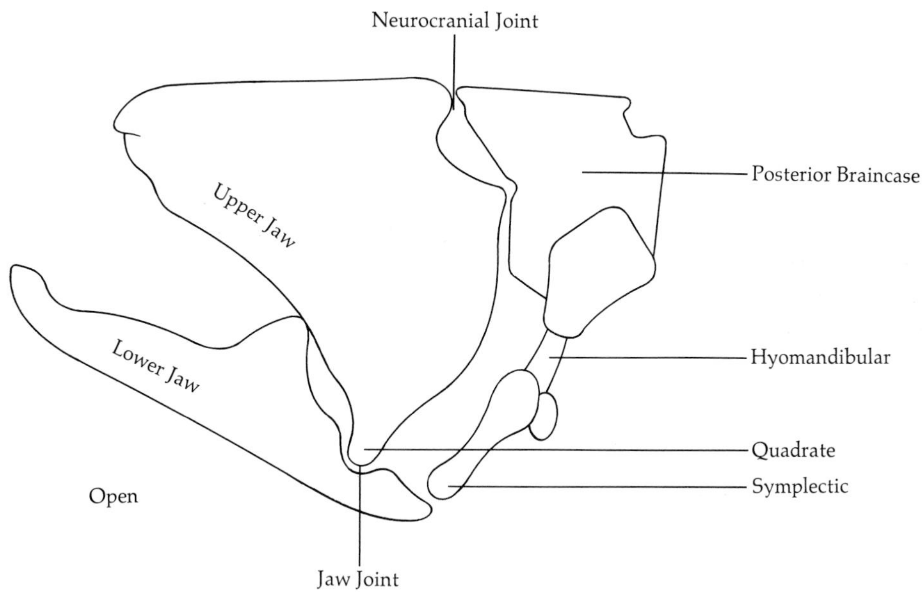

Fig. 4-9. Diagramatic representation of the skull kinesis in the coelacanth, Latimeria. *After K. S. Thomson (1969). "The biology of lobe-finned fishes."* Biol. Rev. **44,** *91–154. By permission of Cambridge Univ. Press.*

This arrangement also puts the nares above water when the mouth is open, making possible aerial as well as gill respiration. Here then is a kinetic skull which is an adaptation not for the active pursuit of large game, but for surface feeding.

The crossopterygian splanchnocranium also differs from that of actinopterygians, although not as much as do the other parts of the skull.

The mandibular arch is well formed and frequently ossified at several points. Dorsally, it forms the quadrate posteriorly, at the jaw joint, and the suprapterygoids more anteriorly; one of these, quite far forward, is the epipterygoid element, which tightly adheres to the anterior part of the neurocranium at the level of the orbit. Ventrally, Meckel's cartilage is well ossified posteriorly as the articular bone, but its more anterior portions are mostly resorbed during life, and the anterior part of the functional lower jaw is all derived from dermatocranium. The hyomandibular is a stout element, helping to brace the upper jaw against the neurocranium and dermatocranium. The ventral parts of the hyoid push forward behind the lower jaw, providing skeletal support for the floor of the mouth and the rudimentary tongue. The more posterior gill arches are typical.

AMPHIBIA

The earliest amphibians are placed in the extinct subclass Labyrinthodontia. Their skeleton was similar to that of the crossopterygian fishes from which they evolved, even including indications of the characteristic intracranial joint. They differed from the crossopterygians in a loss of several bony elements and in the extent of the skull posterior to the orbits. The dermal intertemporal, opercular, subopercular, and (except in the earliest) preopercular bones were lost. Those posterior dermal bones which remained—the supratemporals, postorbitals, tabulars, postparietals—were greatly shortened so that the orbit tended to lie closer to the back of the skull. Simultaneously the skeletal elements anterior to the orbits—the prefrontals, lacrimals, frontals, nasals, and maxillas—became longer, resulting in a longer jaw and a more prominent facial region. This lengthening would necessarily decrease the mobility of any intracranial joint and probably played an important role in the neurocranium's fusion and loss of kinesis in amphibians.

The three orders of living amphibians comprise the subclass Lissamphibia. Their skulls are characteristically flattened; orbits are very large; there is a tendency for further loss of dermal bones and for parts of the neurocranium to remain unossified (Fig. 4-10). The jaw joint is formed from the mandibular arch of the splanchnocranium: above, the posterior part of the palatoquadrate ossified as the quadrate bone; in the lower jaw, an articular bone ossified from the caudal part of Meckel's cartilage. The rest of the jaw is formed by dermal bones: maxillas and premaxillas in the upper jaw and splenials and dentaries in the lower jaw, along with small surangular and prearticular elements. In fossil amphibians the anterior part of the palatoquadrate cartilage ossified as the epipterygoid bone and contributed to the wall of the braincase in the orbital region. In living amphibians, however, the palatoquadrate does not contribute to the braincase.

The tadpoles (larvae) of living amphibians have gill respiration. During metamorphosis the gills are lost, but most of the supporting visceral cartilages remain. The more anterior of these support the tongue; the more posterior lie in the floor of the pharynx and are spe-

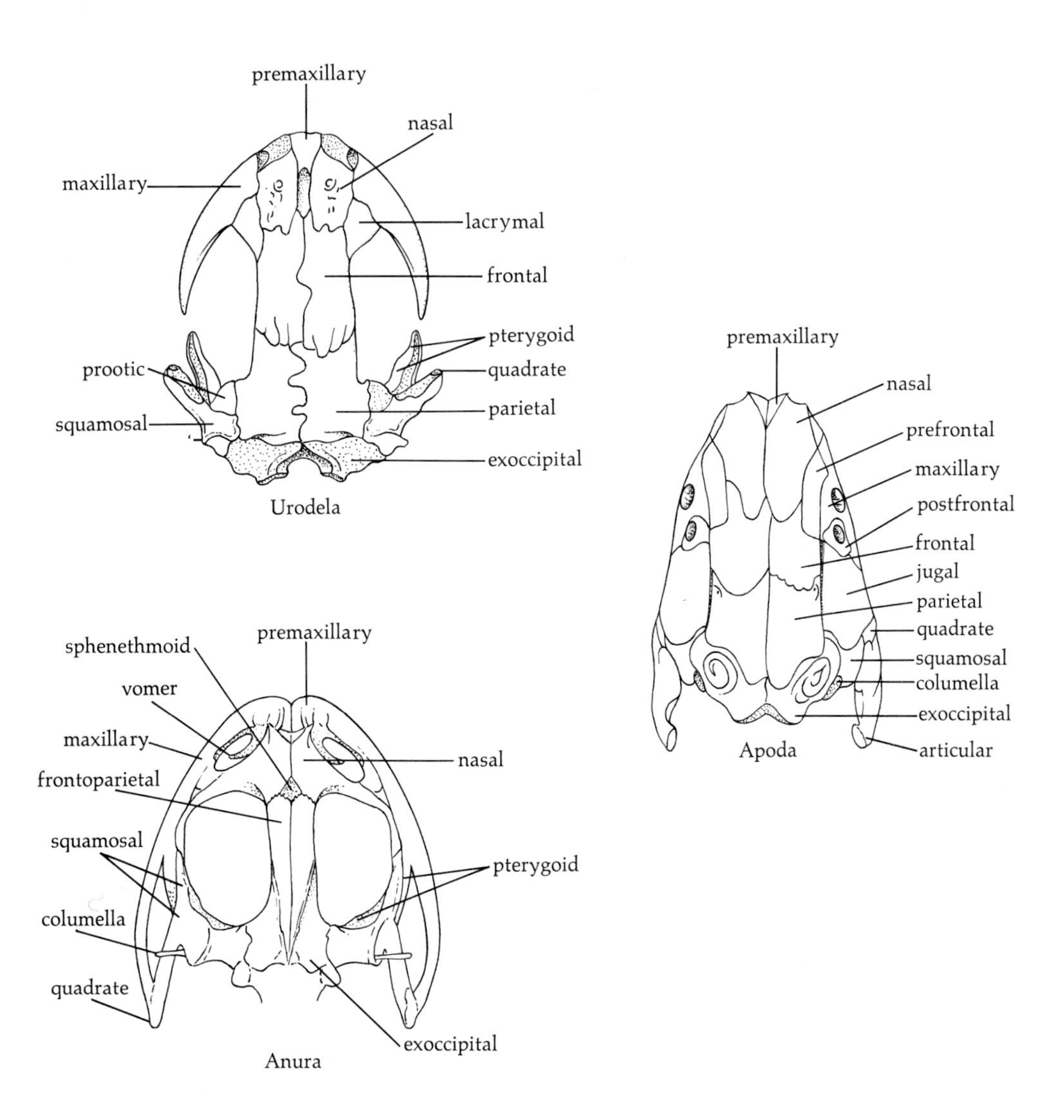

Fig. 4-10. Dorsal views of amphibian skulls.

cialized to guard the glottal chamber (the space leading to the lungs)—these are called glottal or arytenoid cartilages.

Most adaptations for food gathering in Lissamphibia concern the skeleton only indirectly. The tongue, whose base is supported by the basihyal, has evolved into a long and efficient structure with which frogs, toads, and many salamanders pluck insects from the air.

REPTILIA

The first reptiles, Cotylosauria, evolved from Labyrinthodontia in the very late Carboniferous period, and, of course, inherited many labyrinthodont characteristics. For one, the intertemporal bone was absent. For another, the trend for reduction of the posterior part of the dermatocranium was continued: there was reduction and sometimes complete loss of supratemporals, postparietals, and tabulars.

During the adaptive radiation of reptiles there occurred one of the most massive remodeling jobs yet performed by natural selection. The phenomenon is known as fenestration (from Latin *fenestra*, "window") and has to do with the placement of the muscles of mastication, which close the jaws. In cotylosaurs, amphibians, and fishes with dermatocrania, these muscles lie posteriorly on each side of the skull between the dermatocranium and the neurocranium. The muscles extend ventrally to attach to the lower jaw in front of the articular bone. In reptiles, however, there was a newly evolved neck. The lighter the head at the end of the neck, the less metabolic energy would be required to hold and move the head. If the head could become lighter with no sacrifice of function, it would have great adaptive value. It is probably for this reason that fenestration occurred: the dermal bone of the cheek (temporal) region was thinned and lost, opening "windows" in the skull. Not surprisingly, this occurred at points adjacent to the attachments of masticatory muscles, since bone tends to thicken where it is stressed and to become thinner where it is not.

The amount and location of fenestration varied in different groups of reptiles, and in some lines did not occur at all. For instance, the turtles have no fenestration; this skull condition is called anapsid. In the long-necked, paddle-finned Plesiosauria and the fishlike Ichthyosauria, both Mesozoic reptiles, there evolved a temporal fenestra just dorsal to the suture separating the postorbital and squamosal bones; this condition is called euryapsid. The situation was reversed in the mammal-like reptiles (from which mammals evolved) that had a temporal fenestra just ventral to the suture between postorbital and squamosal; this is the synapsid condition. Skulls with temporal fenestrae both dorsal to and ventral to this suture, called diapsid, developed in the lines leading to the great archosaurian groups (which includes today's Crocodilia) and independently in the line leading to the Lepidosauria (today's lizards, snakes, and rhynchocephalian) (Fig. 4-11).

Fenestration does more than just lighten the skull. Since muscles can pull most strongly when their fibers attach to bone obliquely, the fenestrae increase the potential force of muscles by providing more edges for them to pull upon. In many lines the fenestrae enlarged to become confluent with the orbit; in some of the diapsid forms, such as lizards and snakes, they became so large that the junction of the postorbital and squamosal, which normally separates the superior and lateral temporal fenestrae, was eliminated.

The food-gathering adaptations of reptiles include the snakes' extreme development of cranial kinesis (Fig. 4-12). The marginal and cheek bones are loosely and movably attached to the braincase. The elongated quadrate is suspended from the equally elongated supratemporal. Muscles pull these bones forward, opening the mouth and

Fig. 4-11. The earliest amniotes (cotylosaurs) possessed no temporal fenestrae. However, the parallel evolution of fenestration (shown by hatching) in different lines of evolution demonstrates the adaptive value of fenestration in amniotes.

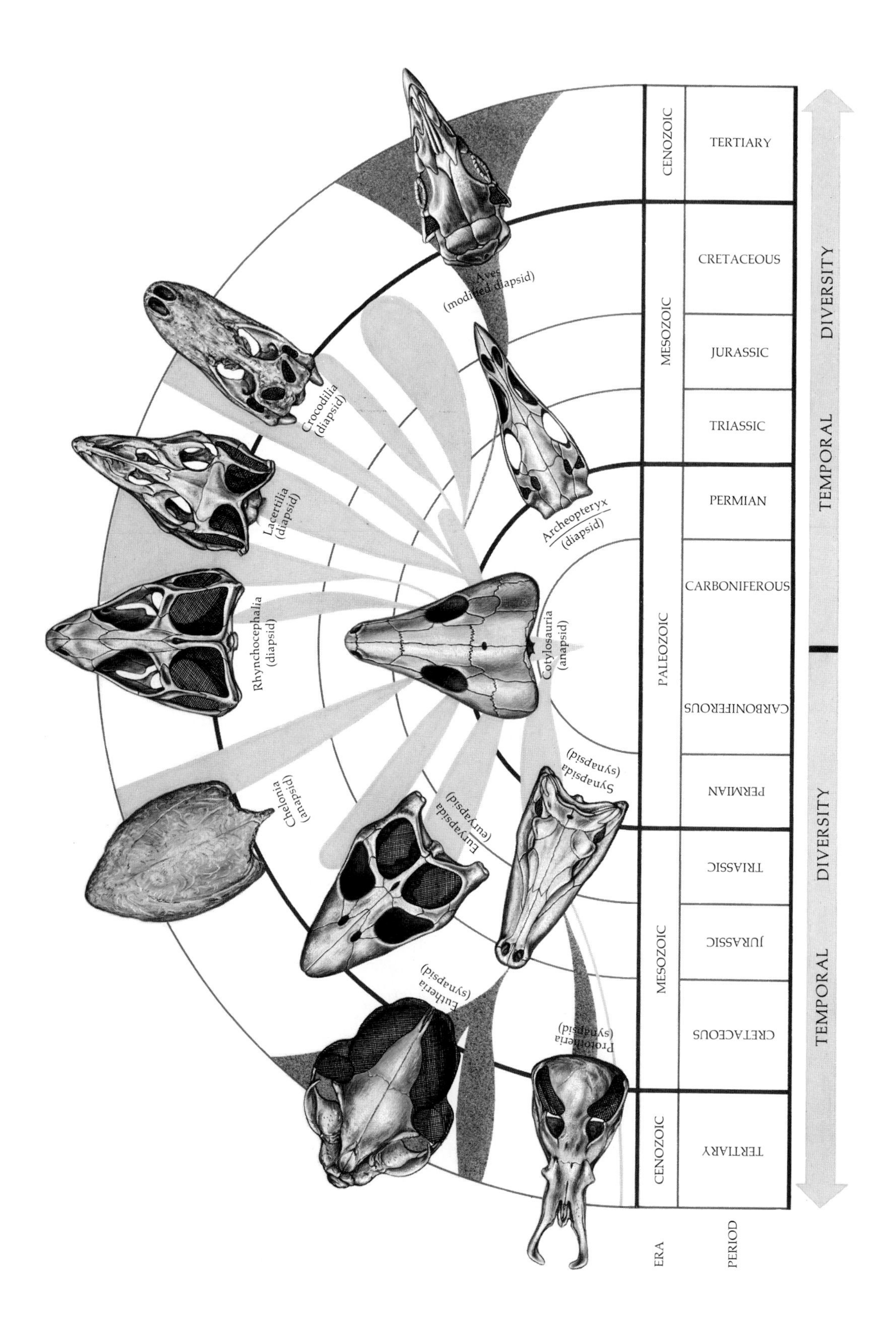
ERA
PERIOD
CENOZOIC
TERTIARY
MESOZOIC
CRETACEOUS
JURASSIC
TRIASSIC
PALEOZOIC
PERMIAN
CARBONIFEROUS
CARBONIFEROUS
PERMIAN
MESOZOIC
TRIASSIC
JURASSIC
CRETACEOUS
CENOZOIC
TERTIARY
TEMPORAL DIVERSITY
TEMPORAL DIVERSITY
Aves
(modified diapsid)
Crocodilia
(diapsid)
Lacertilia
(diapsid)
Rhynchocephalia
(diapsid)
Archeopteryx
(diapsid)
Cotylosauria
(anapsid)
Chelonia
(anapsid)
Synapsida
(synapsid)
Euryapsida
(euryapsid)
Eutheria
(synapsid)
Prototheria
(synapsid)

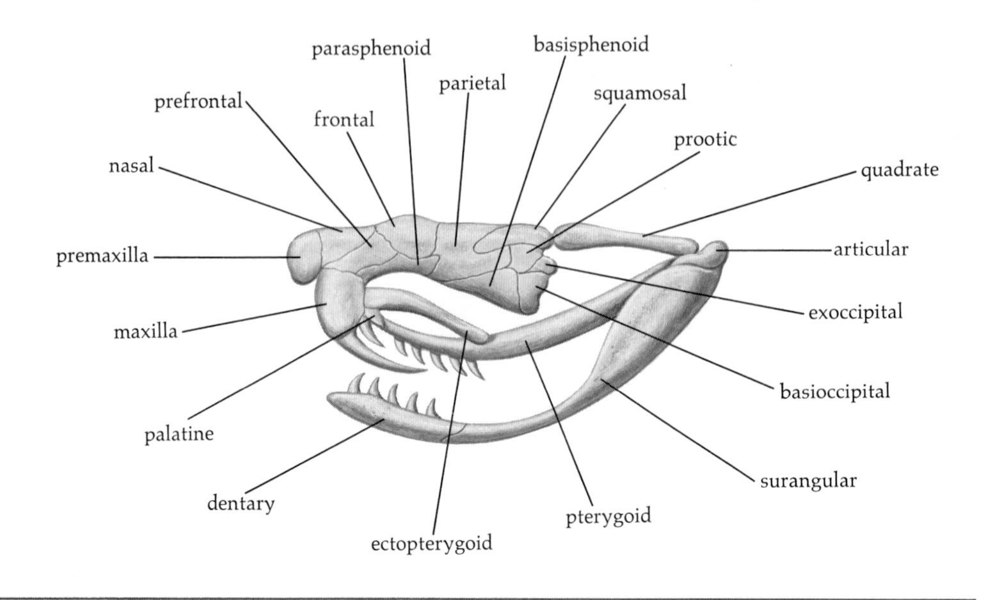

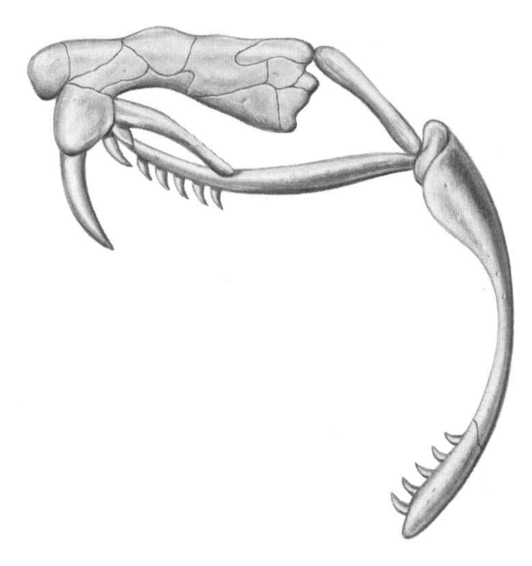

Fig. 4-12. Perhaps the most dramatic kinetic skull is that of the rattlesnakes, whose wide gape allows them to swallow prey of great diameter.

extending the wide open jaws forward. In the upper jaw the pterygoids, palatines, ectopterygoids, maxillas, and premaxillas are all involved in this kinesis. The two halves of the lower jaw are connected by an elastic ligament which allows them to separate and to move independently of one another. The mouth can gape very widely and capture prey many times larger than the snake's diameter. This often does not kill the prey, however, which must be firmly held while it is swallowed. This is accomplished by the independent movement of first one side of the jaw and then the other: while the right half is holding the prey the left half retracts, pulling the prey further in since the teeth point caudally. In this way the prey gradually enters the snake's elastic alimentary canal.

Other modifications also facilitate this unique feeding mechanism. The brain is enclosed in bone, which physically protects it during the ingestion of large prey. A protrusible glottis enables the snake to breathe while swallowing. The various methods of killing the prey before it is swallowed (e.g., constriction, venom) are probably secondary modifications which give added selective value to the snakes' primary food-gathering mechanism.

AVES

Birds, derived from a group of archosaurian reptiles, have extremely lightened skulls which are necessary for long-necked, flying animals. The diapsid condition is expanded so that the lateral fenestra is confluent with the orbit, having obliterated the postorbital–squamosal arch. Many adult bones are fused and their suture lines obliterated; there is also much fusion of dermatocranial, splanchnocranial, and neurocranial elements.

The brain is much larger relative to body size in birds than in reptiles. All three skull components contribute to the fused braincase. Part of the braincase is formed by the splanchnocranial epipterygoid element; its roof and parts of its walls are formed by dermatocranial elements, particularly by flanges that grow ventrally from the parietal and frontal bones and meet the epipterygoid and orbitosphenoid elements. There is a very thin zygomatic arch made up of quadratojugal, jugal, and maxilla elements. The lower jaw is made up of a splanchnocranial

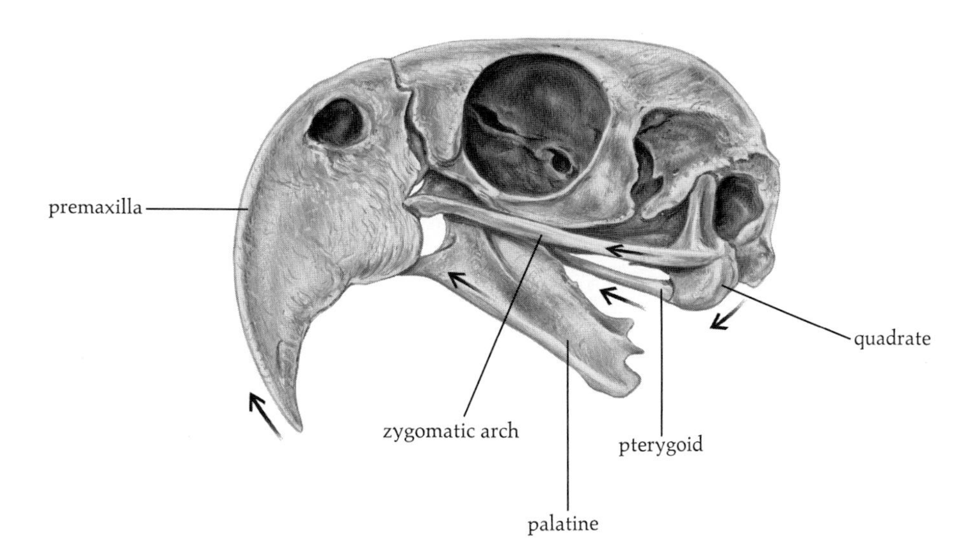

Fig. 4-13. An exaggerated kinetic skull as seen in the cockatoo. Arrows indicate how the skull moves to raise the upper jaw. After A. d'A. Bellairs and C. R. Jenkins (1960). The Skeleton of Birds. In *"The Biology and Comparative Physiology of Birds" (A. J. Marshall, ed.), Vol. 1, pp. 241–300. Academic Press, New York.*

articular bone and five dermal bones: the prearticular, surangular, angular, splenial, and dentary.

Some birds have a kinetic skull. In these birds the quadrate is delicately suspended from the squamosal bone. The joint between quadrate and articular is such that as the lower jaw is depressed, the ventral portion of the quadrate is pushed forward; this movement is passed on to the zygomatic arch laterally and to the pterygoids and palatine bones more medially. This in turn results in raising the upper jaw significantly, which, of course, necessitates a movable joint at the base of the upper beak between the nasal and frontal bones (Fig. 4-13).

The extent of kinesis is variable, being greatest in parrots and their relatives. It is absent in birds of great jaw strength like the hawfinch. This small (55-gram) bird breaks cherry and olive stones in its beak; such a feat, requiring pressures of 106–159 pounds/foot^2, would not be possible with a jaw having the intricate articulations required by cranial kinesis.

Woodpeckers have a different and even more remarkable food-gathering adaptation which involves a posterior part of the splanchnocranium, the hyoid apparatus (Fig. 4-14). From its midline ventral area the two horns of the hyoid wrap around the ventral, the posterior, and then the dorsal part of the skull and are inserted into the woodpecker's right external nostril. Along their length these skeletal elements are wrapped with hyomandibular arch skeletal muscles which, when rapidly contracted, shoot the tongue rapidly in and out. The woodpecker's powerful beak drills holes into which the tongue is jabbed, catching the insect pupae and larvae which have been uncovered.

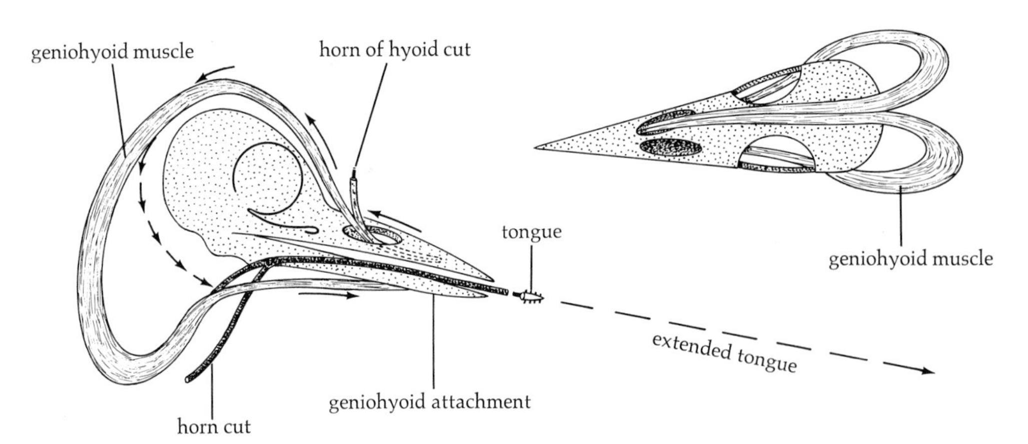

Fig. 4-14. The woodpecker's bizarre hyoid apparatus and associated musculature allows its tongue to be rapidly extended and retracted. After A. d'A. Bellairs and C. R. Jenkins (1960). The Skeleton of Birds. In *"The Biology and Comparative Physiology of Birds" (A. J. Marshall, ed.), Vol. 1, pp. 241–300. Academic Press, New York.*

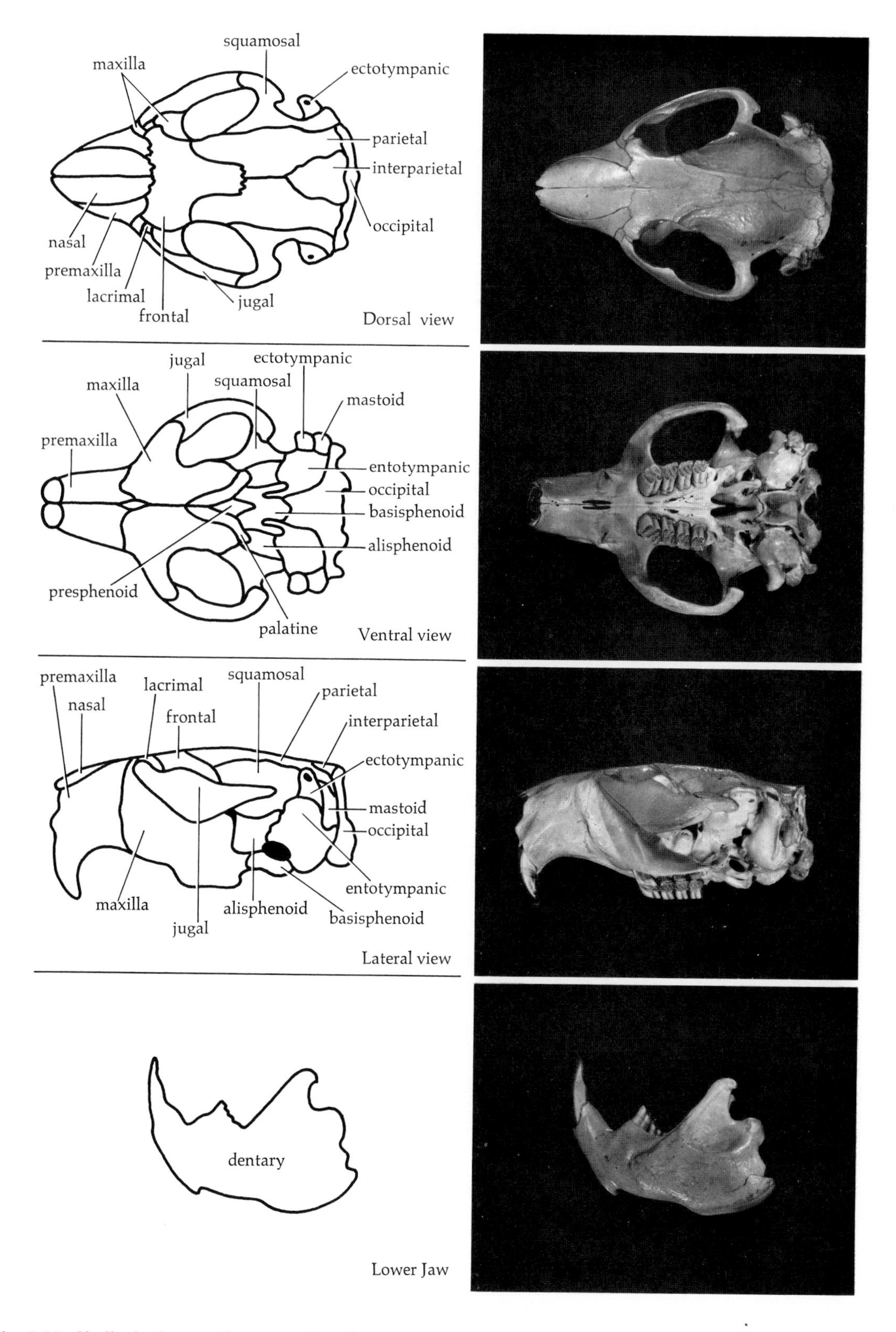

Fig. 4-15. Skull of a beaver demonstrating the arrangement of skeletal elements in a mammal. Outlines indicate suture lines between bones.

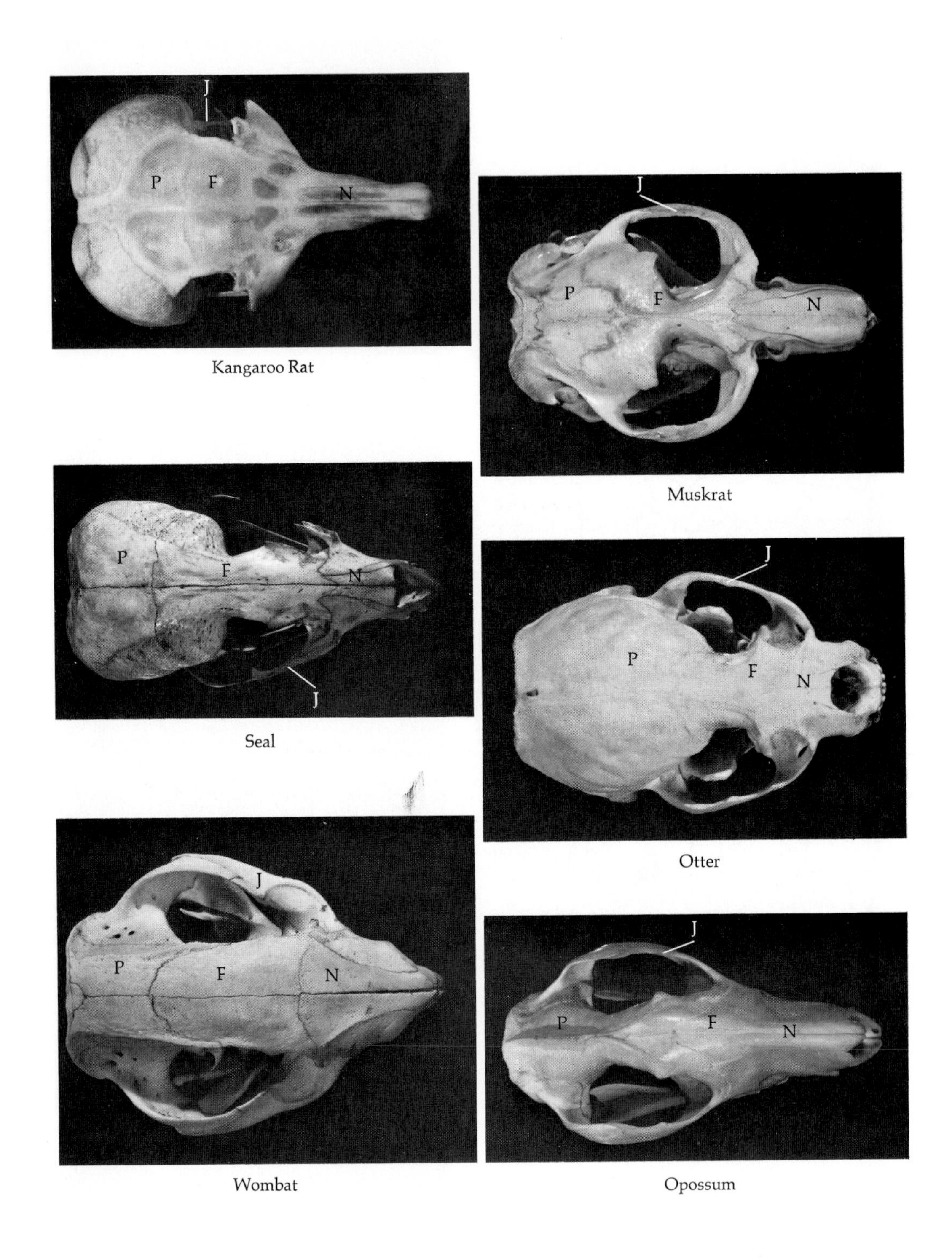

Fig. 4-16. Dorsal view of the skulls of diverse mammals, reproduced in the same size to emphasize comparisons. Although similar bones are found in all, their shapes and proportions vary to such an extent that each skull is distinct. F, frontal; J, jugal; N, nasal; P, parietal.

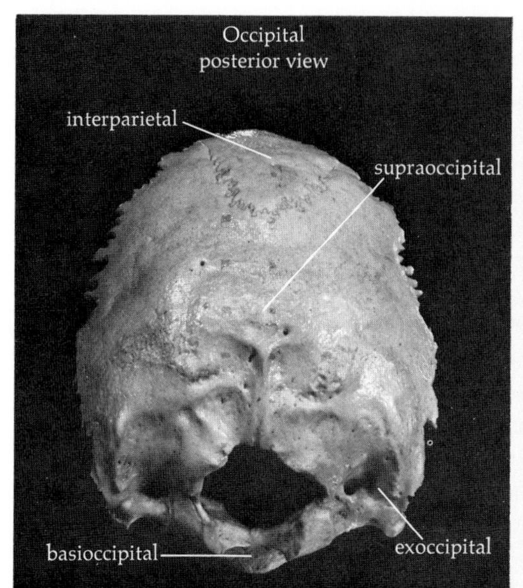

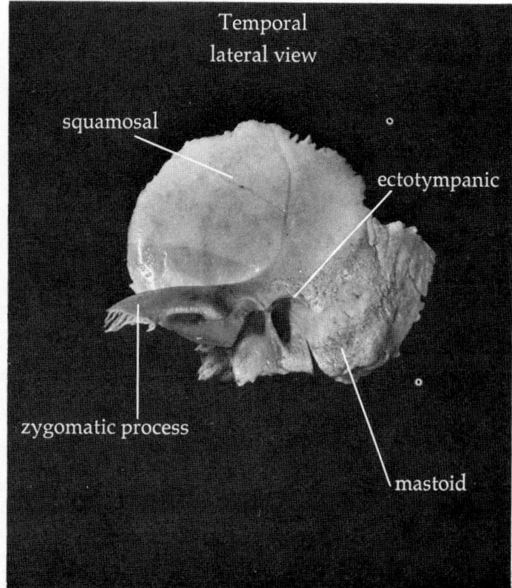

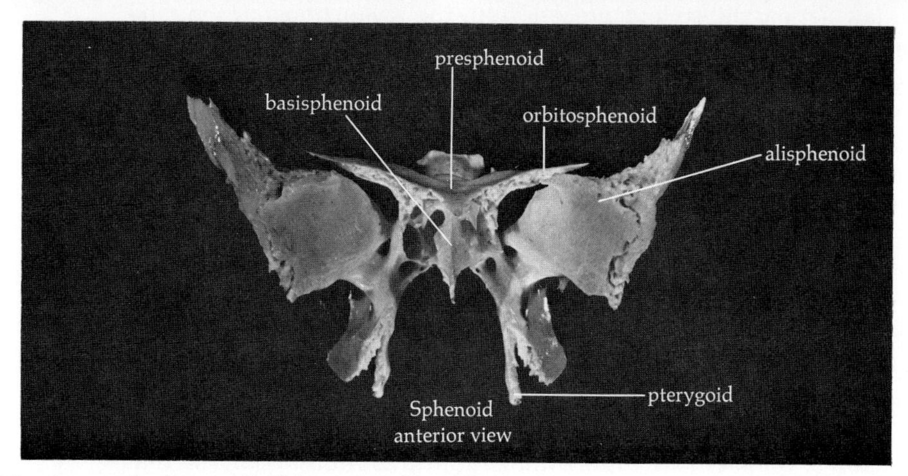

Fig. 4-17. The three major human skull bone complexes. Some elements of the temporal bone appear only in a medial view and so are not seen here.

MAMMALIA

The evolution of mammals from synapsid reptiles was accompanied by several fundamental changes in the skull (Fig. 4-15). There was a great expansion of the lateral temporal fenestrae, the development of a new jaw joint, reduction in the number of bones and fusion between remaining bones to form bone complexes. Particularly dermal bones were lost from both the circumorbital group (prefrontals, postfrontals, postorbitals) and the cheek series (tabulars, supratemporals). Other dermal bones were transformed. To provide the necessary larger braincase, medial flanges of dermal bones grew ventrally, meeting the orbitosphenoid and alisphenoid bones and forming the dorsal roof and much of the lateral walls of the braincase (Fig. 4-16). (The alisphenoid is the homologue of the reptilian epipterygoid bone and, thus, is of palatoquadrate origin.)

The only bone of the mammalian lower jaw is a dermal bone, the dentary, which together with the squamosal portion of the temporal bone (upper jaw) forms the new jaw joint, anterior to where the quadrate–articular joint formed in reptiles. These elements are not lost in mammals, however. In the early mammalian embryo a typical quadrate–articular jaw joint forms. These bones then lose contact with the jaws and become incorporated into part of the middle-ear transmitting mechanism.

There are three major bone complexes in mammalian skulls — the temporal bone, the sphenoid bone, and the occipital bone (Fig. 4-17). The extent of each varies with the species: for instance, in some mammals the squamosal is not incorporated into the temporal bone, while in others it is. In its most complex form, the temporal bone includes the opisthotic and prootic from the otic capsule of the neurocranium; the squamosal from the dermatocranial cheek series; the ectotympanic, homologous with the angular from the dermatocranial lower jaw series; the styloid process from the splanchnocranium; and the entotympanic, an endochondral element with no homologue in other vertebrates. In addition the temporal bone encloses the three auditory ossicles (malleus, incus, stapes) from the splanchnocranium.

The sphenoid bone may include the midline neurocranial elements basisphenoid and presphenoid, the paired neurocranial orbitosphenoids, the paired splanchnocranial alisphenoids, and the paired dermal pterygoids. The occipital includes from the neurocranium the two exoccipitals, the midline basioccipital, and the supraoccipital; occasionally it also includes the dermal interparietal.

There is no kinetic skull development in mammals. As we have seen, the evolution of a kinetic skull has usually been adaptive in seizing, holding, and swallowing prey. In most mammals, especially predatory mammals, there are competent forelimbs and claws for seizing and holding. The teeth are differentiated for different functions so that small pieces of food can be bitten off and chewed. Thus there is no adaptive value to be served by a kinetic skull, and none has evolved.

SUGGESTED READING

Beecher, W. J. (1962). The biomechanics of the bird skull. *Bull. Chicago Acad. Sci.* **11,** 10–33.

DeBeer, G. R. (1937). "The Development of the Vertebrate Skull." Oxford Univ. Press, New York.

Frazzetta, T. H. (1968). Adaptive problems and possibilities in the temporal fenestration of tetrapod skulls. *J. Morph.* **125,** 145–151.

Thomson, K. S. (1969). The biology of lobe-finned fishes. *Biol. Rev.* **44,** 91–154.

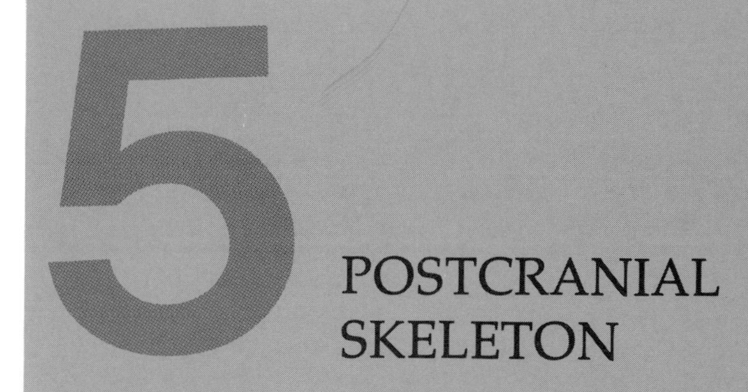

5 POSTCRANIAL SKELETON

5

The two portions of the postcranial skeleton are the axial (vertebral column, notochord, ribs, sternum) and the appendicular (girdles and intrinsic skeletons of the appendages). Both are primarily of endochondral bone (cartilage in Chondrichthyes) with dermal bone limited in most vertebrates to a portion of the pectoral girdle; a few (e.g., turtle, armadillo) also have an extensive protective armor of dermal bone.

AXIAL SKELETON: THE VERTEBRAL COLUMN AND NOTOCHORD

The vertebral column and associated notochord form the main axis of the postcranial skeleton. In fishes these structures prevent telescoping of the trunk and tail and provide points of attachment for the large muscle masses which produce lateral undulations. These muscles are similar throughout the body as is the amount of support provided by the surrounding water, and there is little or no regional differentiation in the axial skeleton.

Such differentiation has evolved in terrestrial vertebrates, which are usually supported on two pairs of appendages and whose axial skeletons must therefore withstand different stresses in different regions. The trunk portion of the vertebral column between the appendages functions as a girder and must withstand both compressional and tensional forces. It must also be flexible enough for both lateral and vertical bending. Even greater flexibility is required of the cervical portion, anterior to the front appendages. Posterior to the hind appendages, in the caudal region, the vertebrae have diverse functions. In fishes they absorb the stress from the power stroke of the tail. During tetrapod evolution tails of various animals have become adapted in diverse ways: for example, they act as a balancing organ in arboreal forms such as squirrels, a supporting third limb in bipedal forms such as kangaroos, and a swimming aid in secondarily aquatic forms such as beavers and whales.

VERTEBRAL DEVELOPMENT

Very early in development there is segmentation of both the myotome and the epimeric sclerotome just medial to it. In each myotome segment there is a single condensation of mesenchyme; this forms the anlage of a segmental muscle mass. In each sclerotome segment there are two mesenchymal condensations, one in front of the other, and from these develop the anlagen for the vertebrae. The posterior half of

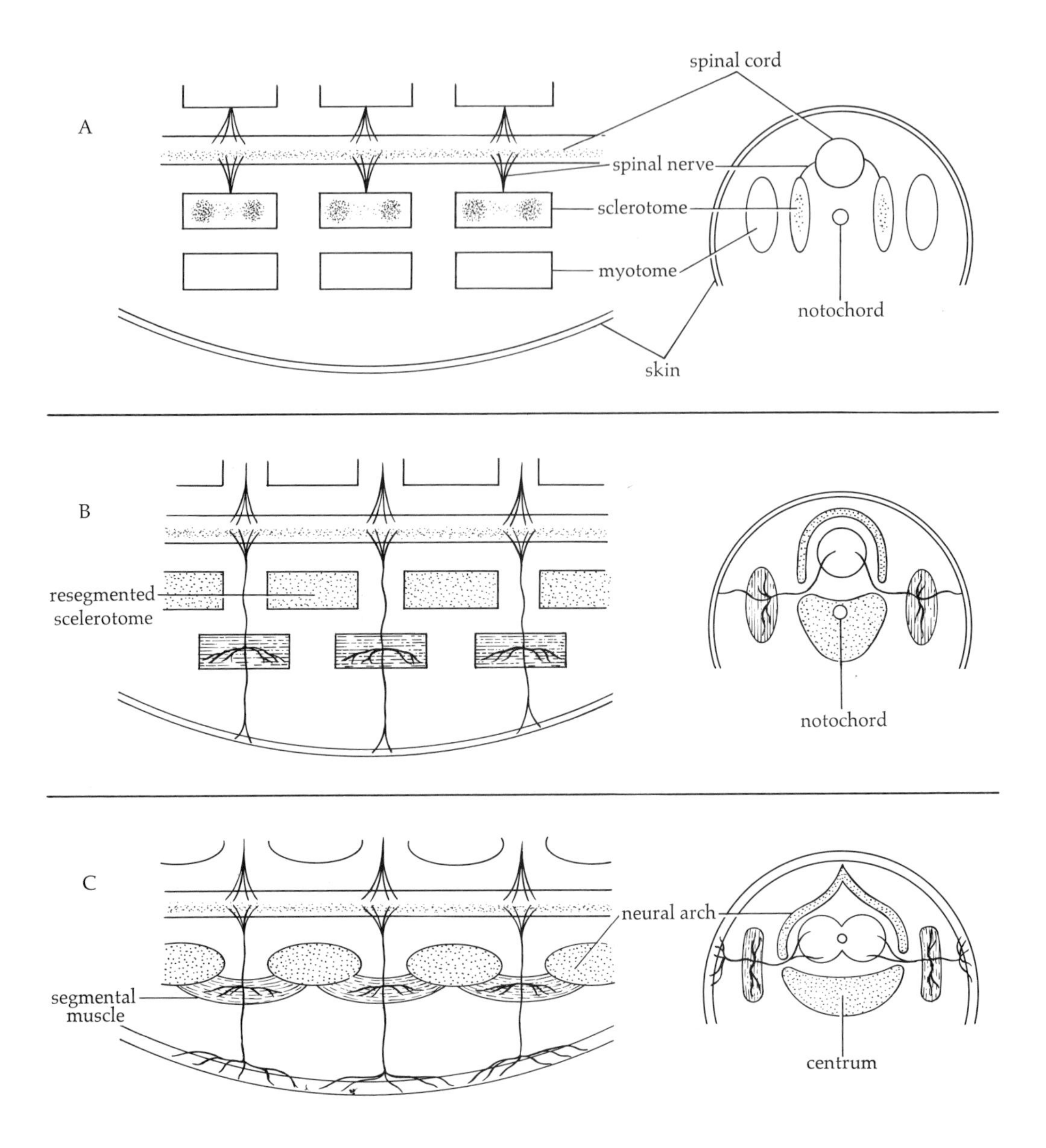

Fig. 5-1. Diagrams of three stages in the development of vertebrae in relation to surrounding structures. To the left are horizontal–longitudinal sections through the trunk, and to the right, transverse sections through the trunk. (A) Initial segmentation of the epimere, with sclerotomes and myotomes at the same levels. The double condensations of mesenchyme in each sclerotome indicate their potential resegmentation. Nerve fibers are just beginning to extend peripherally. (B) Sclerotomal resegmentation has occurred; each sclerotome now lies in an intersegmental position relative to myotomes and spinal nerves. Spinal nerves extend out to myotomes and skin. (C) Vertebrae, muscles, nerves, and skin have differentiated, and functional relationships can be seen.

one sclerotomal segment and the anterior half of the segment just behind it join together, so that each presumptive vertebra forms on a level overlapping two somites (Fig. 5-1). This resegmentation facilitates the development of the adult condition in which axial muscles attach to adjoining vertebrae, making possible bending and twisting movements of the vertebral column.

In further development the right and left sclerotomal mesenchyme differentiate into cartilage, and the right and left sides join dorsally around the developing spinal cord and ventrally around the notochord. The dorsal portion surrounding the spinal cord forms the neural arches and neural spines, and the ventral portion forms the centra (bodies) of the vertebrae. In the tail region, ventral extensions from the centra form hemal arches surrounding the major artery and vein of the tail.

Centrum formation varies greatly among vertebrates. It may encircle the notochord, replace the notochord, or incorporate notochordal tissue as part or all of the vertebra. Moreover, two centra per vertebra may form; if so one is usually larger than the other.

DIVERSITY OF VERTEBRAL STRUCTURE

In living cyclostomes, Dipnoi, and the coelacanth (*Latimeria*), the adult notochord is unconstricted by the vertebral column and is the primary axial skeleton, preventing telescoping and allowing lateral undulations. There are some intersegmental cartilages and bones in *Latimeria*. Throughout the vertebral column *Latimeria* has both neural arches and neural spines attached directly to the heavy, thick notochordal sheath. The anterior neural arches are partially ossified. In the trunk region there are small nubbins of cartilage ventrally; in the caudal region there are hemal arches and spines.

In other fishes each centrum encircles the notochord. This prevents further expansion of the notochord at that level but not between the vertebrae, where the notochord continues to grow. The centrum also continues to grow, enlarging anteriorly and posteriorly and surrounding these expanded intervertebral portions of the notochord. The result is a centrum with concave anterior and posterior surfaces and a tiny hole in the center filled with notochord. This amphicelous centrum, characteristic of Chondrichthyes and Actinopterygii, allows limited movement between centra. The entire vertebral column is jacketed by connective tissue and thus gives general support to the body.

In tetrapods the developing centrum encroaches further upon the intervertebral swelling of the notochord and the intervertebral space in diverse ways (Fig. 5-2). Frequently a ball and socket joint is formed. In urodeles and in the neck region of hoofed mammals the posterior surface of each centrum is concave and the anterior surface convex; this centrum shape is called opisthocelous. The reverse is true in anurans and in the majority of reptiles, where the anterior face of the vertebra is concave and the posterior convex, a condition known as procelous. In mammals, with the exception of the neck region of hoofed mammals, both anterior and posterior surfaces of the centrum are flat,

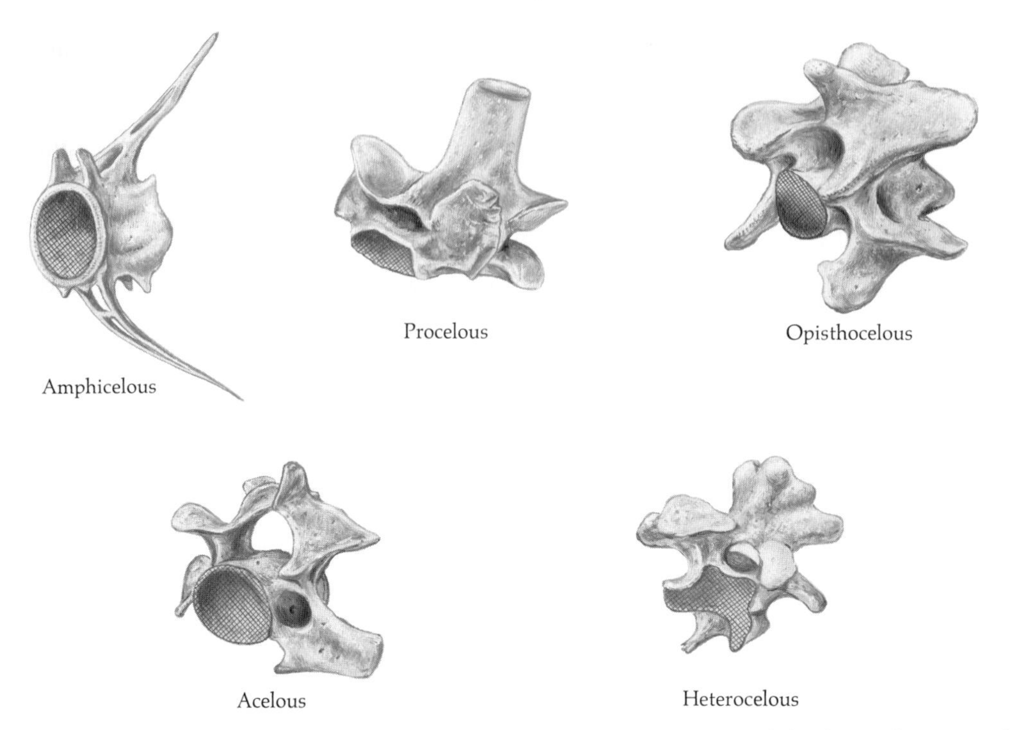

Fig. 5-2. Anterolateral views of the five major types of centra found in vertebrates. The surface of each centrum is hatched.

a condition known as acelous or amphiplatyan. In contrast, the centra faces are complex in birds, being roughly saddle-shaped; considerable movement is possible between these heterocelous vertebrae, particularly in the neck region.

The centrum's spatial diversity suggests that the biconcave amphicelous centrum was constant in fishes and that the diversity of centra shape evolved independently and often parallelly in different tetrapod groups. However, a closer look at living forms makes one suspect a more complex temporal diversity, a suspicion which is confirmed by fossil evidence.

Among fossils and a few living vertebrates there is a diplospondylous condition, that is, vertebrae with parts of two centra per neural or hemal arch. Among living vertebrates this is clearly found in lungfish and in the caudal region of the bowfin (Fig. 5-3). In mammals the peripheral part of the intervertebral disc is probably a remnant of a second centrum (the central part being a notochordal remnant). Among fossils the condition was more pronounced and more widespread: most of the early fossil reptiles, amphibians, and crossopterygian fishes, and probably the primitive tetrapod stock, had two distinct centra—called an intercentrum and a pleurocentrum—per neural arch; in the evolution of different lines one or the other has persisted. The centrum of modern amniotes is probably the pleurocentrum, except in some reptiles where there is also a second, smaller intercentrum. The embryology of modern amphibians gives no indication of whether the centrum is an intercentrum or a pleurocentrum;

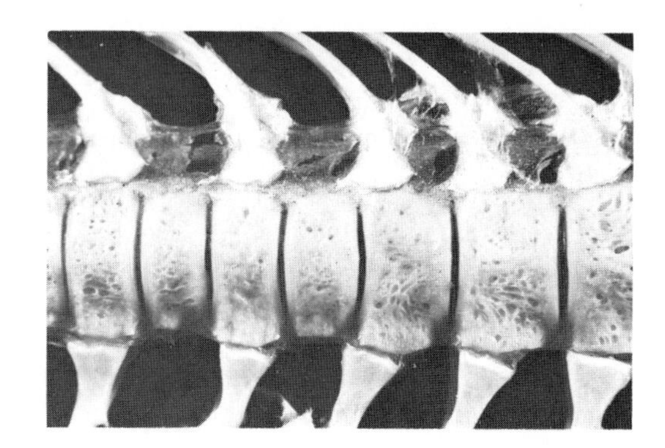

Fig. 5-3. Lateral view of last trunk and first caudal vertebrae of the bowfin. Note the diplospondylous condition of the caudal region.

it is likely that during lissamphibian evolution there was a reorganization of sclerotomal embryology resulting in the modern amphibian vertebra, which would therefore be independent of these categories.

REGIONALIZATION OF THE VERTEBRAL COLUMN

Recall that the vertebral column (1) forms the axial support for most of the body, (2) allows movement of the body, and (3) protects the delicate spinal cord from both body stress and external injury.

Fig. 5-4. A caudal vertebra (left) and a trunk vertebra (right) of a teleost (salmon). Note the amphicelous centra and prominent neural arches and spines. In the caudal region the hemal arch is larger than the neural arch.

Cyclostomes

The notochord carries out all these functions in cyclostomes, and neither the small cartilages which form around it nor the notochord itself shows any regionalization.

Jawed fishes

Trunk and caudal vertebrae are differentiated in the jawed fishes (Fig. 5-4). The trunk vertebrae have ventrolateral processes (basipophyses) to which ribs articulate. There may also be extra protective elements among the trunk neural arches, such as the intercalary arches of chondrichthyeans or the similarly placed ossified supradorsals of many teleosts. The neural arches are usually smaller in the caudal vertebrae, and instead of basipophyses there are hemal arches surrounding the caudal artery and vein.

Amphibians

The movement between the amphibian skull and the first vertebra is possible because of modifications of their articular surfaces. The first vertebra, therefore, is a distinct type, or region—the neck or cervical region.

Caudal to this vertebra are trunk vertebrae, in varying number, each bearing a pair of ribs. The ribs are bicipital, or two-headed, and the heads are articulated (or sometimes fused) with separate processes of the vertebrae—the dorsal head with the diapophysis, the ventral head with the parapophysis.

Further posteriorly, a single sacral vertebra is immovably joined to the pelvic girdle by sacral ribs, fused between the vertebra's centrum and the girdle. The sacral vertebra is the third type or region of the amphibian vertebral column.

In urodeles many caudal vertebrae follow this single sacral vertebra, each with a small neural arch and a small hemal arch. In anurans, however, all vertebrae posterior to the sacrum are fused into a single, solid structure called the urostyle, which acts as a counterbalance during the animal's terrestrial locomotion.

Reptiles

In reptiles there is usually a long and quite mobile cervical region with varying numbers of vertebrae. The first two, the atlas and the axis, are greatly modified to allow movement of the head independent of the neck. The remaining cervical vertebrae are modified to a lesser extent, but enough to permit greater movement than is possible more posteriorly. Except for the atlas and axis, each cervical vertebra has a small rudimentary rib as well as a neural arch and spine (Fig. 5-5).

What are called trunk vertebrae in amphibians are further differentiated in reptiles into thoracic and lumbar vertebrae. Each thoracic vertebra has articulated to it a pair of long, well-defined ribs usually meeting ventrally at the sternum. Like those of amphibians, these ribs

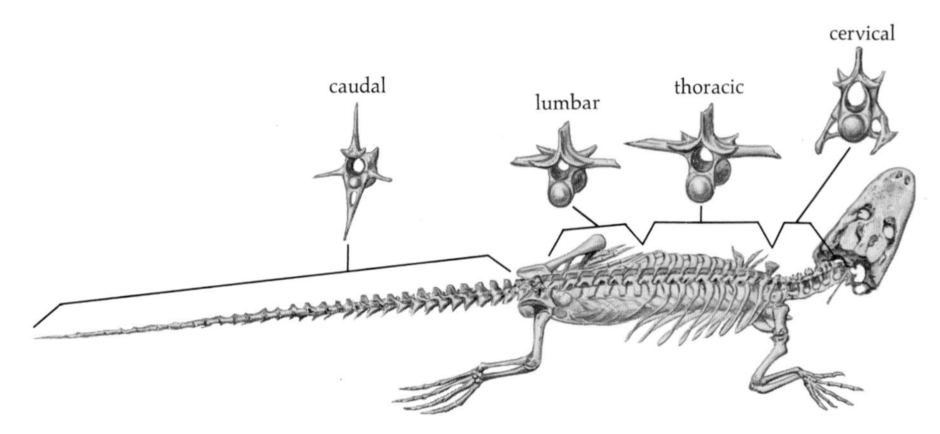

Fig. 5-5. Note the vertebral column of this young alligator skeleton. The inserts show details of its regional specializations.

are bicipital and articulate with a dorsal diapophysis and a ventral parapophysis. Just caudal to the thoracic vertebrae in most reptiles is a variable number of lumbar vertebrae with very short ribs called pleurapophyses fused onto their diapophyses.

Just behind the lumbar region is the sacral region, which in reptiles is typically composed of two large vertebrae with strong sacral ribs fused to the pelvic girdle. These aid in the efficient tetrapod locomotion of the typical reptile. Snakes, however, have undergone a secondary loss of sacral vertebrae and of the entire pelvic girdle. There is no distinction between their thoracic and lumbar vertebrae, and there are fully developed ribs running all the way to the caudal region.

The caudal region is of variable length. In most reptiles the hemal arches are not fused onto the centra but are separate bones, called chevron bones, articulating with the centra.

Throughout the entire reptilian vertebral column, the vertebrae articulate by processes called zygapophyses (Fig. 5-5). Each vertebra bears an anterior and a posterior pair. These interlock between vertebrae, with the articular surface of the anterior zygapophysis facing upward and that of the posterior zygapophysis facing downward. These articulating processes produce a firmer, stronger vertebral column than is possible when adjacent centra are held together by connective tissue alone.

Birds

A highly specialized vertebral column is one of the bird's many adaptations to flight. There are more cervical vertebrae than in any other group, and they are quite elaborate with many processes as well as complex heterocelous centra. This gives great flexibility to the neck, which is particularly important since birds use their beaks to do many things that other tetrapods do with their forelimbs.

In dramatic contrast is the rest of the body. In the thoracic region the neural spines are either fused into one long, platelike neural spine, or are held together by very strong ligaments. To each thoracic vertebra is

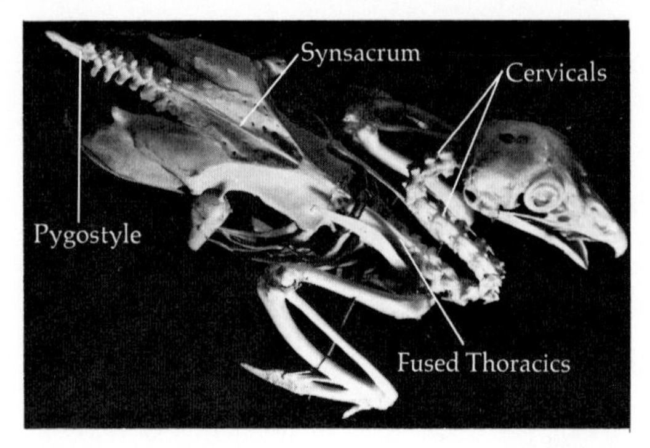

Fig. 5-6. Vertebral column of a chicken, showing regions and relations to other skeletal elements.

articulated a bicipital rib. A prominent structure called the synsacrum is formed from a massive fusion of the most posterior thoracic vertebrae, all lumbar vertebrae, both sacral vertebrae, and the anteriormost caudal vertebrae (Fig. 5-6). Together, the fused thoracic neural spines and the synsacrum make the trunk region rigid and inflexible, so that muscular energy is not required to keep the vertebral column from sagging during flight. The synsacrum also provides a long, strong connection between the pelvic girdle and the vertebral column. This helps the resting bird to balance on its legs without toppling forward, an important consideration since most of its body weight is anterior to the pelvic appendage and of course not supported by the pectoral appendage. It is interesting that a birdlike synsacrum was also present in some of the bipedal dinosaurs, presumably serving the same function.

The bird's caudal vertebrae are greatly reduced in both size and number and fused together to form the pygostyle. This is the skeletal support for the tail feathers, and its size is proportional to the size of and stress on the tail feathers.

Mammals

The cervical vertebrae number seven in all mammals, from the squat-necked whale to the improbable giraffe, with but four exceptions: the three-toed sloth (9), the lesser anteater (8), and the two-toed sloth and the manatee (6 each). The first two cervical vertebrae are highly modified as the atlas and the axis, with the joint between skull and atlas modified for vertical movement, and the joint between atlas and axis modified for rotational movement. As part of this modification the centrum of the atlas does not fuse with the rest of the atlas, but remains separate. It does fuse onto the centrum of the axis and is called the odontoid process of the axis. In this way it forms an efficient rotational joint with the atlas (Fig. 5-7).

Fused to each of the other five cervical vertebrae is a very short bicipital rib which forms a small foramen for the vertebral artery, one of the major arterial supplies to the brain.

The numbers of thoracic and lumbar vertebrae vary in different species. Each type has prominent neural spines, with those in the

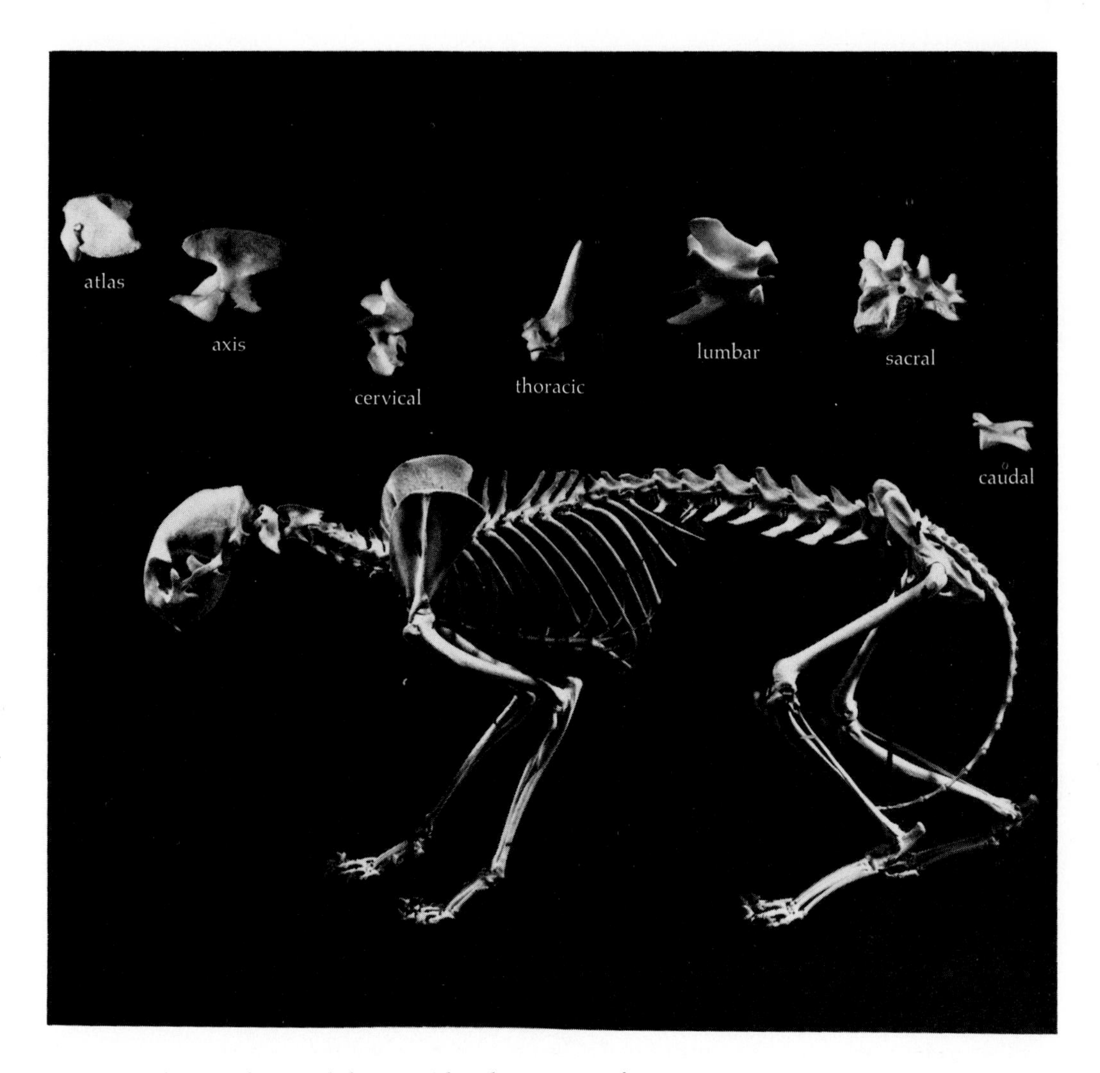

Fig. 5-7. Lateral view of a cat skeleton, with enlargements of an individual vertebra from each region and of the axis and atlas.

lumbar region being somewhat broader. Each thoracic vertebra has a bicipital rib whose dorsal head articulates directly on the centrum and whose ventral head articulates between the anterior and posterior centra. The lumbar vertebrae have pleurapophyses—short, fused ribs.

Characteristically, there are three sacral vertebrae fused to the pelvic girdle, but in some species there are as many as seven or eight. Following these are the small caudal vertebrae, articulating at least anteriorly with chevron bones (Fig. 5-7).

AXIAL SKELETON: RIBS

Ribs may develop in the hypomeric mesoderm anywhere from the posterior edge of the skull through the tail region. If they lie between

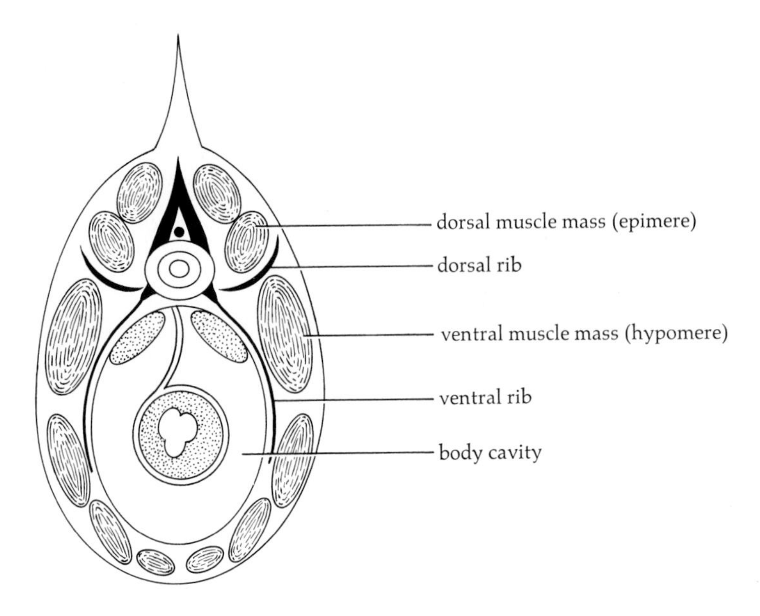

Fig. 5-8. Diagramatic transverse section through the trunk of a teleost showing relative positions of dorsal and ventral ribs.

the dorsal and ventral muscle masses they are called dorsal ribs; if between the ventral muscle mass and the body cavity, ventral ribs (Fig. 5-8). Amphibians have only dorsal ribs, chondrichtheans only ventral ribs. Teleosts often have both, with one pair of each per body segment; it is the dorsal ribs, lying between muscles, which cause their human predators some difficulty.

The exact nature of the amniote rib is not clear. It would seem a simple matter to determine whether it develops embryologically between muscle masses or next to the body cavity. However, the rib develops with one surface against the dorsal muscle mass and the other against the body cavity, while the ventral muscles develop between the ribs. Therefore whether amniote ribs are homologous with the dorsal ribs of teleosts, the ventral ribs of teleosts, or possibly both (which could account for their bicipital nature) is not known.

In some vertebrates the ribs have evolved in spectacular ways. In turtles they are inseparably fused with the dermal bones of the carapace. In snakes their ventral edges are attached to the ventral scales so that the intercostal muscles, which move the ribs, are also the primary muscles of locomotion. In one lizard, *Draco volans,* the ribs pierce through the body wall and support a large flap of skin that aids the animal in gliding from tree to tree. Although in most tetrapods the ventral portion of the ribs remains cartilaginous, in birds both dorsal and ventral portions ossify, and a joint is formed between them. Adjacent ribs are overlapped and held together by the small uncinate processes. This firm rib cage, movable at the dorsoventral joint, is adaptive in respiration, as will be described in Chapter 15.

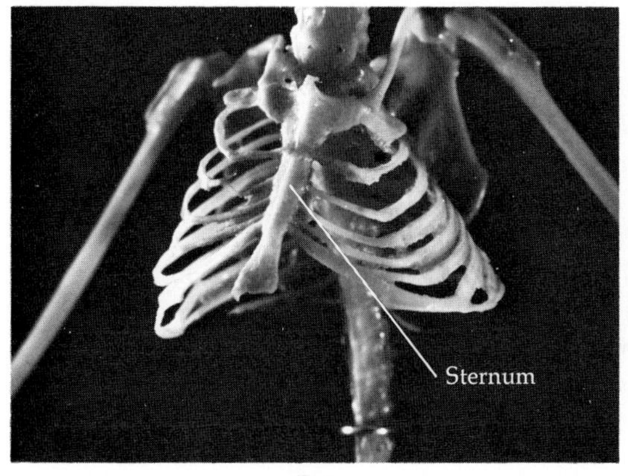

Bat

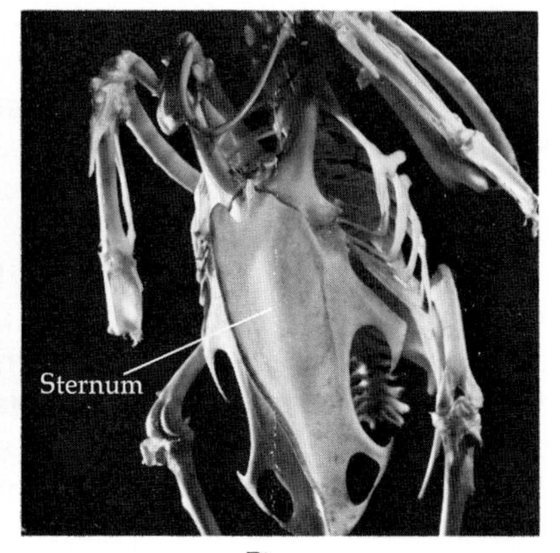

Pigeon

Fig. 5-9. Sternum and surrounding structures of a bird (pigeon) and a bat.

AXIAL SKELETON: STERNUM

Functionally the sternum gives skeletal continuity between the ventral portions of the ribs on the right and left sides of the body. Its phylogenetic origin is unclear; it probably evolved independently in several groups whose environmental situations gave significant adaptive value to this skeletal connection. It is present in only a few fishes and in a limited number of amphibians. It is largest in birds, but is also prominent in modern flying mammals (bats) as it was in the flying reptiles (Pterosaurs of the Mesozoic) (Fig. 5-9).

The sternum does two things for a flying animal: it makes the thoracic cavity more rigid and it provides a large surface area for the attachment of powerful flight muscles. So important are these functions, in fact, that the surface area of the bird's sternum—also called the keel—is to a great extent proportional to its flying ability, being relatively largest in the hummingbird and absent in flightless ratite birds such as the ostrich.

APPENDICULAR SKELETON

The appendicular skeleton can be divided into intrinsic and extrinsic parts. The intrinsic skeleton is that within the appendage itself. The extrinsic skeleton, or girdle, is the skeletal support within the body's main frame to which the intrinsic skeleton articulates. Bones and cartilages of the pectoral and pelvic girdles are readily homologized among vertebrates of widely divergent classes, but the intrinsic skeleton develops completely differently in fishes and tetrapods, and only the broadest homologies can be determined.

EXTRINSIC SKELETON—THE PECTORAL GIRDLE

In Devonian placoderms the dermal bones of the skull continued over the anterior trunk and formed the dorsal portion of the pectoral girdle. Articulating with this was the ventrolateral portion of endochondral bone. In most living vertebrates there is a similar arrangement, although, except in Osteichthyes, the attachment to the skull has been lost. Because the pectoral girdle is the only part of the postcranial skeleton in which dermal and endochondral elements combine and because it nevertheless involves relatively few elements, it makes an interesting demonstration of how structural spatial diversity relates to temporal diversity (Fig. 5–10, see pages following 177).

Chondrichthyes

The cartilaginous fishes, however, having lost all dermal bone, have an entirely endochondral pectoral girdle. Paired scapulocoracoid cartilages lie embedded in trunk musculature along the lateral surface just deep to the skin: the scapular portion is dorsal and the coracoid ventral. At their junction is a depression called the glenoid fossa which receives the pectoral appendage (Fig. 5-11). This fossa is significant in vertebrate locomotion and habitat, and its position shifts throughout temporal and spatial diversity. Ventrally, the right and left scapulocoracoids are joined by a midline element, probably derived from the coracoid. Often there is also another, separate element, the suprascapular cartilage.

In the flattened skates and rays (Batoidea), locomotor energy is provided by huge pectoral fins rather than caudal trunk musculature. The pectoral girdle, therefore, is unusually large, but even so its most prominent component is the scapulocoracoid cartilage. The glenoid fossa, located as in other Chondrichthyes where the scapular and coracoid portions meet, is enlarged to give a broader articulation with the pectoral fins.

Osteichthyes

In most living Osteichthyes the dermal bones of the pectoral girdle are continuous with the skull through posttemporal bones which form synarthrotic articulations—with the skull anteriorly and with other dermal elements posteriorly and ventrally. These dermal elements when fully present include broad plates such as the supracleithrum, cleithrum, and one or more postcleithra, and, more ventrally and medially, the clavicle. However, the series is often not complete; for instance, the clavicle is absent in most Teleostei and Holostei (Fig. 5-11).

The endochondral elements are a dorsal scapula, a ventral medial procoracoid, and either a single scapulocoracoid or two separate bones or cartilages. As in Chondrichthyes, the glenoid fossa lies at the junction of scapula and procoracoid (or of scapular and coracoid portions of the single element).

Among the sarcopterygians there are striking differences between the pectoral girdles of the living lungfish, Dipnoi, and the living coelacanth, *Latimeria*. In the lungfish there is a ligamentous, very loose attachment between the skull and the dermal elements of the girdle. The dorsal dermal elements are greatly reduced, with a small supracleithrum and cleithrum but no postcleithra. Ventromedially there is a prominent clavicle. Between these, articulating dorsally with the cleithrum and ventromedially with the clavicle, is the endochondral scapulocoracoid cartilage and its glenoid fossa.

In *Latimeria*, however, there is no attachment between the skull and the dermal elements of the girdle. A large supracleithrum and cleithrum are present dorsally, as well as an external cleithrum of dubious homology. Ventrally there is a large clavicle. The external cleithrum and the clavicle articulate with the scapulocoracoid, which is primarily cartilaginous with a small ossification; as usual, the scapulocoracoid includes the glenoid fossa.

Amphibians

In living Amphibia the girdle's dermal elements are considerably reduced or lost, particularly the cleithral elements. Among anurans, the bullfrog has a small cleithrum slightly separated from the rest of the girdle; other living amphibians have no cleithrum. Anurans also have an anterior ventral clavicle and a midline ventral dermal bone, the interclavicle, that is found in no living fishes. Most of the anuran pectoral girdle, however, is of endochondral bone, composed of a dorsal scapula and a ventromedial procoracoid with the glenoid fossa lying at their junction.

In urodeles, which have undergone a great reduction of skeletal elements, all dermal bones have been eliminated from the pectoral girdle. The endochondral portion of the girdle consists of a large, mostly cartilaginous scapulocoracoid, ossified only in the area immediately surrounding the glenoid fossa; this area of ossification is partially from the scapula and partially from the procoracoid (Fig. 5-11).

Reptiles

The pectoral girdles of *Sphenodon* and of lizards are similar. Their smaller, dermal portions are the paired clavicles and a midline interclavicle. Their larger, endochondral portions are a dorsal scapula and a ventromedial procoracoid. There is typically full ossification at and around the glenoid fossa, but otherwise both scapula and procoracoid remain largely cartilaginous (Fig. 5-11).

The snakes (Ophidia) have lost paired appendages, and, not surprisingly, the pectoral girdle as well.

The pectoral girdle is bizarrely modified in turtles (Chelonia). Its endochondral portion, which is entirely retained within the rib cage, is triradiate, with a large two-pronged scapula and a single, long procoracoid; the glenoid fossa lies where these two bones articulate. There are no dermal elements in the turtle's functional pectoral girdle, but embryological studies have demonstrated homologies between dermal elements of the pectoral girdle in other tetrapods and the anterior ele-

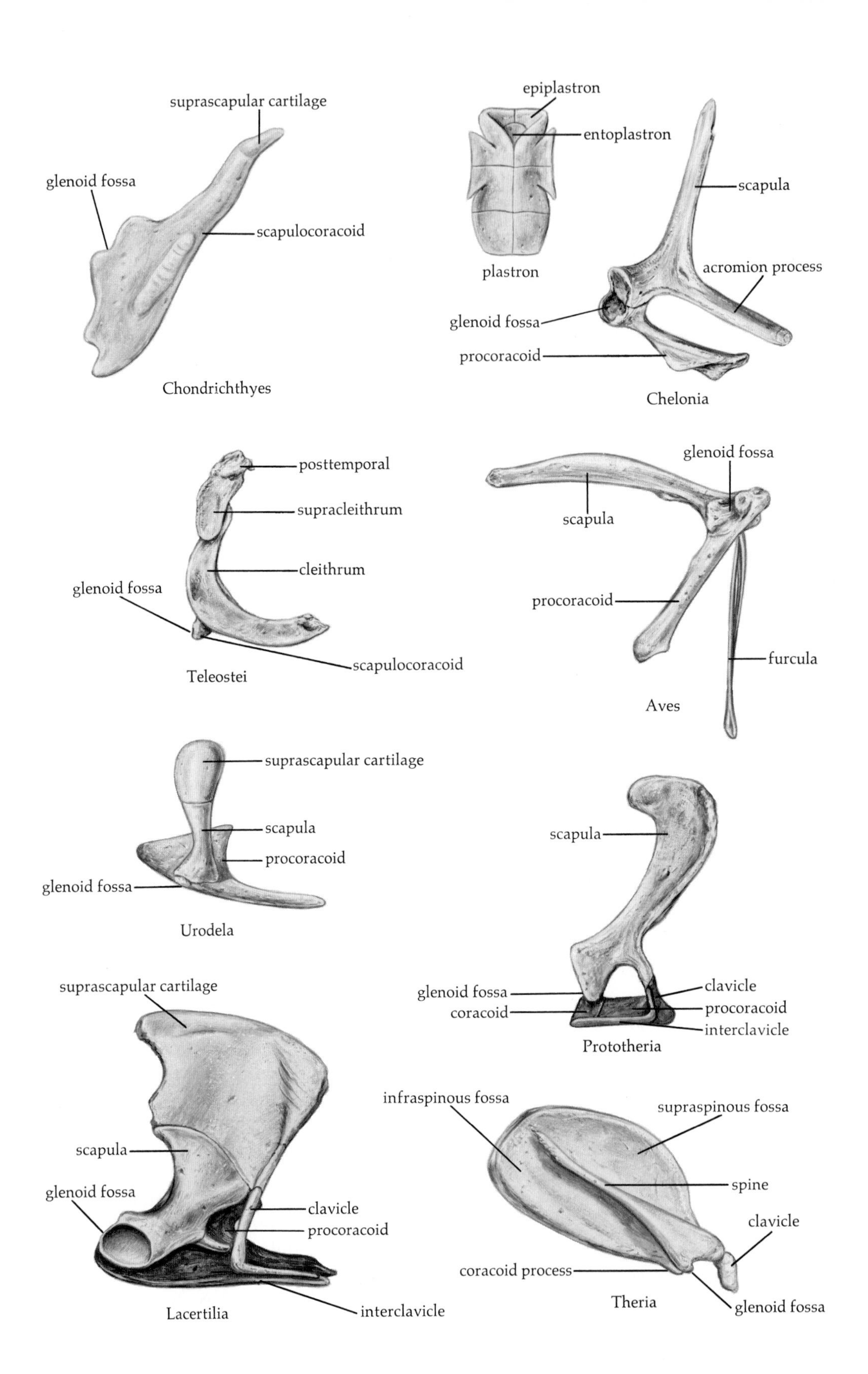
suprascapular cartilage
glenoid fossa
scapulocoracoid
Chondrichthyes
epiplastron
entoplastron
scapula
plastron
acromion process
glenoid fossa
procoracoid
Chelonia
posttemporal
supracleithrum
cleithrum
glenoid fossa
scapulocoracoid
Teleostei
glenoid fossa
scapula
procoracoid
furcula
Aves
suprascapular cartilage
scapula
procoracoid
glenoid fossa
Urodela
scapula
glenoid fossa
coracoid
clavicle
procoracoid
interclavicle
Prototheria
suprascapular cartilage
scapula
glenoid fossa
clavicle
procoracoid
interclavicle
Lacertilia
infraspinous fossa
supraspinous fossa
spine
clavicle
coracoid process
Theria
glenoid fossa

ments of the turtle's ventral shell, or plastron (Fig. 5-11). The paired, most anterior bones, called the epiplastra, are homologous to the clavicles. The midline bone, the entoplastron, is homologous to the midline interclavicle.

In Crocodilia the girdle is primarily endochondral with a large dorsal scapula and a ventromedial procoracoid; the glenoid fossa lies at their point of articulation. The only dermal element is a midline interclavicle which actually functions more like a sternum.

Birds

The pectoral girdle is firmly articulated with the bird's very large sternum by way of the endochondral procoracoid. A narrow, bladelike scapula articulates with the dorsal portion of the procoracoid at the level of the glenoid fossa. Anteriorly and ventrally there is a midline dermal bone, the furcula, which results from the fusion of paired clavicles and a midline interclavicle; this is the bone commonly called the wishbone (Fig. 5-11).

Mammals

In the prototherian Monotremata (the duck-billed platypus and spiny anteater) the dorsal element is a large, simple, endochondral scapula. There are two ventromedial elements: anteriorly a procoracoid and posteriorly a new element, a separate coracoid bone. The glenoid fossa lies between scapula and coracoid rather than between scapula and procoracoid. A clavicle extends anteriorly from the scapula, going ventrally and medially. There is also a very large midline interclavicle (Fig. 5-11).

Among the therian mammals, both marsupials and placentals, the primarily endochondral pectoral girdle is made up of either one or two bones on each side of the body. The largest, most dorsolateral, is the scapula; it bears a fused coracoid process which is homologous to the coracoid bone in prototherians. There is no procoracoid. The scapula typically has a large spine laterally, separating a supraspinous fossa from an infraspinous fossa. As for dermal elements, there is no interclavicle. The more generalized therians have a clavicle, extending ventromedially and articulating with the anterior portion of the sternum. In most carnivores the clavicle articulates with no other bone and is simply embedded in the pectoral muscle mass (Fig. 5-11). In many mammals, particularly hoofed mammals, the clavicle is lost.

In the foregoing brief review one sees certain general characteristics shared by most or all pectoral girdles, such as an endochondral scapular element and various dermal elements. In aquatic vertebrates the primary dermal elements are cleithra, supracleithra, and postcleithra; and in tetrapods, clavicle and interclavicle. Since cleithral elements are so ubiquitously present in bony fishes (including sarcopterygians) and

Fig. 5-11. Lateral views of the right pectoral girdles of diverse vertebrates.

even in anurans, one would expect to find them at some embryologic stage in amniotes even though they are lacking in the adult form. That such is not the case demonstrates that ontogeny does not repeat phylogeny, for there are cleithral elements in amniote phylogeny—in fossil synapsid reptiles, cotylosaurs, and labyrinthodonts.

Because of the generally conservative nature of evolution, a structure which is no longer needed is usually remodeled into another structure with a new function rather than being lost. But this too is shown not always to be true. In addition to the loss of the dermal cleithral elements, the endochondral procoracoid is lost during therian evolution, and the coracoid, which first appears in synapsid reptiles, remains as the only ventral endochondral element of the therian pectoral girdle.

The fossil record is of little or no help in determining the phylogeny of soft structures. One method that can be used is to study spatial diversity and then make logical inferences about temporal diversity. How reliable is this method? In the case of the pectoral girdle, we have seen that even if the fossil record were not available the broad appearance of a cleithrum in so many living vertebrates would suggest that it must have been present in many of the fossil forms including the early amphibians and reptiles. This could be discerned only from the fossil record or from a study of spatial diversity.

However, embryology should not be ignored. For instance, in the ontogeny of the therian scapula the portion which becomes the infraspinous fossa develops first; then an anterior spine develops, and finally the plate that forms the supraspinous fossa develops. Furthermore, the coracoid element is developmentally a separate ossification that does not fuse with the rest of the scapula until quite late in ontogeny. This all indicates that the loss of major ventral endochondral elements in therian mammals was accompanied by the anterior growth of the scapula and the development of its spine—inferences which are drawn from embryological study rather than from spatial diversity and which, in fact, correspond very well with the fossil record.

In order, therefore, to determine the evolution of organ systems, especially soft organ systems with no fossil record, it is essential to study both embryology and spatial diversity.

EXTRINSIC SKELETON—THE PELVIC GIRDLE

The pelvic girdle is of a basically simpler structure than the pectoral girdle, being completely endochondral and having no more than four elements on each side (Fig. 5-12). In fishes it is not articulated with any other part of the skeleton; in tetrapods it is strongly articulated dorsally with the vertebral column at the level of the sacral vertebrae. In most Chondrichthyes and Osteichthyes it is made up of a single puboischiac element on each side, often with a dorsally extending iliac process; the acetabulum, which is analogous to the glenoid fossa, is in a ventrolateral position. In Chondrichthyes, the puboischiac elements are united ventrally into a single puboischiac bar which forms both right and left pelvic girdles.

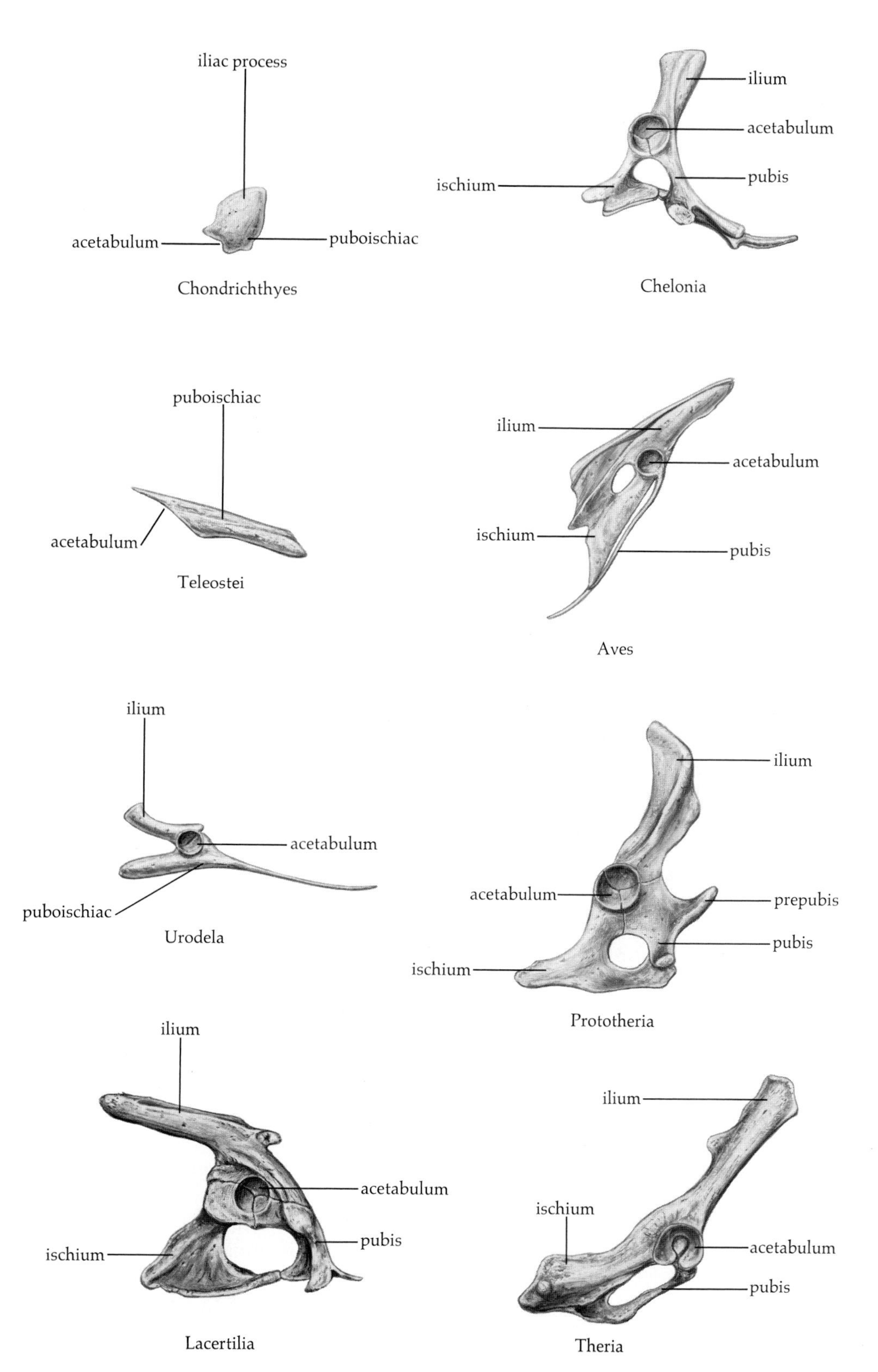

Fig. 5-12. Lateral views of the right pelvic girdles of diverse vertebrates.

Among teleosts the pelvic girdle varies not so much in shape as in position. In many it has migrated anteriorly and lies just behind and between the posterior parts of the gill apparatus, anterior to the pectoral girdle although quite ventral and medial.

In tetrapods, the typical pelvic girdle has three skeletal elements on each side—ventrally the anterior pubis and posterior ischium and dorsally the ilium. The acetabulum lies where the three bones articulate. In most (but not all) tetrapods, the pubes and ischia articulate at the ventral midline, thus joining the right and left pelvic girdles. Dorsally they articulate with the intervening sacral vertebrae, forming a pelvic canal through which the lower alimentary canal and urogenital tracts pass before leaving the body.

In birds and in bipedal archosaurs the right and left pubes and ischia do not articulate with one another, and the pelvic "canal" is left open. In some birds, and in bipedal archosaurs, the pubis extends anteriorly and has also a long posterior process paralleling the ischium; this gives more room for the attachment of the muscles which maintain balance in the perching bird.

In prototherian and metatherian (but not eutherian) mammals there is a fourth skeletal element on each side, called the marsupial or prepubic bone. It articulates ventrally with the pubis (in monotremes it is a fused process on the pubis) and extends anteriorly just deep to the ventral skin of the abdomen. Although having no obvious function in the males of these groups, the marsupial bone does support the brood pouch in female marsupials. However, among monotremes, the female's brood pouch is only rudimentary (spiny anteater) or absent (duck-billed platypus). The presence of this bone in both monotremes, where it is apparently functionless, as well as in metatherians is particularly confusing since there is no indication of a marsupial bone in synapsid reptiles.

Although the number of pelvic girdle elements is both constant and few throughout vertebrate classes, the shape of these elements varies considerably, for they have been modified by natural selection according to the requirements of each species, particularly in relationship to environment (Fig. 5-12).

INTRINSIC SKELETON—UNPAIRED APPENDAGES

Fishes have unpaired appendages located on the midline. There is a dorsal fin, or in some fishes, two, one in front of the other. Caudally there is a tail fin, or caudal fin; ventrally, just posterior to the anus, is an anal fin. The dorsal fin and dorsal parts of the caudal fin are supported either by elongations of neural spines which reach up into the fin or by separate bones or cartilages which develop in longitudinal skin folds. Similarly, hemal arch extensions frequently run into the anal fin and the more ventral parts of the caudal fin; in some fishes the skeletal supports for these fins form from separate skeletal elements differentiating in ventral skin folds.

Functionally, only the caudal fin is much involved in locomotion; it gives a broader surface area to be pushed against the water during lateral undulations. The anal and dorsal fins, usually capable of very little movement, act primarily as stabilizers.

INTRINSIC SKELETON—PAIRED APPENDAGES

In addition to the unpaired median fins, fishes (like most vertebrates) have one pair of pectoral appendages and one pair of pelvic appendages. The exceptions to this generalization have evolved from forms that never had paired appendages (cyclostomes, from Ostracodermi) or that have secondarily lost them during their phylogenies. The latter has occurred in two classes: Amphibia, in the caecilians (Apoda), and Reptilia, in two groups of Squamata—the ophidians (snakes) and some lacertilians (burrowing lizards). All other vertebrates, however, have two pairs of appendages—ample indication of their selective value. The intrinsic skeletons of pectoral and pelvic paired appendages are basically similar throughout each class or subclass.

Fishes

In Actinopterygia and Chondrichthyes the muscles of the very thin, paired fins attach only to the most basal part of the intrinsic fin skeleton, near the girdle. The skeleton itself is made up of a number of radiating cartilages or bones that reach up into the fins and are distally continuous with collagenous fibers of stout connective tissue. Various names have been given to these skeletal elements but homologies are uncertain. It seems that the potential existed for a large number of radiating skeletal elements giving stiffness to the fins and that they evolved differently, that is, not homologously, in different groups.

A distinctive modification is found in the pelvic appendage of male chondrichthyeans in which accessory cartilages have evolved as the skeletal portion of an elongate clasper organ through which sperm is passed into the female during copulation. The clasper cartilages, often quite intricate, will be discussed in the chapter on reproduction.

The sarcopterygian fin is more lobate and contains muscles. Those of Dipnoi and *Latimeria* are similar, with a central series of bones from which radiate several skeletal elements. However, in the rhipidistians (crossopterygian ancestors of amphibians) the fin skeleton was more nearly comparable to that found in tetrapods. It had but a single proximal element—the humerus in the pectoral fin or the femur in the pelvic fin. This articulated distally with two elements—the radius and ulna in the forelimb or the tibia and fibula in the hindlimb. Variable numbers of skeletal elements reached out more distally into the fins.

Tetrapods

Despite the different uses to which it is put, the tetrapod limb skeleton is a conservative structure, pentadactyl (five-toed) in design, and basically similar in living as well as many extinct forms (Fig. 5-13). The forelimb has a single, proximal humerus in the upper arm and an anterior radius and posterior ulna in the forearm. In the wrist is a group

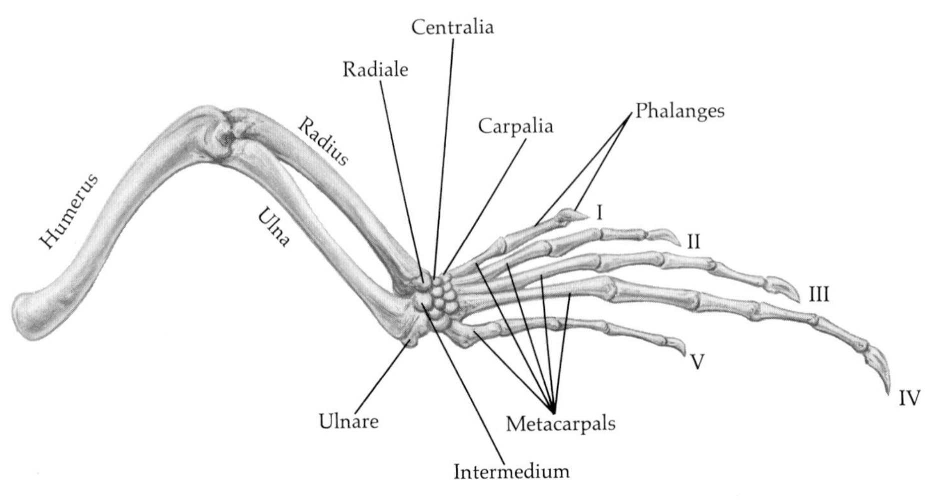

Fig. 5-13. Generalized pentadactyl limb.

of small carpal bones, variously named in different taxonomic groups and anatomical nomenclatures. For our purposes we will divide the carpals into three rows—proximal, intermediate, and distal. The proximal row is comprised of a small bone at the end of the ulna called the ulnare, another at the end of the radius called the radiale, and between these two a third, the intermedium. The intermediate row is a group of up to four centrale elements, numbered from the most anterior (on the radial side) to the most posterior (on the ulnar side). The distal row contains the carpalia, numbered one through five for the five digits. Articulating with the carpalia are elongate "hand bones," the metacarpals, again numbered one through five. Distal to this are the digits, each composed of a number of short "finger bones," or phalanges. In the generalized tetrapod there are two phalanges for the first digit, three for the second, four for the third, five for the fourth, and four for the fifth. A shorthand way of designating the number of phalanges per digit is called the phalangeal formula, written 2,3,4,5,4 for the generalized tetrapod.

The intrinsic elements of the hindlimb are differently named but similarly arranged. The proximal element is the femur. The shank of the leg contains the anterior tibia and the posterior fibula. The ankle bones, called tarsals, include in the proximal row the tibiale, intermedium, and fibulare; in the intermediate row the centralia; and in the distal row five tarsalia. Distal to this are the bones of the foot, the five metatarsals; most distal are the phalanges, with the same phalangeal formula for the generalized tetrapod of 2,3,4,5,4.

Modifications of this basic pentadactyl limb are generally correlated with environment and, as we have come to expect, occur more frequently and more dramatically in distal than in proximal portions. They usually involve a loss or a fusion of elements rather than the evolution of new ones. The bird limb is a good example (Fig. 5-14). The humerus of its forelimb is unremarkable except that, like many bird bones, it tends to be hollow and thus quite light. The radius and ulna are slightly shortened but are otherwise conservative. In the carpal region of the 8-day-old embryo there are thirteen distinct elements; in the adult there are only three—the radial carpal, composed of fused first centrale, radiale, and intermedium; the ulnar carpal, composed of fused ulnare and the pisiform (a sesamoid bone); and distal to this the carpometacarpus, composed of fused centralia two and four, carpalia two, three, four, and five, and metacarpals two, three, and four. Several phalanges are lost; the phalangeal formula is 0,1,2,1,0.

Greater modification of distal compared to proximal portions has also occurred in the bird's hindlimb. The femur is unremarkable. The distal end of the tibia, however, is fused with the tibiale, intermedium, and one centrale, forming a large tibiotarsus; the fibula is extremely small and vestigial. The remaining tarsal elements are fused with the metatarsal elements of digits 1, 2, 3, and 4; digit 5 is lost. The exact number of phalanges varies according to species.

There is the same emphasis on distal modification in mammals. The fibula is frequently either fused with the tibia or lost, or present only as a small, almost functionless splint bone. The greatest distal modifications occur in ungulates, such as the horse whose only functional digit is the third in both fore- and hindlimbs. The horse's third metacarpal (or metatarsal) is greatly elongated, giving it what amounts to an additional long bone and joint which greatly facilitate its running

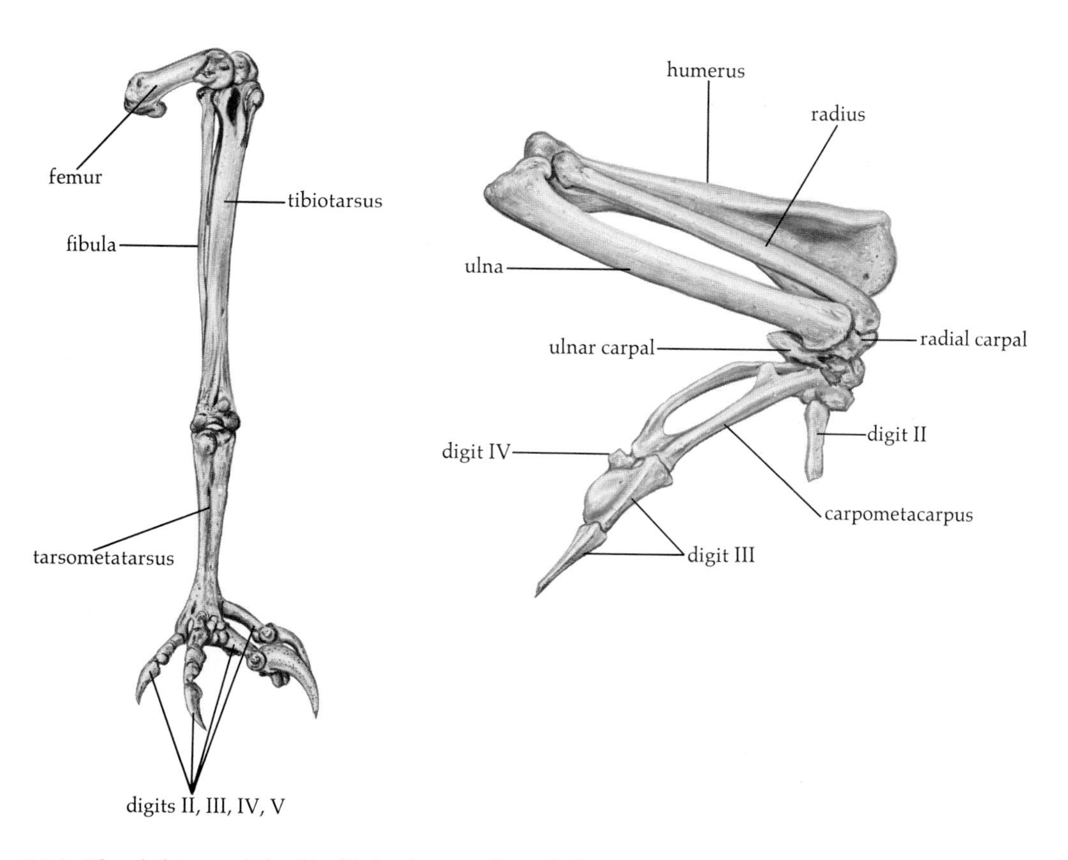

Fig. 5-14. The skeletons of the hindlimb of an eagle and the forelimb of a turkey.

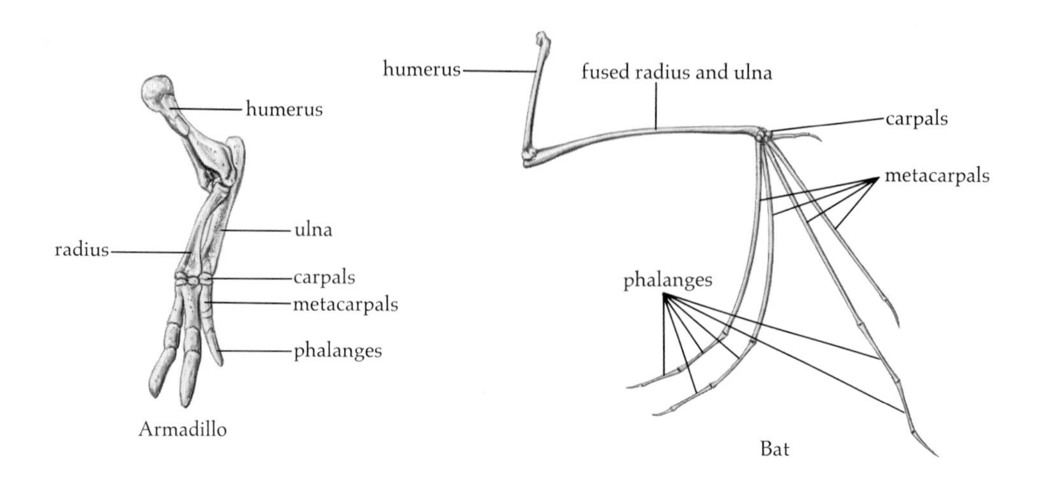
humerus
ulna
radius
carpals
metacarpals
phalanges
Armadillo
humerus
fused radius and ulna
carpals
metacarpals
phalanges
Bat

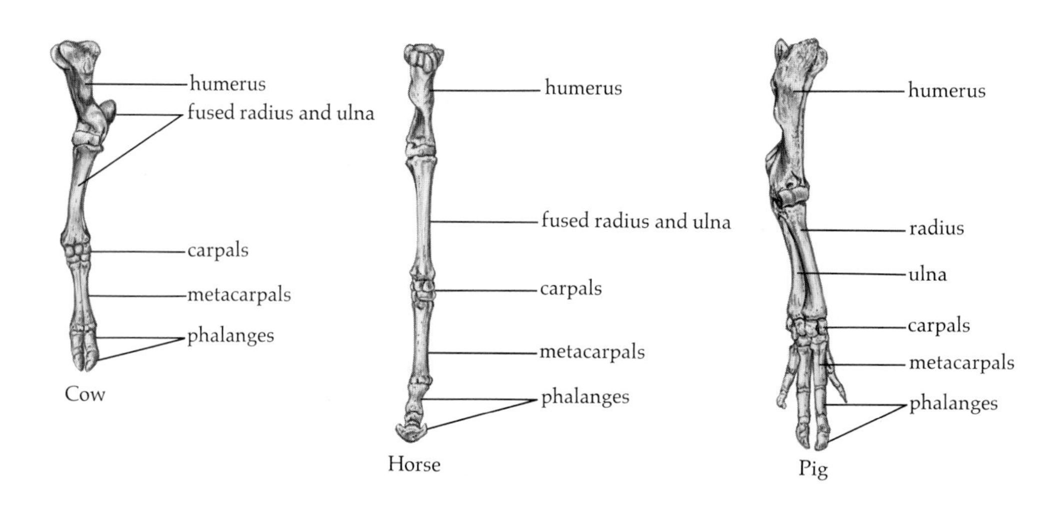
humerus
fused radius and ulna
carpals
metacarpals
phalanges
Cow
humerus
fused radius and ulna
carpals
metacarpals
phalanges
Horse
humerus
radius
ulna
carpals
metacarpals
phalanges
Pig

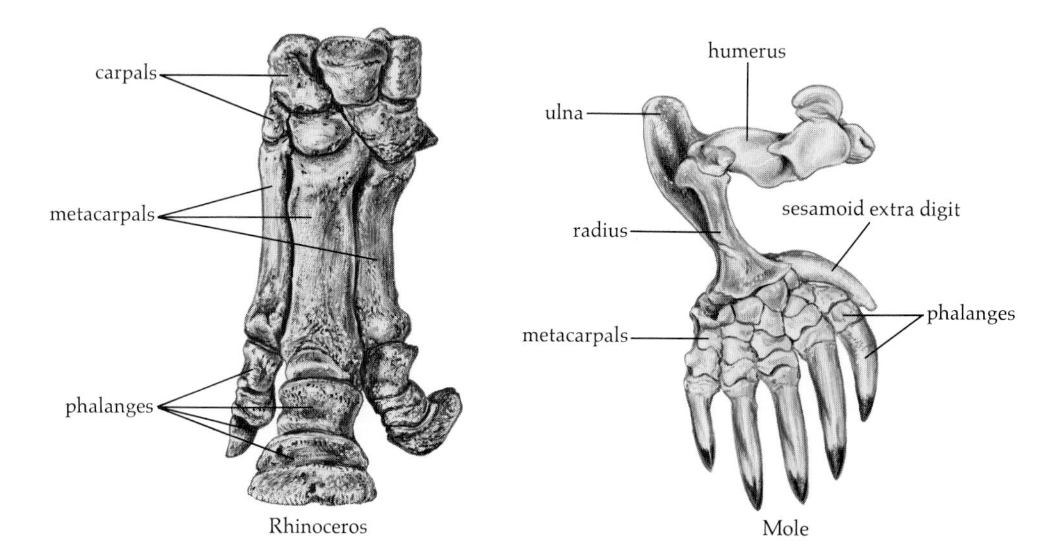
carpals
metacarpals
phalanges
Rhinoceros
humerus
ulna
radius
sesamoid extra digit
phalanges
metacarpals
Mole

gait. The three small phalanges of the third digit are enlarged; the distal one is completely covered by the dermal modification called the hoof (Fig. 5-15).

A different type of limb modification is seen in an animal such as the mole, whose habitat is subterranean. The forelimbs are short and stout, and the broad hand (the digging tool) is made even broader by a large sesamoid element which is a functional sixth digit.

ADAPTIVENESS OF THE TETRAPOD APPENDICULAR SKELETON

The diverse modifications of the appendicular skeleton include both general and specific adaptations. Among the general adaptations is the firm attachment of the pelvic girdle to the axial skeleton so that the power stroke of locomotion, coming from the pelvic limbs, is passed directly to the axial skeleton. Another adaptation is the loose or nonexistent skeletal connection of the pectoral girdle to the axial skeleton, which allows the forelimbs and pectoral girdle to act as shock absorbers during locomotion. This freedom of the forelimbs also facilitates greater movement and manipulation.

Specific adaptations occur primarily in the distal portions of the limbs; examples are the reduced toes and lengthened metacarpals in horses and other ungulates; the grasping ability of the digits in man, opossum, and raccoon; and the short, stout, powerful digging appendages of moles.

In the appendicular skeleton, therefore, there are general adaptations proximally and specific adaptations distally. In fact, the structural and functional adaptations of the distal portion of the limb—that portion in most intimate contact with the environment—are an effective mirror of the environment.

SUGGESTED READING

Bodemer, C. W. (1968). "Modern Embryology." Holt, Rinehart and Winston, Inc., New York.

Goodrich, E. S. (1930). "Studies on the Structure and Development of Vertebrates." The Macmillan Co., London. (Reprinted in two volumes by Dover, 1958.)

Gray, J. (1953). "How Animals Move." Cambridge Univ. Press, New York.

Hildebrand, M. (1968). How Animals Run. *In* "Vertebrate Adaptations," pp. 30–37. W. H. Freeman and Co., San Francisco, California.

Romer, A. S. (1966). "Vertebrate Paleontology." Univ. of Chicago Press, Chicago.

Fig. 5-15. Several skeletal modifications of the mammalian forelimb. Note the shortened distal elements in digging forms (mole and armadillo) and the elongated distal elements in flying forms (bats) and in ungulates.

Smith, J. M., and Savage, R. J. G. (1956). Some locomotory adaptations in mammals. *J. Linn. Soc. London, Zool.* **42,** 603–622.
Williams, E. E. (1959). Gadow's arcualia and the development of tetrapod vertebrae. *Quart. Rev. Biol.* **34,** 1–32.

MUSCULAR TISSUES

6

Except for the streaming of cytoplasm and the actions of cilia and flagellae, all body movements of vertebrates, whether grossly visible or not, are caused by the contractions of muscles. Microscopic study reveals three types of muscle cells—smooth, striated, and cardiac—each containing the contractile proteins actin and myosin (Fig. 6-1). Smooth muscle is incorporated into every organ system except the skeletal, the nervous, and the muscular system itself; it controls the movement of food through the alimentary canal, the secretions of glands, the contractions of the urinary bladder, and even the fine movements of hairs in the skin. In general, the smooth musculature regulates a vertebrate's internal environment. It acts more slowly than other muscles, but can sustain contraction for long periods.

Skeletal muscles, also called striated muscles, form distinct organs themselves and are referred to as "the muscular system." Most of these muscles move the skeleton; others insert on the skin and cause it to move. In general, striated muscles move large parts of the body in response to and in relation to the external environment. They act much more quickly than do smooth muscles, with rapid, strong contractions, and they fatigue more readily.

Cardiac muscle is found only in the vertebrate heart. It is characterized by its rhythmic activity, which can be modified by neural or hormonal stimulation. In general its structural and functional characteristics are intermediate between those of smooth and skeletal muscles. For instance, cardiac muscle contracts more rapidly and with a shorter delay after stimulus than smooth muscle, but less rapidly than skeletal muscle.

SMOOTH MUSCLE

CELL STRUCTURE

The smooth muscle cell is relatively long and tapered, with a central oval nucleus. Little more than this can be seen with the light microscope, although with the electron microscope one finds typical mitochondria, Golgi apparatus, a granular endoplasmic reticulum, and, most prominently, fine parallel myofilaments running the length of the cell and occupying much of its space. Biochemical analysis shows that there are large amounts of actin and myosin in smooth muscle cells. These proteins must be principal constituents of the myofilaments rather than of the sarcoplasm (as muscle cell cytoplasm is called), for the myofilaments occupy such a large part of the cell that there is but a scant amount of sarcoplasm—not enough to contain large amounts of anything.

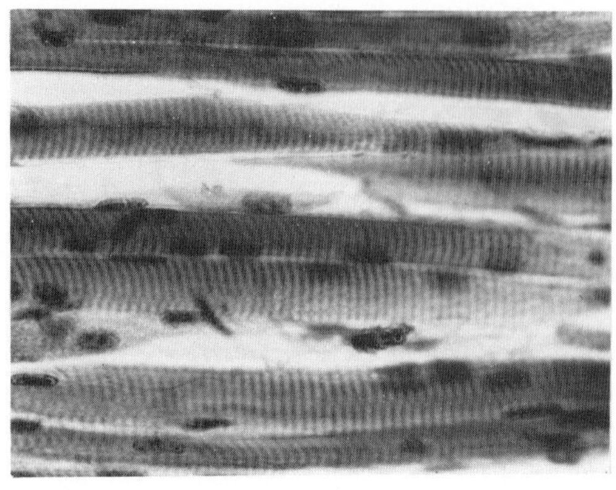

striated

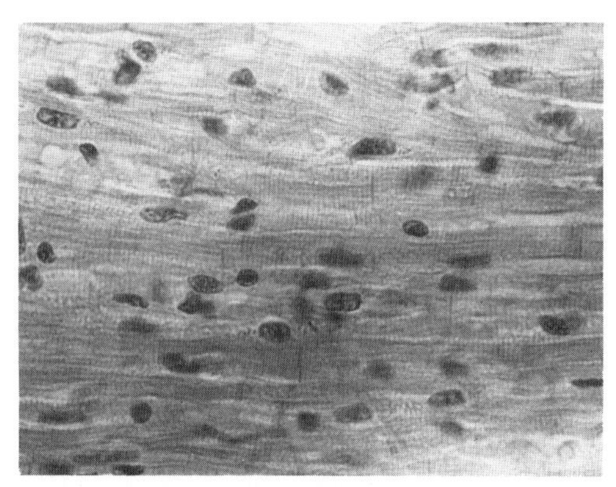

cardiac

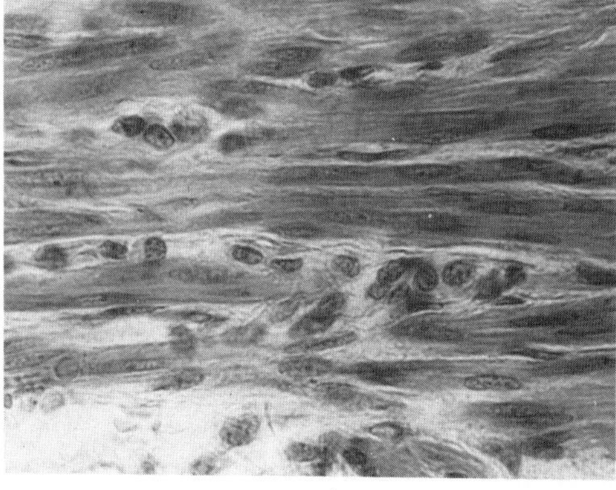

smooth

Fig. 6-1. The three types of muscle cells found in vertebrates. Note the intercalated discs in the cardiac muscle.

Groups of smooth muscle cells

Smooth muscle cells are almost always grouped, in either large sheets or small fascicles. For instance, small groups of smooth muscle cells form subunits of the eye, some groups controlling the dilation and constriction of the pupil and others controlling the roundness of the lens. In the alimentary canal and in blood vessels, smooth muscles form sheets around the lumen and can contract to change its diameter or length (Fig. 6-2).

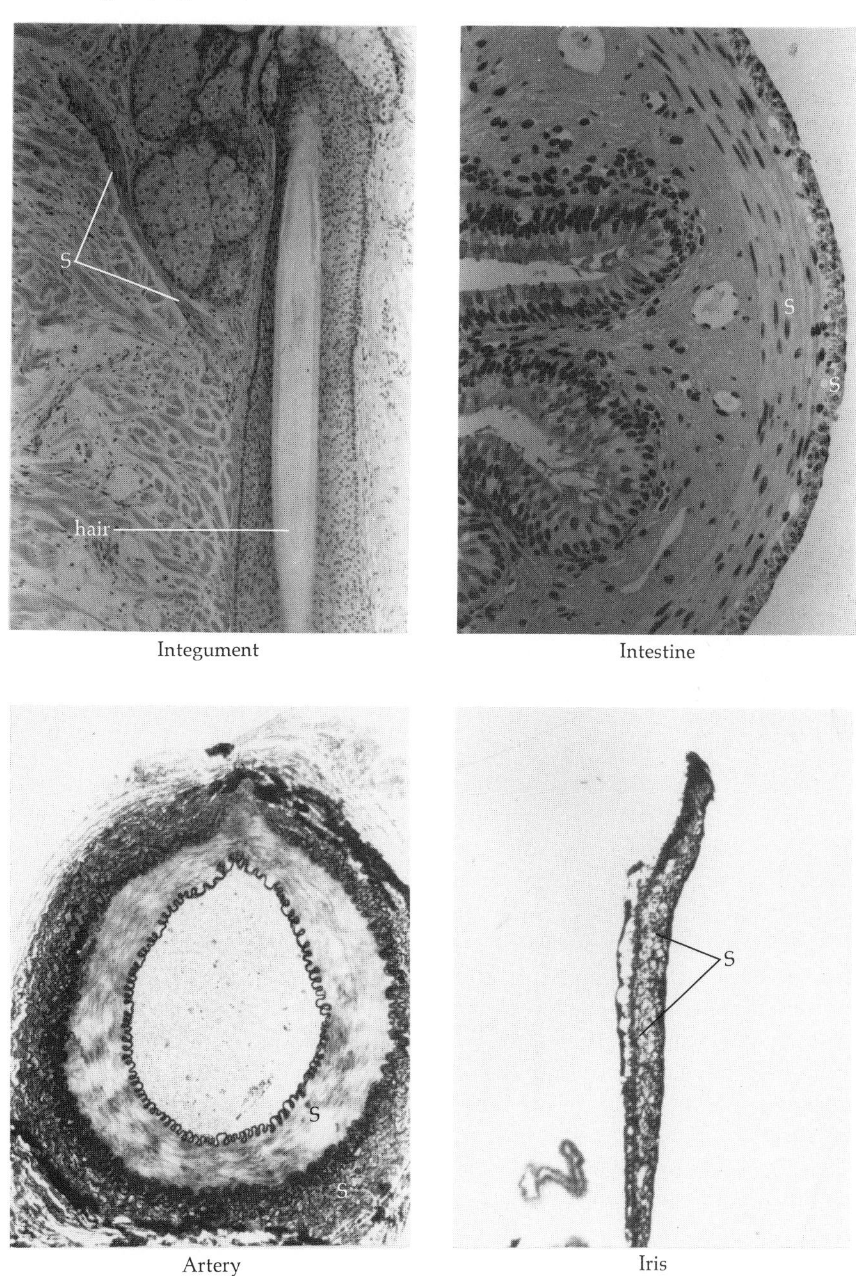

Fig. 6-2. Smooth muscles (S) occur in many organs of the body, some of which are represented here. In the mammalian integument, arrector pili muscles raise individual hairs; in other organs smooth muscles move food through the intestine and control the diameters of blood vessels and of the pupil of the eye.

The individual smooth muscle cells are surrounded and held together by connective tissue fibers, through which blood vessels and nerves pass. Smooth muscles generally have a uniformly distributed but scant capillary supply; only a few smooth muscle cells have direct nervous innervation.

FUNCTIONAL CHARACTERISTICS

Functionally there are two quite distinct types of smooth muscles. Most are visceral muscles (e.g., those of the alimentary canal) which have a diffuse innervation and are slow to react but also very slow to fatigue. They have spontaneous rhythmic contractions, the rate of which can be altered by neural or hormonal stimulation. In addition they readily contract in response to mechanical stimulation, for instance, when the smooth muscles of the alimentary canal are stretched during feeding.

The multi-unit smooth muscle, on the other hand, has a larger innervation. It contracts only upon neural or hormonal stimulation, and then somewhat more rapidly than do the visceral smooth muscles. The muscles of the iris, the muscles controlling lens curvature, and the muscles surrounding blood vessels are multi-unit smooth muscles.

SKELETAL MUSCLES

CELL STRUCTURE

When skeletal muscle is cut in longitudinal sections and studied with the light microscope, its characteristic striations appear as alternating light and dark bands. The light bands are called I (isotropic) bands because these portions of the cell are nonbirefringent; the dark bands are called A (anisotropic) bands and are birefringent.

With greater magnification under the light microscope subdivisions of these bands are apparent. In the center of the I band is a thin, dark line called the Z line; in the center of the A band is a pale streak, the H line. Studies combining electron microscopy, biochemistry, and X-ray diffraction have shown that the bands and lines are caused by the macromolecular structure of the contractile mechanism of these muscles.

The A bands of a skeletal muscle are the areas where the contractile protein myosin occurs in "thick" filaments (80–100 angstroms [Å] in diameter). The I bands lack myosin, but have "thin" filaments (about 50 Å in diameter) of the other contractile protein, actin. These actin filaments are not only in the I band but also extend into the A band, running between myosin filaments. The H line is a small area in the center of the A band where there is no actin but only myosin filaments. The Z line, in the center of the I band, is the point of origin and attachment of the actin filaments (Fig. 6-3). During muscular contraction the actin and myosin filaments slide past one another, so that the thin actin filaments overlap the myosin filaments to a greater extent

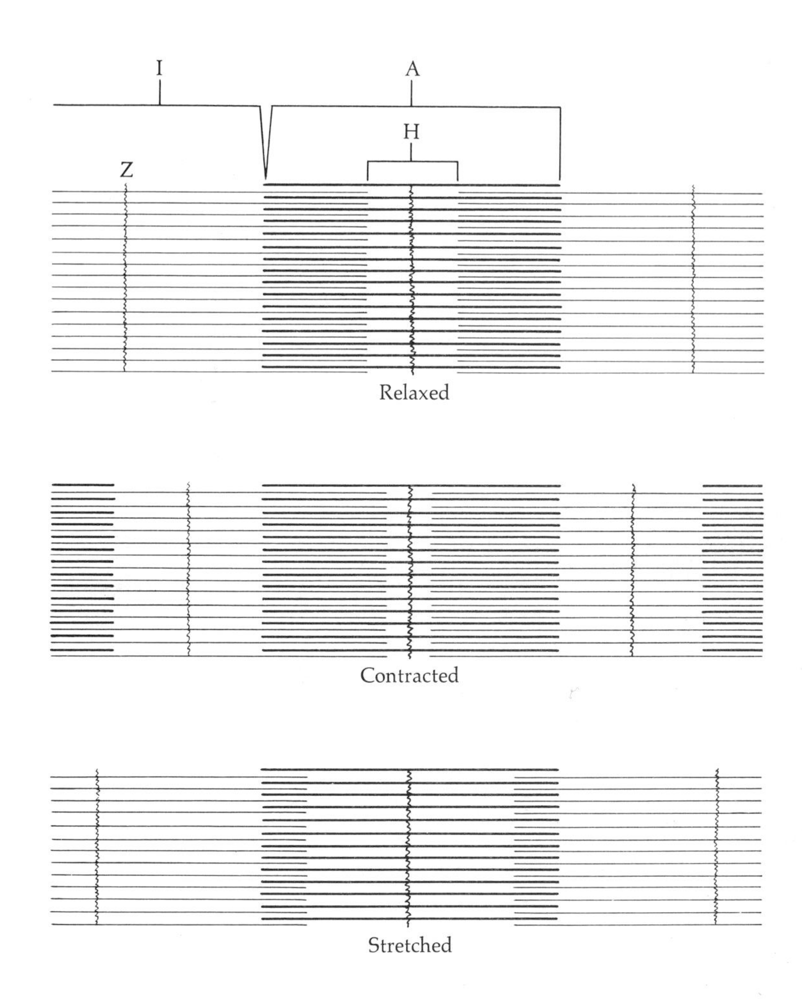

Fig. 6-3. Diagrams showing the arrangements of actin (thin) and myosin (thick) filaments in skeletal muscles while relaxed, contracted, and stretched.

than usual. This can be observed microscopically in a contracted muscle. The precise molecular mechanism causing this sliding action is not known; it is thought that weak chemical bonds are changed when a muscle is stimulated to contract.

OTHER STRUCTURAL FEATURES OF SKELETAL MUSCLE CELLS

There are usually many elongated mitochondria lying just beneath the cell membrane (called sarcolemma in muscle cells) among the nuclei, and more mitochondria deep in the cell between fibrils. In resting muscles there are prominent glycogen particles between fibrils. These glycogen granules are sparse in fatigued muscles because most of them have been depleted by energy production.

The skeletal muscle cell has an intricately organized network of endoplasmic reticulum called sarcoplasmic reticulum. There is a group of longitudinal tubules running between fibrils and extending the length of the muscle cell. At right angles to the longitudinal system, at the level of the Z line, is a transverse system called the T system. Pores in the T system bring the extracellular fluids surrounding the muscle cell into the myofibrils at the Z line. The current theory, which has substantial experimental justification, suggests that the T system aids rapid, almost simultaneous stimulation of all the myofibrils of a single muscle cell. In other words, it appears to be the T system tubules that cause the necessary coupling between the nerve impulse received by the cell and the contraction of the entire cell in one unified response.

TYPES OF SKELETAL MUSCLE FIBERS

Skeletal muscle fibers vary in length and in diameter, but their greatest difference is a structural one which is indicated by their color; there are red fibers and white fibers, and the percentage of each in any gross muscle will make it redder or whiter (e.g., the "light meat" and "dark meat" of poultry).

Red muscle fiber gets its color from a globular protein, myoglobin. Functionally myoglobin has many of the characteristics of hemoglobin, and it loosely binds oxygen. Thus red fibers can store oxygen to be used when needed. Red fibers have slightly more sarcoplasm, and less glycogen, than do white muscle fibers; they are of smaller diameter, have a greater blood supply, and are slightly slower both to contract and to fatigue. On the other hand, white fibers have little myoglobin and sarcoplasm, and a considerable amount of glycogen; they are of rather large diameter, have a small blood supply, and both contract and fatigue rapidly.

Most gross muscles contain some of both kinds of fibers, and take their name and character from whichever kind predominates. Tetrapods have more red muscles than do fishes, and tetrapod muscles which must sustain a high tonus to counteract gravity (e.g., the limb extensors) are especially high in red fibers. On the other hand, muscles which must be activated rapidly for escape or predation (e.g., the limb flexors) have more white fibers. Within any taxonomic group there are more red muscles in species which characteristically undergo sustained locomotor activity, and more white muscles in species which rely on sudden spurts of great speed. For instance, although fishes generally have many more white muscles than do tetrapods, the salmon's muscles look pink; its long migrations necessitate a relatively high percentage of red fibers. Even so, the salmon's red, or pink, muscles have less myoglobin and fewer red fibers than those of any mammal.

EXCITATION OF SKELETAL MUSCLES

In contrast to smooth muscles, every skeletal muscle fiber is directly innervated. Each motor nerve fiber can branch and innervate few or many muscle fibers in the gross muscle; the exact number is called the motor unit, and the smaller this unit, the finer the muscle's control. Muscles which move the eyeball, for instance, have a motor unit of 1, while the motor unit for a major extensor of a large animal's leg may be 100 or more.

The area of the sarcolemma in which nerve endings are received is called the synaptic membrane. The nerve impulse causes a release of acetylcholine onto the synaptic membrane, which causes its brief depolarization. If the depolarization is great enough the stimulus is said to be at threshold or above, and a muscle action potential occurs which spreads over the entire muscle fiber. This action potential causes the muscle fiber to contract, which shortens it by approximately one-half. Skeletal muscle fibers either contract fully or not at all to a given stimulus—this is called the all or none phenomenon.

Although it is known that the spread of a muscle action potential over the sarcolemma always precedes contraction, the precise cause-and-effect mechanisms are not known. However, as described above, the T system is implicated.

STRIATED MUSCLES AS ORGANS

Around each gross muscle is a dense sheath of connective tissue, called the epimysium. At each end of the muscle, the epimysium is continuous with the muscle's tendon, or aponeurosis as it is called if it is in a broad, flat sheet. The tendon or aponeurosis itself is formed of similar but even tougher connective tissue, which is continuous with the periosteum of a bone or perichondrium of a cartilage, as the case may be.

The epimysium is also continuous with a looser and finer connective tissue, the perimysium, which separates bundles (or fascicles) of muscle fibers within the muscle, and through which nerves and blood vessels travel to the muscle. Like the epimysium, the perimysium is continuous with the tendon or aponeurosis.

The arrangement of muscle fibers and their relationships to the perimysium and to the tendon or aponeurosis vary in different muscles (Fig. 6-4). Some muscles are strap-shaped, with long, parallel muscle fibers which run most of the length of the muscle and attach to the perimysium and epimysium where the tendon or aponeurosis begins. Contraction of these strap-shaped muscles causes maximum movement with minimum force.

In many cases the tendon extends far into the muscle and has short fibers obliquely attached to it. If these fibers all enter obliquely into just one side of the tendon, the muscle is called a unipennate muscle, whereas if diagonal fibers run to both sides of the tendon deep within the muscle, it is a bipennate muscle.

Within other muscles, the tendon splits into several tough connective tissue subunits, to each of which very short fibers are obliquely attached. Such a multipennate muscle, with a great many short muscle fibers, has the greatest strength because of the large number of fibers that can be brought into play for a movement. However, it can move for only a short distance.

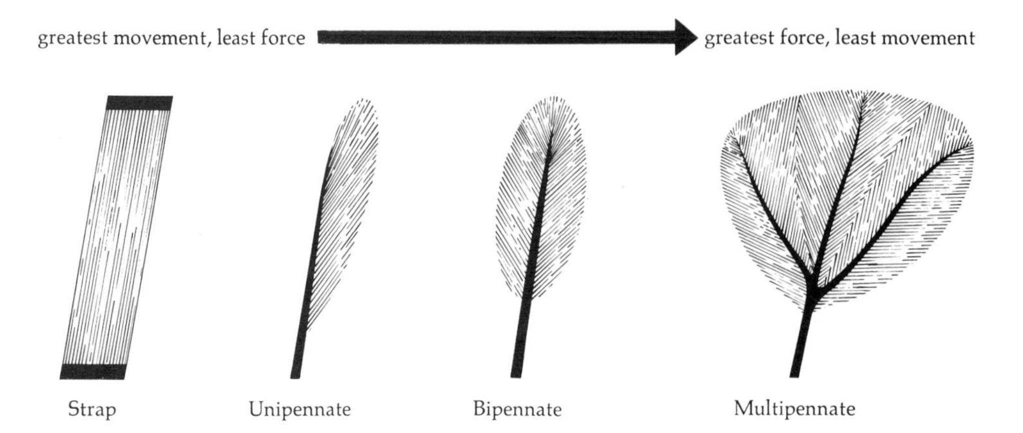

Fig. 6-4. The relative positions of muscle fibers and tendons vary with the functional characteristics of the muscles.

Whereas an individual muscle fiber will contract either completely or not at all, a gross muscle exhibits a graded contraction depending upon the percentage of fibers contracted at any one time. Under normal physiological conditions, even a resting muscle has some fibers in contraction. This gives the muscle "tone." The further contraction or relaxation of the muscle increases or decreases its tonus. A muscle in which no fibers are contracting is without tone and is called flaccid; this abnormal condition usually occurs only under deep anesthesia.

The muscle is usually attached to two skeletal elements. Contraction causes the belly of the muscle (the mass between its two ends) to swell and one or both of its ends to move. The end attached to the more distal skeletal element is usually the one that moves; if so, it is called the insertion, and the proximal end, which usually remains fixed, is called the origin. However, these are operational terms and sometimes are reversed. For instance, the large muscle of the human forearm, called the biceps, runs between the shoulder girdle and the forearm. If the shoulder joint is held rigid while the biceps is contracted, the elbow bends. In this case the insertion is the distal end of the muscle and the origin is the proximal end. However, if the elbow is kept rigid and the shoulder muscles are relaxed so that the shoulder joint can move, the contraction of the biceps raises the upper arm. In this case, the insertion is the proximal end of the biceps, and the origin is the distal end.

The gross movement of the muscle is called its action, which can be further characterized in one of several ways. If the muscle action decreases the angle between two bones, it is called flexion; if it increases the angle, it is extension. Adduction is a movement toward the midline of the body, and abduction is a movement away from it. A depressor is a muscle which lowers, and a levator is one which raises; for instance, there are depressors and levators of the lower jaw. A rotator, as the name implies, twists something. There are two different types of rotators in the forearm: the pronator rotates the arm toward the midline, and the supinator rotates it away from the midline. Finally, a dilator makes a spherical opening larger, and a constrictor makes it smaller; for instance, dilators and constrictors control the pupil of the eye.

The process of naming, dissecting, and describing muscles as separate entities tends to obscure the fact that, to an even greater extent than other organs, skeletal muscles work together rather than individually. For any given movement several muscles are involved. First there is the prime mover, which does most of the work. Another muscle which potentially would do the opposite action is called the antagonist. During movement the prime mover increases its tonus and the antagonist decreases its tonus, or, to put it another way, the prime mover contracts while the antagonist actively relaxes. Other muscles function as joint-fixers, stabilizing parts of the skeleton which are not to be involved in the movement and thus limiting and controlling the movement. Still other muscles act as secondary movers and secondary antagonists, assisting the action of either the prime mover or the prime antagonist.

The involvement of several—even as many as two dozen—muscles in one smooth, coordinated movement requires a tremendous amount of integration, which is controlled by the nervous system. We will discuss how this happens in a later chapter.

Because muscles are attached to skeletal elements by tendons, which are sometimes quite long, the action can occur at some distance from the muscle fibers. Thus it is possible to have delicate, nonbulky structures, such as the bird's foot or the human hand, in which muscles located elsewhere exercise complex, discrete control.

CARDIAC MUSCLES

The muscles of the heart are striated as are skeletal muscles, but the muscle cells are different. Individual cardiac muscle cells are branched, with two or three side branches from each fiber. The cells are shorter than those of skeletal muscle, and where two cells join end-to-end there are sizable thickenings called intercalated discs. Electron microscopic studies have demonstrated that these intercalated discs lie at the level of a Z line and are actually the connections between two muscle cells, the thickening being caused by an interdigitation of the two sarcolemmas which helps to hold the adjacent cells tightly together.

Cardiac muscle cells undergo automatic, rhythmic contractions, without external stimulation. The rate of this rhythm, however, can be modified by either hormonal stimulation or the autonomic nervous system.

Heart muscles are red, with a great deal of myoglobin; unlike most red muscles they also have considerable glycogen. These two features, combined with some specific anatomical characteristics (to be discussed in the chapter on the circulatory system), enable the heart to fulfill its function of continual contraction and relaxation throughout life.

SUGGESTED READING

Gans, C., and Bock, W. (1965). The functional significance of muscle architecture: A theoretical analysis. *Ergeb. Anat. Entwicklungsgesch.* **38,** 115–142.

Gordon, M. (1972). "Animal Physiology: Principles and Adaptations," 2nd ed. Macmillan, New York.

Huxley, H. E. (1956). Muscular contraction. *Endeavour,* **15,** 177–188.

Patt, D. I. and Patt, G. R. (1969). "Comparative Vertebrate Histology." Harper and Row, New York.

MUSCULAR SYSTEM

Skeletal muscles maintain the vertebrate's position and cause its movements. Additionally, about 80% of their energy is dissipated as heat that is circulated by the blood to all parts of the body. In homeothermic ("warm-blooded") vertebrates this helps maintain a constant body temperature; in poikilotherms ("cold-blooded" vertebrates) it provides a changing but warmer than ambient temperature which facilitates the chemical reactions of metabolism.

The dermal skeleton, primarily protective in function, involves few muscles. No doubt this contributes to its diversity, since during evolution many dermal elements have changed and even been lost without requiring comparable changes in the muscular system. By contrast, most endochondral bones have strong and frequently complex muscles, their number being roughly proportional to the complexity of the endochondral skeleton. Altogether skeletal muscles comprise about one-third to one-half of the animal's bulk.

ONTOGENY OF SKELETAL MUSCLES

AXIAL MUSCLES

Most skeletal muscles are derived from somatic mesoderm; the axial muscles (those concerned primarily with the axial skeleton) come from the myotome of the epimere, each segment (somite) of which spreads into a dorsal, epaxial portion and a ventral, hypaxial portion. The epaxial portion, innervated by the dorsal rami (branches) of spinal nerves, gives rise to epaxial musculature which in the adult lies mainly along the dorsolateral body wall adjacent to the vertebral column. The hypaxial portion migrates ventrolaterally along the body wall deep to the integument and gives rise to the ventrolateral musculature of neck, trunk, and tail.

MUSCLES OF THE LIMBS (APPENDICULAR MUSCLES)

In Chondrichthyes, ventral extensions from the somites grow down into the developing limb buds and there differentiate, giving rise to the limb muscles.

In all other limbed vertebrates the limb bud mesenchyme dorsal to the developing skeletal tissue differentiates into the limb's extensor muscles, and the mesenchyme ventral to the developing skeletal tissue differentiates into limb flexor muscles. Outgrowths of the hypaxial musculature contribute to some of the girdle muscles.

GILL ARCH MUSCLES (BRANCHIOMERIC)

Unlike other skeletal muscles, those of the gill arches form from anterior visceral mesoderm; the remainder of this visceral mesoderm contributes to smooth muscles, connective tissue, and blood vessels. Because of this visceral derivation, the gill arch muscles are innervated by special visceral, rather than by somatic, nerves. These muscles are the branchiomeric muscles.

PATTERNS OF LATER MUSCULAR DIFFERENTIATION

Before individual muscles and muscle fibers differentiate to the point of being identifiable, two major events occur. First, most skeletal elements differentiate, at least in cartilage. The muscles then develop in relationship to the skeletal elements and are best understood in that context. Second, the muscle masses become innervated. This innervation is retained despite even large migrations: in mammals, for instance, cervical myotomes migrate caudally and give rise to the muscles of the diaphragm which, although separating the abdomen from the thorax, is still innervated by cervical nerves.

The patterns by which large muscle masses differentiate to form individual muscles vary in different parts of the body and in different species, but can be divided into a few general categories. Delamination forms flat muscles running parallel with the surface of the body and occurs, for instance, in tetrapod hypaxial muscles. Other muscle masses split longitudinally and form individual muscles running parallel to an axis, as, for instance, in tetrapod appendages.

Either of these processes can combine with fusion between adjacent myotomes: for example, in the development of tetrapod epaxial musculature individual myotomes fuse with one another and then split longitudinally into long, cablelike muscles adjacent to the vertebral column. Fusion can occur quite early in development, as in this example, or later, when muscles are already partially differentiated. In the latter case the adult muscle frequently has transverse layers of connective tissue (myoseptae) between the parts derived from different myotomes (Fig. 7-1).

DETERMINING MUSCLE HOMOLOGIES AND PHYLOGENIES

It is difficult to be certain of muscle homologies, particularly in vertebrates of widely separate groups. Muscles of the same shape and location are not necessarily homologous, for they may have evolved convergently. Although embryological development can indicate homologous muscles in closely related forms, in those that are more distantly related embryological development can establish only general, or field, homologies rather than specific homologies.

Since it is relatively easy to establish skeletal homologies, a similarity in origin, insertion, or both has often established supposed muscle homologies, but this method may be misleading: in axial musculature,

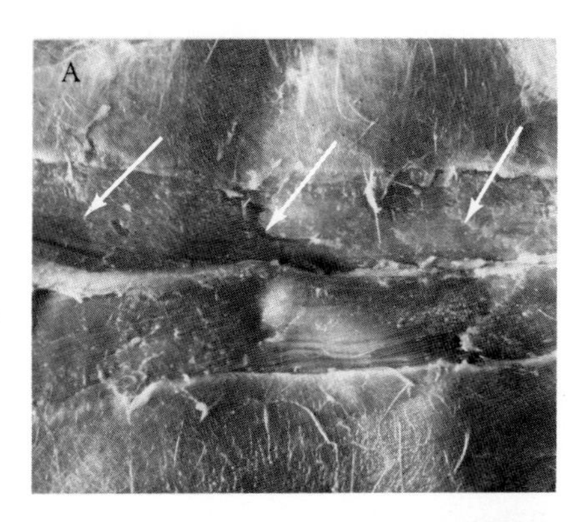

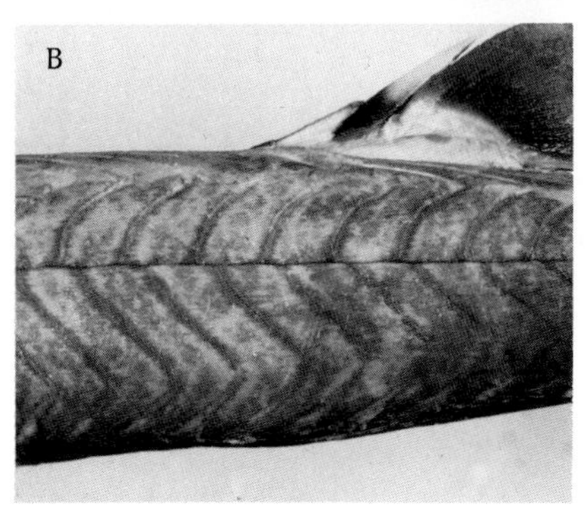

Fig. 7-1. Myoseptae separate the segmental axial muscles of fish, as in the dogfish (B). Although in mammals most axial muscles are not segmented, there are traces of myoseptae such as in the rectus abdominus muscles of the rat (A) (arrows indicate myoseptae).

the origin is more conservative between species; in appendicular musculature, the insertion is more conservative. As for innervation, there is no real reason why different innervations could not have evolved for two homologous muscles in different vertebrates, or why, if a muscle were lost in one vertebrate, its nerve could not then innervate a newly evolved, nonhomologous muscle. Furthermore, peculiar problems arise when one relies on innervation: does a bird's last cervical nerve (which may be number 13 or 14) innervate a muscle that is homologous to the one innervated by a mammal's last cervical nerve (which is number 8)?

In determining muscle homologies, then, all these methods must be considered. If two muscles have similar embryology, similar origins or insertions, and similar innervations, then they are undoubtedly homologous; if not, the most one can be certain of is a field homology for a group of muscles. In any case, the differentiation from the same general muscle mass to specific muscles often is completely different in two forms, particularly in two different classes, and field homologies are much more common than specific homologies.

Phylogenies can be inferred only after homologies are determined, and they are limited by the extent and precision of the homologies.

When homologies are known in diverse living forms, one can make solid inferences about the temporal diversity of a muscle or group of muscles; this will be attempted later in this chapter.

Classification of Skeletal Muscles

It is possible to classify the muscular system into subgroups in several different ways, but to the evolutionary morphologist this is probably most reasonably done on the basis of their embryonic origin. By this criterion, there are two major groups of skeletal muscles, somatic and branchiomeric, as shown in Table 7-1.

Table 7-1. Vertebrate Skeletal Muscles

	muscles	*derivation*	*innervation*
SOMATIC		SOMATIC MESODERM	SOMATIC MOTOR NERVES
AXIAL	1. EXTRINSIC OCULAR	PREOTIC HEAD MYOTOMES	SOMATIC MOTOR CRANIAL NERVES III, IV, VI
	2. HYPOBRANCHIAL	POSTOTIC HEAD MYOTOMES	OCCIPITAL NERVES (ANAMNIOTES), HYPOGLOSSAL NERVE (AMNIOTES)
	3. TRUNK		
	EPAXIAL	DORSAL PORTION OF MYOTOMES	DORSAL RAMI OF SPINAL NERVES
	HYPAXIAL	VENTRAL PORTION OF MYOTOMES	VENTRAL RAMI OF SPINAL NERVES
APPENDICULAR	1. INTRINSIC	LIMB BUD MESENCHYME	VENTRAL RAMI OF SPINAL NERVES
	2. EXTRINSIC	LIMB BUD MESENCHYME AND/OR MYOTOMES	VENTRAL RAMI OF SPINAL NERVES
BRANCHIOMERIC		ANTERIOR SPLANCHNIC MESODERM	SPECIAL VISCERAL MOTOR CRANIAL NERVES V, VII, IX, X, XI

SOMATIC MUSCLES: AXIAL

EXTRINSIC OCULAR MUSCLES

The most anterior somatic muscles are the extrinsic ocular muscles, which move the eyeball (Table 7-2). These are phylogenetically the most conservative muscles of the vertebrate body. They are derived from the three preotic myotomes, and, except for the relatively simple condition of cyclostomes, six of them are present and are the same in all vertebrates. These are the four rectus and two oblique muscles; their origins are in various parts of the bony or cartilaginous orbit, and their insertions are onto the wall of the eyeball (Fig. 7-2). Working in coordination, they can move the eyeball freely and smoothly.

Table 7-2. Extrinsic Ocular Muscles

muscle	*myotome*	*cranial nerve and no.*
SUPERIOR RECTUS INFERIOR RECTUS MEDIAL RECTUS INFERIOR OBLIQUE	FIRST	OCULOMOTOR (III)
(LEVATOR PALPEBRAE SUPERIORIS)	FIRST	OCULOMOTOR (III)
SUPERIOR OBLIQUE	SECOND, DORSAL PART	TROCHLEAR (IV)
EXTERNAL RECTUS	SECOND AND THIRD, VENTRAL PARTS	ABDUCENS (VI)
(RETRACTOR BULBI)	SECOND AND THIRD, VENTRAL PARTS	ABDUCENS (VI)

Two others, the retractor bulbi and the levator palpebrae superioris, are variably present. The retractor bulbi pulls the eyeball directly into the orbit, a protective maneuveur for animals with slightly protruding eyes. It has no antagonist; instead, the elasticity of a large fat pad at the back of the orbit pushes the eyeball toward the surface when the retractor bulbi relaxes. The levator palpebrae superioris inserts on the upper eyelid and raises it. Its antagonist is not a somatic muscle but a branchiomeric muscle, the orbicularis oculi, which attaches also to the eyelid and lowers it.

HYPOBRANCHIAL MUSCLES

The dorsal part of the third myotome plus the entire fourth, fifth, and sixth myotomes never differentiate in vertebrates, but are resorbed

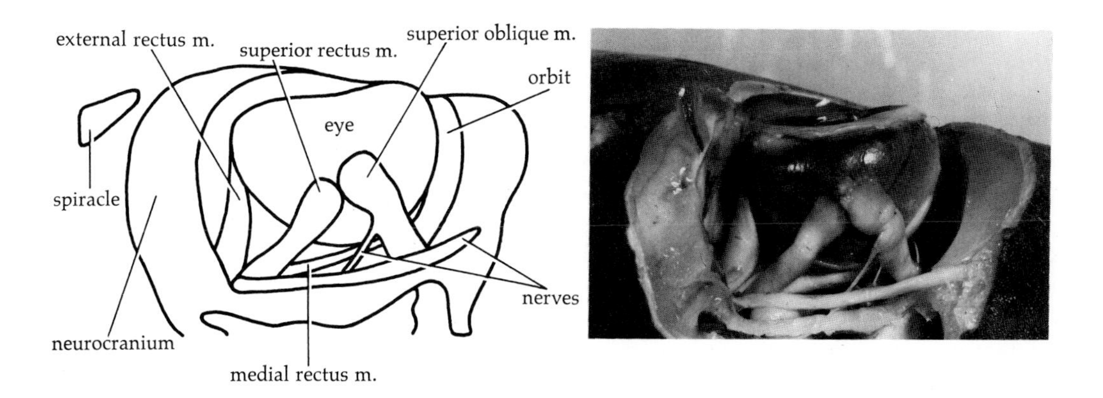

Fig. 7-2. Dorsal view of the extrinsic ocular muscles as seen in the spiny dogfish.

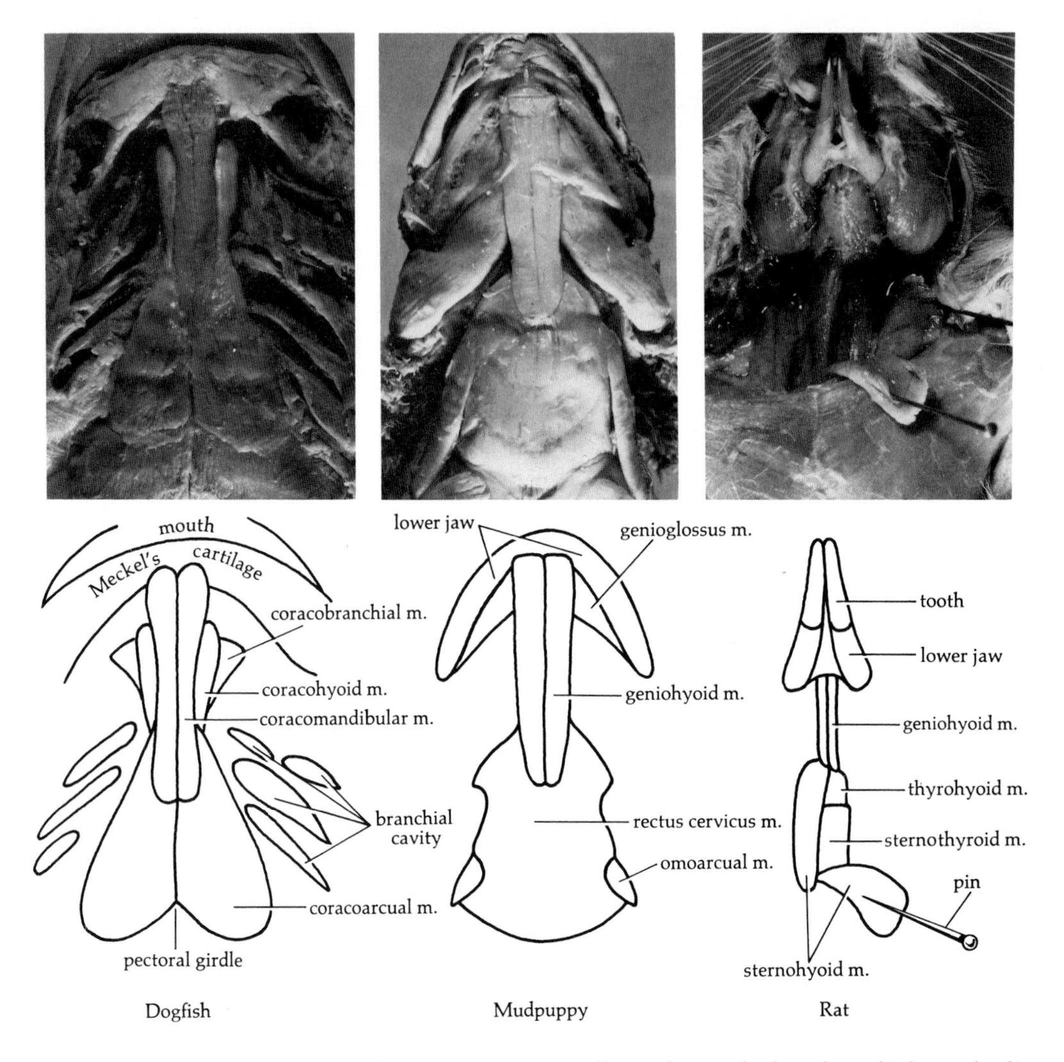

Fig. 7-3. In all vertebrates the hypobranchial muscles lie ventrally between the level of the pectoral girdles and lower jaws.

during development of the otic capsule. Postotic myotomes form just behind this region and migrate ventrally, medially, and finally anteriorly, coming to lie ventrally between the gill arches and the floor of the mouth–pharynx. These myotomes then differentiate into the hypobranchial muscles. In anamniotes they are innervated by the occipital nerves, and in amniotes, by the hypoglossal nerve (cranial nerve XII) which is a coalescence of the occipital nerves.

In fishes the hypobranchial muscles insert on the ventral parts of the gill arches and the jaw apparatus and play a major role in depressing them. In Chondrichthyes they have their origin on the coracoid (ventral) portion of the scapulocoracoid. The most superficial pair extends all the way to Meckel's cartilage and forms the coracomandibular muscles. Deep to this is a pair of coracohyoid muscles inserting on the basihyoid. Deeper still are coracoarcuals and coracobranchials going to the more posterior gill arches (Fig. 7-3). In teleosts, although the hypobranchial muscles are more diverse, they still go primarily to the jaw apparatus and the gills and have generally depressor functions.

In living amphibians the hypobranchial muscles can be divided into a prehyoid and a posthyoid group, as outlined in Table 7-3.

In amniotes, as outlined in Table 7-4, the hypobranchial musculature forms the major intrinsic and some extrinsic musculature of the tongue, and also some of the superficial ventral neck muscles controlling gross movements of the hyoid and thyroid apparatus. In contrast to anamniotes, amniotes have no pectoral girdle attachments of hypobranchial muscles (Fig. 7-3).

Table 7-3. Hypobranchial Musculature of Lissamphibians

muscle	*position*
PREHYOID	
GENIOHYOID	SYMPHYSIS OF LOWER JAW TO HYOID
GENIOGLOSSUS	SYMPHYSIS INTO TONGUE
POSTHYOID	
RECTUS CERVICUS	INSERTING CAUDALLY ONTO HYOID
OMOARCUALS	CORACOID PORTION OF PECTORAL GIRDLE
PECTORISCAPULARIS	SCAPULA

Table 7-4. Hypobranchial Musculature of Amniotes

muscle	*position*
HYOGLOSSUS	HYOID APPARATUS TO TONGUE
GENIOHYOID	SYMPHYSIS TO HYOID
GENIOGLOSSUS	SYMPHYSIS TO TONGUE
STERNOHYOID	ANTERIOR PART OF STERNUM, TO BASIHYAL
STERNOTHYROID	STERNUM TO THYROID CARTILAGE
THYROHYOID	THYROID CARTILAGE TO BASIHYAL

TRUNK MUSCULATURE

Fishes

In fishes the postcranial axial musculature remains segmented in the adult, with connective tissue myoseptae between adjacent myotomes and a horizontal septum between epaxial and hypaxial musculature. The myotome runs dorsally to ventrally in a zigzag line whose angles are greatest in the most active fishes. Further, the myotomes are slanted sharply anteriorly from the body surface to the vertebral column, with the fibers converging toward the centra of the vertebrae. Thus a transverse section through the trunk musculature of an active fish cuts across several myoseptae, which appear as concentric circles as the myotomes taper in toward their insertions: the myotomes are "stacked" like cones with their apices pointing forward (Fig. 7-4). With this arrangement a larger percentage of the muscle fibers exert

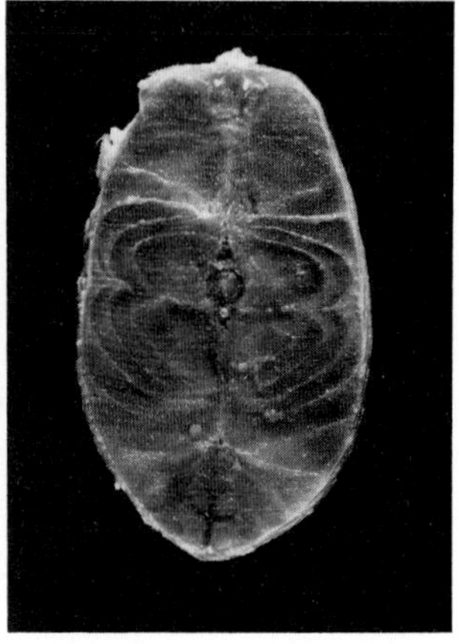

Fig. 7-4. Lateral view of a perch with all but a single caudal myotome dissected away. Both the zigzag surface pattern and the deeper portion of the myotome can be seen. On the right is a transverse section cut through the caudal musculature of a perch, showing several myotomes cut in section.

their force on or close to the vertebrae, efficiently producing the characteristic lateral undulations of fish locomotion.

Tetrapods

Tetrapods that locomote on four limbs sustain quite different stresses in the axial region and have undergone much more regional differentiation of the axial musculature—both anterior to posterior and dorsal to ventral. In general, the epaxial (dorsal) musculature, rather than being segmented, forms long, cablelike muscles. These prevent the vertebral column from sagging between the appendage girdles, allow freedom for the tail, and give strong support to the head. The hypaxial (ventral) musculature differentiates as laminae of broad, flat muscles which protect the body cavities and, fore and aft of the girdles, support the neck and tail. The stresses differ in the epaxial and hypaxial regions, and the two regions have evolved quite separately.

TETRAPOD EPAXIAL MUSCULATURE (Fig. 7-5)

Amphibians. The epaxial musculature of urodeles is quite fishlike in some respects. Even in the adult, myoseptae separate individual myomeres, and in the tail there are typical myotomes producing lateral undulations. In the trunk region, however, the epaxial part of the myomere is divided into a very small medial portion, the interspinalis, and a large lateral portion, the dorsalis trunci.

In anurans the transverse myoseptae are lost and long, cablelike muscles extend the length of the trunk. The medial muscle, here named the longissimus dorsi, extends all the way from the ilium to the posterior part of the skull. The lateral muscle, here called the iliolumbaris, runs from the ilium onto the transverse processes of more anterior vertebrae. The caudal vertebrae are fused to form a single bony rod, the urostyle. A single dorsal muscle, the coccygial iliacus, goes from the ilium caudally and obliquely to the urostyle, extending the urostyle when it contracts.

Reptiles. Since the turtles' vertebral column is fused to the carapace, bending of the trunk is impossible. Not surprisingly, there has been an almost complete loss of axial trunk muscles.

Among other living reptiles (Squamata, Rhynchocephalia, Crocodilia) several strong, cablelike epaxial muscles are formed in the trunk region. Most medial is the transversospinalis, connecting the neural spines and arches of adjacent vertebrae. Lateral to this there is the broader longissimus dorsi muscle, running from the ilium and sacrum to the transverse processes of a variable number of more anterior vertebrae. Lateral to the longissimus dorsi is the iliocostalis, arising from the lateral part of the ilium and adjacent fascia and inserting on the ribs. In the cervical region the epaxial musculature is more complex, with many muscles controlling the movements of the neck and head. The most intricate of these movements are caused by the occipital muscles, derived from the transversospinalis. More superficially, the longissimus dorsi partially inserts onto the spines of cervical vertebrae, and a separate muscle, the cervicus capitus, runs from the neural spines to the posterior part of the skull and controls the gross movements of head and neck.

Mammals. In generalized mammals, the epaxial musculature is similar to that of the more generalized reptiles, although split into more parts anteriorly (Fig. 7-5).

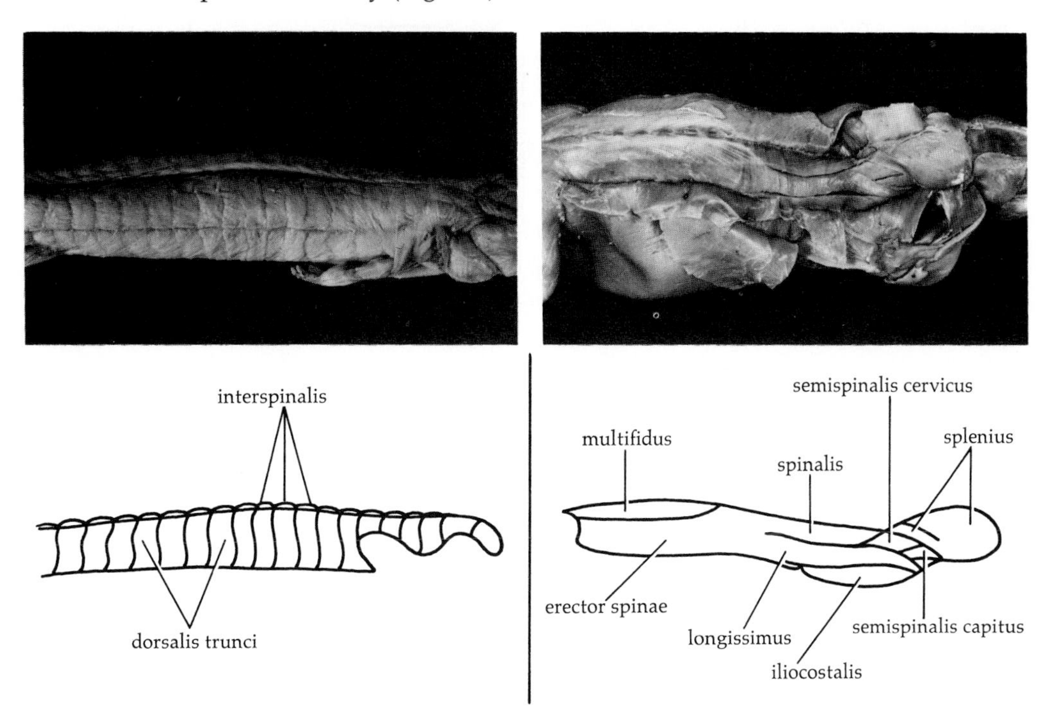

Fig. 7-5. Dorsolateral views of the epaxial musculature of a mudpuppy (left) and a rat (right).

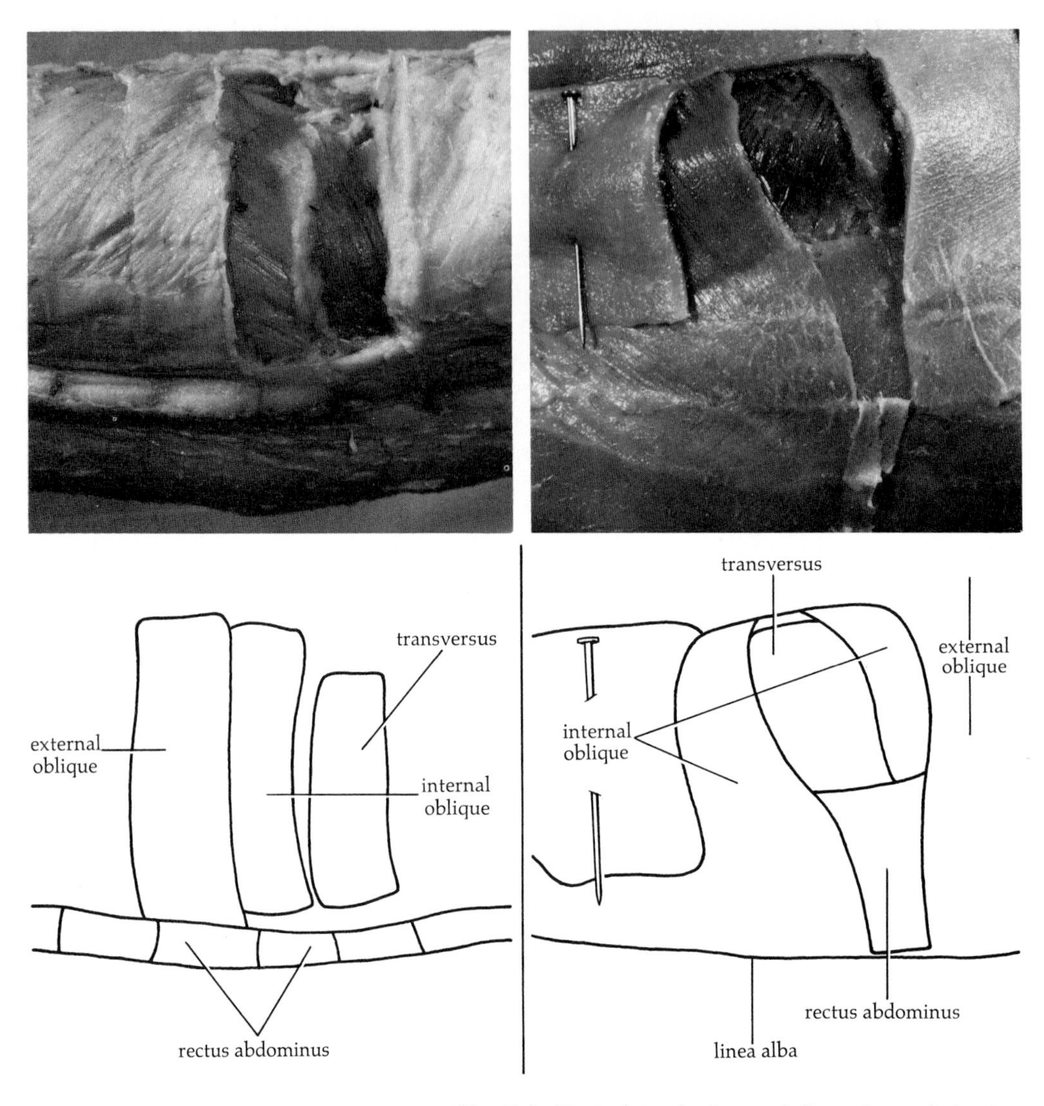

Fig. 7-6. Ventrolateral views of dissections of the hypaxial musculature of a mudpuppy (left) and a rat (right).

Birds. There is little epaxial musculature of the trunk in birds, for the fusion of thoracic, lumbar, sacral, and anterior caudal vertebrae into one rigid avian structure has eliminated much of the selective pressure for it. In the cervical region, however, where complex and elaborate movement is necessary, the epaxial musculature includes many fine, small muscles that control movement between individual vertebrae and between vertebrae and skull; these muscles will not be discussed here.

TETRAPOD HYPAXIAL MUSCULATURE (Fig. 7-6)

Amphibians. In urodeles the myoseptae between myotomes persist, and in the caudal region the myotomes are "typical." In the trunk hypaxial region, however, each myotome differentiates into four muscles. Superficial fibers, forming the external oblique muscle, run at right angles to intermediate fibers, forming the internal oblique muscle; the former runs anterodorsally to posteroventrally, and the latter caudodorsally to ventroanteriorly. Deep to these two is a third, very thin muscle, the transversus, whose fibers run transversely.

Midventrally there is a line of connective tissue called the linea alba. On either side, the rectus abdominus muscle runs from the pubis anteriorly to the cartilages of the pectoral girdle. This segmented muscle is the fourth trunk hypaxial muscle.

Anurans retain more of the myoseptae in the hypaxial than in the epaxial musculature. For instance, distinct myoseptae divide the rectus abdominus into segments. The other two hypaxial muscles, however, are broad and sheetlike with no segmentation: the external oblique, superficially, and the transversus deep to it. There is no internal oblique.

Reptiles. In reptiles (except Chelonia) the hypaxial musculature differs regionally. In the lumbar region there are typical sheetlike, unsegmented muscles—the external oblique, internal oblique, and transversus; ventrally and medially is the segmented rectus abdominus. In the thoracic region, between the prominent ribs that reach ventrally to the sternum, are two sets of muscles whose fibers run at right angles—the external intercostals and, deep to them, the internal intercostals. A third group lies deeper, running between the ribs and the transverse processes of vertebrae—the levatores costarum muscles. In the cervical region fine muscles run from the cervical vertebrae to the skull and depress the head. Other muscles lower the neck; chief among these is the longissimus coli which is attached along the ventral border of the centra from the thoracic through the cervical region.

Birds. The avian axial musculature is that of a modified reptile. Because the trunk is fused and the sternum and appendicular musculature expanded, the hypaxial musculature is much reduced in the trunk region. It is, of course, extremely elaborate in the cervical region.

Mammals. In mammals the abdominal hypaxial musculature is almost reptilian, consisting of the external and internal obliques, the transversus, and the rectus abdominus. In the thoracic region the external and internal intercostals are developed further than they are in reptiles, with the external intercostals pulling the ribs forward and the internal intercostals pulling them backward. Variable numbers of supercostals and subcostals aid in respiration, as does a migrated cervical hypaxial muscle, the diaphragm; these will be discussed in a later chapter. Finally, a pair of longitudinal thoracic hypaxial muscles, the scalene muscles, run from the ventrolateral portion of the ribs up to the cervical vertebrae; depending upon which skeletal part is fixed, these can either bend the neck or raise the rib cage.

SOMATIC MUSCLES: APPENDICULAR

Fishes

In Chondrichthyes and apparently also in teleosts the muscles of the paired fins arise as direct buds from the hypaxial myotomes. Connec-

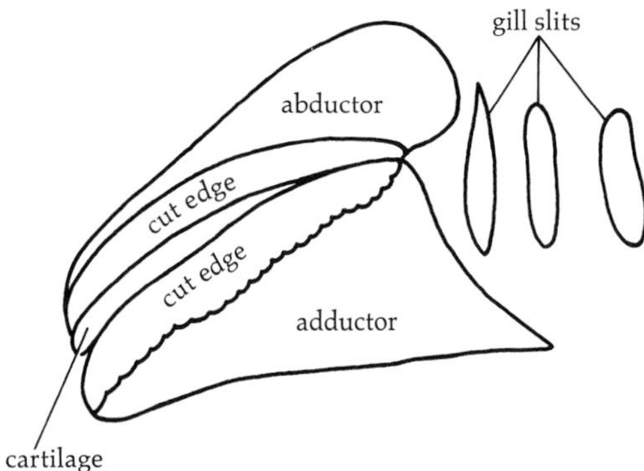

Fig. 7-7. Lateral view of the pectoral musculature of a dogfish. The fin has been cut near its base.

tive tissue holds the girdles in position on the trunk. Small muscles running from the girdle to the proximal part of the fin form its extensors and flexors: dorsally the extensors and ventrally the flexors (Fig. 7-7).

In fishes there are few muscles which are completely contained within a fin, and movement is usually of the fin as a whole rather than of its parts. Since the primary function of paired fins is stabilization rather than locomotion or manipulation, fine movements of parts are unnecessary. The male chondrichthyean pelvic fin provides significant exceptions, for, modified as the intromittent clasper organ, it has complex muscles and elaborate movements.

Skates and rays, however, are flattened dorsoventrally and hence are incapable of lateral undulations. Their primary locomotor organs are huge pectoral fins which have very large muscles continuous with the myotomes of the hypaxial region. Nervous coordination permits rhythmic, anterior to posterior contractions which provide the impulses for locomotion.

Tetrapods

In tetrapods there are extrinsic and intrinsic appendicular muscles. The extrinsic muscles relate the girdles to both the axial structure and the limbs. A functional category, it includes muscles derived from axial musculature, branchiomeric musculature, and limb musculature, which itself is derived from limb bud mesenchyme. The intrinsic muscles, however, are totally derived from limb bud mesenchyme; they are defined as muscles whose origins and insertions are both within the limb itself. The extrinsic limb musculature, involved with proximal parts of the skeleton, is conservative compared to the more distal, intrinsic musculature which, like its associated skeleton, has undergone gross modifications in different tetrapod groups. Generally, intrinsic appendicular muscles are simpler in living amphibians than in reptiles and most complex in birds and mammals, and, of course, they are correlated with the degree of variation in the distal part of the skeleton and the degree of fine movement of parts.

A detailed examination of all the individual appendicular muscles in each class is beyond the scope of this book. Instead, we will concentrate on the tetrapod shoulder musculature as seen in a therian mammal, a lizard, and a salamander, and follow this limited study of spatial diversity with an analysis of temporal diversity.

SHOULDER: TETRAPODS (Fig. 7-11, see pages following 177)

In those tetrapods whose clavicle and sternum are not articulated, the pectoral appendage has no skeletal connection to the rest of the animal's skeleton but is held to it by an elaborate muscle sling. This is advantageous for two aspects of the animal's behavior. First, the girdle can be moved relative to the trunk, and, therefore, the limb is more mobile. Second, the muscles that form the interface between the pectoral and axial skeletons act as shock absorbers when the animal lands on its front legs, as during locomotion or a predatory leap.

In mammals, three major hypaxial muscles contribute to this pectoral girdle sling—the levator scapulae ventralis, the rhomboideus, and the serratus ventralis (Table 7-5, Fig. 7-8). The levator scapulae ventralis is a bandlike muscle running from the posterior part of the skull to the spine of the scapula. The rhomboideus is a heavy muscle coursing from the base of the neural spines to the dorsal border of the scapula. A slip of this muscle, the rhomboideus capitus, arises from the back of the skull. The serratus ventralis is the largest of the three; many separate fascicles originate on the ribs and then converge, inserting on the dorsomedial surface of the scapula. Lizards lack a rhomboideus muscle but have a large levator scapulae and a serratus ventralis that is somewhat smaller than that of mammals. Salamanders have the same two muscles, but the latter is called the thoraciscapularis muscle.

Branchiomeric muscles also contribute to the pectoral girdle sling (Table 7-5). In mammals these are the trapezius complex and the sternocleidomastoid complex (Fig. 7-9). The trapezius muscles originate from the back of the skull and from the neural spines as far caudally as the thoracic region. The anterior portion of the complex inserts on the clavicle and the more posterior portions on the spine and fascial surface of the scapula. The cleidomastoid runs from the clavicle to the mastoid process of the temporal bone; the sternomastoid, as its name

Table 7-5. Major Tetrapod Shoulder Muscles

embryonic origin	*mammal*	*lizard*	*salamander*	*labyrinthodont*
		TRUNK TO GIRDLE = "SLING"		
HYPAXIAL	LEVATOR SCAPULAE VENTRALIS	LEVATOR SCAPULAE	LEVATOR SCAPULAE	LEVATOR SCAPULAE
	SERRATUS VENTRALIS	SERRATUS VENTRALIS	THORACISCAPULARIS	THORACISCAPULARIS
	RHOMBOIDEUS	—	—	—
BRANCHIOMERIC	TRAPEZIUS COMPLEX STERNOCLEIDOMASTOID COMPLEX	TRAPEZIUS AND CLEIDOMASTOID	CUCCULARIS	CUCCULARIS
HYPOBRANCHIAL	—	—	PECTORISCAPULARIS	—
	—	—	OMOARCUAL	—
		TRUNK TO HUMERUS		
LIMB BUD	LATISSIMUS DORSI TERES MAJOR CUTANEOUS MAXIMUS (PART)	LATISSIMUS DORSI	LATISSIMUS DORSI	LATISSIMUS DORSI
	PECTORALIS MAJOR PECTORALIS MINOR PECTOANTIBRACHIALIS XIPHIHUMERALIS CUTANEOUS MAXIMUS (PART)	PECTORALIS	PECTORALIS	PECTORALIS

Table 7-5. (Continued)

embryonic origin	mammal	lizard	salamander	labyrinthodont
		GIRDLE TO HUMERUS		
	DELTOID { CLAVODELTOID	CLAVODELTOID	PROCORACOHUMERALIS LONGUS }	DELTOID
	DELTOID { SPINODELTOID, ACROMIODELTOID }	SCAPULODELTOID	SCAPULODELTOID }	DELTOID
	TERES MINOR	SCAPULOHUMERALIS ANTERIOR	PROCORACOHUMERALIS BREVIS	SCAPULOCORACOHUMERALIS
	SUBSCAPULARIS	{ SUBCORACOSCAPULARIS, SCAPULOHUMERALIS POSTERIOR }	SUBCORACOSCAPULARIS	SUBCORACOSCAPULARIS
	SUPRASPINATUS, INFRASPINATUS }	SUPRACORACOIDEUS	SUPRACORACOIDEUS	SUPRACORACOIDEUS

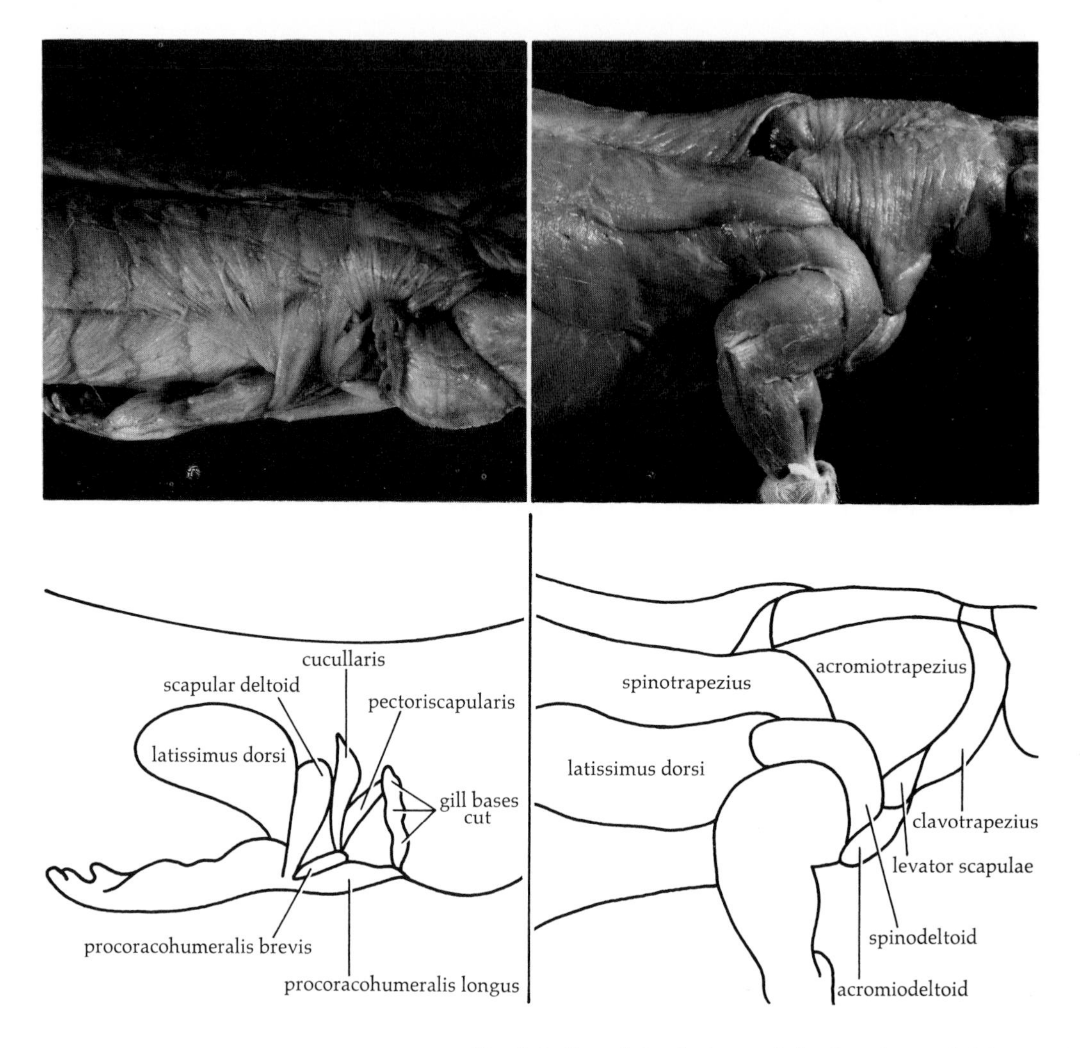

Fig. 7-8. Dorsolateral views of the dorsal superficial musculature of the pectoral girdle of a mudpuppy (left) and a rat (right).

implies, does not attach to the pectoral girdle. In lizards the cleidomastoid and trapezius are a continuous muscle originating on the back of the skull and neural spines and inserting onto the clavicle and scapula. In urodeles these muscles are represented as a small muscle, the cuccularis, inserting onto the front of the scapula (Fig. 7-9).

As mentioned (Table 7-3), two hypobranchial muscles are part of the shoulder musculature of amphibians (Table 7-5).

With this exception, the remaining shoulder muscles are all derived from limb bud mesenchyme and go from either the trunk fascia or the pectoral girdle to the humerus (Table 7-5). Those from the trunk to the humerus are, in all tetrapods, two large superficial muscles—the latissimus dorsi and the pectoralis complex.

The latissimus dorsi, which has its origin from the dorsolateral portion of the trunk behind the pectoral girdle, is a major muscle pulling the humerus caudally. It is small in urodeles, larger in reptiles, and still larger in mammals (Fig. 7-10). In mammals two other muscles are derived from the same embryonic mass as the latissimus dorsi: these are the teres major which runs from the ventral border of the scapula to the humerus and a portion of the cutaneous maximus which inserts onto the skin of the trunk and acts to twitch it.

The pectoralis complex, which has its origin from either the sternum or midventral fascia (the linea alba), courses medially to insert onto the humerus (and sometimes the forearm); thus it is a major adductor of the forelimb (Fig. 7-10). Like the latissimus dorsi, the pectoralis complex is smallest in urodeles, larger in reptiles, and largest in mammals. In most mammals this complex is divided into separate muscles; in the cat, these are the pectoralis major, pectoralis minor, pectoantibrachialis, and xiphihumeralis. The complex also gives rise to the portion of the cutaneous maximus not contributed by the latissimus dorsi.

Those muscles derived from limb bud mesenchyme and going from the pectoral girdle to the humerus include the deltoids, teres minor, subscapularis, and supra- and infraspinatus (Table 7-5).

The deltoid musculature undergoes considerable variation both within and between classes (Fig. 7-11, see pages following 177). In mammals there may be a single deltoid originating from the clavicle, scapular spine, and acromian process and inserting on the humerus (as in the human), or there may be three separate muscles, the clavodeltoid, spinodeltoid, and acromiodeltoid (as in the cat). Lizards usually have a clavodeltoid (clavicle to humerus) and scapulodeltoid (scapula to humerus). Salamanders also have a scapulodeltoid, but since they lack a clavicle, the other deltoid muscle runs from the

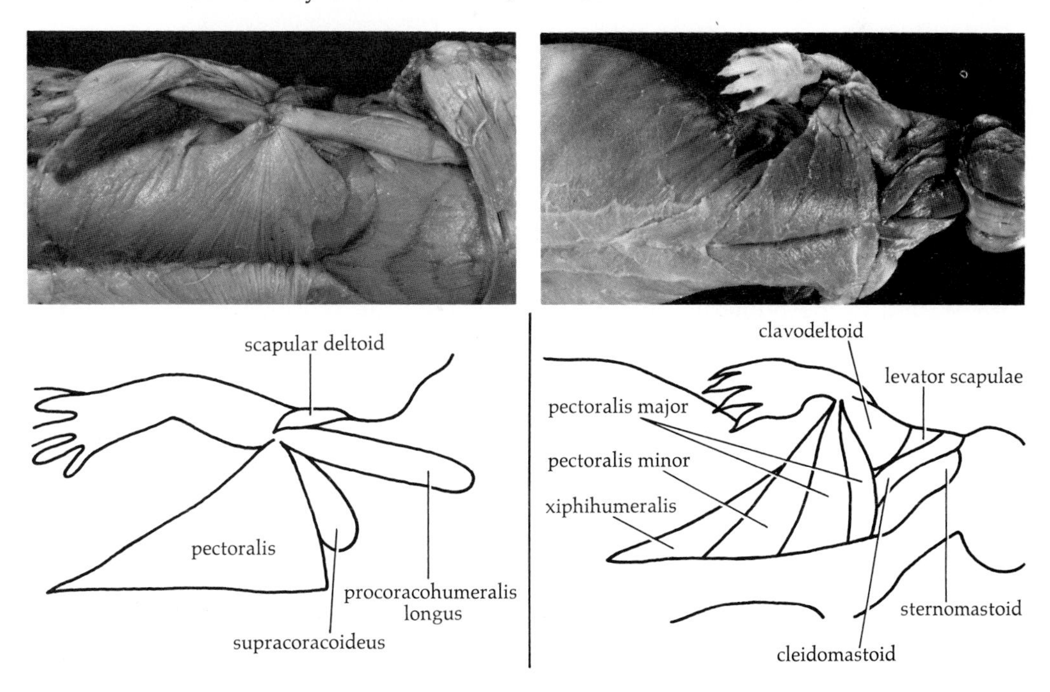

Fig. 7-9. Ventral views of the ventral superficial musculature of the pectoral girdle of a mudpuppy (left) and a rat (right).

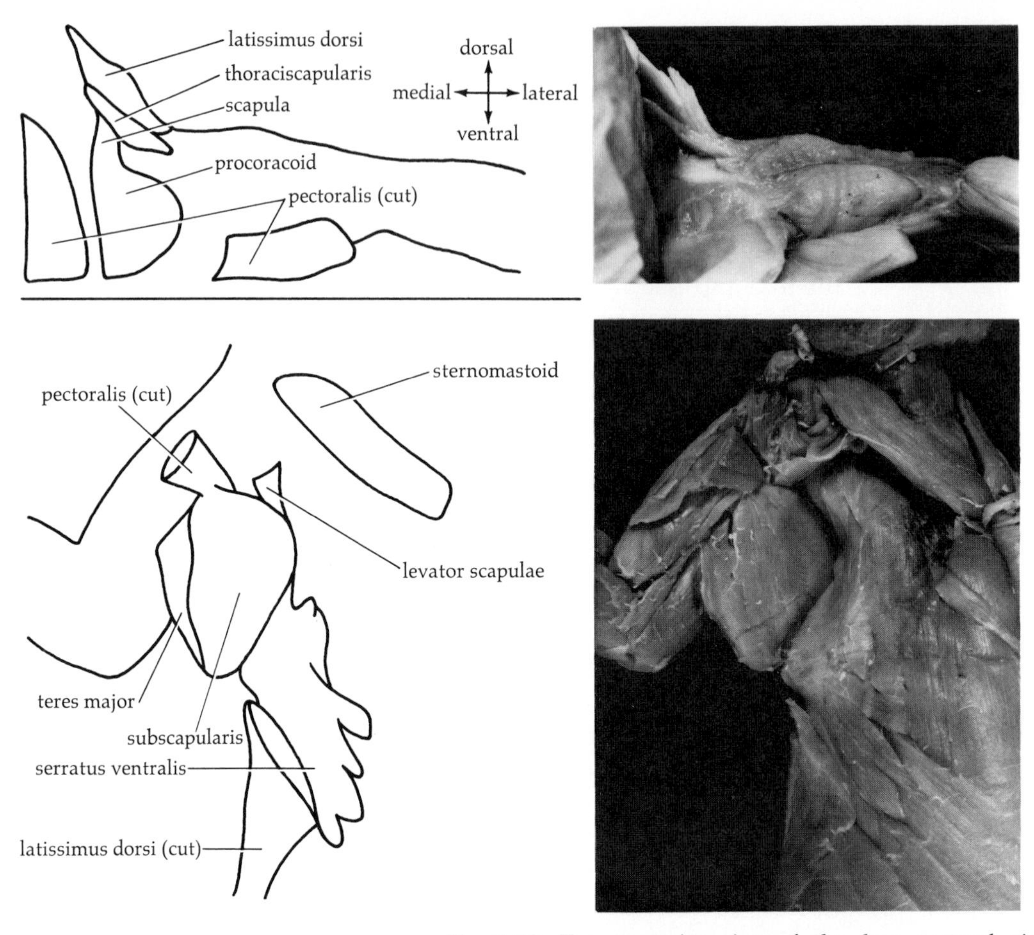

Fig. 7-10. Top: posterior view of the deep pectoral girdle musculature of a mudpuppy. The pectoralis has been cut and the limb and girdle pulled forward. Bottom: ventral view of the deep pectoral girdle musculature of a rat. The pectoralis and latissimus dorsi have been cut and the limb and girdle pulled dorsally.

procoracoid to the humerus; this muscle is named the procoracohumeralis longus.

In mammals there is a small muscle deep to the teres major called the teres minor, which runs from the scapula to the humerus. The same muscle is relatively larger in lizards and is called the scapulohumeralis anterior. In urodeles this muscle runs from the procoracoid to the humerus and is called the procoracohumeralis brevis.

In mammals the large subscapularis is a strong adductor. It originates from most of the medial surface of the scapula and inserts on the medial surface of the proximal portion of the humerus (Fig. 7-11, see pages following 177). In lizards two muscles form from the same embryonic anlage—the subcoracoscapularis and the scapulohumeralis posterior. As with the mammalian subscapularis, each runs from scapula to humerus. Urodeles have a small subcoracoscapularis but no scapulohumeralis portion.

Finally, in mammals there are two adductor muscles which lie on the lateral surface of the scapula, just above and below the scapular spine, and are named the supraspinatus and infraspinatus, respectively (Fig. 7-11, see pages following 177). Both insert on the proximal end of the humerus. Embryologically these muscles form in the ventral portion of the limb bud and then migrate dorsally. In the lizard and salamander a single muscle, the supracoracoideus, is homologous to these two; not surprisingly it is located on the ventral surface of the shoulder girdle, the procoracoid, and runs to the humerus.

SHOULDER: LABYRINTHODONT (Fig. 7-11, see pages following 177)

The common ancestors of today's living mammals, lizards, and salamanders were the stem amphibians, Labyrinthodontia. From the fossil record we know that the labyrinthodont pectoral girdle was composed of an endochondral scapula and procoracoid and a dermal cleithrum, clavicle, and interclavicle. Faint "muscle scars" on these fossil bones indicate some of the points of muscle attachment. Putting this together with our knowledge of its living descendents, inferences can be made about its shoulder musculature as shown in Table 7-5 and Figs. 7-8, 7-9, 7-10, and 7-11. Although most of these inferences are self-evident, some require comment.

With the loss of gills in adult amphibians the muscles which had acted as levators of the gill apparatus lost their adaptive value. At the same time, as amphibians began to walk on land, there was a strong adaptive value for anything that would strengthen the shoulder girdle. Thus it is not surprising that the branchiomeric gill levators became attached to the pectoral girdle. This apparently happened quite early: muscle scars on the labyrinthodont clavicle and cleithra are probably from the cuccularis.

The fact that among living vertebrates only amphibians have hypobranchial muscles involved in shoulder movement (the pectoriscapularis and omoarcuals) suggests that these evolved after the lissamphibians had split off from the evolutionary group giving rise to the amniotes.

The diversity of deltoid musculature in living tetrapods strongly suggests that they are derivatives of a single, broad deltoid muscle, not unlike that in humans, and that they evolved differently in independent tetrapod lines.

PELVIC MUSCULATURE

In all tetrapods the pelvic girdle is firmly attached to the axial skeleton by way of the sacrum; therefore it cannot move independently of the axial skeleton, and so needs no "sling" muscles. The extrinsic pelvic muscles, therefore, run only from the girdle and associated vertebrae onto the femur. There are also intrinsic muscles. These complex pelvic muscles are best studied in the laboratory.

BRANCHIOMERIC MUSCLES (Fig. 7-12)

Pharyngeal pouches form embryologically in all vertebrates, and the splanchnic mesoderm between the pouches differentiates into a mass of skeletal muscles which form the branchiomeric muscles. Homologies are easy to determine, since in all classes each branchial pouch (and hence each muscle) is innervated by a specific, special visceral nerve from the brain. On the basis of their ontogeny and innervation these muscles can be considered in groups, as indicated in Table 7-6, with each group composed basically of levators and dorsal and ventral constrictors.

Table 7-6. Branchiomeric Muscles

muscles	*pouch derivation*	*innervation*
MANDIBULAR	MANDIBULAR, OR FIRST PHARYNGEAL	TRIGEMINAL NERVE (FIFTH CRANIAL, OR 'V')
HYOID	SECOND	FACIAL NERVE (VII)
GLOSSOPHARYNGEAL	THIRD	GLOSSOPHARYNGEAL NERVE (IX)
VAGAL	FOURTH-SIXTH	VAGUS NERVE (X)

Fishes

In the dogfish, the jaws are formed by the mandibular arch and are moved by its muscles. The levator is the levator palatoquadrati, running from the neurocranium to the upper jaw; as its name implies it raises the upper jaw. The dorsal constrictor muscles are a large adductor mandibulae mass, which is broken up into a small preorbitalis and quadratomandibularis; these close the mouth. The ventral constrictors are the paired intermandibularis muscles, running from the linea alba to Meckel's cartilage; they function in unison to raise the floor of the mouth.

The hyoid arch has a very small levator, the spiracularis; when contracted it does not raise the hyoid arch but constricts the spiracle, which is this gill arch's small gill slit. The dorsal constrictor is the epihyoidean; it is the main muscle raising the hyoid, and it also helps push the jaws forward. The ventral constrictor is the interhyoideus, lying just deep to the intermandibularis and inserting on the ventral hyoid structures.

The third through the sixth arches have superficial dorsal and ventral constrictors, which are numbered as the arches (third, fourth, etc.). Their levators are fused into a single muscle, the cuccularis, raising the gill arches.

In the quite independent actinopterygian evolution the branchiomeric muscles, like the head and gill arch skeleton, have become much more complex and are intimately involved in the mechanism of kinetic jaws. In the adult many originate and insert not on the splanchnocranial elements with which they are associated in early embryology, but on dermatocranial elements; for example, the dermal bones of the teleost operculum are moved by branchiomeric muscles.

Amphibians

Since adult amphibians do not respire by gills, their branchiomeric muscles differ from those of fishes, but they can still be recognized by their innervation and embryology. As in fishes, the dorsal constrictor of the mandibular arch closes the jaws; in amphibians the muscles derived from the dorsal constrictor are the levator mandibuli muscles. The levators of the mandibular arch are fused with and into these muscles. Ventrally the intermandibularis muscles are quite similar to those in Chondrichthyes in both position and function.

In the second gill arch the dorsal constrictor and levator are also fused and together are called the depressor mandibuli. This muscle runs from the back part of the skull to the very back part of the lower jaw, behind its articulation. When it contracts, therefore, it pulls the back part of the jaw upwards, depressing the rest of the lower jaw and opening the mouth. The interhyoideus (ventral constrictor) runs from the midline to the hyoid apparatus. Its posterior part, the sphincter coli, becomes attached to the integument rather than to the hyoid arch and acts to tighten the skin in the ventral throat region (and, therefore, is an integumentary muscle).

The constrictor muscles of arches three through six are primarily concerned with the larynx and voice production. The levators are fused as the cuccularis, which functions as an extrinsic shoulder muscle.

Amniotes

In amniotes the mandibular arch is greatly reduced; only one branchiomeric muscle retains its splanchnocranial connections, and that only in mammals. This is the tensor tympani, a very small muscle from the dorsal part of the mandibular arch musculature. When the tensor tympani contracts it pulls on a modified splanchnocranial element, the malleus of the middle ear, and thus tightens the tympanic membrane.

With this one exception the amniote mandibular muscles are involved with the dermatocranium. The fused levator and dorsal constrictor, in all amniotes, are the major muscles of mastication, closing the jaws and often allowing for lateral chewing: these are the masseter, the temporalis, and the pterygoideus internus and externus.

The mandibular arch ventral constrictor is the intermandibularis in birds and reptiles; in mammals it is divided into two elements. Anteriorly there is the mylohyoid, which runs to the lower jaw much as the intermandibularis does in other vertebrates. The posterior part has become attached to part of the ventral constrictor of the second gill arch, thus forming a two-part muscle called the digastricus. Because of its embryonic origin, the digastricus has a double innervation, the anterior part by the trigeminal nerve and the posterior part by the facial

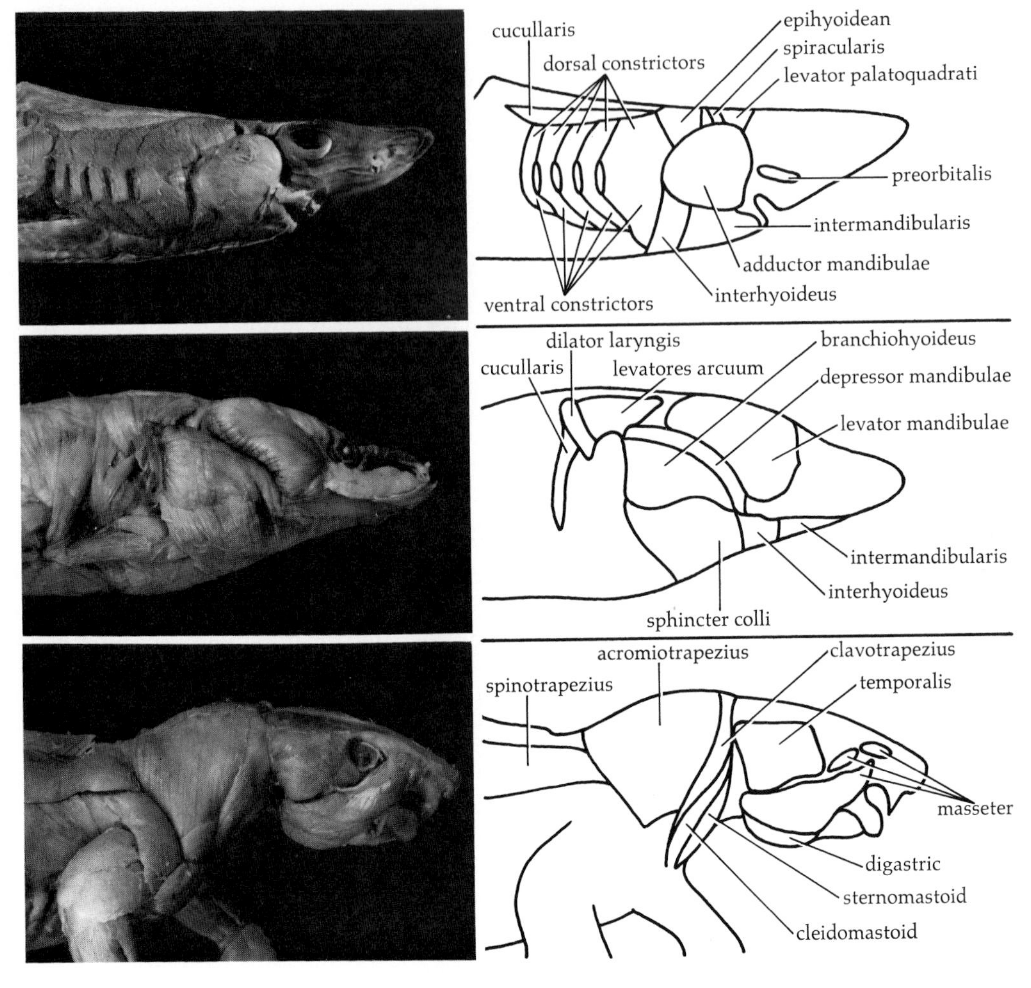

Fig. 7-12. Lateral views of the superficial branchiomeric musculature of a dogfish (top), mudpuppy (middle), and rat (bottom).

nerve. The digastricus muscle is the primary muscle opening the mammalian jaw.

In reptiles and birds (as well as amphibians) the levator and dorsal constrictor of the second gill arch give rise to the depressor mandibuli which opens the mouth. In mammals, whose jaws are opened by the digastricus, the levator and dorsal constrictor of the second gill arch are present as the tiny stapedius muscle. This muscle is attached to a hyoid arch derivative, the stapes of the middle ear, and acts to tighten the chain of tiny bones in the middle ear, thus protecting the ear from overstimulation.

In adult reptiles and adult birds, as in amphibians, the ventral constrictor is present as the interhyoideus and sphincter coli. In mammals, it is present as the posterior portion of the digastricus and the muscles of facial expression, so finely developed in humans. In the neck region this musculature is called the platysma, and its more specialized regions around the eyes, nose, and mouth are named according to the part of the facial skin they move, e.g., the orbicularis oculi closes the eye.

The musculature of arches three through six becomes involved primarily with the intrinsic muscles of the pharynx and larynx and with some large muscles of the shoulder (trapezius muscle), as well as muscles running from the sternum and clavicle to the mastoid at the rear of the skull (sternomastoid and cleidomastoid muscles, respectively). Thus the loss of functional gills in tetrapods frees the muscles which in early embryology were related to gill pouches, and they then become involved in facial expression, jaw movements, swallowing, vocalizing, hearing (in the middle ear), and movement of the shoulder. Despite these diverse developments, their homologies can be accurately traced by their embryology and innervation (Fig. 7-12).

SOME MUSCLE SPECIALIZATIONS

Birds (Fig. 7-13)

One of the most extreme adaptations of the muscular system is seen in the flight muscles of the class Aves. Flight requires powerful adduction (when the wings are brought down in the power stroke) and then rapid abduction (when the wings are raised in preparation for another power stroke). Although small epaxial muscles give fine control and stability to the wing dorsally, the mass of both its adductors and abductors are located ventrally, along the enlarged sternum (unlike other vertebrates, whose abductors are located dorsal to the girdle). The largest of these flight muscles is also the largest muscle in the bird's body (often providing as much as one-fifth of the total body weight). This is the pectoralis muscle, whose origin is from the very large sternum and its keel, from the ribs, and from the furcula. The pectoralis muscle fibers all converge to insert onto the proximal part of the humerus, so that its contraction adducts the wing powerfully and rapidly.

Deep to the pectoralis lies the supracoracoideus, sometimes called the deep pectoral muscle. It originates on the sternum beneath the pectoralis muscle. Its tendon passes up parallel to the procoracoid to its dorsal aspect and then passes through the foramen triosseum, which is formed by the articulation of the scapula, procoracoid, and furcula of the pectoral girdle. After passing through this foramen, the tendon inserts on the dorsal part of the proximal portion of the humerus. When the supracoracoideus contracts, the foramen triosseum acts as a pulley and the wing is abducted.

In this way both the abductor and adductor add weight only ventral to where the wing articulates onto the pectoral girdle. Thus the bird's center of gravity is below the axis of the wings and stability is added to its flight.

Bats (Fig. 7-13)

The only flying mammals are bats, order Chiroptera, which have quite different adaptations. The sternum has only a small keel. The pec-

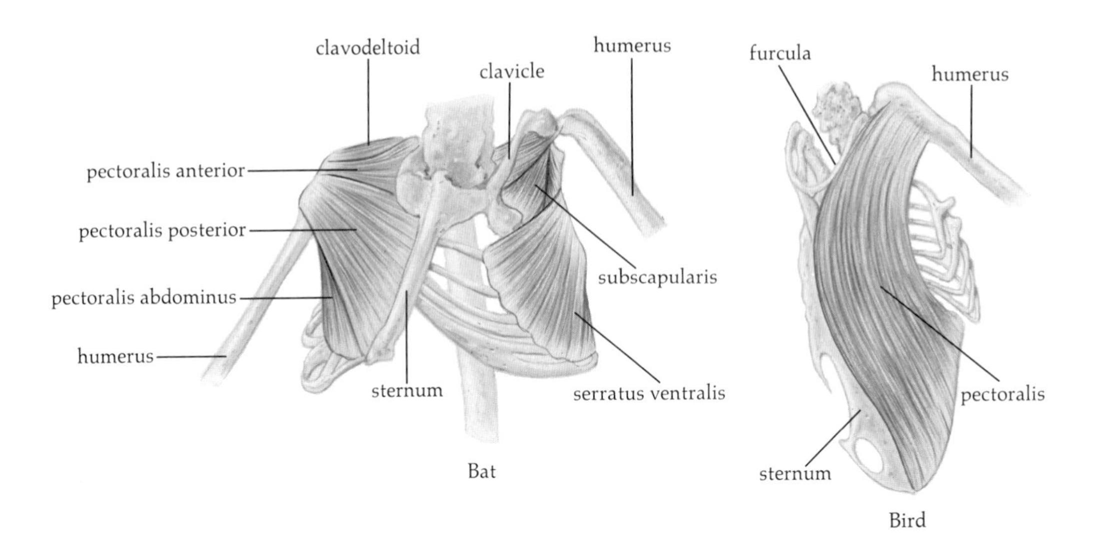

Fig. 7-13. The power stroke of flight (adduction of the humerus) is caused primarily by a single large muscle in birds; in bats it requires the coordinated contractions of six muscles.

toralis musculature is divided into three parts, the pectoralis anterior, posterior, and abdominus, all of which insert on the proximal end of the humerus and provide much of the power stroke for adduction of the wing. The clavodeltoid, posterior division of the serratus ventralis, and subscapularis provide additional power and also give fine angle adjustments. Abduction involves several muscles, including the supra- and infraspinatus, the acromio- and spinodeltoids, and the teres major; although all of these are located dorsally, none is large enough to make the bat "top-heavy." There are elaborate intrinsic muscles that control the patagium (wing membrane) and the digits which support it. One such muscle originates from the back of the skull and inserts distally on the wing, supporting its anterior portion.

Moles

Among fossorial (burrowing) animals there are several different adaptations; in the case of moles it is once again an adaptation of the forelimb, particularly in the shoulder region. The teres major is the largest and most powerful muscle in the complex, and the one that does the work. However, the humerus and the pectoral girdle are rotated approximately 90 degrees, so that when the teres major contracts it does not rotate the limb but pulls it directly posteriorly. The other muscles, such as the pectoralis and triceps, fix the joints.

Whales and Porpoises

Quite a different locomotor adaptation is found in the wholly aquatic whales and porpoises, order Cetacea. Their hindlimbs have been secondarily lost; the forelimbs act as stabilizers for steering much as do the paired appendages of fishes. Also as in fishes, the caudal musculature produces the power stroke for the cetacean's swimming, but instead of lateral undulations they are dorsoventral undulations. The tail is moved dorsally by very large dorsal iliocostalis muscles running back and inserting onto the caudal vertebrae; it is depressed by a very large ventral muscle, the ischiocaudalis, which runs from the ischia of the pelvic girdle back to the caudal vertebrae.

It is thus apparent that, particularly in vertebrates with bizarre locomotor mechanisms, certain muscles become very enlarged and others very reduced, but only rarely are there muscles which do not have homologues in the more generalized vertebrates. In this respect, the muscular system is similar to the skeletal system: mutations which are retained by natural selection usually involve modification of existing structures rather than development of totally new structures.

SUGGESTED READING

Cheng, C. C. (1955). The development of the shoulder region of the opossum, *Didelphis virginiana,* with special reference to the musculature. *J. Morph.* **97,** 415–471.

George, J. C., and Berger, A. J. (1966). "Avian Myology." Academic Press, New York.

Gilbert, P. W. (1957). The origin and development of the human extrinsic ocular muscles. *Contrib. Embryol.* **36,** 59–78.

Konigsberg, I. R. (1964). The embryological origin of muscle. *Sci. Amer.* **211** [August], 61–66.

Romer, A. S. (1944). The development of the tetrapod limb musculature—the shoulder region of Lacerta. *J. Morph.* **74,** 1–41.

THE INTEGUMENT

The integument is more than the wrapping around the vertebrate's body. It is a dynamic interface between the animal and its environment and as such has several functions. Most obviously it separates the animal's internal and external environments. These environments have different ionic concentrations, and the integument, acting as a selective barrier, helps maintain this difference. It also helps regulate the animal's internal temperature, by dissipating and absorbing heat. It protects the animal against injury from normal contacts with the external environment. It also acts as a mechanical barrier to various potentially harmful microorganisms.

While separating and protecting the animal from its environment, the integument also provides contact and communication with it. The receptors for heat, cold, pressure, and many other sensations are composed of modified integumentary elements; even more highly modified integumentary elements comprise parts of the distance receptors for audition, olfaction, and vision.

A vertebrate's color and pattern are determined in the integument; colors can blend with the surroundings ("protective coloration") or be boldly patterned to provide a warning (e.g., the skunks), mimic the warnings of other animals (e.g., the harmless coral king snake, which looks much like the poisonous coral snake), or resemble threatening structures such as large horns (as in many African ungulates). In other cases (e.g., birds), bright colors and bold patterns are related to sexual habits and apparently help attract mates and perhaps establish territories.

The integument also gives rise to various other modifications, such as the several forms of claws, hooves, horns, antlers, and even the rattlesnake's warning rattle. The structural and functional variety of these modifications is not surprising in view of the intimate contact between the integument and the external environment, which in turn puts intense selective pressure on the integument.

DEVELOPMENT OF THE INTEGUMENT

The vertebrate integument is a compound organ derived from both ectoderm and mesoderm. Its outer layer, the epidermis, is formed by the ectoderm that remains on the embryo's surface after development of the neural tube. Its deep layer, the dermis (still occasionally called the corium), is formed by mesoderm lying just deep to the surface ectoderm. In the epaxial region the dermis is formed by the dermatome of the epimere, and in the hypaxial region, by the somatic mesoderm of the somatopleure. Throughout the dermis blood vessels and nerves

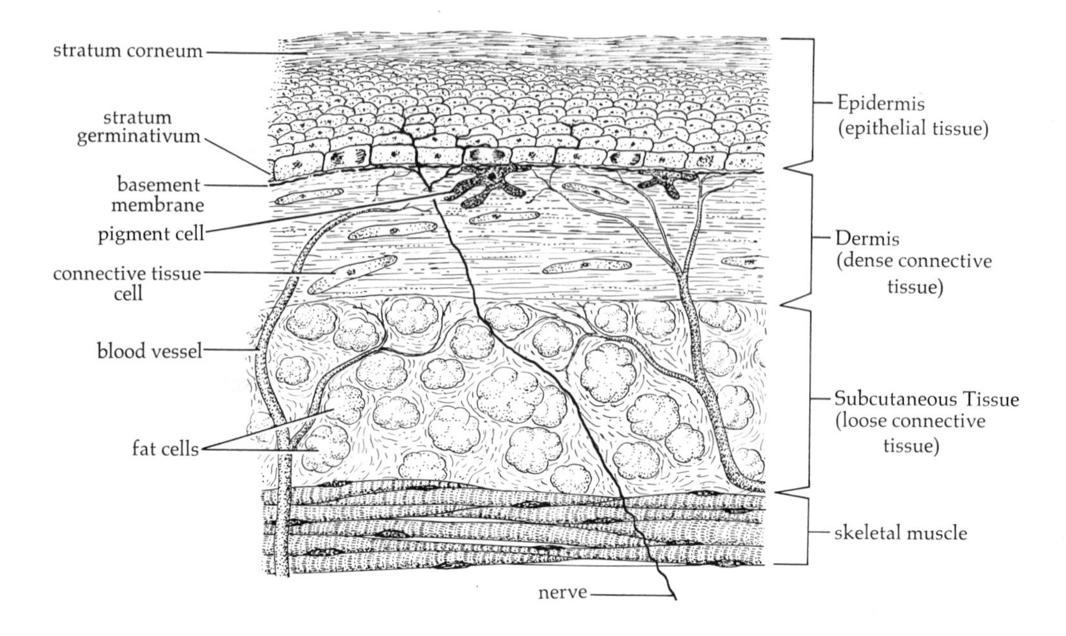

Fig. 8-1. An idealized, generalized vertebrate integument seen in cross section.

grow into it from other parts of the body, and pigment cells migrate in from their neural crest origin.

MICROSCOPIC STRUCTURE

EPIDERMIS

Layers of epithelial cells compose the epidermis (Fig. 8-1). All epidermal cells arise in the deepest layer, the stratum germinativum, whose cuboidal or columnar cells rest on a fine, fibrous basement membrane and undergo mitotic divisions throughout life. After each mitotic division, one of the two daughter cells stays in contact with the basement membrane, but the other is pushed outwards and differentiates according to species and body region.

Those cells just superficial to the stratum germinativum usually are connected to adjacent cells by desmosomes, which probably give the epidermis physical coherence. Cells located more peripherally may develop granules of the structural protein keratin within their cytoplasm. Some cells differentiate into glands or other integumentary derivatives.

The skin's most superficial layer, the stratum corneum, contains dead cells whose most significant component is keratin. Keratin is formed most massively in amniotes and exists in two molecular forms: α-keratin, with filaments 70–80 Ångstroms (Å) wide, and β-keratin, with filaments about 30 Å wide. Although a β-keratin layer is invariably hard, e.g., the surface of avian and crocodilian scales, α-keratin layers may be either soft, as in mammalian flexible skin, or hard, as in a mammalian claw. These dead cells of the stratum corneum are continually being sloughed off and replaced by deeper cells.

The stratum germinativum thus is the mitotic layer for a process which continues throughout life, producing cells which are pushed toward the surface and gradually differentiate, die, and come to rest in the stratum corneum, and finally are lost. Cells or cell debris are lost in tiny groups in most vertebrates; dandruff, for instance, is the sloughing off of stratum corneum. However, in some vertebrates—notably the lepidosaurian reptiles—an entire layer of skin is shed as a unit, after a "new skin" has been developed just below it.

DERMIS

The dermis (corium) is often divided into two portions. The dermis proper is a superficial layer of tough, usually collagenous connective tissue cells and fibers. It also contains blood vessels, nerves, pigment cells, smooth muscles, occasional skeletal muscles, and dermal bone.

Deep to the dermis proper is the subcutaneous tissue (hypodermis or subcorium); it is made up of very loose connective tissue and may contain a great deal of fat. In some vertebrates the fat is layered and can make the subcutaneous tissue extremely thick (e.g., the blubber of whales). Blood vessels and nerves of the integument also pass through the subcutaneous tissue (Fig. 8-1).

HOW THICK IS A LAYER?

The thickness of different skin layers varies considerably between species, between individuals, and between body regions. In areas of mechanical stress the epidermis is usually much thicker, with a partic-

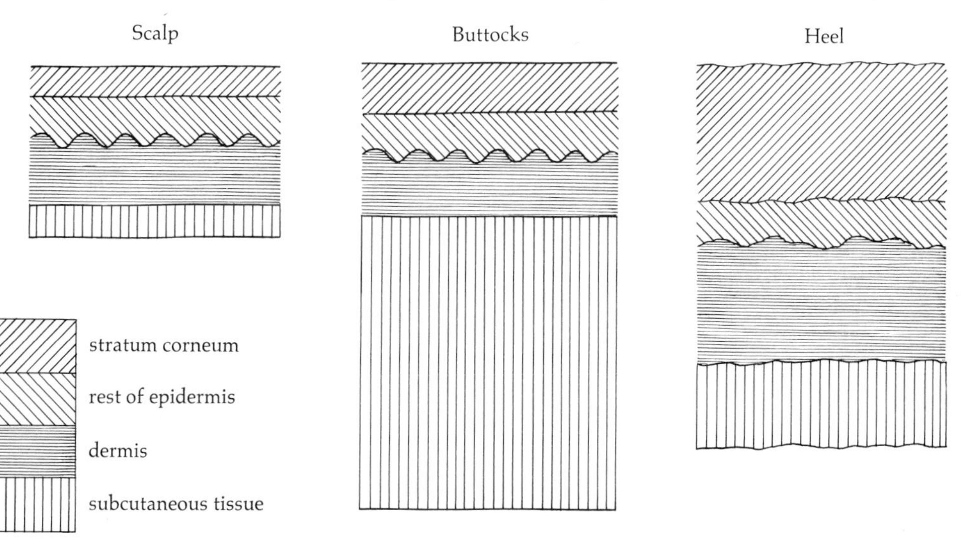

Fig. 8-2. The thickness of the skin layers varies in different body regions, as diagramed here for the human.

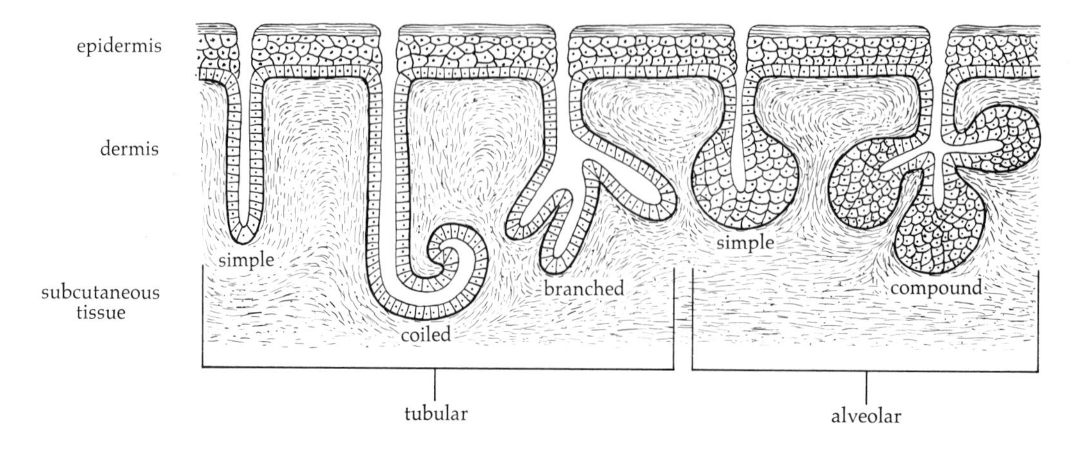

Fig. 8-3. Classification of multicellular glands by their structure.

ularly heavy stratum corneum; callouses, for instance, are layers of stratum corneum that have become thickened in response to pressure or friction. In areas where the skin endures little mechanical stress the epidermis is much thinner. The regional differences in thickness of the stratum corneum are known to be the result of both genetic and environmental influences.

The dermis, too, is relatively thick in some areas and thin in others. The greatest variation, however, is found in the fatty subcutaneous layer. In the human (Fig. 8-2) it may vary from less than 1 mm on the scalp to over 3 cm on the buttocks, and even more in a woman's breast.

Derivatives of the Integument

Both epidermis and dermis give rise to specialized structures of the skin. In addition to dermal bone (as discussed in Chapter 3), these structures include glands, scales, horns and antlers, claws and nails, feathers, and hair (all of which will be dealt with here).

INTEGUMENTARY GLANDS

Almost without exception, integumentary glands are epithelial structures derived from epidermis. Their embryonic development is determined by the underlying mesoderm that induces a particular gland in a particular area of the body, even when the ectoderm above it has been transplanted from another area.

Integumentary glands can be classified by structure as unicellular or multicellular. Multicellular glands, usually an invagination from the stratum germinativum extending down into the dermis, are further classified as tubular or alveolar, and each of these may again be subdivided (Fig. 8-3).

Tubular glands are deeper than alveolar, some extending even into the hypodermis. They are lined by a single layer of epithelial cells; usually only those in the deeper part of the gland do the secreting. If the tube is straight, it is called a simple tubular gland; if the tube coils at the end, it is called a coiled tubular gland; and if after entering the dermis or hypodermis the tube branches into several smaller tubes, it is called a branched tubular gland. Sometimes each branch coils in the hypodermis, forming a branched, coiled tubular gland.

Alveolar glands seldom go as deeply into the dermis and only rarely enter the hypodermis. They have more bulk, however, being several cells thick with a small lumen. If arranged as one large clump of cells, the gland is called a simple alveolar gland; if in several clumps, it is called a compound alveolar gland.

Glands can also be classified by their mode of secretion. If the secretion leaves the cell without visibly damaging it the gland is called a merocrine gland. If the cell must burst to release the secretion, it is called a holocrine gland. In an intermediate type, a small part of the cell is pinched off with the secretion but the rest of the cell survives; this is an apocrine gland.

FISHES

Most integumentary glands in fishes are holocrine or merocrine unicellular glands lying individually throughout the epidermis and secreting mucus, a complex glycoprotein containing an albuminous protein. This mucus forms a protective layer over the body, and, not surprisingly, the glands are more numerous when scales are fewer. In the scaleless hagfish, there are elongated mucous cells ("thread cells") that are so active that one hagfish in a scrub pail full of water, startled by a hammer blow on the pail, is said to secrete enough mucus to jell the water.

The mucus provides more than physical protection, however. It reduces friction between the water and the body, enabling the fish to move with less energy. It carries away microorganisms, since it is continually secreted and lost. It can coagulate mud, which then precipitates; it has been demonstrated that two drops of mucus from an eel will clear a half-pint of muddy water in less than 30 seconds. For fishes which live in continually muddy waters, this is important not only because it improves vision but also because it helps prevent irritation of the gills during respiration.

The concentration of body fluids is greater than that of the surrounding water in freshwater fishes, and less in saltwater fishes. The mucous layer slows the passage of water and ions between external and internal environments and thus helps to maintain osmotic balance.

Mucus has some other, more specialized uses. When the dry season approaches and its home streams diminish, the African lungfish, *Protopterus*, burrows into the mud and secretes a great deal of mucus around itself. This precipitates the mud, forming a quite impervious cocoon in which the lungfish survives the dry period. Sticklebacks construct nests on stream bottoms, holding them together with large amounts of mucus. Paradise fish make more peculiar nests: the male blows mucus-enclosed air bubbles, which float to the surface; the female deposits her eggs on the under surface of this floating nest, and there they develop.

In addition to these unicellular mucous glands, a few fishes have multicellular glands. In several taxonomic groups there are fishes with poison glands: for instance, the stingray among Chondrichthyes, and many catfish among Osteichthyes. These glands are usually associated with the spines of the fins, and they serve primarily a defensive purpose.

The luminescent organs, found mostly in deep-sea fishes, are unusual glands of several types all producing "cold light" (i.e., light with a minimal heat production). Fishes which possess these organs have large prominent eyes, while other deep-sea fishes have small, often nonfunctional eyes. In some, the luminescent organ is not actually a gland but a structure containing light-producing, symbiotic microorganisms. In others, the light is produced by a true gland. The glandular portion, formed by the stratum germinativum, sinks into the dermis as a sphere; the superficial hemisphere forms a transparent lens, and the deeper hemisphere forms the light-producing cells; deep to this the dermis forms reflector fibers and, deeper still, pigment cells. Light produced by the gland is thus reflected out to the environment (Fig. 8-4). The precise biochemistry of light production in deep-sea fishes is not fully understood, but it is evidently not the same sort of enzymatic reaction that occurs in fireflies.

AMPHIBIANS

Most amphibian glands are simple, alveolar mucous glands in the superficial dermis. As in fishes, the mucus serves various functions. It is most pronounced in the tree frogs whose feet bear glands that secrete a sticky mucus of obvious adaptive value.

In addition to mucous glands, several anurans have serous glands whose secretions, although proteinaceous, are more watery than mucus. Most of these serous glands secrete rather mild toxins; those of the common toad can produce irritations (but *not* warts!) in sensitive persons. Others, however—particularly those of some tropical frogs—are extremely powerful. The Kokoe frog of South America, for instance, produces batrachotoxin, which is the most powerful neural poison known.

REPTILES

The reptilian integument contains numerous glands, most of which are relatively inconspicuous and poorly understood. Holocrine alveolar glands are most common. Their structural details and location on the body are often species specific. In some species there are small glands adjacent to the germinal epithelium. In some species, larger glands are located near the cloaca; the still meager functional studies suggest that

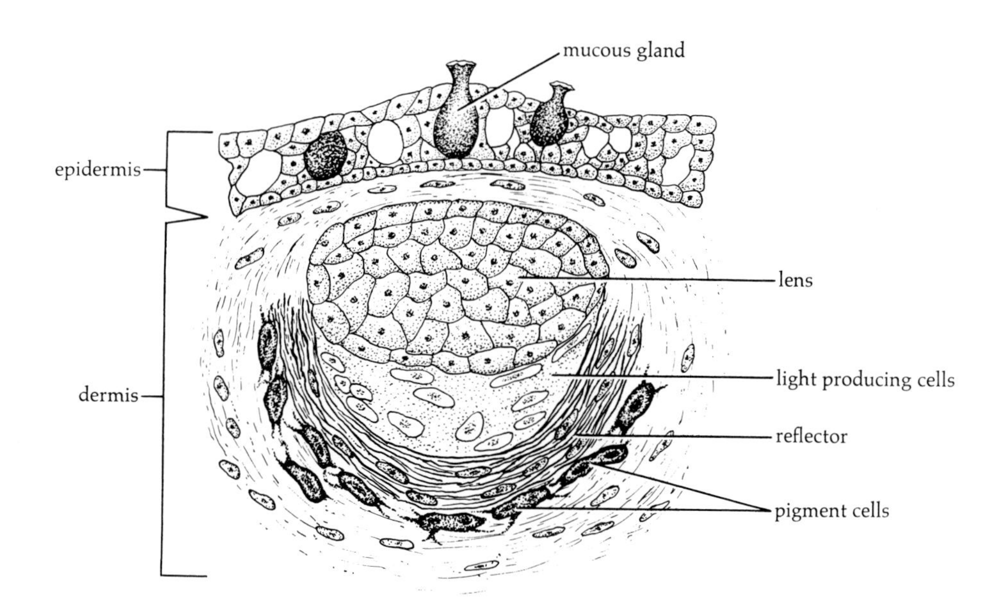

Fig. 8-4. The luminescent organ of the California toadfish, a deep-sea form that comes to the surface only to breed.

their odoriferous secretions may function in intraspecific communications, such as by marking territory or attracting mates.

BIRDS

The only prominent integumentary gland in birds is at the base of the tail, just anterior to the pygostyle, and is called the uropygeal gland. This gland is bilobed, with two ducts to the dorsal surface, one on each side of the pygostyle (Fig. 8-5). It is a holocrine sebaceous gland, producing an oily secretion called promatum which was long considered necessary for grooming the feathers; hence the gland has also been called the preening gland. Its function is now in doubt, however. While it is true that during preening many birds pick up some of the promatum, the preening solution also comes partially from salivary glands (not of integumentary origin). Further, some birds, such as bustards, most pigeons, parrots, and most ratite birds, lack a uropygeal gland, while in others it is only vestigial. However, it has been shown to be at least transiently present during the development of all birds in which this has been studied. Some researchers have suggested that it functions as an odoriferous gland with sexual adaptation. Others have believed it to be a source of vitamin D; some birds develop rickets if it is removed, but not if supplementary vitamin D is given at the same time.

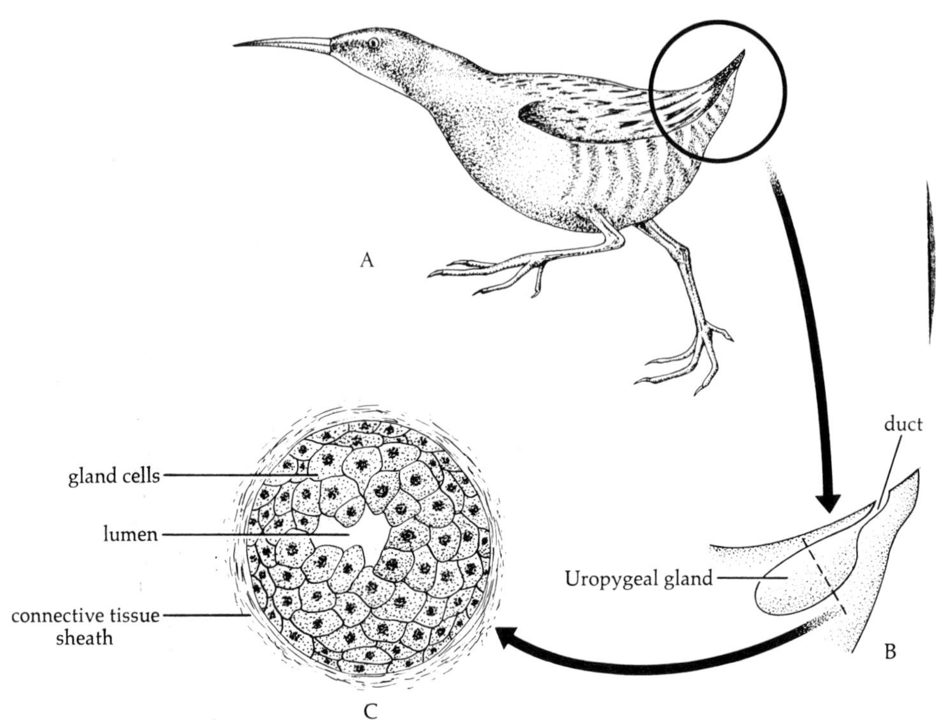

Fig. 8-5. The prominent integumental gland in birds is the preening or uropygeal gland. Its secretion is high in lipids and glycogen.

MAMMALS

Mammals have integumentary glands of great number and diversity, all derived from stratum germinativum (Fig. 8-6).

Sudoriferous (or sweat) glands

Found in no other class of vertebrates, these glands secrete a thin, watery substance containing some salts and urea. The exact composition of this secretion varies with the metabolic state of the animal, and it plays a small role in maintaining proper osmotic balance. The evaporation of the sweat gland secretion causes cooling and thus helps maintain a constant body temperature.

There are two types of sweat glands, eccrine and apocrine; the former has merocrine secretions, and the latter, of course, apocrine secretions. Both are coiled tubular glands (apocrine are also branched); their deeper portions are vascularized and contain myoepithelial cells whose contractions force out the secretions.

Eccrine glands, which secrete sweat, extend well into the dermis and frequently into the hypodermis. Their distribution is quite variable. Whales have none; most mammals have them at least on the soles of the feet and around the lips, genitalia, mammary glands, and anus; some species, such as humans and horses, have them profusely all over the body.

Apocrine glands are larger, extending deep into the hypodermis. Their secretion is thick and milky, with the distinct odor of a sexual attractant. They are found only in the axillary, genital, and anal regions, and around the nipples.

Sebaceous glands

These compound, alveolar holocrine glands secrete an oily substance called sebum, which is a lubricant and grooming agent for hair. Their ducts open into hair follicles.

Meibomian glands

These are found on the edges of the eyelids in most mammals; they produce an oily film which covers the surface of the cornea as a protective and lubricating agent.

Genital glands

Found in conjunction with the genitalia are the preputial glands in males and the vulvar glands in females. Their oily secretions act as lubricants.

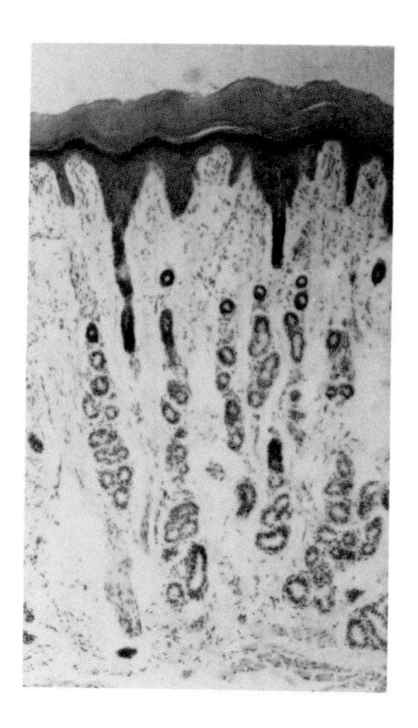

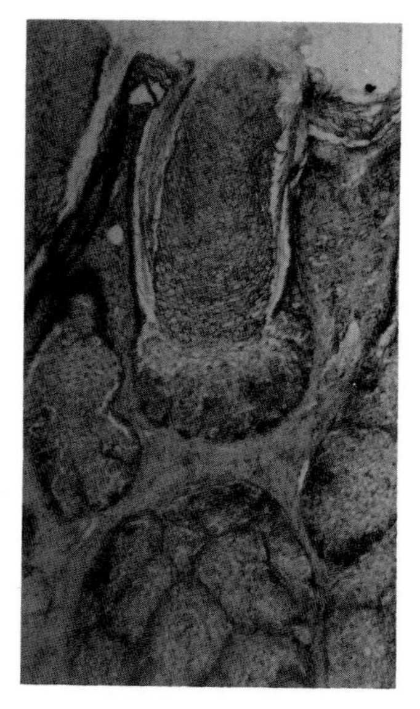

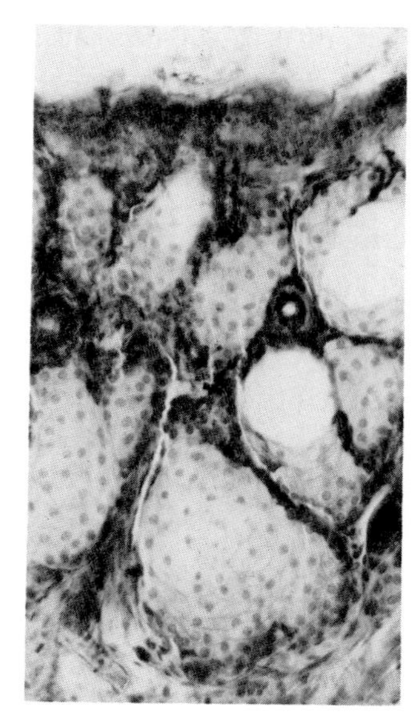

Fig. 8-6. Three different, epidermally derived integumentary glands from one mammal, the kangaroo rat. On the left are coiled, branched, tubular sweat glands found on the palmar surface of the front paws. In the center is a large holocrine gland of unknown function located in the interscapular region. On the right are ceruminous (wax) glands from the external auditory meatus.

Anal scent glands

Many mammals, such as beavers, cats, dogs, and foxes, have large, hormonally controlled, compound, alveolar scent glands in the anal region, in addition to and not to be confused with the apocrine glands serving the same function. In various mammals, the odoriferous secretions of these alveolar glands are used to mark territories and attract mates. In skunks, however, this same gland is further modified to produce the well-known, highly irritating protective agent.

Ceruminous (or wax) glands

These are nearly tubular in structure. They are found only in the mammalian external ear canal, where their secretions act with the hairs of the canal to protect the delicate tympanic membrane from microorganisms and bits of dust or dirt.

Mammary glands

These branched tubular glands are probably highly modified, apocrine sweat glands. Their hormonally controlled secretion, restricted to the female, contains water, fat, sugar, albumen, and calcium and other salts. Its precise composition varies according to species.

In most mammals, the tubules of the mammary gland empty into a collecting duct which comes to the surface at the nipple, a modification of surface epidermis. The prototherian monotremes, however, have no nipple; the several branching tubules, rather than opening into a single collecting duct, open independently and secrete onto the female's ventral body hairs, from which the young suck the milk. Ungulates, on the other hand, have a large storage compartment, the udder, with the glands opening into its center. This is called an inverted mammary gland, as opposed to the more common everted type where the nipple is turned out and the ducts open together at its tip.

Mammary glands are located at various points along two mammary lines, on the ventral side of the trunk from the pectoral to the inguinal region. In some species, such as pigs, they form all along these lines, but arboreal mammals generally have a single pair of pectoral mammae, while the udder of ungulates is in the inguinal region.

SCALES AND DENTICLES (Fig. 8-7, see pages following 177)

Various means of hardening the integument have developed during the evolution of vertebrates (as well as of successful invertebrates such as arthropods). Their adaptive value lies in both the mechanical strength and the protection from external agents which they afford. In vertebrates these hardenings may be bone or tough collagen formed in the dermis, or hard structural proteins (e.g., keratins and various enamel-like proteins) produced by epidermal cells, or a combination of these dermal and epidermal structures. If protection is provided at the expense of flexibility, however, the animal becomes less mobile, as witness the turtles. In the great majority of vertebrates, flexibility is retained because the hardenings are formed discontinuously in folds of the skin and overlap one another. This advantageous "shingle effect" provides hard layers with potential movement between them. Such dermal and/or epidermal hardenings formed in skin folds are called scales. They occur in several forms and are found in all living vertebrate classes except Chondrichthyes and Agnatha. In Chondrichthyes, isolated denticles are formed in a smooth skin and are called appendages rather than scales. Living Agnatha lack skin hardenings.

PLACOID DENTICLES

Elasmobranchs have rows of placoid denticles embedded in the skin over the entire body. Each denticle has a flat basal plate embedded in the dermis and a sharp, curved, protruding spine (Fig. 8-8). The basal plate and most of the spine are derived from the mesoderm of the dermis, where fibroblasts accumulate and dentine is secreted by odontoblasts. In the center of each denticle is a hollow pulp cavity filled with blood vessels and nerves, from which fine canaliculi reach into all parts of the dentine. The spine itself is covered with a thin layer of a hard, enamel-like substance called vitrodentine which is secreted by epidermal cells called ameloblasts. In both embryology and adult structure the placoid denticles are similar to teeth, and, in fact, the teeth and denticles are continuous at the edges of the mouth. In sharks the teeth are simply exaggerated placoid denticles, and both the teeth and the denticles can be replaced if lost.

SCALES OF OSTRACODERMS AND PLACODERMS

These ancient fishes characteristically had heavy, dermal scale bones of three layers: an inner layer of compact lamellar bone, a middle layer of spongy vascular bone, and an outer layer of dentine from which usually protruded tubercles or spines (Fig. 8-7). Although common in Devonian and Silurian ostracoderms and placoderms, these scales occur in no living vertebrates.

SARCOPTERYGIAN COSMOID SCALES

The heavy, bony scales of the ancient crossopterygians and lungfish were called cosmoid scales. They were structurally similar to those of ostracoderms and placoderms, except that instead of dentine over the two bony layers there was a dentine-like cosmine substance containing radiating tubules and pulp cavities; overlying this substance, there was a very thin, extremely hard layer of vitrodentine (Fig. 8-7).

Living sarcopterygians, however, have much thinner scales resembling the cycloid scales of most living osteichthyeans (to be described below).

GANOID SCALES

In these heavy, bony scales the cosmine layer is overlaid not with vitrodentine but with extremely hard, multilayered ganoine (Fig. 8-7). Its many layers and thickenings give ganoine a lustrous, often metallic sheen. True ganoid scales occurred in the palaeoniscoid fishes; among living vertebrates there are slightly modified ganoid scales (Fig. 8-8) in bichirs (genus *Polypterus*) and garpikes (genus *Lepisosteus*).

CYCLOID AND CTENOID SCALES

The evolutionary trend in most groups has been away from heavy armor toward other protective devices and mobility. The cycloid or ctenoid scales of most living fishes have neither cosmine nor ganoine. They are composed of two thin layers: a somewhat bonelike, but acellular osteoid layer and an inner, very dense fibrous layer. They are rather small and flexible, arranged in an overlapping shingle fashion and entirely embedded in the dermis. The very thin, overlying epidermis often contains unicellular mucous glands. These scales occur in varied shapes—particularly in their lamellae ("growth rings")—which are so species specific that they have sometimes been used for taxonomy. The two scale types differ only in that the ctenoid scales have small stiff spines (ctenii) on their free posterior surfaces (Fig. 8-8), while the cycloid scales do not.

AMPHIBIAN SCALES

The stem amphibians, Labyrinthodontia, possessed very heavy, bony, and usually cosmoid scales; during amphibian evolution these have become reduced or lost. Some modern apodans have tiny bony scales of dermal derivation; anurans and urodeles have none.

REPTILIAN SCALES

The scales of reptiles are formed from the epidermis; their hard portion is the structural protein keratin. (There are no enamel-like proteins in any amniote scales.) Reptilian scales can undergo various modifications. In adult turtles and some other reptiles, the scales are underlaid with dermal bones. In the so-called horned toad, actually a lizard, the "horns" are prominent scales on various parts of the body,

Fig. 8-8. Dogfish placoid denticles occur as "appendages" on a smooth skin; the garpike's ganoid scales, the buffalo fish's cycloid scales, and the perch's ctenoid scales, on the other hand, all form as hard, dermal portions of a folded skin and overlap one another. The black dots over the perch scales are melanophores.

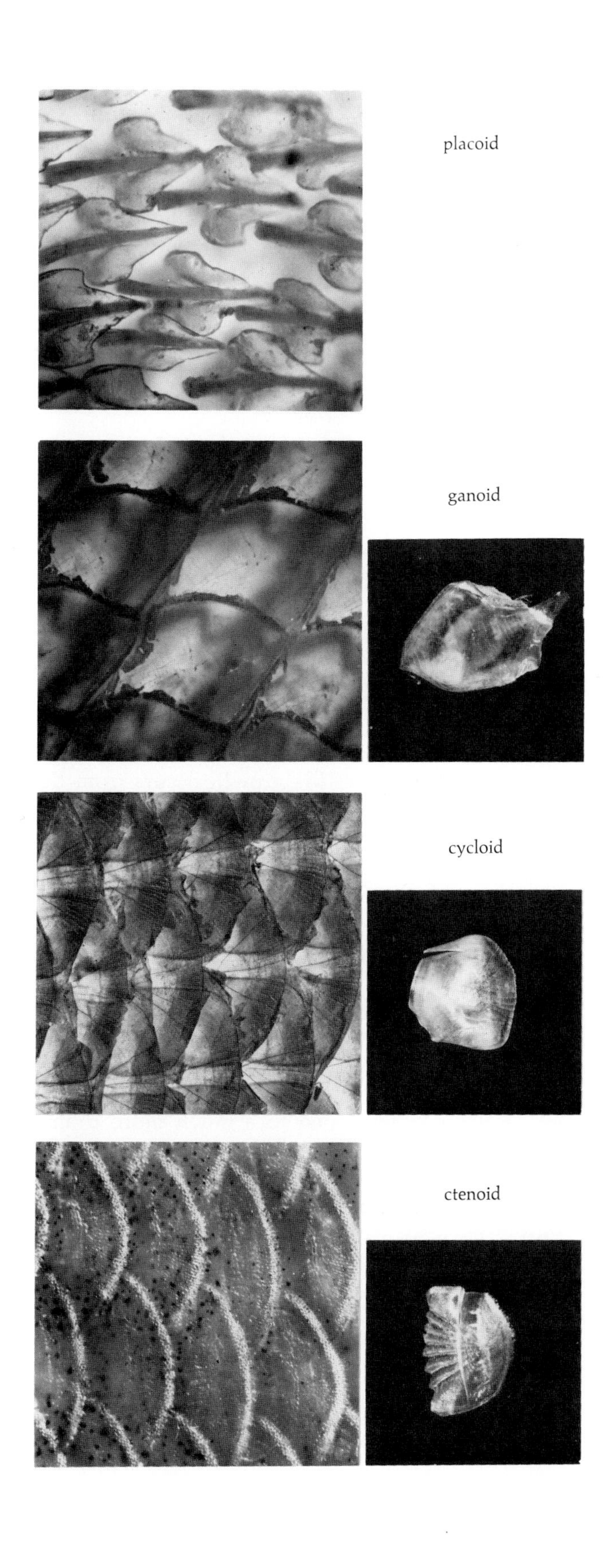
placoid
ganoid
cycloid
ctenoid

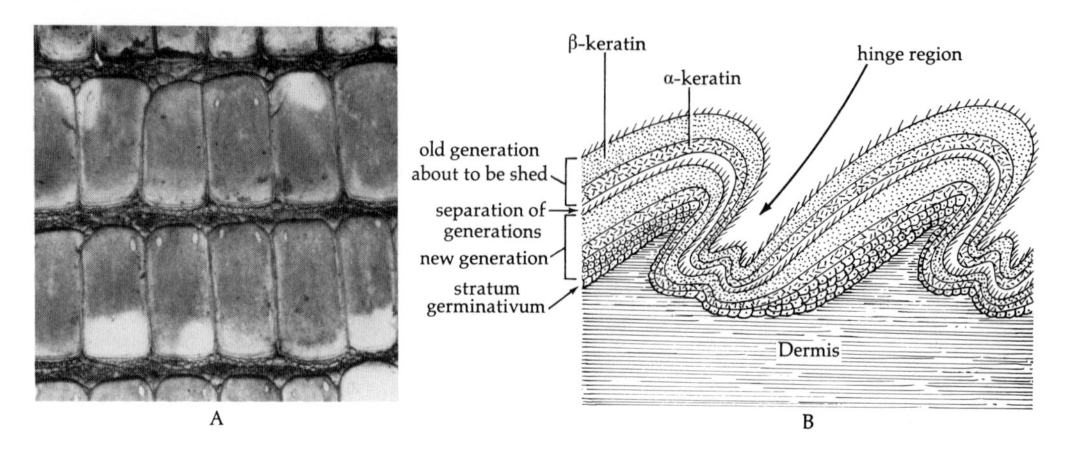

Fig. 8-9. A. Surface view of a freshly shed skin of a tegu lizard. B. Diagrammatic cross section of the skin of a generalized lepidosaurian. Note that the scales are part of the epidermis rather than appendages from it. Note also that new scales are formed in the deeper epidermis before the worn epidermis is shed. The thinning of keratins in the hinge region allows flexibility.

especially dorsally. In lepidosaurian reptiles (snakes, lizards, *Sphenodon*), a complete new epidermis—including its scales—is formed intermittently. This growth occurs from the stratum germinativum, and upon its completion the outer, older epidermis is shed as a unit (Fig. 8-9).

Both α- and β-keratin are involved in reptilian scales (Fig. 8-10).

SCALES IN BIRDS AND MAMMALS

Most birds have typical, epidermal keratin scales on the feet and lower leg and frequently a few small scales at the base of the beak. Most mammals have no scales, but some mammals, particularly rodents, have a few epidermal scales on the tail. In a very few mammals, however, scales are developed to a much more dramatic extent. The armadillo, for instance, has an overlying layer of epidermal horny scales fused to heavy, dermal bony armor similar to the armor of turtles. The termite-eating pangolin of tropical Africa and southeast Asia is almost entirely covered with heavy, sharp-edged dermal scales. When in danger it tucks its head between its forelegs and curls into a ball, exposing only its scales.

Mammals produce only α-keratin, although birds have both α-and β-keratin (Fig. 8-10).

HORNS AND ANTLERS

Among the most bizarre integumentary modifications are horns and antlers. Although present in some extinct reptiles and several extinct, nonhoofed mammals, among living forms they are confined to the ungulates. There are four basic types (Fig. 8-11).

1. The midline keratin fiber horn is found only in the rhinoceros. It is formed by long epidermal strands of keratin held together by a sticky substance. It is an integral part of the skin, is never shed, and can be removed from the dead animal along with the skin. The Indian rhinoceros has one horn, and the African two, one behind the other.

2. Antlers are characteristic of the deer family, Cervidae; although in reindeer they occur in both sexes, they are usually restricted to the male. An outgrowth from the dermal frontal bone forms the antler's core. This is covered by a soft layer of haired skin called the velvet. When the bony portion is fully grown, the vascularization is restricted basally; the velvet then dies and is sloughed off. The bare, fully developed antlers remain throughout the mating season and for some time

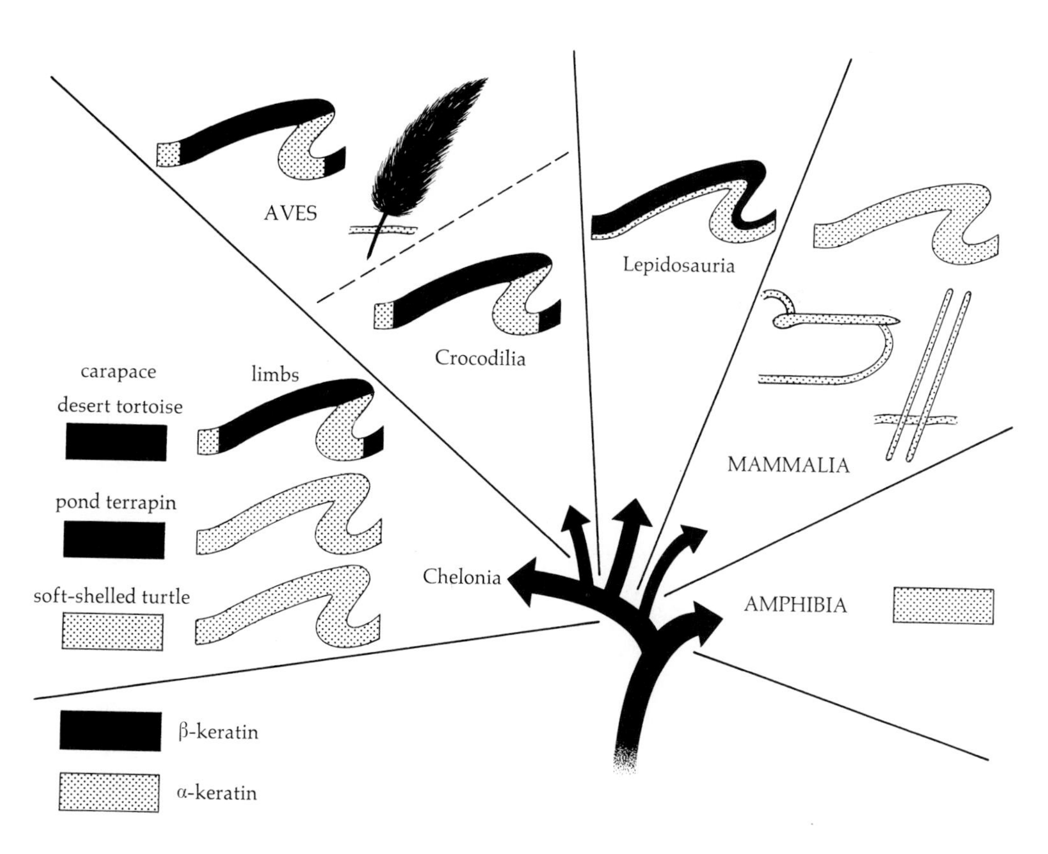

Fig. 8-10. The distribution of the fibrous keratin proteins in the epidermis of tetrapods. Note that β-keratins are found only in reptiles and birds. This distribution suggests that the ability to synthesize β-keratins evolved in early reptiles but after the synapsids were already a separate group. After H. P. Baden and P. F. A. Maderson (1970). Morphological and biophysical identification of fibrous proteins in the amniote epidermis. J. Exp. Zool. **174,** *225–232.*

Fig. 8-11. Diverse horn and antler types among ungulates.

thereafter until they are shed, regrowing the following year. (The exception is the giraffe, whose antlers remain covered with velvet throughout life and are never shed.) The antlers are used chiefly as offensive weapons in intraspecific combat for females and rank. Their growth is hormonally controlled; females and castrated males, which do not normally produce them, will do so if given testosterone.

3. In the male pronghorn "antelope," a North American ungulate that is not truly an antelope, there is an unusual horn. Its bony core also is an outgrowth from the frontal bone, but its outer part is of heavy, hard keratin with a single prong (which is not present on the bony core). The keratin part is shed annually, forced off by a new one forming deep to it; the inner, bony part is not shed.

4. Finally there is the so-called hollow horn that is never shed, found, for example, in both male and female cattle, sheep, and goats. These horns are also bony outgrowths from the frontal bone with an overlying portion of hard keratin derived from epidermis. They are not truly "hollow" unless the keratin portion is removed from the bony core.

DIGITAL TIPS

Without protection the tips of tetrapod digits would be quite vulnerable. The epidermis of the tip forms hard keratin, however, which is modified as claws, nails, and other structures. These not only protect the tips, but are often used for offense and defense as well. They can also improve the hand's manipulative ability (as will be obvious if one tries to pick up a small object with too-short fingernails).

AMPHIBIANS

There are only two examples of modified, keratin digital tips in amphibians. The very aquatic anuran, *Xenopus,* has claws. The spadefoot toads have a hard keratin sheet on the hind foot, which assists in digging.

AMNIOTES

Although structural details differ between classes, all amniotes have generally similar claws made up of two curved, scalelike keratin plates. The larger, dorsal plate is the unguis, and the somewhat smaller, softer, ventral plate is the subunguis; together these form a sheath over the entire distal phalanx of each digit. They are derived completely from epidermis, with a thickened layer of stratum germinativum forming the claw bed in the skin of the distal phalanx.

In birds, claws form mainly on the digits of the hind feet, although the fossil Archeopteryx also had three claws on each wing. In some modern species there are claws on some wing digits in the nestlings; these claws help the young birds move about and are lost before they leave the nest.

Claw shape varies considerably and is related to habits and habitats. Woodpeckers for instance have very sharp, narrow claws with which they anchor themselves firmly into tree bark during drilling. Most fowls have blunt, strong claws, with which they can effectively scratch the ground. Birds of prey have long, sharp, powerful talons.

The mammalian claw has basically the same structure as reptilian and avian claws, with two exceptions: there is usually a thickened area of epidermis and underlying dermis forming a footpad proximal to the subunguis, and the stratum corneum is of greatly thickened, soft (not hard) keratin.

Within this basic mammalian structure, however, are extremes in development (Fig. 8-12). In ungulates, the hoof is formed by the heavy unguis; the subunguis is much softer, and there is a very large, soft keratin pad which forms just behind the subunguis. At the other extreme are the primates, whose "claws" are so greatly modified that they are called nails; these have a very flattened unguis and an extremely small subunguis. Some lemurs have a specialized condition, with nails on all the digits except the second of the hind foot which retains a claw.

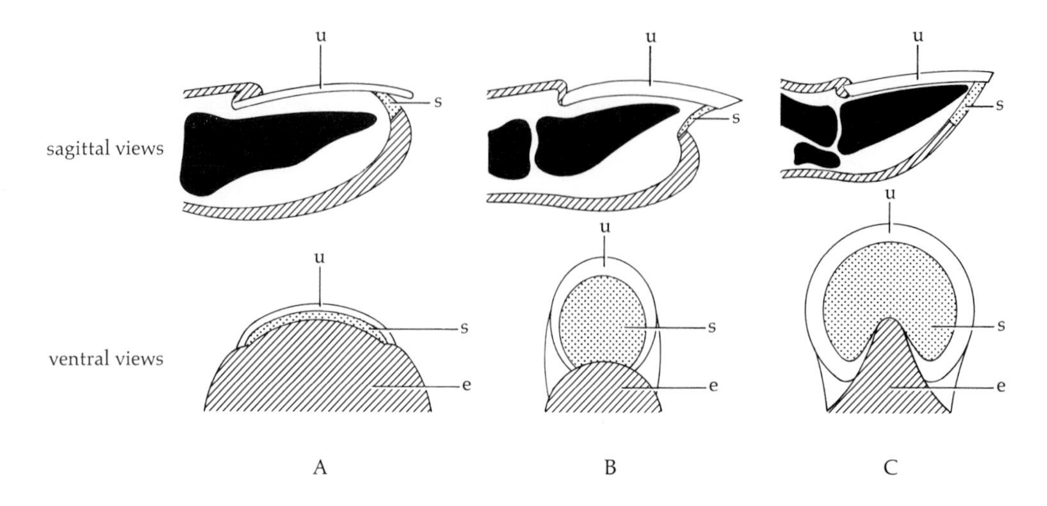

Fig. 8-12. Mammalian epidermal hardenings at digital tips include a hard unguis (u), a softer subunguis (s), and a thickened stratum corneum (e). The extent of each of these varies, as shown in a nail (A), a claw (B), and a hoof (C), but they are all composed of α-keratins. Drawings after J. S. Kingsley (1917). "Outlines of Comparative Anatomy of Vertebrates," second edition. Blakiston, Philadelphia, Pennsylvania.

FEATHERS

Unique to the class Aves, feathers are light, strong, waterproof, and admirably suited to their functions of flight and insulation. They are basically an elaborate modification of β-keratin epidermal scales, for embryologically both are preceded by the dermal papilla, a clumping of mesodermal fibroblasts and concentrated vascularization in the dermis. The overlying epidermis then proliferates rapidly, producing cells which die and form the β-keratin from which the feather develops.

There are three basic types of feathers (Fig. 8-13).

1. The largest and most prominent are contour feathers, or plumae (singular, pluma). These include the quill feathers of the body and the two largest groups of flight feathers (the remiges of the wing and the retrices of the tail).

The shaft of the pluma is composed of the calamus, which is embedded in the feather follicle and protrudes a short distance beyond it, and the rachis, which supports and is actually part of the feather's vane. Deep in the follicle at the tip of the calamus is a hole, the inferior umbilicus, through which the growing feather is nourished (the mature feather is hollow and dead). A second hole, the superior umbilicus, is present at the base of the vane where the calamus becomes the rachis. From the superior umbilicus usually protrudes a "secondary feather," the aftershaft. The aftershaft is small in most birds, but in some, such as the ostrich, it may be as long as the main feather itself.

The vane of the feather is made up of rachis, barbs, barbules, and hooklets (Fig. 8-14). The tapering rachis divides the vane into its two webs. The barbs, each attaching to the rachis, are the largest elements of the web. On each barb there are two rows of many smaller barbules; one row is along the surface of the barb that faces the feather tip, and the other is along the opposite surface. Thus the barbules from the anterior surface of one barb cross at right angles the barbules from the posterior surface of the next barb. The barbules on the side toward the tip have microscopic hooklets which firmly hold the crossing barbules.

Contour feathers are important in flight. During the wing's downstroke the vanes of neighboring wing feathers close together and present a common, continuous, and almost impervious surface to the air. During the upstroke, smooth integumentary muscles attached to the feathers slightly separate the wing feathers, letting air through with less resistance. It is the interlocking barbules with their hooklets that make the web so impervious and yet capable of "instant repair" after the barbs have separated.

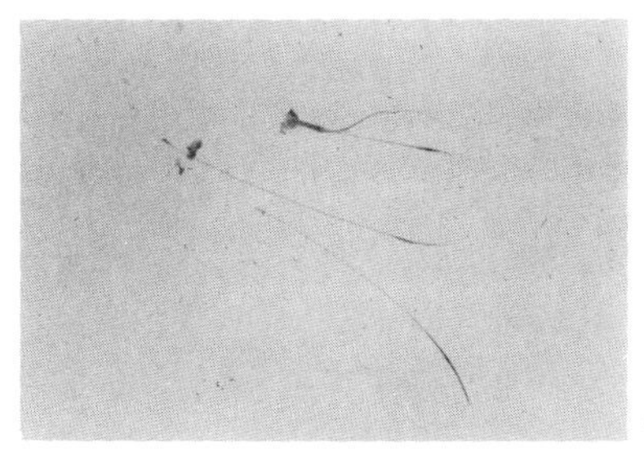

Filoplumes

Plumula

Pluma

Fig. 8-13. The three major types of feathers. The pluma photographed here lacks barbules and hooklets on the distal portions of its basal barbs. This common condition in plumae allows their basal portions to function as insulators, as do plumulae.

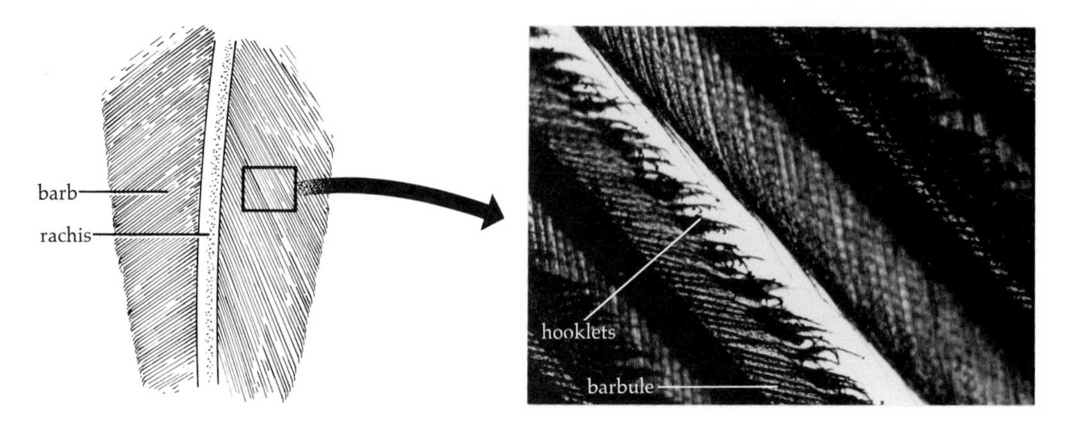

Fig. 8-14. On the left is part of a pluma, showing both rachis and vane; on the right, greatly enlarged, is the portion indicated by the rectangle. Two sets of barbules have been separated to demonstrate the hooklets.

2. Secondly there are the down feathers, or plumulae (singular, plumula), lying between the quill feathers over most of the body. These much softer feathers are particularly abundant on the breast and abdomen, and are important in the incubation of eggs, protection of young, and retention of body heat.

Each plumula has a short calamus embedded in the skin and several barbs from which barbules radiate. The barbules have no hooklets, and, therefore, the plumulae cannot interlock.

3. Thirdly there are the hair feathers, "pin feathers," or filoplumes, which superficially resemble hair. These are made up of a calamus within the skin from which comes a single rachis having either no barbs or just a few at its distal tip. The adaptive function of these feathers, if any, is unknown.

PLUMAGE

The distribution of these three types of feathers constitutes the bird's plumage. The young bird's first plumage, called the nestling down, contains only the fluffy tips of emerging contour feathers. This temporary plumage wears off as the full quill feathers develop. Gallinaceous birds (e.g., chicken) are fully cloaked in nestling down when they hatch, but songbirds and most others develop nestling down after being born almost naked.

The nestling down is replaced by juvenile plumage, the first full coat of feathers that includes quill feathers. It normally lasts through the first winter and is usually less brilliant in color than subsequent plumages.

Shortly before the first mating season the juvenile plumage is replaced by the nuptial plumage. This has the coloration of the adult bird and is often quite brilliant.

A postnuptial plumage develops after the breeding season. Some birds develop both a nuptial plumage and a postnuptial plumage each year; others retain the postnuptial plumage unchanged after the first year of life.

Although from a casual glance it appears that the contour feathers occur over most of the bird's body, they are actually attached in lines, called pterylae, between which are areas called apteriae that have only down and hair feathers. Apteriae also occur immediately around movable joints, where heavy contour feathers would impede movement.

HAIR

Hair is a diagnostic characteristic of mammals and, like the plumage of birds, has several functions. It insulates against cold and protects against rain and mechanical injury from the environment. In por-

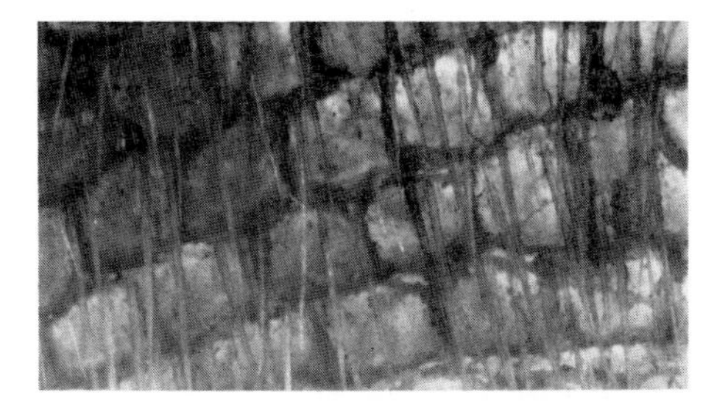

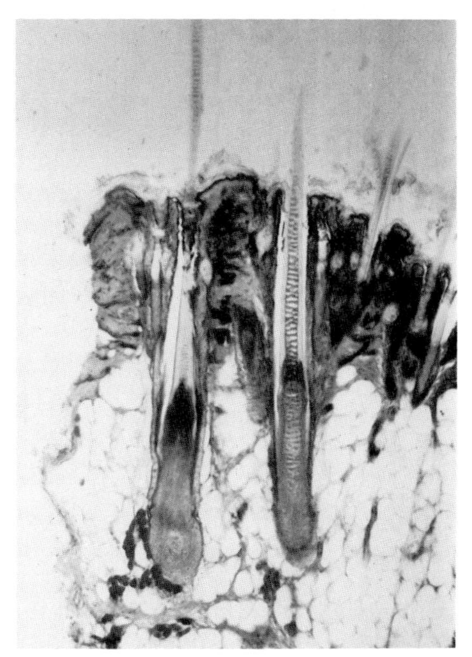

Fig. 8-15. Upper picture: hairs growing out between the epidermal scales of a rat's tail. Lower picture: a section through mammalian skin showing two large and three small hairs. Note that the large hairs extend deep into the subcutaneous tissue.

cupines, hedgehogs, and spiny anteaters it is stiffened into sharp quills and protects against would-be predators as well. The hairs are innervated basally, and thus serve as sense organs; this is particularly true of the specially stiffened hairs called vibrissae ("whiskers") in the facial region. Finally, it is primarily the hair which gives most mammals their color.

The epidermal keratin shaft which projects from the skin is only part of the hair (Fig. 8-15). Its root is sunk into a deep pit, or follicle, extending into the dermis. Opening into the follicle is a sebaceous gland, and below the root, at the follicle's base, is a small dermal papilla. Although they extend into the dermis, both follicle and hair are epidermal structures (except for the dermal papilla).

A small, smooth muscle, the arrector pili, runs from the basement membrane of the epidermis to a deep portion of the follicle. When it contracts, it raises the hair and causes the pelage to become thicker. This is done in response to cold, since it traps more dead air near the skin and increases the insulating layer; interestingly, it is also a response to fear, since it makes the animal appear larger and thus more threatening.

In the armadillo and the tails of some rodents, the hairs grow between epidermal scales in an orderly pattern. In most other mammals, although epidermal scales are lacking, the hairs grow in a similar pattern, suggesting the presence of scales at one time in their phylogeny (Fig. 8-15).

Just as birds have several plumages, mammals have more than one pelage. The first is the lanugo hair, which is frequently developed and usually also lost in the uterus; most mammals are born quite naked except for vibrissae. Next there is the juvenile pelage, often not the same color as the adult pelage. At about puberty, the juvenile pelage is replaced by the adult pelage, often in an orderly molt which starts at one end of the animal, progresses to the other, and is marked along its way by a distinct line between new and old hairs.

An extreme modification of hair is found in the toothless, or baleen, whales. From the mouth's edges hang huge, heavily keratinized, modified hairs which act as strainers. As water is drawn in through this apparatus the microorganisms and plankton are strained out and then swished off with the tongue and swallowed. Thus these largest vertebrates that ever lived do so thanks to minute organisms.

COLOR

An animal's color may be located in its dermis, in its epidermis, or in some epidermal modification such as scales, feathers, or hair. In vertebrates, color is usually caused by a pigment ("pigment color"), but it can also result from a physical structure which reflects or scatters specific wavelengths of white light ("structural color").

Pigment color is caused by specialized cells, called chromatophores; embryonically they originate from neural crest and then migrate into the dermis or epidermis. The most common chromatophores are melanophores, containing the protein pigment melanin which has two forms: eumelanin gives a brown or black coloration, while phaeomelanin gives a yellow-brown coloration. Other chromatophores, called xanthophores, contain carotenoid and pteridine pigments and cause much of the yellow-to-red coloration. Most vertebrates are incapable of synthesizing carotenoids and must obtain them directly from plants; but once a synthesized carotenoid has been ingested it can be modified to produce quite a different color than it does in the plant. The third common type of chromatophore produces structural rather than pigment color. This is the iridophore, filled not with pigment but with tiny crystals of guanine or other purines. These may reflect all wavelengths of light (producing a silvery effect), or just the shorter wavelengths (producing a metallic-blue effect).

Green pigments are very rare in vertebrates; what appears green usually does so because of two or more combined factors. For instance, the green of frogs is caused by the reflection of blue light from iridophores lying deep to yellow xanthophores.

In many parts of the body the pigment granules produced by the chromatophores are incorporated into dead, keratinous structures, for instance, hair, feathers, horns, and claws.

Changes of color are caused primarily by the position of the pigment granules within the highly branched pigment cell. When concentrated around the center of the cell near its nucleus, the pigment forms too small a spot to be perceived by the naked eye. However, when it is dispersed far out into the cell's long processes, it occupies a much larger area and thus can be easily detected. These color changes are controlled by external stimuli, such as light, by hormones, or by a combination of these factors.

COLORATION IN BIRDS

Visually, the most dramatic vertebrate group is birds, whose colors include white, black, brown, and many brilliant hues of blue, green, red, and yellow. Black and brown colorations are produced by melanophores which aggregate at the dermal papilla during feather growth, reach their long cytoplasmic processes into the epithelium that will form the feather, and there deposit melanin granules; thereafter the melanophores die. Red and yellow colorations are caused by carotenoids from plant material, which are chemically changed into specific bird carotenoids and are deposited by xanthophores in the epithelium of the developing feather bud. In addition to the carotenoid reds, there is also an exceptional bright red pigment, called turacin. This pigment is water soluble and is chemically a porphyrin combined with copper. It is known only in the African plantain-eating touraco, which, after a rain, appears dull brown because of the melanins which lie deep to the water-soluble turacin. In the same group of birds is found a closely related green pigment, an oxidized form of turacin called turacoverdin.

White is a structural color, caused by many tiny, air-filled spaces [about 0.4 micrometer (μm) in diameter], primarily in the feather's barbs. These tiny air chambers reflect white light and thus make the

Fig. 8-16. A white feather placed in oil becomes colorless because oil fills the air spaces in the barbs so that they no longer scatter light. In the photograph, oil has traveled along the barbs by capillary action.

feathers appear pure white. If one of these feathers is put into a clear oil having a different refractive index than air, and the oil enters the spaces, the feather will look colorless (Fig. 8-16).

Blue is another structural color. The so-called box cells in the barbs have many tiny canal-like pores (0.10–0.25 μm in diameter) which cause the feathers to reflect only the shortest wavelengths of light, i.e., blue. Originally described by Tyndall, this type of structural blue is called Tyndall blue. This structural color is usually enhanced by melanin lying deep to it in the center of the barb where it absorbs all other wavelengths of light.

Except for turacoverdin, birds' feathers contain no green pigments. The green of parrots and parakeets is caused by a combination of structural blue and carotenoid yellow.

One of the most remarkable color characteristics of birds is the iridescent plumage of many, such as hummingbirds, peacocks, and trogans. Iridescence is a structural phenomenon of the barbules rather than the barbs. These barbules are broad and flat and usually have no hooklets. They contain a dark melanin overlaid by several fine, transparent films, each about 0.4 μm thick. Light passing through these layered films is diffracted into its component parts, and most of it is reflected back. Light which does pass all the way through is absorbed by the black pigment deep to it. It is essentially the same as what one sees when oil has been spilled in a puddle on the road, but it is much more pleasing on the feathers of birds.

One cannot help but marvel at the genetic precision during the development of bird plumages. Structural colors would be vastly altered by a change in pore size of a fraction of 1 μm. Pigment colors must be laid down at exactly the right time during the feather's development in order to produce the intricate patterning by which many birds are recognized. The further fact that these feathers grow from independent follicles to form a single pattern in the adult pelage makes the genetic precision even more remarkable.

THE INTEGUMENT AS AN ORGAN SYSTEM

The integument is often viewed as a single organ which wraps the vertebrate body and which is composed of two tissue types: ectodermally derived epidermis and mesodermally derived dermis and hypodermis. As we have seen, however, many and diverse structures are derived from this organ, such as bones, glands, scales, horns, claws, feathers, and hair. Structurally, therefore, it is more realistic to consider the integument an *organ system.* The same is true functionally: the integument and its derivatives are involved in protection, food storage, heat regulation, sensation, excretion, secretion, respiration, and locomotion. The integument, therefore, is actually one of the more complex and diverse organ systems.

SUGGESTED READING

The following papers comprised a symposium on the vertebrate integument, presented in December, 1971, and published in *The American Zoologist,* Volume 12, Number 1 (1972).

Flaxman, B. Allen. Cell differentiation and its control in the vertebrate epidermis, pp. 13–25.

Moss, Melvin L. The vertebrate dermis and the integumental skeleton, pp. 27–34.

Quevedo, W. C., Jr. Epidermal melanin units: Melanocyte-keratinocyte interactions, pp. 35–41.

Taylor, J. D. Dermal chromatophores, pp. 43–62.

Hadley, M. E. Functional significance of vertebrate integumental pigmentation, pp. 63–76.

Ling, J. K. Adaptive functions of vertebrate molting cycles, pp. 77–93.

Quay, W. B. Integument and the environment: Glandular composition, function, and evolution, pp. 95–108.

Montagna, W. The skin of nonhuman primates, pp. 109–124.

Kollar, E. J. The development of the integument: Spatial, temporal, and phylogenetic factors, pp. 125–135.

Argyris, T. S. Chalones and the control of normal, regenerative, and neoplastic growth of the skin, pp. 137–149.

Maderson, P. F. A. When? Why? and How?: Some speculations on the evolution of the vertebrate integument, pp. 159–171.

COLOR PLATES

Figs. 4-5, 5-10, 7-11, 8-7, 11-10, 16-18

Fig. 4-5. Diversity of vertebrate skulls showing surface patterns of neurocranial, splanchnocranial, and dermatocranial elements.

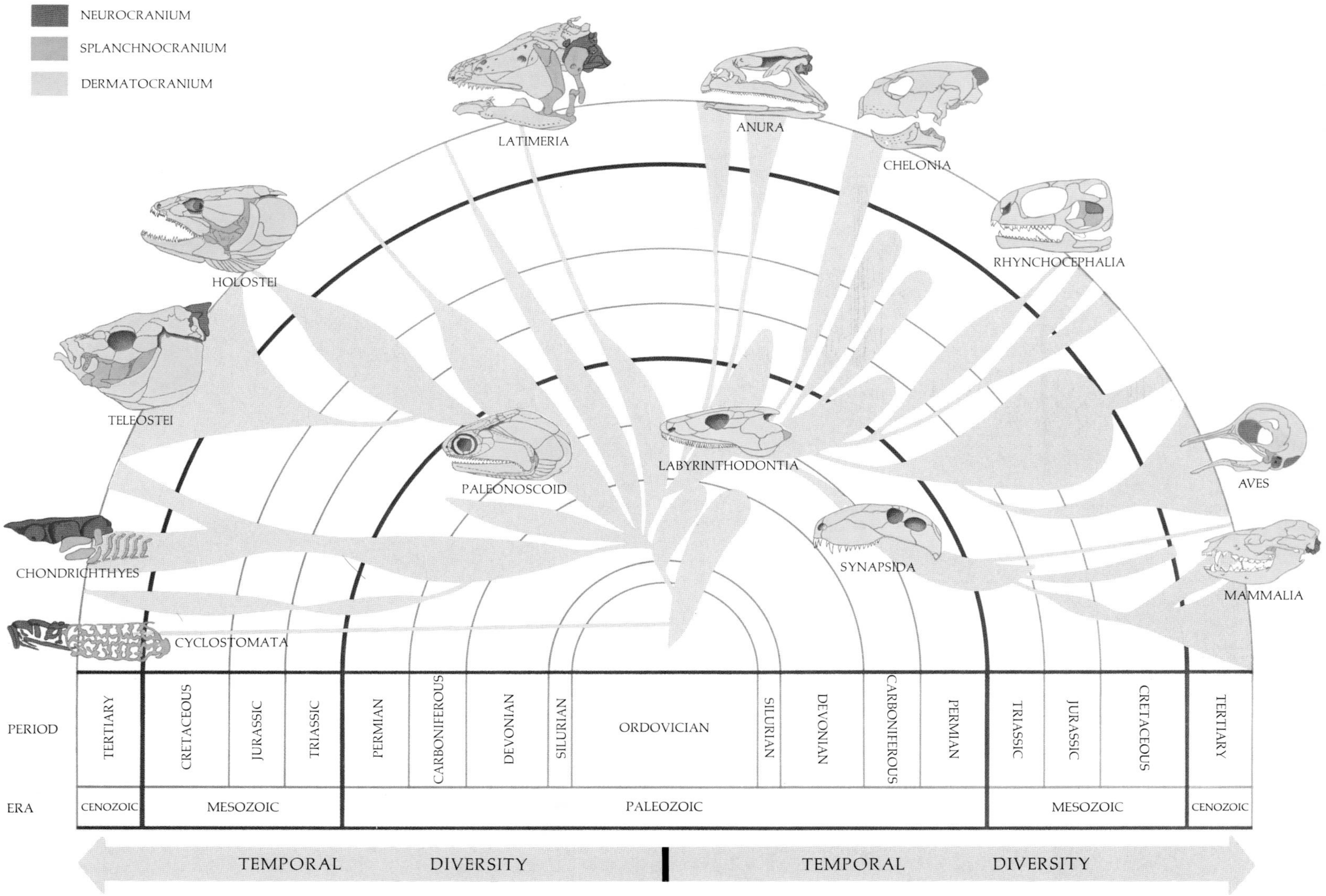

NEUROCRANIUM
SPLANCHNOCRANIUM
DERMATOCRANIUM
LATIMERIA
ANURA
CHELONIA
RHYNCHOCEPHALIA
HOLOSTEI
TELEOSTEI
PALEONOSCOID
LABYRINTHODONTIA
AVES
SYNAPSIDA
MAMMALIA
CHONDRICHTHYES
CYCLOSTOMATA
PERIOD
TERTIARY
CRETACEOUS
JURASSIC
TRIASSIC
PERMIAN
CARBONIFEROUS
DEVONIAN
SILURIAN
ORDOVICIAN
SILURIAN
DEVONIAN
CARBONIFEROUS
PERMIAN
TRIASSIC
JURASSIC
CRETACEOUS
TERTIARY
ERA
CENOZOIC
MESOZOIC
PALEOZOIC
MESOZOIC
CENOZOIC
TEMPORAL DIVERSITY
TEMPORAL DIVERSITY

Fig. 4-5

Fig. 5-10. Spatial and temporal diversities of the vertebrate pectoral girdle. The × indicates location of glenoid fossa, where the forelimb articulates with the pectoral girdle. Stippling indicates cartilage.

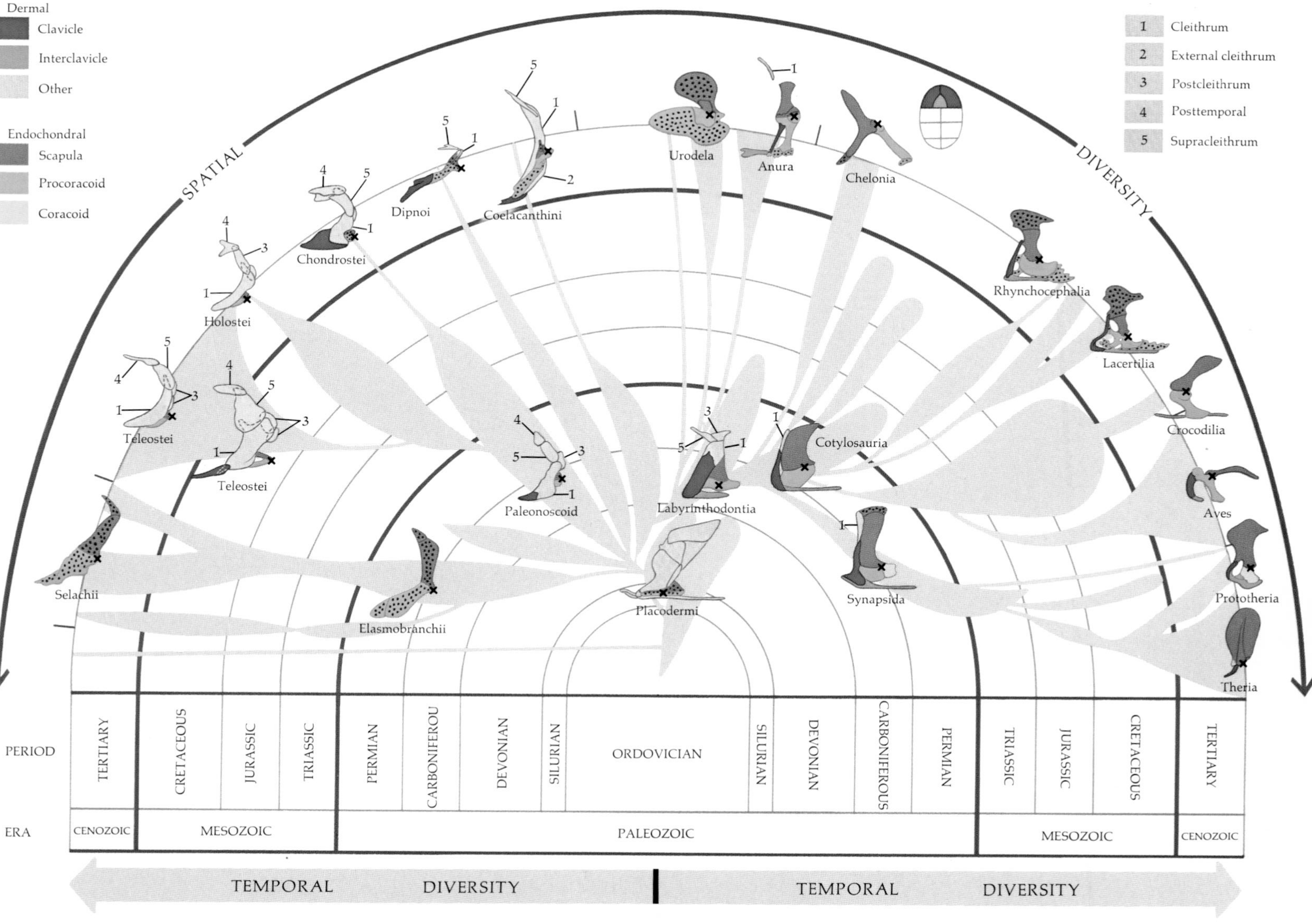
Dermal
Clavicle
Interclavicle
Other
Endochondral
Scapula
Procoracoid
Coracoid
1 Cleithrum
2 External cleithrum
3 Postcleithrum
4 Posttemporal
5 Supracleithrum
SPATIAL
DIVERSITY
Selachii
Teleostei
Teleostei
Holostei
Chondrostei
Dipnoi
Coelacanthini
Urodela
Anura
Chelonia
Rhynchocephalia
Lacertilia
Crocodilia
Aves
Prototheria
Theria
Elasmobranchii
Paleonoscoid
Placodermi
Labyrinthodontia
Cotylosauria
Synapsida
PERIOD
ERA
TERTIARY
CRETACEOUS
JURASSIC
TRIASSIC
PERMIAN
CARBONIFEROU
DEVONIAN
SILURIAN
ORDOVICIAN
SILURIAN
DEVONIAN
CARBONIFEROUS
PERMIAN
TRIASSIC
JURASSIC
CRETACEOUS
TERTIARY
CENOZOIC
MESOZOIC
PALEOZOIC
MESOZOIC
CENOZOIC
TEMPORAL DIVERSITY
TEMPORAL DIVERSITY

Fig. 5-10

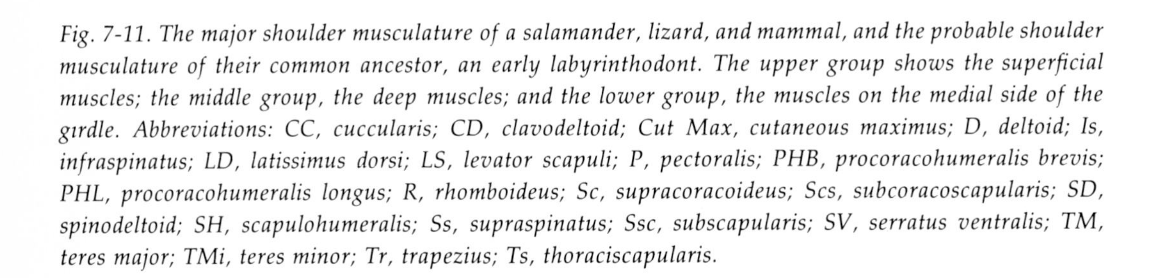

Fig. 7-11. The major shoulder musculature of a salamander, lizard, and mammal, and the probable shoulder musculature of their common ancestor, an early labyrinthodont. The upper group shows the superficial muscles; the middle group, the deep muscles; and the lower group, the muscles on the medial side of the girdle. Abbreviations: CC, cuccularis; CD, clavodeltoid; Cut Max, cutaneous maximus; D, deltoid; Is, infraspinatus; LD, latissimus dorsi; LS, levator scapuli; P, pectoralis; PHB, procoracohumeralis brevis; PHL, procoracohumeralis longus; R, rhomboideus; Sc, supracoracoideus; Scs, subcoracoscapularis; SD, spinodeltoid; SH, scapulohumeralis; Ss, supraspinatus; Ssc, subscapularis; SV, serratus ventralis; TM, teres major; TMi, teres minor; Tr, trapezius; Ts, thoraciscapularis.

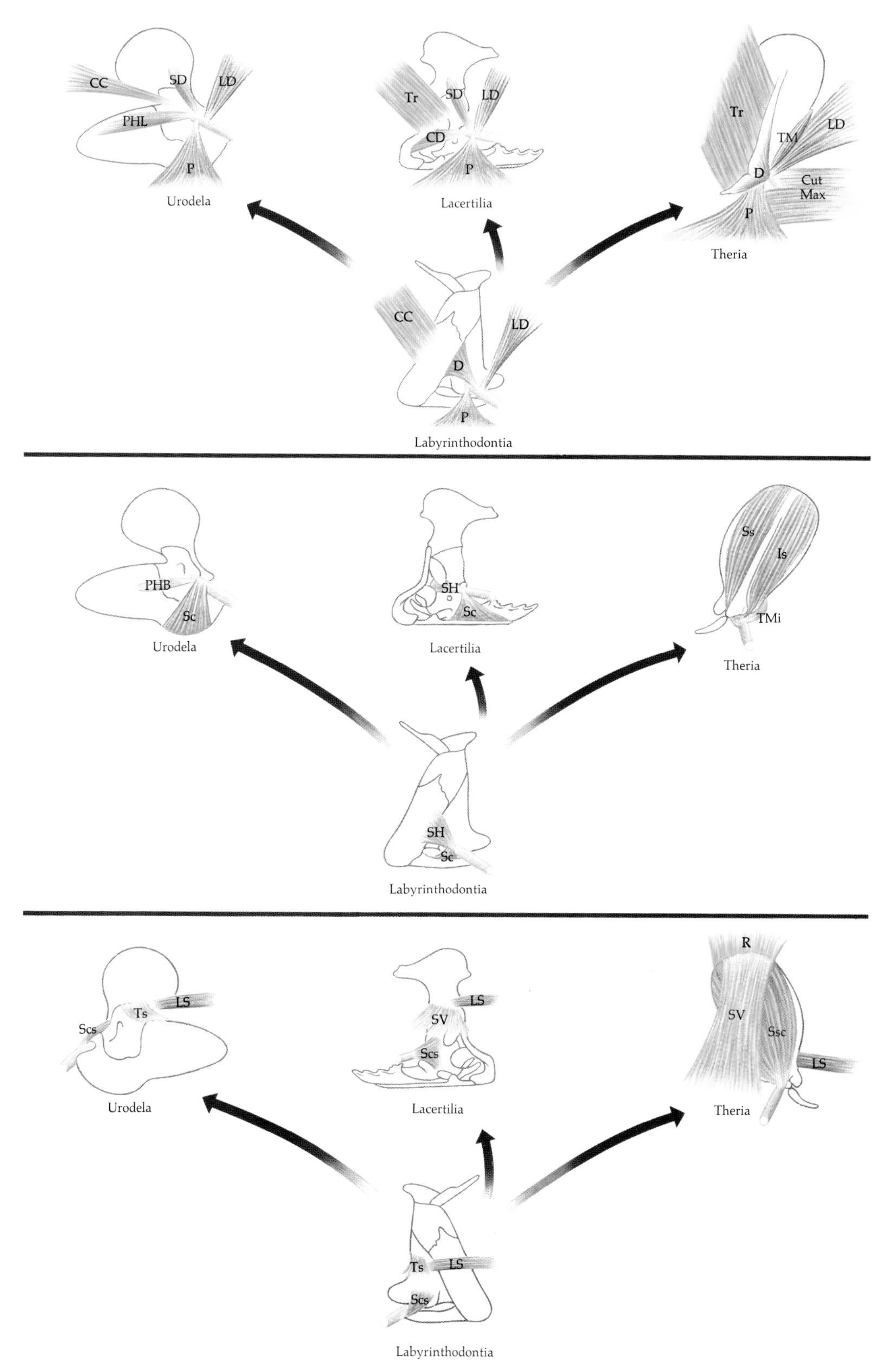
CC
SD
LD
PHL
P
Urodela
Tr
SD
LD
CD
P
Lacertilia
Tr
TM
LD
D
Cut
Max
P
Theria
CC
LD
D
P
Labyrinthodontia
PHB
Sc
Urodela
SH
Sc
Lacertilia
Ss
Is
TMi
Theria
SH
Sc
Labyrinthodontia
LS
Ts
Scs
Urodela
LS
SV
Scs
Lacertilia
R
SV
Ssc
LS
Theria
Ts
LS
Scs
Labyrinthodontia

Fig. 7-11

Fig. 8-7. Spatial and temporal diversities of the most common integumentary derivatives of the major groups of vertebrates.

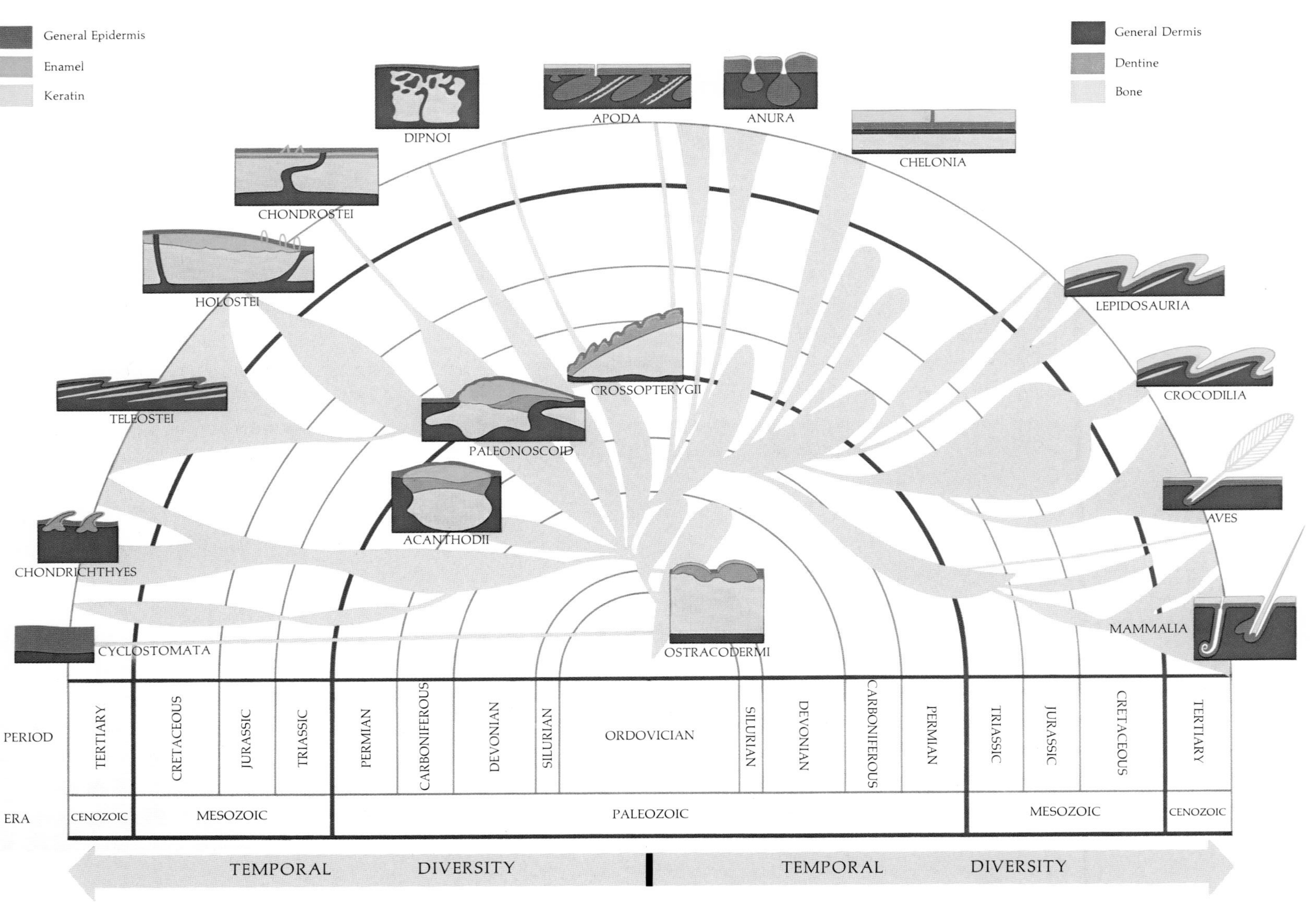

General Epidermis
Enamel
Keratin
General Dermis
Dentine
Bone
DIPNOI
APODA
ANURA
CHELONIA
CHONDROSTEI
HOLOSTEI
LEPIDOSAURIA
CROSSOPTERYGII
CROCODILIA
TELEOSTEI
PALEONOSCOID
AVES
ACANTHODII
CHONDRICHTHYES
MAMMALIA
CYCLOSTOMATA
OSTRACODERMI
PERIOD
TERTIARY
CRETACEOUS
JURASSIC
TRIASSIC
PERMIAN
CARBONIFEROUS
DEVONIAN
SILURIAN
ORDOVICIAN
SILURIAN
DEVONIAN
CARBONIFEROUS
PERMIAN
TRIASSIC
JURASSIC
CRETACEOUS
TERTIARY
ERA
CENOZOIC
MESOZOIC
PALEOZOIC
MESOZOIC
CENOZOIC
TEMPORAL DIVERSITY
TEMPORAL DIVERSITY

Fig. 8-7

Fig. 11-10. Spatial diversity of the vertebrate brain as seen in dorsal view.

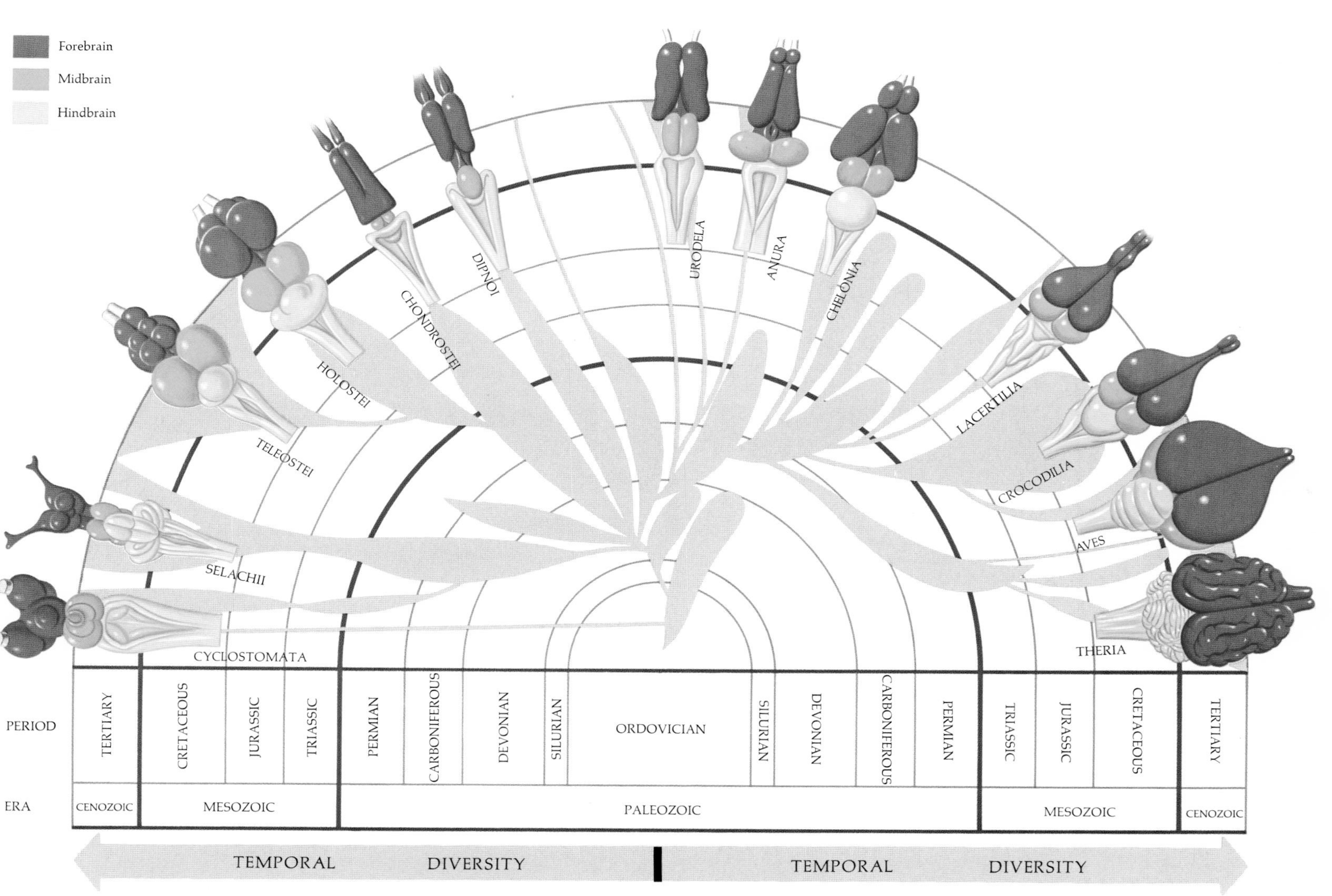
Forebrain
Midbrain
Hindbrain
CYCLOSTOMATA
SELACHII
TELEOSTEI
HOLOSTEI
CHONDROSTEI
DIPNOI
URODELA
ANURA
CHELONIA
LACERTILIA
CROCODILIA
AVES
THERIA
PERIOD
TERTIARY
CRETACEOUS
JURASSIC
TRIASSIC
PERMIAN
CARBONIFEROUS
DEVONIAN
SILURIAN
ORDOVICIAN
SILURIAN
DEVONIAN
CARBONIFEROUS
PERMIAN
TRIASSIC
JURASSIC
CRETACEOUS
TERTIARY
ERA
CENOZOIC
MESOZOIC
PALEOZOIC
MESOZOIC
CENOZOIC
TEMPORAL DIVERSITY
TEMPORAL DIVERSITY

Fig. 11-10

Fig. 16-18. Spatial diversity of the derivatives of the aortic arches. Blue indicates vessels carrying unoxygenated blood and red those carrying oxygenated blood. The discontinuities between the afferent and efferent branchial arteries of fishes represent the capillary beds of the gills. Roman numerals indicate the aortic arches from which the vessels developed. DA, dorsal aorta; EC, external carotid artery; H, heart; IC, internal carotid artery; PA, pulmonary artery. The mammalian arch of the aorta is formed from the left fourth aortic arch and the avian arch of the aorta from the right fourth aortic arch.

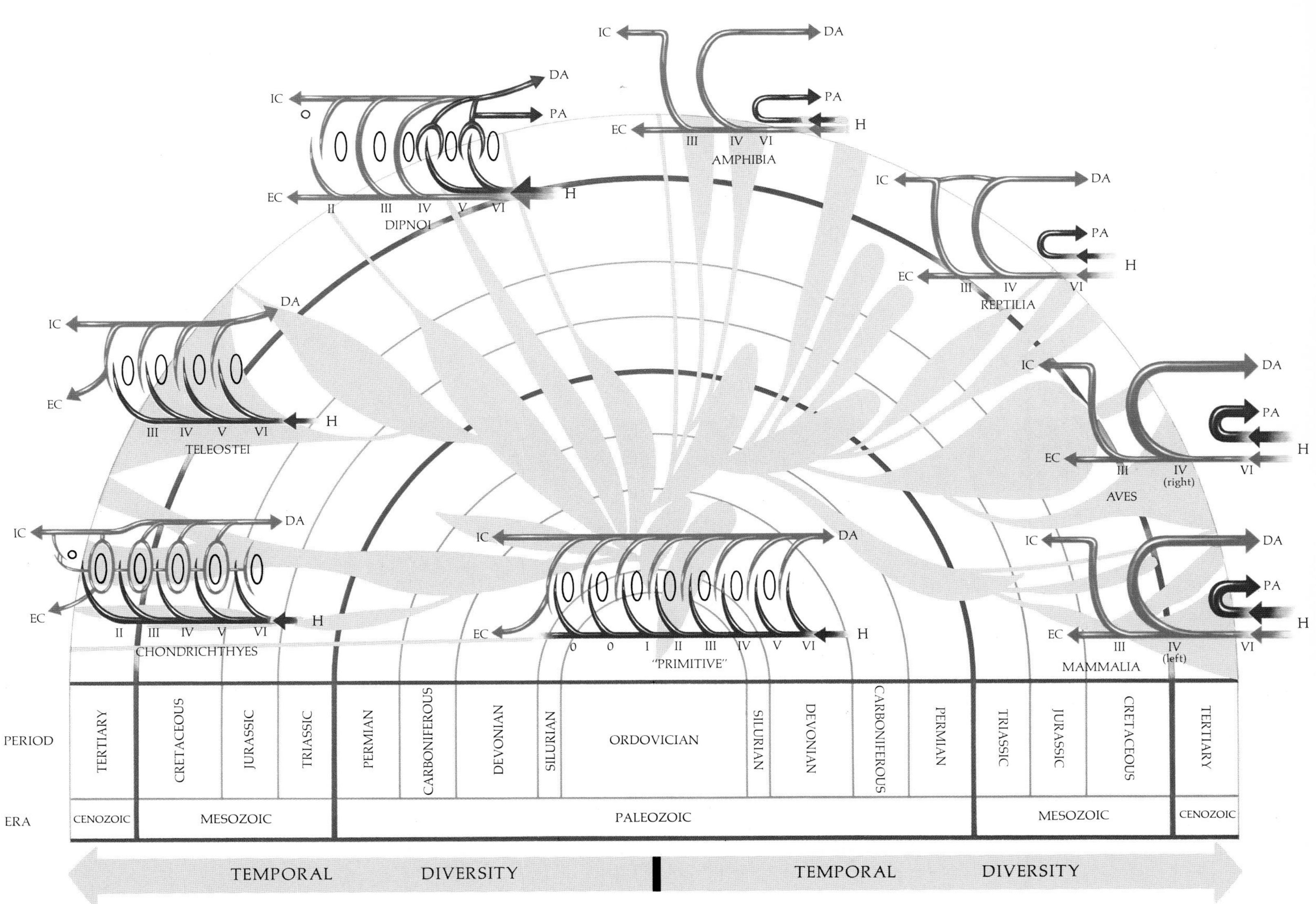
DIPNOI
AMPHIBIA
REPTILIA
TELEOSTEI
AVES
CHONDRICHTHYES
"PRIMITIVE"
MAMMALIA
IC
EC
DA
PA
H
PERIOD
ERA
TERTIARY
CRETACEOUS
JURASSIC
TRIASSIC
PERMIAN
CARBONIFEROUS
DEVONIAN
SILURIAN
ORDOVICIAN
CENOZOIC
MESOZOIC
PALEOZOIC
TEMPORAL
DIVERSITY

Fig. 16-18

9 NERVOUS TISSUES

9

The sense organs and nervous system are responsible for all perceiving, processing, and storing of information and for controlling and coordinating most of the actions of muscles and glands (some being influenced by endocrine organs). The functioning of nervous tissues and of the entire system is, therefore, of critical importance and has been shaped by evolutionary pressures into a finely tuned instrument. For example, a hungry frog that sees an airborne insect darts out its tongue to catch it in midflight. Any malfunctioning in the reception or processing of the information, or the performance of the action, would mean the frog's early demise.

NEURONS

MORPHOLOGY OF THE NEURON

The fundamental morphological unit of the nervous system is the nerve cell, or neuron. Although neurons exhibit tremendous structural diversity, most share certain anatomical characteristics (Fig. 9-1). The

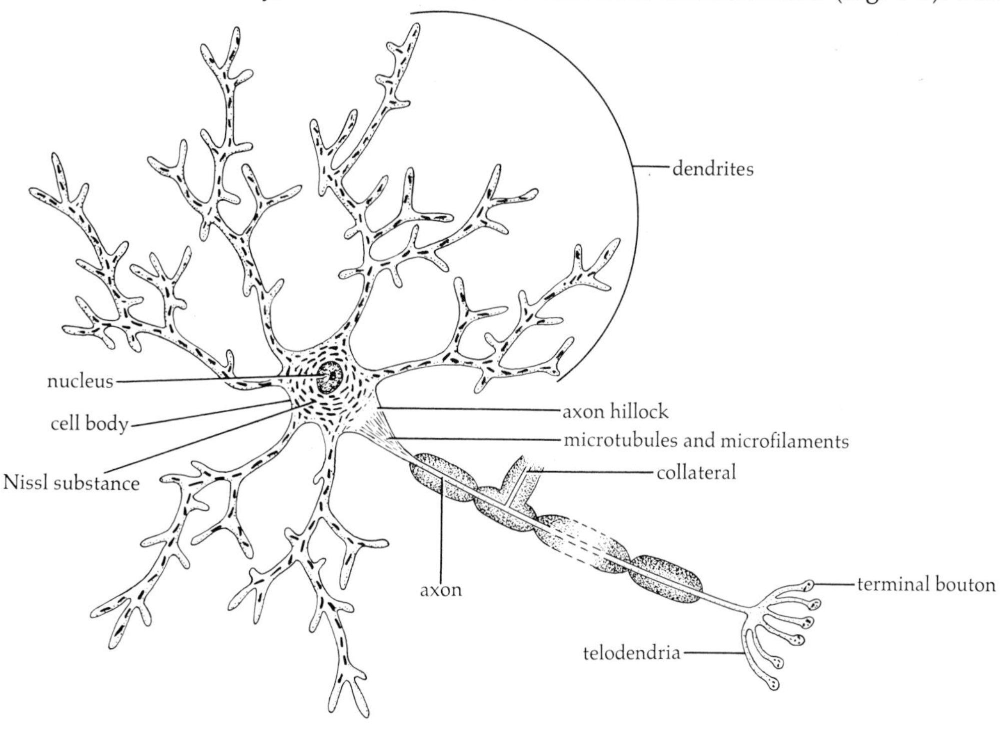

Fig. 9-1. A generalized multipolar neuron.

typical neuron is composed of a cell body (perikaryon) and one or more narrow, usually branched cytoplasmic processes (axon, dendrites).

Perikaryon

The perikaryon contains the nucleus and the same organelles found in most cells. However, it has a far greater concentration of densely folded, rough endoplasmic reticulum (ribonucleoprotein). In nerve cells this is called Nissl substance, and its concentration indicates the capacity for a high metabolic rate, particularly a high rate of amino acid incorporation and protein synthesis.

Axon

Each neuron has only one axon, functionally defined as the process that carries the nerve impulse away from the cell body. With one significant exception (to be described), the axon is the longest process.

The area of the cell body from which the axon emerges, called the axon hillock, lacks Nissl substance, as does the axon itself. The axon contains microtubules, microfilaments, and a few mitochondria. Collaterals branch from the axon, usually at right angles. Distally, the axon (and each collateral) breaks up into many fine branches called telodendria, each of which terminates in a swelling called a terminal bouton. Each terminal bouton contains mitochondria and submicroscopic, membrane-bound spheres called synaptic vesicles.

Dendrites

A neuron may have no dendrites or as many as thirty. Dendrites are usually shorter and more branched than axons. Their internal structure is similar to that of the perikaryon, and they frequently contain Nissl substance in their initial portions. A neuron with no dendrites is termed unipolar, for it is polarized by the axon. A single dendrite makes a bipolar neuron, and two or more dendrites, a multipolar neuron (Fig. 9-2).

The Nerve Fiber

Usually an axon, but in one case a dendrite, the nerve fiber is the process of the neuron which propagates the nerve impulse. Fibers gathered together form nerves or tracts.

PHYSIOLOGY OF THE NEURON

Inside a neuron there is a high concentration of potassium ions and a low concentration of sodium ions; in the interstitial fluids outside the cell the concentrations are reversed. Instead of osmotically equilibrating, this imbalance of cations is maintained by the cell membrane's physical characteristics plus an enzyme system, called the

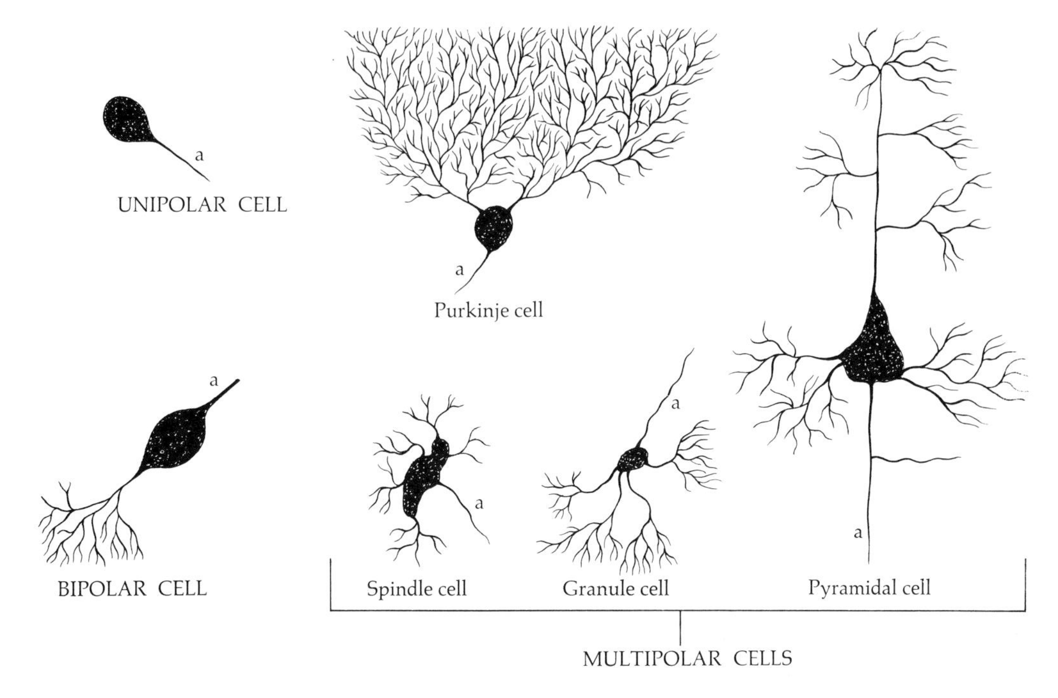

Fig. 9-2. Diversity of neuron types as seen in the number and extent of their dendrites. Only the proximal portions of their axons (a) are shown.

sodium pump, which actively removes sodium from inside the cell. Owing in part to this imbalance, a neuron in its "resting" state has an internal electric potential which is negative relative to the interstitial fluid; this *resting potential* is as much as −70 millivolts (mV) in a large neuron, and less in smaller neurons.

A large (for convenience's sake) nerve fiber can be stimulated in the laboratory by a small electric shock. This causes depolarization (also called excitation) and instantaneous reversal of the potential, from about −70 to +40 mV in a large neuron and less in smaller neurons. This "wave of external negativity," occurring in about 1 millisecond (msecond) and then traveling rapidly the length of the fiber (see Fig. 9-3), is the nerve impulse or *action potential.*

Essentially the same thing happens in the body (*in vivo*) except that the stimulus is picked up by either the dendrites or the cell body. The stimulus can be any physical distortion of the cell membrane, caused even by certain chemicals, which will change its permeability and cause depolarization. If the depolarization is less than 10–15 mV, it does not generate an action potential; its stimulus is called subthreshold. If the depolarization becomes as great as 10–15 mV, however, it triggers an action potential at the axon hillock of the cell body. The action potential is then propagated down the axon.

Any nerve cell can produce action potentials of only one amplitude; for large neurons this is about 110 mV (from −70 to +40 mV). The action potential is thus called an all-or-none phenomenon: it occurs at full strength or not at all.

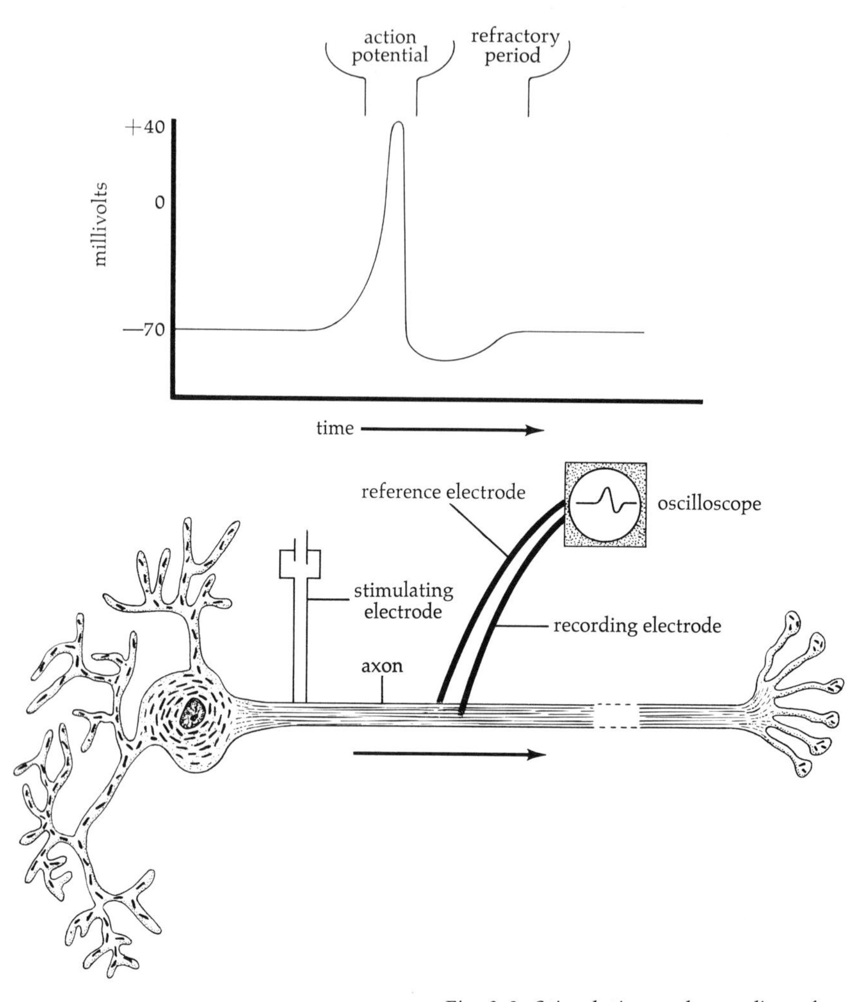

Fig. 9-3. Stimulation and recording of an action potential from a large neuron.

The termination of the action potential coincides with a transient increase in the resting potential, which amounts to as much as 90 mV for large neurons. During this *refractory period,* from $\frac{1}{2}$ to 4 mseconds in duration, no further action potentials can be generated. After the refractory period, the cell's potential returns to about −70 mV (for large neurons), and the process can begin again. All of this occurs in as little as 2 mseconds (Fig. 9-3).

The action potential is the only type of message that most axons can transmit. In most neurons this message is sent from time to time without external stimulus in what are called spontaneous firings. When these neurons are excited, the spontaneous rate of firing increases; inhibition causes the spontaneous rate to decrease. Some neurons, however, do not generate spontaneous action potentials; in these, inhibition results in a greater stimulus being required to generate an action potential.

MYELIN SHEATH

Nerve fibers are protected from the extracellular fluids through which they pass by a close wrapping of other cells: glial cells in the brain and spinal cord and Schwann cells in the peripheral nervous system. Often these cells contain a fatty substance, myelin, which wraps around the fiber somewhat like many layers of insulation around a wire. At points between glial or Schwann cells there are small unmyelinated areas called the nodes of Ranvier.

Although protective cells enclose all nerve fibers, the myelin sheath is variable; thus there are heavily myelinated, lightly myelinated, and even unmyelinated fibers (Fig. 9-4). In general, the larger the diameter of the axon and the thicker its myelin sheath, the more rapidly it will conduct nerve impulses. The primary function of the myelin sheath thus appears to be to increase the propagation speed of the action potentials.

There is an unusual situation in neurons carrying sensory information from the periphery. Their cell bodies are located just outside the cord or brain, in groups called spinal ganglia or cranial ganglia depending on their location. The fiber leading to the cell body must often be extremely long; a sensory fiber in a man's toe, for instance, runs up the leg to the lower back region, where its cell body is located

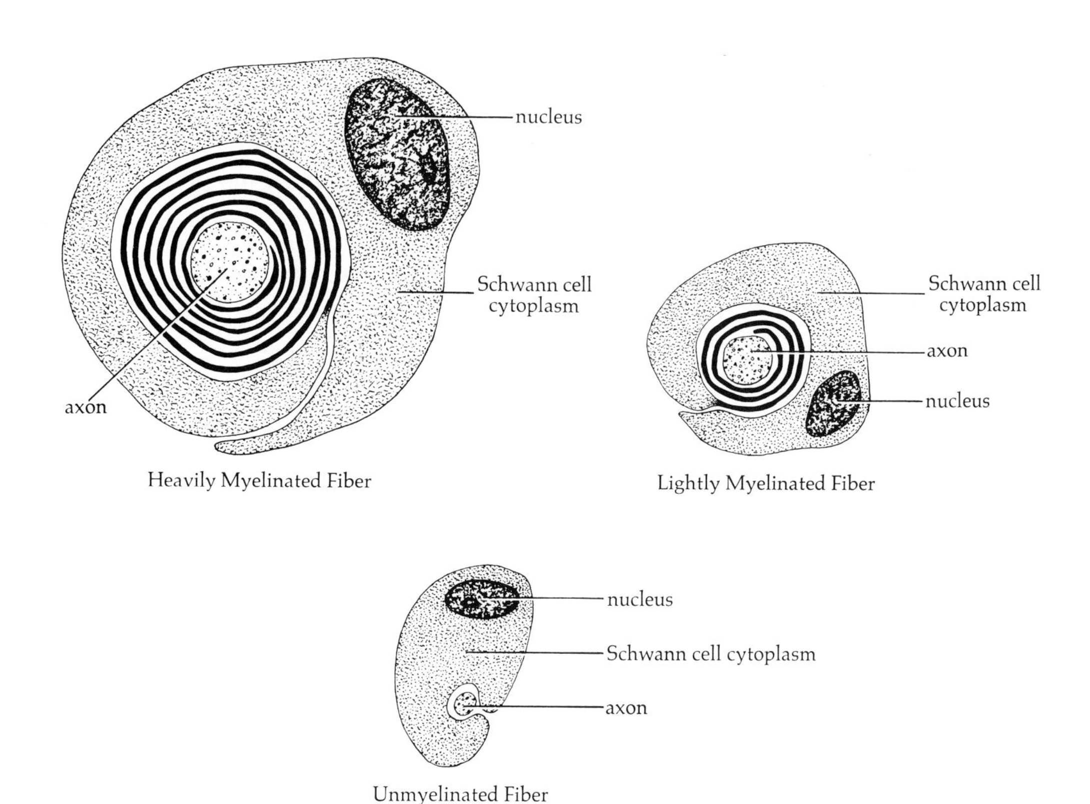

Fig. 9-4. Variations in the extent of Schwann cell protection of peripheral fibers and the amount of myelin (shown in black).

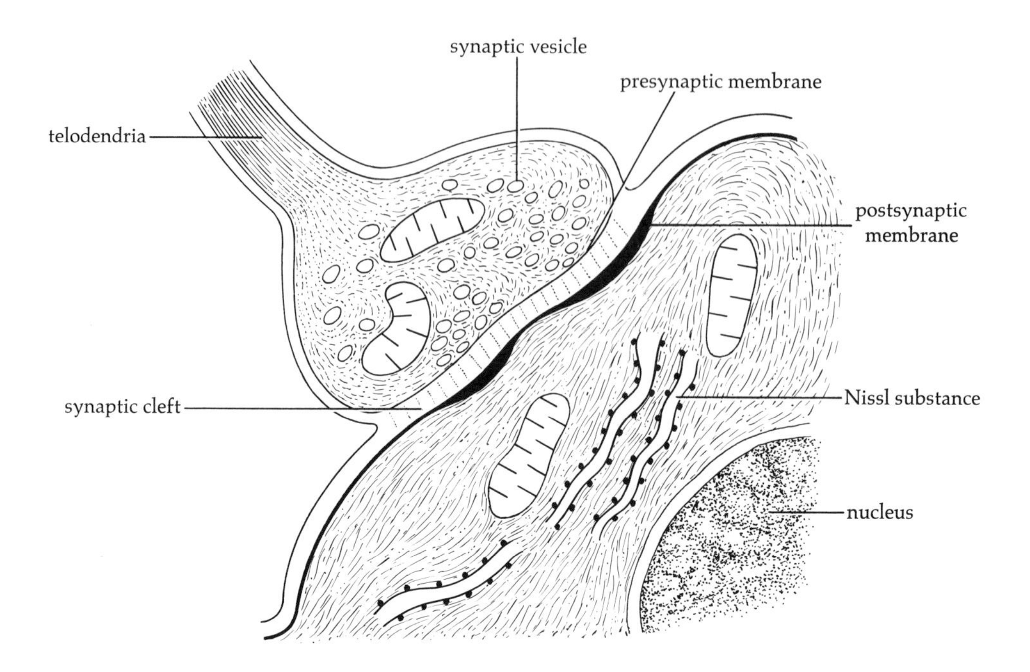

Fig. 9-5. Diagram of an axosomatic synapse as it would be seen with the electron microscope.

in a spinal ganglion. The axon leading from the cell body is frequently much shorter, for it passes into the spinal cord and may terminate immediately (or may first travel some distance in the cord). This is the one exception to two generalizations: that the axon is the neuron's longest process, and that the axon is the only "fiber", that is, the only process which propagates the nerve impulse. Furthermore, these dendrites resemble axons in their internal structure, length, and the wrapping of Schwann cells that frequently produce myelin. Except in this particular case of peripheral sensory nerves, however, dendrites do not propagate nerve impulses and have no myelin sheath.

THE SYNAPSE

Nerve cells pass neural information (action potentials) from one to another, in one direction only, across a narrow synaptic cleft [150–200 Ångstrom (Å) units wide]. A synapse may be axosomatic or axodendritic, or sometimes axoaxonic, that is, between the terminal boutons of one axon and the next cell body, or its dendrites, or sometimes its axons (Fig. 9-5).

When an action potential reaches the terminal bouton of the axon it causes the release of a neural humor, or chemical transmitter, that is stored in the synaptic vesicles of the bouton. (Of several neural humors, the one most studied is acetylcholine; it is found at all neuromuscular junctions and at many other synapses.)

Release of the neural humor into the synaptic cleft causes a change in the cell membrane on the other side of the synaptic cleft, on which the axon is said to terminate. This membrane is called the postsynaptic membrane, and its cell, the postsynaptic cell. Depending on the specific cell and on the neural humor involved, the change in the postsynaptic membrane can be either hypopolarizing or hyperpolarizing. A hypopolarizing change produces an excitatory postsynaptic potential (EPSP), making the resting potential of the postsynaptic cell slightly less negative (i.e., more positive). A hyperpolarizing change has the opposite effect and is called an inhibitory postsynaptic potential.

An EPSP lowers the threshold required for generating an action potential in the postsynaptic cell. If the EPSP reaches 10–15 mV, an action potential is generated in the postsynaptic cell and travels down its axon. An inhibitory postsynaptic potential raises the threshold of the postsynaptic cell, which inhibits its firing. Inhibitory mechanisms are necessary for such phenomena as the active relaxation (= decreased tonus) of a muscle whose antagonist is contracting.

TERMINOLOGY

The nervous system may be divided into the central nervous system (CNS), made up of the brain and spinal cord, and the peripheral nervous system, made up of neurons leading to and from the brain (= cranial nerves) and spinal cord (= spinal nerves). Within these large divisions are many smaller ones.

Neurons having similar functions are usually grouped together. In the CNS, such an aggregate of cell bodies is called a nucleus, not to be confused with the nucleus of an individual cell. In the peripheral nervous system, such an aggregate is called a ganglion.

In the CNS, nerve fibers are arranged in functional groups: such a group is called a tract, or a specific type of tract such as a lemniscus or fasciculus. In the peripheral nervous system, such a group includes fibers of different functional types; these form a morphological unit which is called a nerve—a gross anatomical structure that can be seen in dissection. Nerves have specific names, for instance, the radial nerve of the arm, or the sciatic nerve of the leg.

In an unstained section the myelin sheaths are a glistening white; therefore tracts and nerves are called the white matter. Cell bodies, which lack myelin, appear a dull gray and are called the gray matter. The white matter typically overlies the gray matter in both the cord and brain. Where a broad surface of the brain is composed of layered gray matter superficial to white matter, as in the cerebrum and cerebellum, this external gray matter is called cortex.

Nerves or tracts carrying information from sense organs, viscera, etc., to the CNS are called sensory nerves outside the cord or brain, or sensory tracts inside; those carrying information from centers in the brain and cord toward muscles or glands are called motor tracts or nerves.

Between sensory and motor centers in the CNS are thousands of association (internuncial) neurons and tracts.

The nerves, tracts, ganglia, and nuclei associated with the somatic portions of the body (primarily skin and muscles) constitute the somatic nervous system. Those concerned with the more vegetative portions (viscera, gills, gill arches, and their derivatives) comprise the visceral nervous system. These are not distinct entities, but ways of thinking about the organization of the vertebrate nervous system.

EXPLORING THE NERVOUS SYSTEM

In studying the nervous system homologies are as important as they are in any other system, yet their determination is particularly difficult. The embryology is so diffuse and complex that conventional means scarcely reveal which groups of adult cells have common origins. A recent advance is the use of radioautography: radioactive tracers are introduced at various stages throughout ontogeny, and their incorporation into specific sites in the developing embryo is plotted against time. Another method, particularly useful in closely related vertebrates, involves the microscopic comparison of nuclei whose cells may be defined either by perikaryon morphology or dendritic patterns. This method can also be useful in studying a structure such as the cerebellar cortex, which varies little from group to group.

Homologies can also be indicated by the position of a nuclear group or fiber tract relative to surface or internal landmarks. Histochemistry can sometimes demonstrate the same chemical structure in different nuclear groups. Electrophysiology can show whether similar nuclei or tracts in different vertebrates respond to similar electrical or mechanical stimulation. Behavior after either stimulation or damage to specific parts can, to some extent, indicate homologies. Since all these methods are subject to confusion and problems of interpretation, as many of them as possible should be used before homologies are stated with confidence. The currently most important method, however, is to determine the fiber connections between nuclear groups.

These connections consist of miles of intermingled fibers connecting millions of neurons with an intricate complexity. Neither in structure nor in function are they completely understood. In the late nineteenth and early twentieth centuries (when the study of descriptive comparative morphology was advancing most rapidly), there were no accurate methods for studying nervous pathways. Dissection can reveal only the very largest and most prominent pathways. The microscope shows such an elaborate network of fibers that following a single one for any distance is impossible. Therefore, instead of these classical descriptive methods, it is necessary to use experimental methods in charting the nervous system. Of these, there are presently two possibilities: physiology and anatomy.

Physiological techniques usually involve recording the action potentials elicited by a specific stimulus. For instance, one can stimulate an animal with light and, by "scanning" with electrodes, "map out" the areas in the brain where action potentials are elicited; these areas of stimulation are the visual centers. One can also stimulate groups of

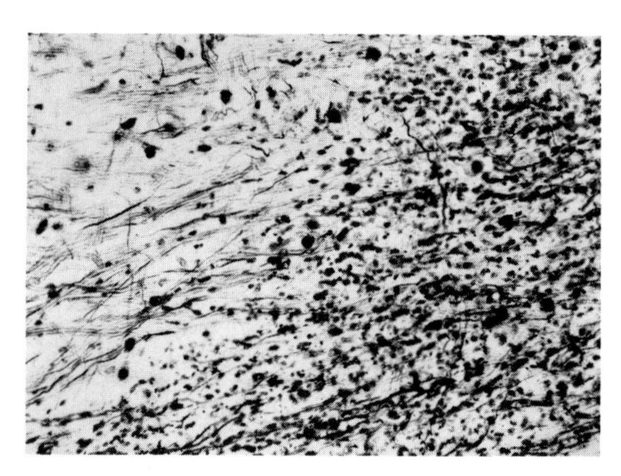

Fig. 9-6. Special silver-staining techniques selectively demonstrate degenerating fibers and terminals. The upper left portion of the photo is normal; the rest contains degeneration.

fibers or cells in the brain and note the muscles or glands that are activated.

Only experimental anatomy, however, can provide an actual tracing of nerves and central pathways. In these procedures, a specific nucleus or ganglion is surgically destroyed and the animal is allowed to survive long enough for the axons coming from that nucleus or ganglion to start degenerating. The animal is then killed and the pertinent parts of the nervous system processed to show degenerating fibers and terminal boutons (Fig. 9-6). Although this is a long, painstaking process, yielding information with frustrating slowness, it is the most accurate way to trace the "wiring diagrams" of the nervous system. Reliable versions of these techniques have been available for mammals only since the 1950s and for nonmammals since the 1960s. Not surprisingly, many pathways are still unknown.

Most of the work using these two experimental methods has been done on mammals. In addition, attempts to understand and treat human neurological disorders have led to many important advances, but usually in primates. As a result, we now have a fairly comprehensive view of the fiber pathways in a few mammalian brains, but our knowledge is much less complete in important groups of nonmammals. The question of homologies, despite its importance, remains frequently unanswered.

REFLEXES

A good place to start studying the anatomical organization of the nervous system and how it functions is with the fixed motor patterns which result from certain specific sensory stimulations. These largely unchanging motor patterns are called reflexes. They can be quite simple, involving two or only a few elements, or quite complex, involving many muscles and a long series of events.

In the simplest possible reflexes there are only two neurons: a sensory neuron from the sense organ to the CNS, which synapses with a

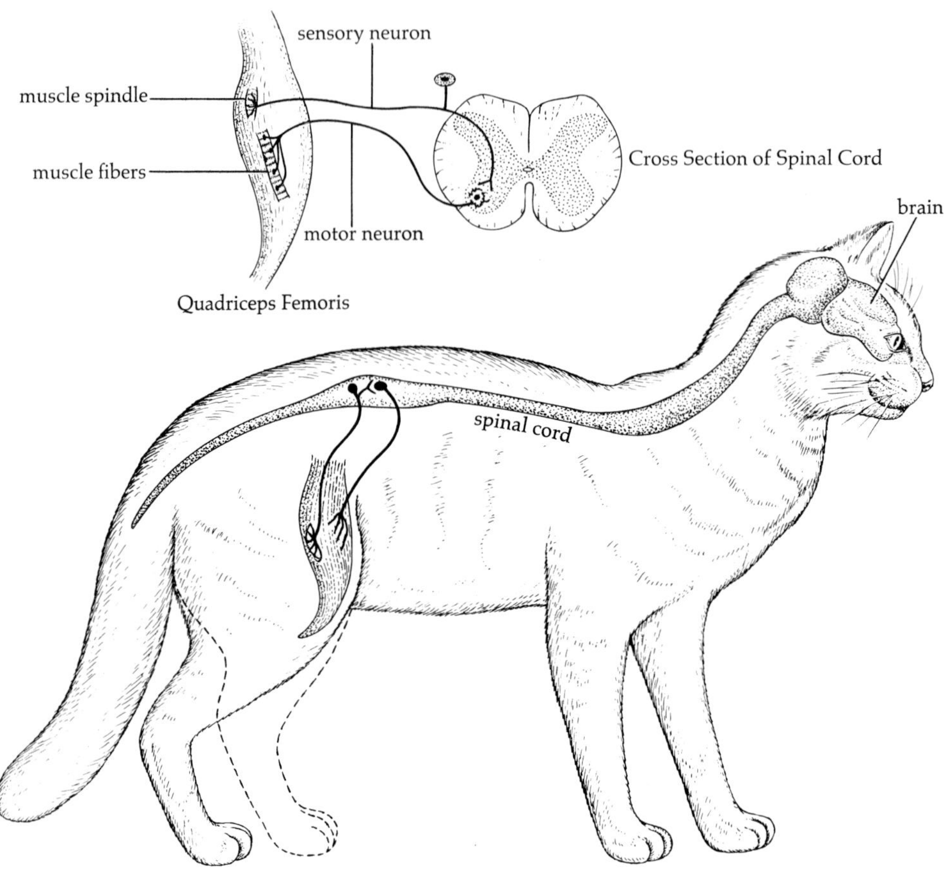

Fig. 9-7. A two-neuron reflex arc. Stretching of the quadriceps femoris stimulates sensory fibers coming from the muscle spindle sense organs. These sensory fibers synapse with spinal cord motor cells whose fibers innervate the quadriceps femoris muscle. The resulting contraction extends the shank of the leg, which also counteracts the stretching of the muscle. Note that the CNS is involved only in a local portion of the spinal cord; the rest of the cord and the brain are not affected by this reflex.

motor neuron from the CNS to the contracting muscle. All two-neuron reflexes are "stretch reflexes," so far as is known. The most familiar is the patellar or "knee-jerk" reflex. When the tendon just below the kneecap is tapped, the muscle attached to that tendon is stretched. A proprioceptive sense organ within the muscle, the muscle spindle, is excited by this stretching; it generates action potentials that pass up the sensory fibers to the cell body, located in a spinal ganglion just outside the spinal cord. From the cell body the nerve impulse travels through the neuron's axon to motor cells in the ventral part of the spinal cord gray matter. Neural transmitters, released from the terminals of this sensory axon, cause excitation of the motor neurons in the spinal cord. Action potentials then travel from the motor neurons down their axons to the muscle, causing the release of acetylcholine at the neuromuscular junction. This excites the skeletal muscle cells enough for them to contract, and when the muscle contracts the leg jerks forward (Fig. 9-7).

Most reflexes are more complicated, involving many more neurons and sometimes also a graded response that is related to the strength of the stimulus. In the relatively simple flexion reflex, an appendage is withdrawn when a noxious stimulus, such as a pinch, is applied to its distal end. If a pinch on a person's finger is too weak to cause this reflex, the stimulus is said to be below threshold. If the stimulus is somewhat stronger the fingers or hand will be pulled away. If the stimulus is very strong the whole arm and shoulder are pulled away; with an even stronger stimulus, the whole body may leap away. This increased response is called irradiation of the reflex.

In this more complicated reflex, sensory neurons pass from the fingertip up to the spinal ganglia; their axons pass into the cord and synapse with association neurons, which in turn synapse with many motor neurons, some in different levels of the cord. Since most of these motor neurons innervate muscles controlling the specific finger being stimulated, a stimulus barely above threshold causes excitatory postsynaptic potentials just adequate to generate action potentials in the motor nerves of that finger. If the stimulus is stronger, however, the rate of action potentials in the sensory nerves is faster and more and more motor neurons are recruited by way of the association neurons; more muscles are then stimulated and the gross response is stronger.

Thus the flexion reflex, although more complicated than the stretch reflex, is, nevertheless, a strictly spinal reflex. Information about what is happening may be passed on to the brain, but the entire reflex will occur even if the connection between brain and cord is severed.

Many reflexes, however, involve cranial centers as well; an example is the placing reflex. If a cat is blindfolded and an object brought gently against the dorsal side of one foot, the cat will raise the foot and place it precisely on top of the object. If a specific area of the cerebral cortex is destroyed this placing reflex is abolished, but the flexion and stretch reflexes are not. Thus the placing reflex is shown to be not merely a cranial (rather than spinal) reflex but to involve the highest cranial centers (Fig. 9-8). Knowledge of which central nervous structures are involved in particular reflexes is a valuable diagnostic tool in delineating damage to the CNS.

These are but a few examples of the many fixed motor patterns which respond to specific sensory stimuli. They provide shortcuts through the network of information reception and decision making to ensure quick and proper responses to certain important types of stimuli regardless of the physiological state of the animal.

Despite the value of reflexes, however, the behavior of vertebrates is more importantly characterized by plasticity, that is, by the ability to modify behavior rather than always responding in the same way to the same stimulus. The degree of this plasticity varies with the structural complexity of the nervous system. Nevertheless, it is reasonable

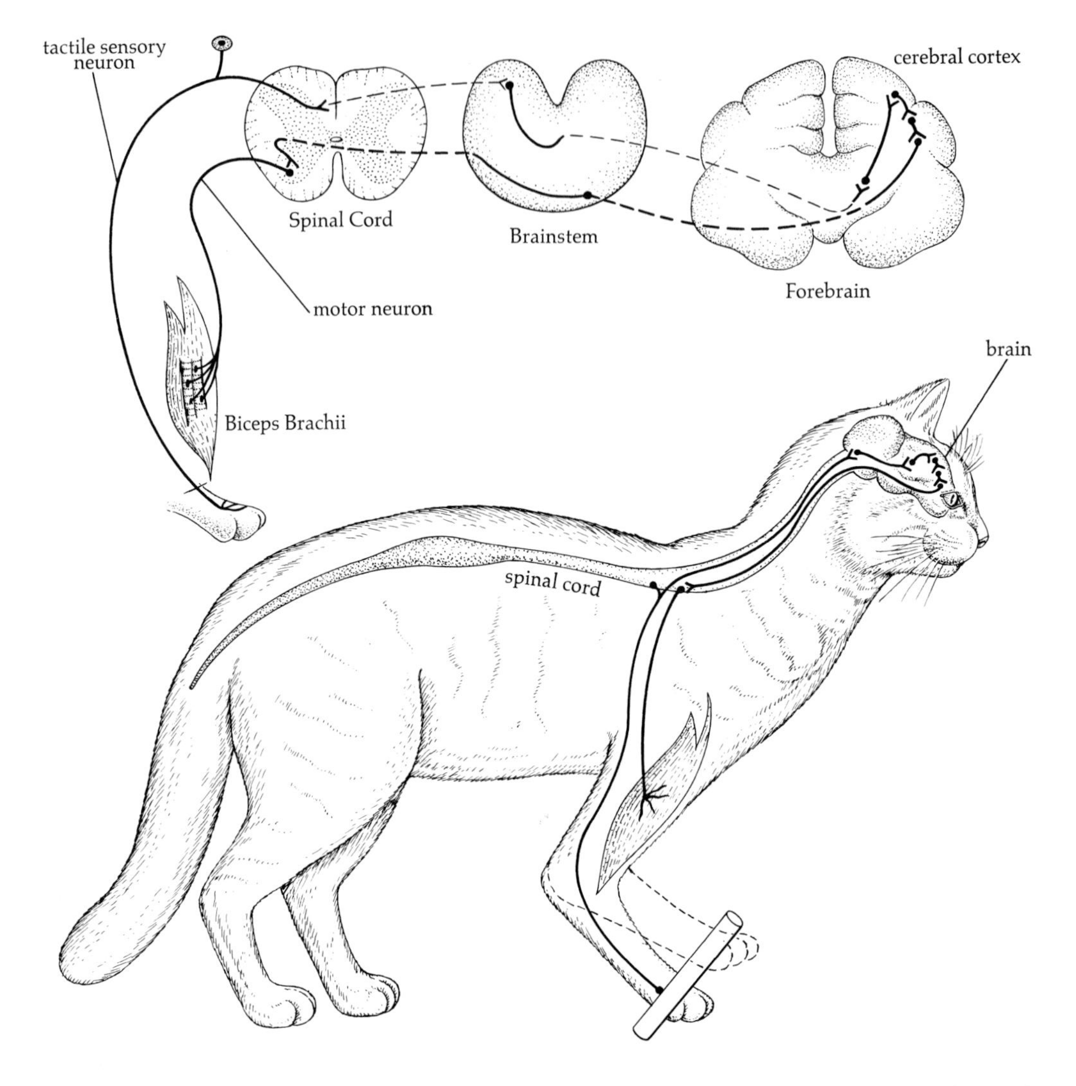

Fig. 9-8. A multisynaptic reflex, the placing reflex. When the dorsum of the paw receives tactile stimulation, the paw is raised and precisely placed on the stimulating object. Here the reflex involves the spinal cord and brain, including the cerebral cortex. (The biceps brachii is only one of several muscles involved in this reflex.)

to ask just how fixed are these motor patterns and whether they can be changed.

Various experiments have studied this question. Consider again the example at the beginning of this chapter: a frog, seeing a passing fly, darts out its tongue and catches the fly in midflight. This requires very detailed, precise perception and processing of sensory information and a motor output of equal precision. If a frog's eyes are surgically rotated 180 degrees, so that each eye is upside down but otherwise normal, its tongue will dart out in a direction that is opposite to the fly's position; the frog will never learn to compensate for this handicap.

Similar experiments have been done with mammals, either surgically or with special lenses which reverse the image. A difficult period of adjustment ensues, during which the subject sees upside down. Unlike the amphibian, however, the mammal will adapt to this situation, compensating by his behavior for the differences between his perception and reality. In effect, he undergoes a central adaptation and learns to see things as if they were reversed. So great is this adaptation that when the special glasses are removed or the surgery reversed, the subject must make another adjustment to bring perception into line with reality.

However, mammals also have some relatively inflexible motor patterns. If the flexor and extensor muscles of the laboratory rat's hindlimb are cut and their tendons interchanged, so that the extensor muscles flex the limb and the flexor muscles extend it, the rat will respond to a noxious stimulus by extending its leg instead of flexing it. Furthermore, this reflexive extension will continue even if the foot becomes badly damaged. The same thing happens if the muscles are undisturbed but their nerves are switched, so that the extensor nerves go to the flexor muscles and vice versa. If both the nerves and the muscle insertions are switched, there is a "double negative" and the rat responds normally. This is a fixed motor reflex which in the laboratory rat is inflexible, even if it injures the animal.

If similar surgery is done on a monkey, its first reaction, like that of the rat, is to extend rather than flex the leg in response to a noxious stimulus. However, after a considerable "learning period," the monkey will learn to flex the leg after first extending it, evidently exerting cerebral control over the reflex but never learning to bypass the reflex-controlled extension.

It therefore becomes apparent that certain reflexes are genetically fixed and can be modified only slightly, if at all, by learning. It is certain, however, that despite the large number of vertebrate reflexes and their adaptive value, there is an even greater amount of vertebrate behavior which is not fixed but learned and a great deal of central nervous plasticity which allows behavior to be changed in the CNS.

SUGGESTED READING

Heimer, L. (1971). Pathways in the brain. *Sci. Amer.* **255** [July], 48–60.
Ochs, S. (1965). "Elements of Neurophysiology." Wiley, New York.
Sperry, R. W. (1968). Plasticity of neural maturation. *Devel. Biol.* Suppl. **2,** 306–327.
Van der Kloot, W. G. (1968). "Behavior." Holt, New York.

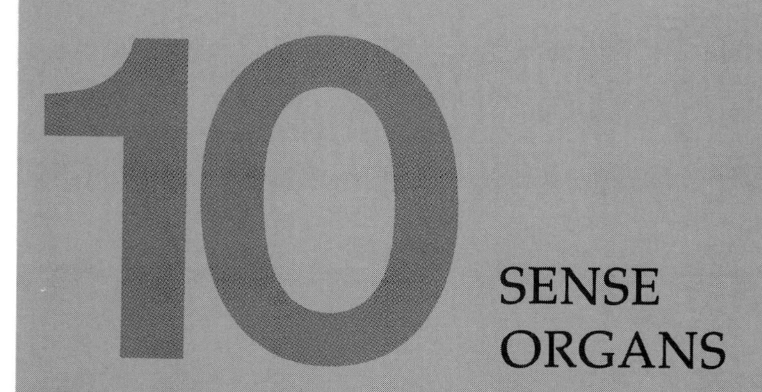

10 SENSE ORGANS

Vertebrate Hearing

Events both without and within an animal are received by sense organs and transduced into nerve impulses. Thus an animal receives information about both its external and internal environments. The cliche gives us five senses: sight, hearing, taste, smell, and touch; there are many others equally important, such as balance, cold, heat, pain, hunger, thirst, and so on. In nonhumans there are still others: for instance, some fishes have electroreceptors capable of detecting very fine electrical fields. Each sense is distinct, distinguishable from other senses, and related to a specific sense organ.

Since each sense organ is adapted for specific information, they are of diverse structure. Yet all sense organs share some basic characteristics and can be understood as elaborations of a basic pattern.

FUNDAMENTAL CHARACTERISTICS OF SENSE ORGANS

It is advantageous for the neurons of sense organs to be extremely sensitive so that they can respond to stimuli of very low intensities. This is accomplished by bare nerve endings, that is, one or more dendrites in each sense organ that have no surrounding protective myelin sheath or Schwann cells. In fact, the simplest sense organs are bare nerve endings intertwined with epithelial or connective tissue cells in such a way that mechanical deformation of the non-nerve cells is transferred to the bare nerve endings, which then initiate nerve impulses.

Such bare nerve endings would be even more sensitive if they were located on the body surface. However, they would require protection from damage or dessication. They would also require a means for achieving or enhancing specificity, since if they were equally accessible to all types of stimuli they would respond equally to all types, and their messages to the CNS would simply say "something has happened," rather than telling what it was.

The requirements for protection and for specific information have selected for the evolution of both simple and complex superstructures. These superstructures surround the bare nerve endings, protecting them from damage and from all stimuli except the one modality to which the sense organ is adapted. In most sense organs the superstructure also facilitates a very low threshold for the specific modality (Fig. 10-1).

In spite of their superstructures, however, sense organs will respond to a stimulus of any modality if the stimulation is sufficiently intense. A blow on the head mechanically stimulates the eye's receptor cells, initiating nerve impulses that the brain registers as "light."

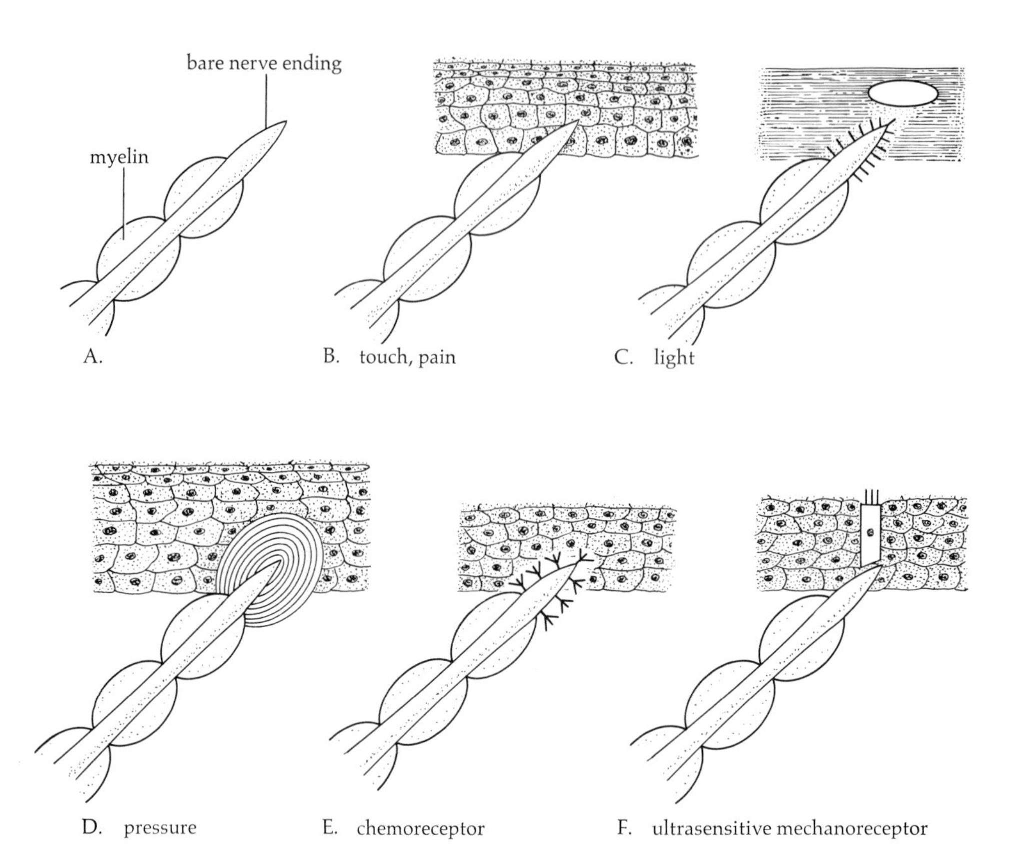

Fig. 10-1. Any bare nerve ending (A) is a sense organ in that a mechanical, chemical, or light stimulus of sufficient intensity will cause it to generate action potentials. In B epithelial cells surround the ending and make it a touch receptor (if the ending is deep) or a pain receptor (if near the surface). In C there are light-sensitive molecules on (sometimes in) the ending and a lens; these make it a light receptor. The concentric membranes surrounding the ending in D collect force from a large area and focus it onto the ending, making it a pressure receptor. In E the ending is modified by a moist permeable epithelium and chemoreceptor membrane molecules, and forms smell and taste receptors. Extreme mechanical sensitivity is produced by the modification in F, where another cell with sensitive processes is interposed between the ending and the surface; this hair cell is found in the acousticolateralis system.

An intense sound or electrical stimulation, if they could reach the photoreceptors, would have the same effect. Normally, however, the eyeball's complex structure protects the bare nerve endings (photoreceptor cells) from all stimuli except a particular modality (light) and, at the same time, helps lower the threshold for that modality by making it more accessible to nerve endings (in this case, by focusing light onto the photoreceptor cells).

CUTANEOUS RECEPTORS

Some types of sensation are easier to detect than others, however, and there are some sense organs which have no superstructure or only simple ones. In these sense organs, the bare nerve endings are free or encapsulated.

FREE NERVE ENDINGS

Some sensory nerve fibers split up into many terminations of bare nerve endings, each contacting an epithelial cell or ending in the interstitial space (Fig. 10-2). The morphology of these terminations varies; usually they are bulb-shaped, flattened, or complexly branched (flower-spray) endings. These free nerve endings are found in the deeper layers of the epidermis, in the underlying dermis, in the walls of the alimentary canal and other viscera, and in the lining of body cavities. All have a relatively high threshold. The simplest have the highest thresholds; these are pain fibers which send to the CNS nonspecific information of general pain at a given point.

Some free nerve endings, particularly in the dermis, are associated with integumentary derivatives which greatly lower their thresholds: the large stiff vibrissae ("whiskers") and, to a lesser extent, hairs and feathers are examples. The free nerve endings around the lower portions of these structures are stimulated when the base of the hair or

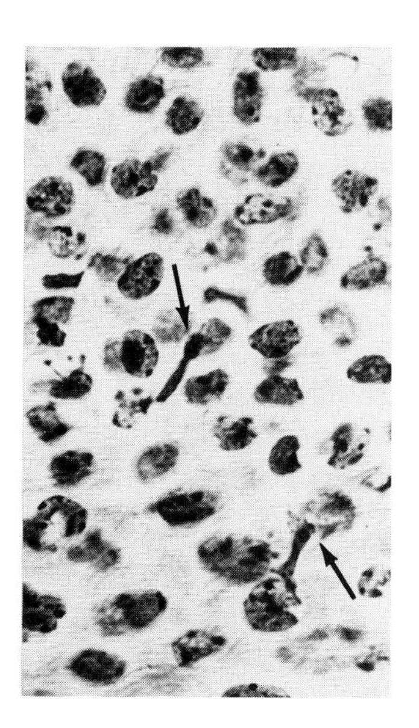

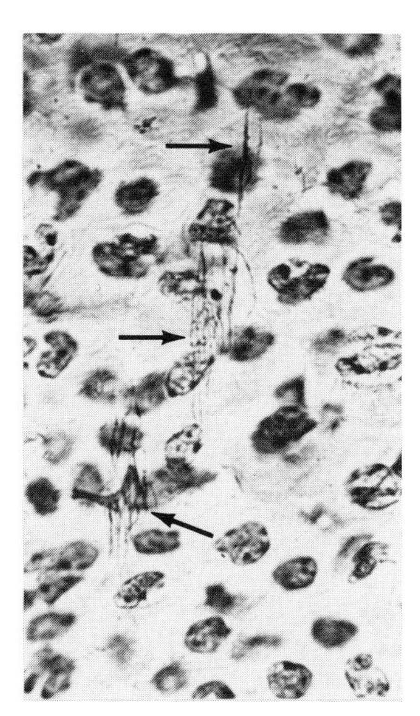

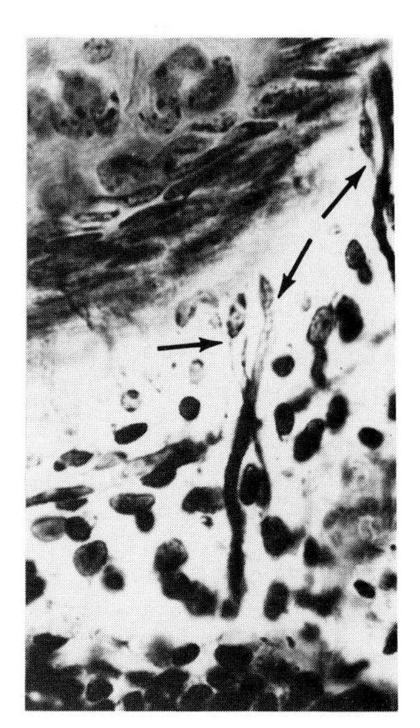

Fig. 10-2. Varied types of silver-stained free nerve endings in the outer epidermal cells of a vibrissa follicle. Endings are indicated by arrows. On the left are large blunt endings; in the center are flower-spray endings; and on the right are fine terminals on specialized cells called Merkle cells.

feather moves. The hair or feather can be compared to a lever: the "long lever arm" sticks out of the skin, the "fulcrum" is its connection with the skin, and the "short lever arm," where the force is greatest, occurs at the site of transduction where free nerve endings encircle the hair or feather base. This amplification of stimulus force in effect increases the sensitivity of the free nerve endings: a small force outside causes a large force at the nerves. A fine, light, specific sense of touch results.

ENCAPSULATED NERVE ENDINGS

In amniotes, many of the bare nerve endings in dermis, alimentary canal, and body cavity linings are not free but are encapsulated by connective tissue. Thus they are protected from many types of stimulation and at the same time made specifically sensitive to others (Fig. 10-3).

Meissner's corpuscle, usually lying in the most superficial dermis, is composed of a thin connective tissue capsule enclosing epithelial cells on each of which rests a flattened, bare nerve ending. These corpuscles are particularly sensitive to touch, but are protected from most changes of temperature. On the other hand, the end bulbs of Ruffini—flattened, bulblike endings within a fine, connective tissue network—respond to temperature increases. The end bulbs of Krauss are similar but are more spherical bulbs of connective tissue with knoblike neural endings; these bulbs respond to decreases in temperature.

The most studied encapsulated nerve ending organ is the pressure transducer called the Pacinian corpuscle. The central core of this sense organ is a single, slightly branched dendrite which terminates in a slight bulb. Myelin surrounds the proximal portion, where the dendrite leads toward the CNS. Tough connective tissue surrounds the bare nerve ending and the distal myelin sheath in onionlike layers (lamellae), which enlarge the Pacinian corpuscle until it is often up to 4 mm long. Surrounded by these lamellae, the nerve ending is protected from fine touch, heat and cold, and from pain unless the lamellae themselves are cut. Pressure, however, received by the outer layer and passed down to deeper but smaller spheres, is magnified and concentrated. When it has reached the core, even slight pressure has been amplified considerably. This stimulates the bare nerve ending which depolarizes and initiates a nerve impulse starting, in this case, at the first node of Ranvier.

PROPRIOCEPTORS

Proprioceptors are sense organs in muscles, tendons, ligaments, and joints that give information concerning position, movement, and tension. The most studied proprioceptor is the muscle spindle, or stretch receptor, several of which lie within each tetrapod skeletal muscle. Each muscle spindle is made up of a fusiform connective tissue sheath containing three to twelve special, small-diameter in-

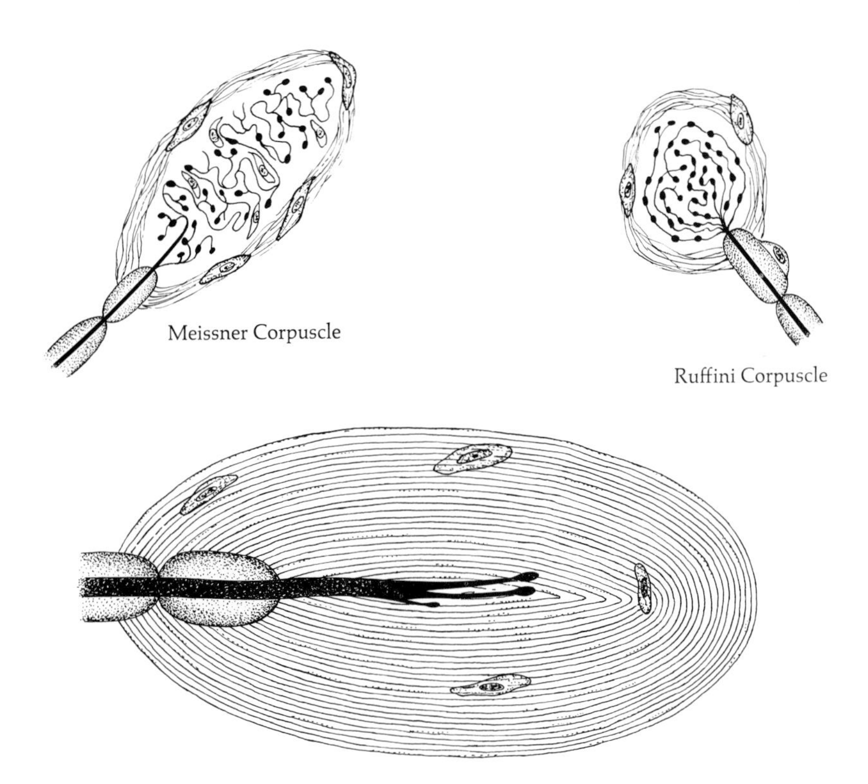

Fig. 10-3. Three types of encapsulated nerve endings.

trafusal muscle fibers. Each intrafusal fiber is striated only at its two ends; its central portion contains nuclei but no contractile elements. Each end of the intrafusal fiber attaches to the connective tissue sheath. A large, bare nerve ending, the annulospiral ending, wraps around the central, nonstriated, nuclear bag portion of the intrafusal fiber. Smaller sensory nerve endings (flower-spray endings) terminate around the connective tissue and the distal striated portion of the intrafusal fiber. In addition, small motor fibers called gamma motor fibers terminate at regular neuromuscular junctions on the striated portions of the intrafusal fiber (Fig. 10-4).

Stretching of the muscle spindle initiates firing in both the annulospiral ending and the flower-spray endings. The annulospiral endings have a larger diameter and transmit impulses more rapidly, while the flower-spray endings propagate slower nerve impulses.

Nerve impulses from the gamma motor fibers (either reflexive or initiated in the CNS) cause the intrafusal fibers to contract, which in turn stretches the nuclear bag and thus excites the annulospiral fibers but inhibits the flower-spray fibers which end on the striated portions of the intrafusal fibers. Contraction of the skeletal muscle fibers around the spindle constricts and thus inhibits both the annulospiral and the flower-spray endings. This inhibition is partially counteracted when the intrafusal fibers contract; if they contract enough they can still stretch the nuclear bag and excite the annulospiral fibers.

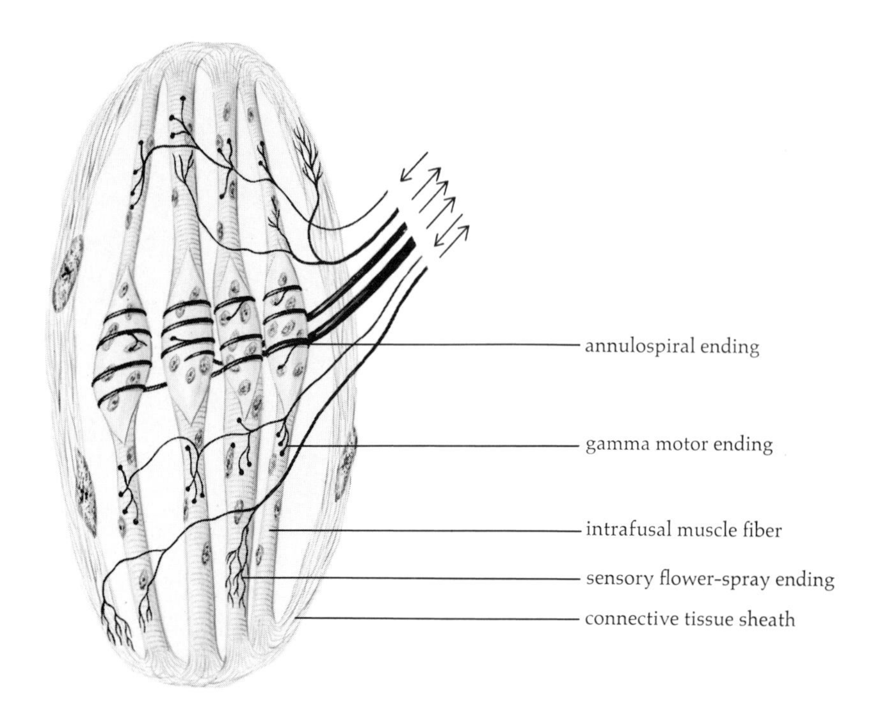

Fig. 10-4. Diagram of a muscle spindle. The entire spindle is surrounded by large skeletal muscle fibers, not shown. Arrows indicate the direction the nerve impulses travel in different fibers.

The muscle spindle is the sense organ of the stretch reflex that is important in maintaining the proper tonus of the skeletal musculature. It is a "dynamic" sense organ, since its sensitivity and firing can be controlled by the CNS through the gamma motor fibers.

There are also sense organs in tendons, called tendon organs, which are made up of flower-spray free nerve endings with club terminations on the heavy collagen fibers of the tendons. These flower-spray endings respond to increased tension on the tendon. Both the tendon organ and the muscle spindle fire in response to stretch; the tendon organ also fires in response to muscle contraction. Neither gives information about position or movement.

Several kinds of sensory endings, including tendon-type flower-spray endings, Pacinian corpuscles, and Ruffini-type endings, are found in ligaments and joints. These proprioceptive sense organs send information to the CNS regarding the positions of skeletal elements at all joints.

CHEMORECEPTORS

For chemoreception the stimulating chemical must dissolve on a moist membrane containing cells which are sensitive to it. The stimulating chemical probably forms a weak bond with the receptor cell membrane, causing a temporary depolarization of the sensory cell. In vertebrates both olfactory and taste organs are chemoreceptors.

OLFACTORY ORGANS

All vertebrates have olfactory areas in the brain and olfactory sense organs peripherally, either adjacent to the external nostrils (e.g., fishes), or deep to them (e.g., mammals).

In fishes, except lungfish, the olfactory epithelium lies in blind nasal sacs into and out of which water is pumped. In living lungfish and all tetrapods, external nares open into a nasal cavity, and internal nares lead from the nasal cavity into the mouth and pharynx. The nasal cavity is narrow in amphibians, but greatly enlarged in amniotes, and, except in turtles, contains complex folds called conchae, usually with skeletal supports (e.g., mammalian turbinate bones; Fig. 10-5). The conchae increase the surface area of the nasal cavity and its vascularized epithelium. Although no gas exchange takes place here, the increased size and surface area of the nasal cavity no doubt evolved for its adaptive value to respiration rather than to olfaction as explained below.

There are three types of epithelium in the nasal cavity. Most of the epithelium is respiratory—ciliated columnar epithelium containing

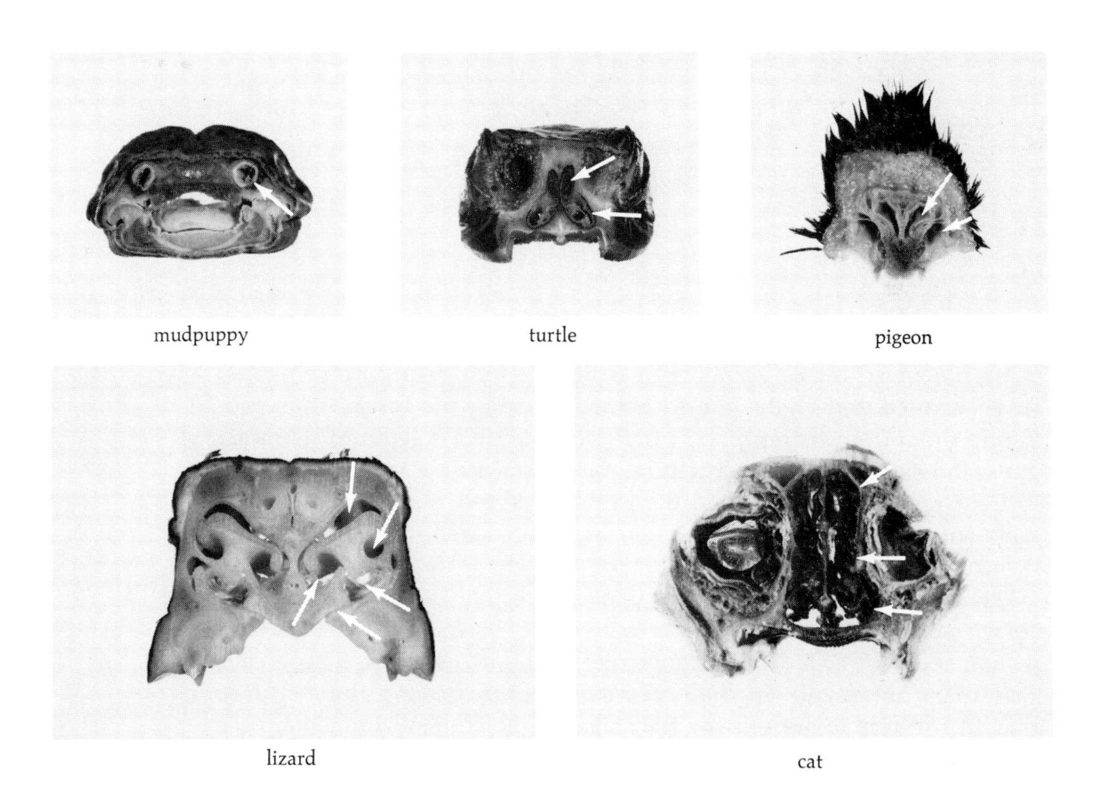

Fig. 10-5. Cross sections through the anterior part of the head region of diverse tetrapods, showing the complexity of the nasal chambers (indicated by arrows). Note that the mudpuppy and turtle lack conchae. In the lizard, the duct from the vomeronasal organ to the mouth is indicated by a double arrow.

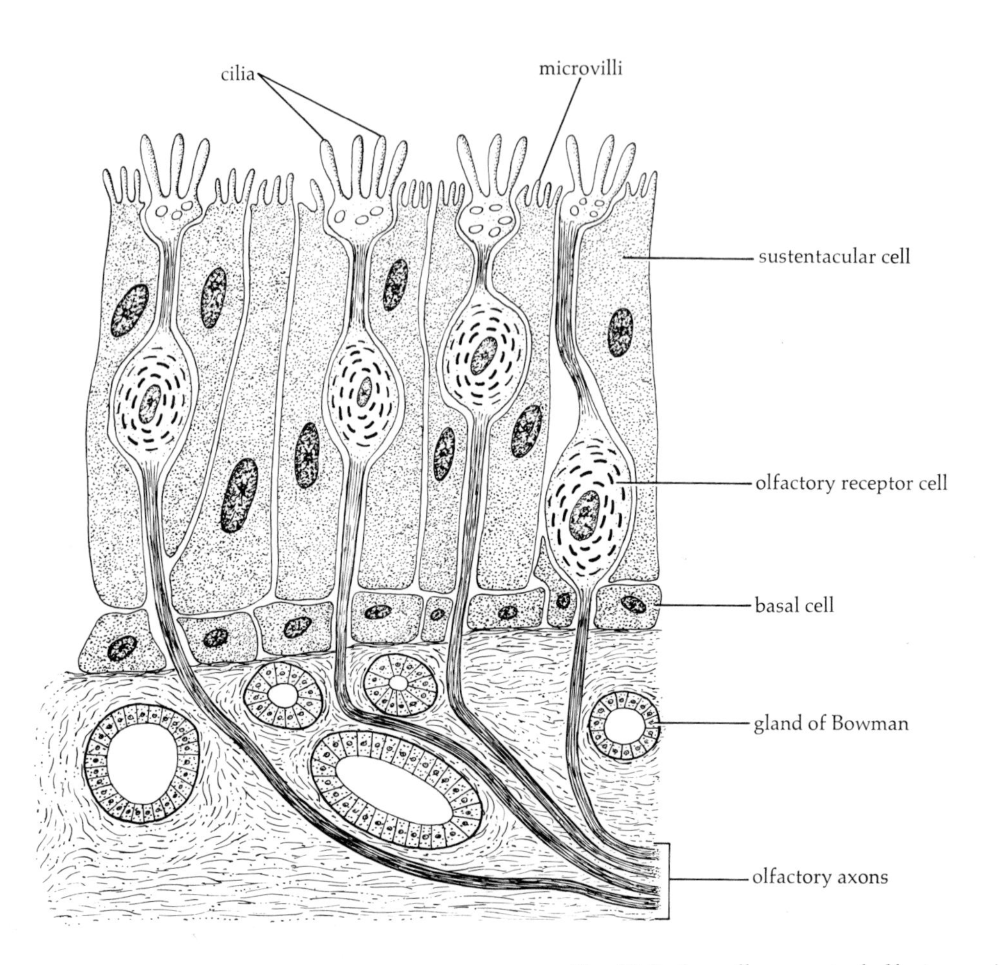

Fig. 10-6. A small segment of olfactory epithelium.

many mucus-secreting goblet cells which benefit respiration. Inhaled air is warmed and made humid before reaching the lungs, thus protecting them from irritation. Air being exhaled is cooled and its moisture condensed on the epithelium, thus conserving heat and water. Particulate objects in inhaled air can be filtered out and prevented from entering the lungs and later expelled, often violently by the sneeze reflex.

The second type of epithelium is olfactory (Fig. 10-6); it lies in the dorsal, posterior portion of the nasal cavity. Its most numerous cells are not receptor cells but sustentacular (supporting) cells. These cells have many microvilli on their free surfaces, increasing their surface area. In addition to supporting the sensory cells, the sustentacular cells secrete a mucus which protects and moistens the entire olfactory epithelium. Along with occasional, small basal cells, these sustentacular cells rest on a thin basement membrane below which is loose connective tissue. Deep to the basement membrane are small, branched tubular glands of Bowman which also secrete mucus through small ducts onto the olfactory epithelium.

Among the sustentacular cells lie the receptor cells which receive and transduce olfactory stimuli. Human beings can distinguish an estimated 10,000 or more olfactory nuances; other vertebrates can presumably distinguish even more, since their peripheral and central olfactory areas are more prominent. Yet there is but one type of receptor cell—a bipolar neuron with an axon to the olfactory bulb of the brain and short dendrites that reach the epithelial surface where each forms an expansion with several nonmotile cilia (Fig. 10-6). It is probably on these cilia that the receptor sites actually lie, and it may be that their specificity is at the molecular level.

THE VOMERONASAL ORGAN OF JACOBSON

The third type of epithelium in the nasal cavity is that of Jacobson's organ. This apparently olfactory sense organ is present in most amphibians, reptiles, and mammals; it is formed and then resorbed during development in birds and crocodilians. It lies just dorsal to, and in reptiles communicates with, the oral cavity; its nerve runs adjacent to (but separate from) the olfactory nerve. The epithelium of the organ of Jacobson is similar to olfactory epithelium, except that the bulblike endings of the dendrites of the bipolar neurons lack cilia, having instead microvilli at the epithelial surface.

The specific function of the vomeronasal organ of Jacobson is still in doubt. Experimental and morphological evidence indicate that in snakes and lizards, at least, it facilitates the smelling of food which is in the mouth.

TASTE ORGANS

Vertebrate taste organs, called taste buds, lie within surface epithelium as ovoid clusters of two types of fusiform cells: supporting cells and gustatory (taste) receptor cells (Fig. 10-7). From the receptor cell at the surface of the epithelium a prominent apical process protrudes into a tiny hole, the taste pore. This apical process bears microvilli which greatly increase its surface area and probably provide the receptor sites for the chemicals dissolved on the epithelium.

Whereas in olfactory epithelium the receptor cell itself sends an axon to the brain, this is not the case for gustatory cells. Instead, fibers of taste in the facial, glossopharyngeal, and vagus nerves make synaptic contact with the gustatory receptor cells. Their sensory dendritic endings with terminal boutons surround the gustatory receptor cells in a complex wrapping. The receptor cell, stimulated by a chemical, excites the nerve cell endings which carry the information to the brain. In this respect it is more sophisticated than olfactory organs.

The distribution of taste buds on the body is variable. In fishes they are usually on the epithelium of the oral cavity and pharynx, including the gill apparatus. Some scaleless teleosts, such as catfish, have taste buds all over the body. Experimental evidence indicates that these chemical sense organs are used in detecting and locating food. In tetrapods, taste buds are located on the tongue and often in the pharynx; there are none on the general body surface.

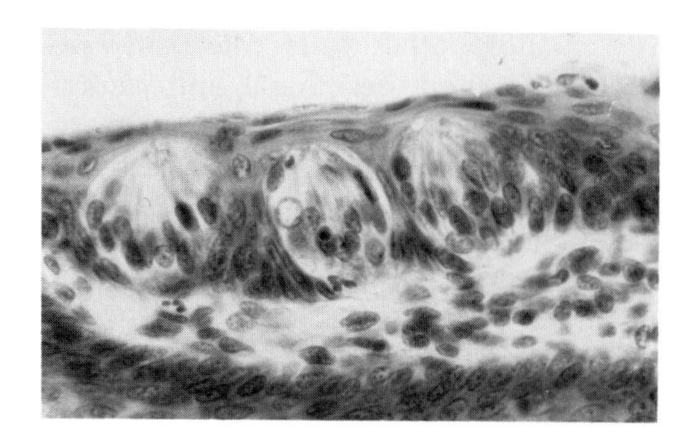

Fig. 10-7. A row of three taste buds on a foliate papilla of a rabbit's tongue.

As common experience indicates, olfaction and taste strongly interact. Often when one smells something one can also taste it; but in humans, since there is no functional connection between the vomeronasal organ of Jacobson and the mouth, this phenomenon does not involve that organ. On the other hand, the sense of taste is altered by experiments which temporarily obliterate olfaction without affecting taste per se, or, as we all know, when the olfactory mucosa is affected by a head cold.

PHOTORECEPTORS

There are two types of photoreceptors in vertebrates: lateral eyes, which apparently were present even in the earliest fossils, and median eyes, which are possibly even older.

LATERAL EYES

Rods and cones

The receptor cells of lateral eyes contain photosensitive pigments made up of a protein (one of several opsins) in combination with a pigment which is altered by the presence of light. In rods, this is a carotenoid derived from vitamin A (11-*cis*-retinaldehyde). Light striking this chemical causes it to break down into all-*trans*-retinaldehyde plus opsin, and this photoreaction triggers a series of events which can lead to the generation of action potentials. In cones the chemistry is not understood.

The two types of light receptor cells, rods and cones, have a number of developmental, structural, and functional characteristics which allow them to be regarded (equally legitimately) either as modified nerve cells or as epithelial receptor cells. For our purposes we shall consider them modified nerve cells.

They differ in their morphology and specific photopigments (Fig. 10-8). The rods have a lower threshold of stimulation. Their visual pigment is called rhodopsin, or porphyropsin in freshwater vertebrates (including amphibian tadpoles but not amphibian adults). This pigment is sensitive to all wavelengths of visible light; therefore the rods sense "light" or "no light," but cannot distinguish colors. Each rod has an outer and an inner segment, designated by their distance from the center of the eyeball. The inner segment contains mitochondria, some Nissl substance, and two centrioles. The outer segment is composed of cylindrical stacks of membranes, or lamellae, which contain the photopigment. There is a short connecting piece between the outer and inner segments that is, structurally, the proximal portion of a cilium. It appears, in fact, that the entire outer segment is an elaborately modified distal portion of a cilium. Therefore, as in the olfactory and gustatory receptors, cilia contain the "receptor sites."

Cones generally are smaller and have fewer stacks of lamellae, which are arranged as cones rather than as a cylinder; there are similar inner segments and connecting pieces. Each cone contains one of three different visual pigments; these pigments are sensitive to different portions of the visual spectrum, and therefore the cones distinguish colors. Although the absorption spectra of the pigments are known, their exact chemical structures are not.

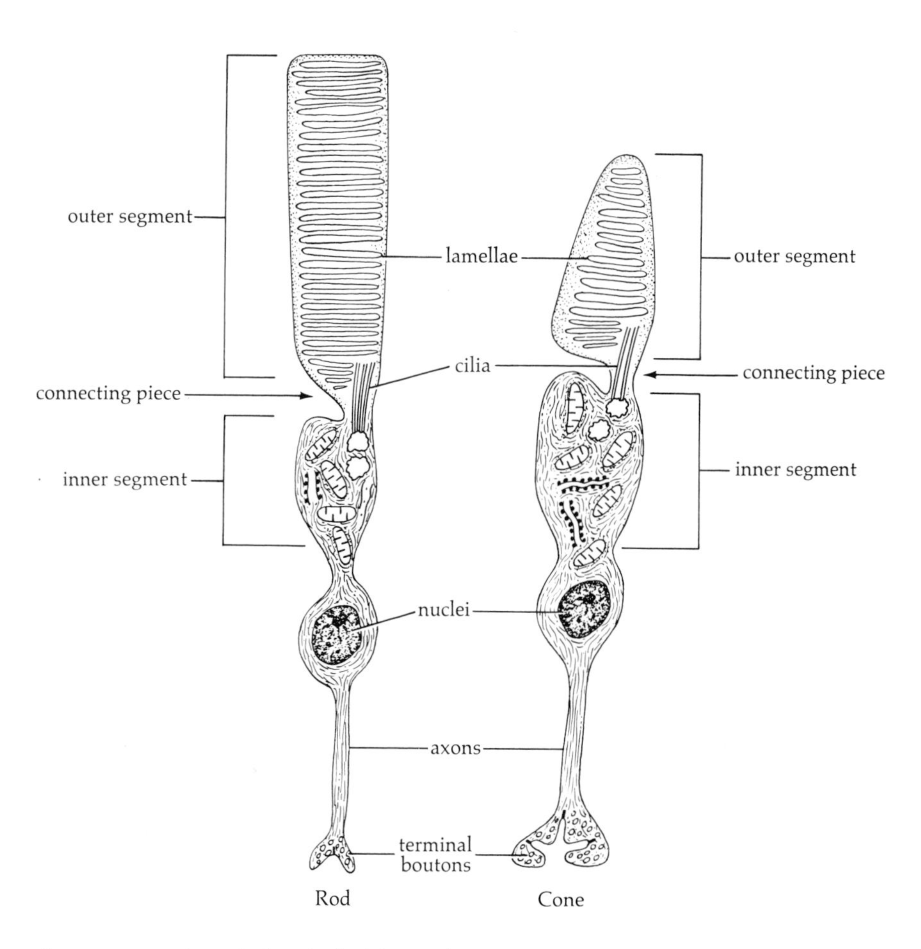

Fig. 10-8. Diagramatic representation of the similarities and differences in rod and cone morphology.

Because cones have a much higher threshold of stimulation than rods, colors can be seen only in bright light; in dim light only the rods function and vision is restricted to black and white. Many vertebrates, including humans, have both rods and cones, the rods giving greater sensitivity (because of their lower threshold) and the cones giving color vision. However most sharks and nonprimate mammals have retinas that contain only rods, giving these animals their night vision. A few strictly diurnal mammals, such as ground squirrels, have retinas that contain only cones, giving these animals improved color perception in good light, but presumably making them functionally blind at low light levels.

Superstructure

If the rods and cones are themselves impressive, the eye as a whole is indeed an elegant structure, controlling the amount of light and how it is focused onto the rods and cones and protecting the retina from other stimuli.

Even the eye's embryology is remarkable. During the brain's early development, two outpocketings, the optic vesicles, evaginate from its lateral anterior portion and extend toward the surface ectoderm. The part of the vesicle nearest the brain narrows, forming a thin optic stalk between the brain and the optic vesicle; it is this optic stalk which becomes the adult optic nerve. The external hemisphere of the vesicle then invaginates toward the inner hemisphere, forming the optic cup. The layer of the cup nearest the brain forms the pigmented retinal epithelium; the layer nearest the surface ectoderm forms the neural retina which includes the rods and cones as well as other neurons to be described. During the development of the optic cup the overlying surface ectoderm thickens, forming a lens placode. This placode then invaginates; the invaginated portion differentiates into the lens and the remaining surface ectoderm differentiates into the corneal epithelium. Mesenchyme condenses around the optic cup and the lens, forming the eyeball's outer tough connective tissue layer (the sclera) and its vascular coating (the choroid). Thus several areas contribute to the eyeball (Fig. 10-9): the brain (retina), the surface ectoderm (lens and part of the cornea), and the mesoderm (outer layers).

The sclera

This outer part of the eyeball is a tough capsule of collagenous connective tissue that protects the eyeball from external damage and stabilizes its considerable internal pressure. Cartilage or bone frequently develops within the sclera; in most reptiles and birds there are scleral ossicles in a ring around the pupil and iris.

The cornea

Anteriorly the sclera gives way to the transparent cornea, formed from surface ectoderm and a deeper layer of mesenchymal cells. In order for the cornea to be transparent it must be avascular (lacking vasculariza-

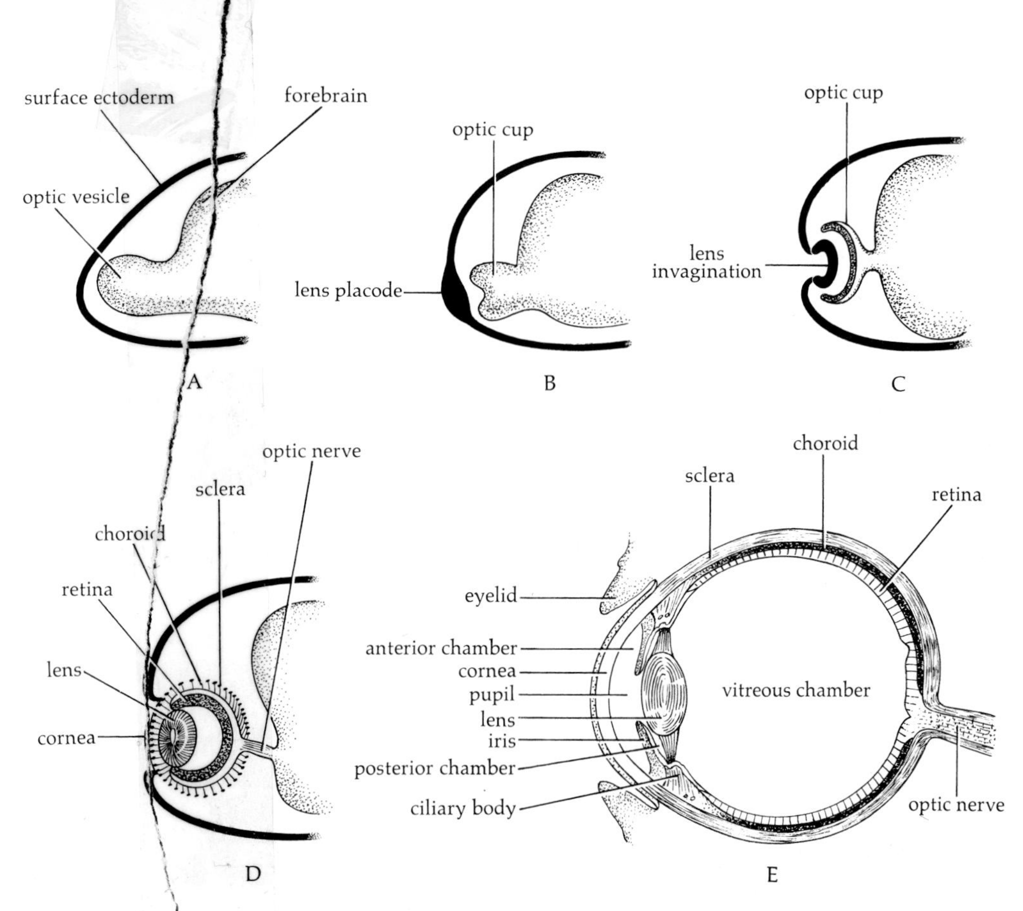

Fig. 10-9. Development (A,B,C,D) and adult structure (E) of the eye, as seen in sections through its axis.

tion); therefore, it is nourished primarily by the lacrimal secretions (tears) in which it is bathed. In cyclostomes it is somewhat cloudy, and its two layers of ectoderm and mesoderm remain distinct. In many teleosts the cornea has a light yellow tint because of its carotenoid pigments.

The refractive index of the cornea matches that of water in fishes; in tetrapods the refractive index of the cornea differs from that of air. This is important in accommodation, to be discussed below.

The choroid

The richly vascularized and usually pigmented choroid layer lies just deep to the sclera, except deep to the cornea and lens where the choroid is incomplete. In many vertebrates the choroid contains a reflective element called the tapetum lucidum, which causes light passing through the retina to be reflected back into the eyeball. Most of the vertebrates with this structure are nocturnal—it is the "eyeshine" of cats, dogs, and many wild animals seen by flashlight or car headlights. Their vision is made more sensitive because the reflected light stimulates the rods and cones a second time, but sharpness of focus is sacrificed.

There are three types of tapeta. The most common is fibrous, composed of shiny collagenous, connective tissue fibers. This type is found in some teleosts, most marsupials, whales and dolphins, elephants, hoofed mammals, and many nocturnal monkeys. The second type, found in lemurs and carnivores, is cellular; each cell contains many reflective purine rods, about 1 micrometer (μm) in diameter and 3–4 μm long. The third type is composed of a purine, guanine, occurring in extracellular plates with a silvery luster; it is found in many teleosts and some chondrosteans and has its most complex structure in elasmobranchs, whose tapetum is occlusive. This is possible because over each guanine plate lies a melanophore. In dim light the pigment is retracted into the center of the melanophore, and the tapetum functions. In bright light the melanin disperses in the melanophore, covering the guanine plate and absorbing the light, and, therefore, there is no functional tapetum.

The iris

Surrounding the pupil is the pigmented iris. Its origin is complex: the outer portion develops from mesoderm of the choroid, the inner, from ectoderm of the edge of the optic cup. The iris of tetrapods contains smooth musculature which acts reflexly in response to light: radial fibers contract in dim light and dilate the pupil; circular fibers contract in bright light and constrict the pupil. In this way the retina is provided with proper light in a broad range of environmental conditions.

The pupil, or hole in the center of the iris, is spherical in most vertebrates, but slit-shaped in cats and many snakes and rectangular in many ungulates.

The lens

The transparent lens is formed by an invagination of surface ectoderm and is covered by an elastic capsule. Because the refractive index of the lens is different from that of the surrounding air or water, the lens plays a role in focusing the light rays on the retina; this role is greatest in fishes and least in tetrapods.

The ciliary body

The lens is suspended in the front of the eyeball by the ciliary body which is composed of fibrous connective tissue and usually some muscular tissue. Structure of the ciliary body is diverse, as described below.

Accommodation

Light rays come to the eye from near and far and then must be bent to focus on the retina. How much bending is necessary depends on the distance from object viewed to eye. Adjustments must be made in the eyeball in order to focus to different distances, and this is called accommodation. These adjustments may involve the lens, ciliary body, and cornea, and it is here that there is the greatest structural diversity in lateral eyes.

In fishes, three different methods of accommodation have independently evolved. Cyclostomes have no ciliary body; the lens is held in position by the humors of the eyeball, and a unique corneal muscle flattens the cornea and changes its diffractive index for accommodation. Elasmobranchs have a ciliary body, but it contains no muscles. The eye normally is focused for far vision; a small, unique protractor muscle pulls the lens forward to accommodate for near objects. In teleosts the opposite is true. The eye normally sees near objects, and a small, unique retractor muscle pulls the lens backward for distance vision. Amphibians, like elasmobranchs, normally have distance vision, but in amphibians it is the ciliary muscles which pull the lens forward for accommodation to near vision.

Amniotes also use ciliary muscles, but change the curvature of the lens rather than its position. Reptiles and birds have the largest and most muscular ciliary bodies, with striated muscles which run from the scleral–corneal junction to the lens; these muscles respond rapidly with strong contractions that pull the lens forward, thus squeezing it into a smaller area and making it thicker; this thickening provides accommodation for near focus. Snakes are an exception; the entire eyeball is compressed, and the lens is pushed forward by the vitreous humor, resulting in accommodation for near focus.

In mammals the ciliary muscles are small, smooth (rather than striated) muscles which are slower in responding. The lens is normally held tautly and distended by the connective tissüe portion of the ciliary body. When the ciliary body is pulled forward by the ciliary muscles, there is less tension on the lens, and it thickens by its own elasticity—a slower process of accommodation than that in birds and reptiles.

Fluid spaces of the eyeball

In front of the lens is a space filled with a watery fluid called the aqueous humor. This space is divided into the anterior chamber (between cornea and iris) and the posterior chamber (between iris and lens). A much larger space lies between lens and retina and contains a very viscous fluid, the vitreous humor.

Pigmented retinal epithelium

Between the choroid and the rods and cones is the thin layer of pigmented retinal epithelium. In animals without a tapetum lucidum, this pigmented retinal epithelium contains heavy deposits of melanin and absorbs light that has passed through the retina. Despite its name it is colorless in animals with a tapetum lucidum, although if the tapetum is composed of guanine, the pigmented retinal epithelium may also contain guanine crystals.

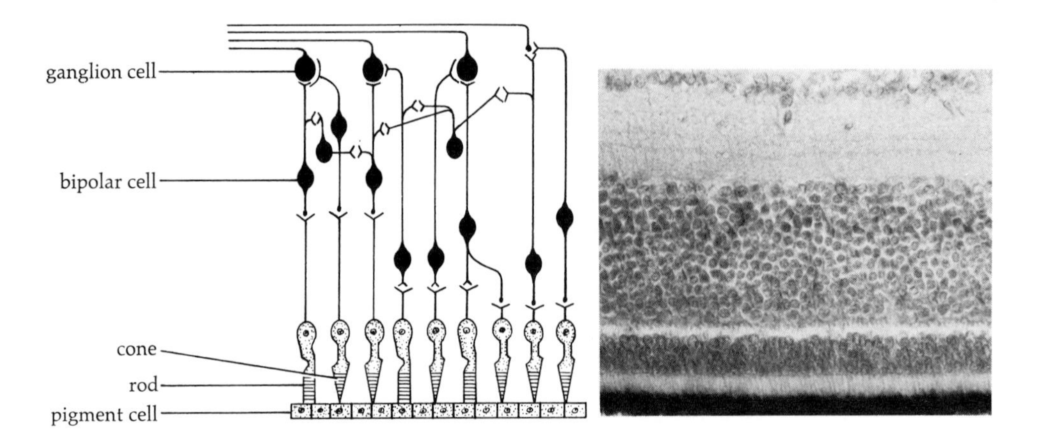

Fig. 10-10. The structure of the neural retina of a mammal, as photographed through the light microscope, is shown on the right; on the left the neural connections are diagramed.

Neural retina

Both the pigmented retinal epithelium and the neural retina form from the forebrain, but only the neural retina differentiates into nerve cells (Fig. 10-10). The neural retina is relatively thick and has three major layers of cells. Its outermost layer (furthest from the center of the eyeball) contains the rods and cones, whose outer segments abut against the pigmented epithelium. The next layer (toward the center) contains small ("midget") and large ("flat") bipolar cells with which the receptor cells make synaptic connections. Either a single cone or several rods will synapse on a single, small bipolar cell. Thus the cones, in addition to detecting colors, give more precise pattern discrimination and sharper vision. On the other hand, because rods synapse in groups on one bipolar cell they have a lower threshold but less visual acuity.

In addition to bipolar cells this middle layer contains two types of cells whose axons run parallel to the curvature of the eyeball. These are the horizontal cells and the small amacrine nerve cells that connect distal portions of the bipolar cells (Fig. 10-10).

Bipolar cells make synaptic connections directly on the soma of the ganglion cells which compose the innermost neuron layer of the retina. Although they are virtually without dendrites, the ganglion cells have long axons which course along the inner layer of the neural retina and then pierce completely through neural retina, pigmented retina, choroid, and sclera and pass thence as the optic nerve carrying the visual information to the brain. At the point where they pierce through the retina there are no rods and cones, but a blind spot. Yet such is the plasticity of our nervous system that humans at least normally "fill in" the missing information and are not aware of a discontinuity in vision.

At a different place on the retina there is a great concentration of photosensitive cells—in diurnal animals, mostly cones, in nocturnal animals, rods. This is called the area centralis, located by an imaginary line drawn directly through the center of the cornea and lens and extended to the retina. In the center of the area centralis is the area of most sensitive vision, a particularly compact region of photosensitive cells called the fovea. The bipolar and ganglion cell layers are very thin over the fovea, thus giving less interference to the incoming light. Because the light has to pass through the cornea, lens, aqueous and vitreous humors, and three layers of the neural retina before reaching the photoreceptor cells, this thinning helps to increase the fovea's sensitivity; for although all these layers are transparent they nevertheless absorb some of the light energy.

MEDIAN EYES

In addition to lateral eyes many living fishes, amphibia, and reptiles also have one or two eyes located dorsally on the midline, called the pineal eye and the parietal eye. They may be phylogenetically older than lateral eyes, for they have a less complex structure and there are openings for them (as well as for lateral eyes) in many fossil forms, including the oldest fossils (ostracoderms). Cyclostomes have two median eyes, a pineal eye developing from the dorsal part of the brain slightly on the right and a parietal eye developing from the dorsal part of the brain slightly on the left. Most other living vertebrates with median eyes have retained either the pineal or the parietal eye, but not both.

Median eyes develop in much the way that lateral eyes develop. A dorsal evagination from the brain comes to lie just under the surface ectoderm, which then develops a cornea. However in median eyes the optic vesicle retains its vesicular shape. The outer part forms the lens, and the inner part forms the retina (Fig. 10-11); therefore the receptor

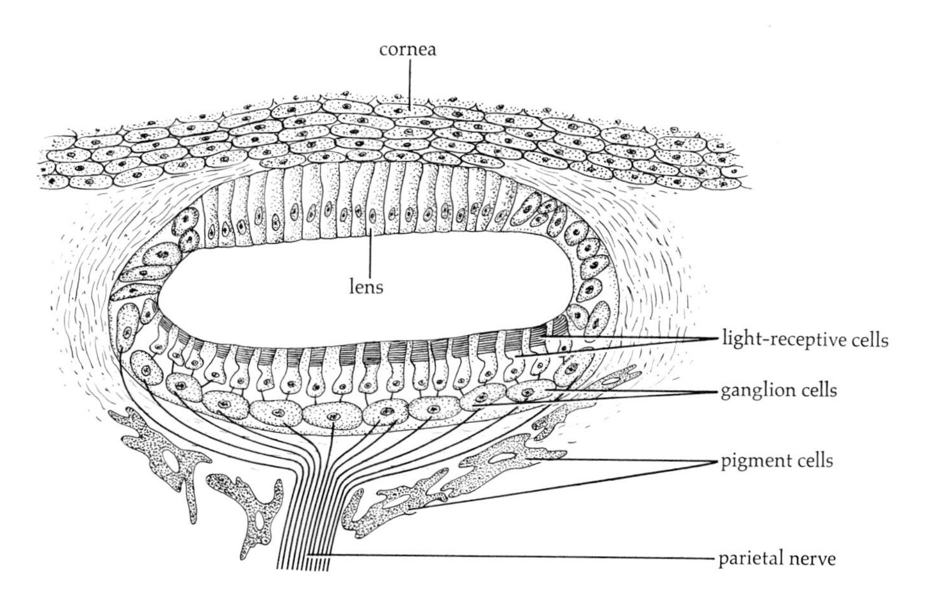

Fig. 10-11. Diagramatic cross section through the median eye of a lizard.

cells face toward the lens rather than away from it as in lateral eyes. In all vertebrates at least one such vesicle develops; in those without median eyes it never reaches the body surface but differentiates instead into the pineal gland.

The photoreceptive cells of median eyes contain photopigments that are similar to those of lateral eyes. Their ultrastructure includes rows of discs (lamellae) and cilia. The stalk connecting the median eye to the brain becomes its nerve and is called either the pineal or the parietal nerve.

The physiology of median eyes has been studied most extensively in lizards and amphibians. Although they are distinctly light sensitive, they form little or no image. Most recent evidence suggests that diurnal rhythm (the pattern of activity and inactivity over a 24-hour period) is triggered by the amount of light received by the median eye.

OPHIDIAN PIT ORGANS

The pit vipers (including rattlesnakes and their allies) and some of the boid snakes (pronounced bō-ĭd) (including pythons and their allies) have unique sense organs called pit organs which are sensitive to longer wavelengths than visible light and, therefore, are essentially radiant heat receptors.

The pit vipers have a single pit on each side of the head, between the nostril and the eye. The pit is covered by an extremely thin membrane filled with unmyelinated nerve endings containing mitochondria (Fig. 10-12). In the boids, there is a series of pits in the labial scales lining the mouth. The membrane floor of the pits is richly innervated with free nerve endings packed with mitochondria.

These pit organs are so sensitive that they can detect a radiant heat energy change of 0.003°C. Behavioral experiments suggest that they are the primary organs for the detection and localization of the small warm-blooded mammals on which these snakes usually prey.

The transduction mechanism of pit organs is not understood. It is known that heat stimulation alters the morphology of the mitochondria. Possibly the mitochondria themselves receive heat energy and transduce it into neural stimulation.

THE ACOUSTICOLATERALIS RECEPTORS

The vertebrate inner ear contains two sense organs: the organ of balance, called the vestibular apparatus, and the organ of hearing, called the auditory apparatus. A related group of sense organs called the lateral line is located in the integument of fishes and larval amphibians. As different as they may seem, these three sense organs of the acousticolateralis system are similar in some important respects. All have similar embryonic origin. All are mechanoreceptors, and in all the receptor cell is the hair cell, which is the body's most sensitive mechanoreceptor.

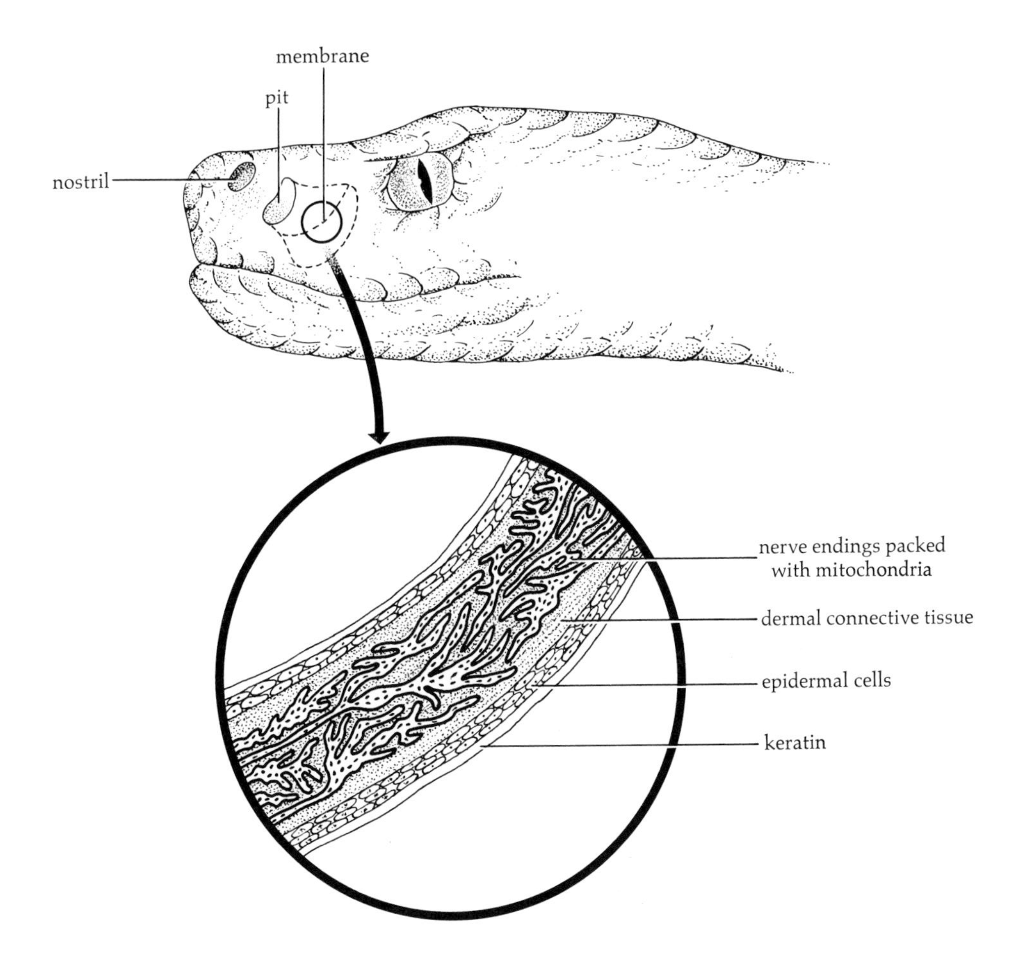

Fig. 10-12. Diagram of the lateral view of a rattlesnake head showing location of the pit organ and how the pit expands deep to the surface. The pit membrane divides the pit into anterior and posterior chambers.

EMBRYONIC ORIGIN

Thickened placodes form in the surface ectoderm and invaginate to deeper portions of the body. In the inner ear they become dissociated from the surface ectoderm; in the lateral line they sink only as far as the deep portions of the skin and maintain contact with the surface through pores.

THE HAIR CELL (Fig. 10-13)

The hair cell varies from pear shaped to cylindrical. The cell has a prominent nucleus and the normal cellular organelles, such as mitochondria, endoplasmic reticulum, and Golgi apparatus. Its most notable organelles however are the two types of cilia ("hairs") at its apical end. There are from 40 to 200 long, thin cytoplasmic processes,

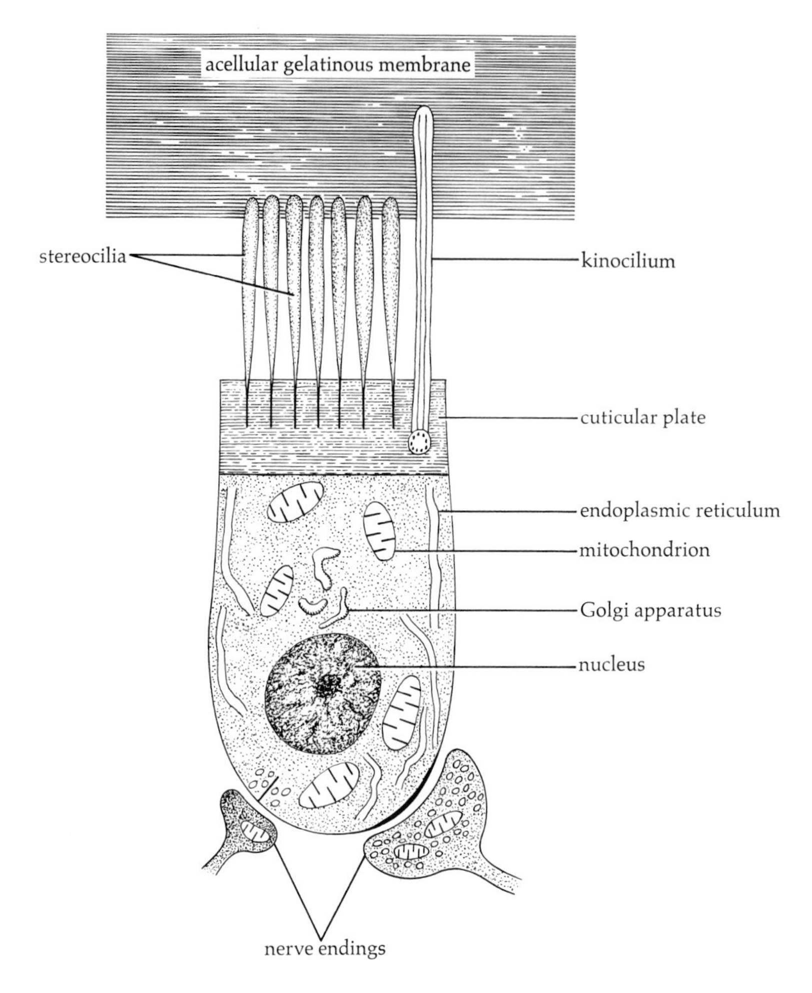

Fig. 10-13. Diagram of a generalized hair cell.

the stereocilia. They have little internal structure except for an electron-dense rootlet embedded in the hair cell. There is also (usually) a single, eccentrically placed kinocilium, which typically possesses nine pairs of peripheral fibrils and two central fibrils. The hair cell is thus polarized, both across its apical surface (by the kinocilium) and basally to apically.

At the base of the cilia the cytoplasm is particularly dense and is called the cuticular plate. This surface of the hair cell lies against the extracellular fluid space. The tips of the cilia extend into this space and are embedded in an acellular gelatinous membrane, which may or may not contain inorganic crystals. At its basal end, bare nerve endings form synapses with the hair cell.

These bare nerve endings are excited when the cilia are bent. The degree of force required to do this, and its nature and source, vary in different portions of the acousticolateralis system; however, the force is usually applied to the gelatinous membrane.

THE LATERAL LINE SYSTEM (Fig. 10-14)

The lateral line system of fishes and larval amphibians is comprised of a complex network of fluid-filled canals in the head skin and, in the trunk, a single long canal from head to tail at about the junction of the epaxial and hypaxial musculature. Along these canals are small pores opening at the surface, and groups of sensory epithelia called neuromast organs. In each neuromast organ are both a group of supporting (sustentacular) cells and a group of hair cells identically oriented with respect to the kinocilium. Any disturbance of the water near the fish causes movements of the canal fluids; these movements bend the hairs of the hair cells and that stimulates the neuromast organs.

For a long time the lateral line system was believed to have an auditory function. We now know that it is instead a "distance–touch receptor." Its adaptive value is twofold: it detects disturbances in the water ("near field sound"); and it gives the fish information about its

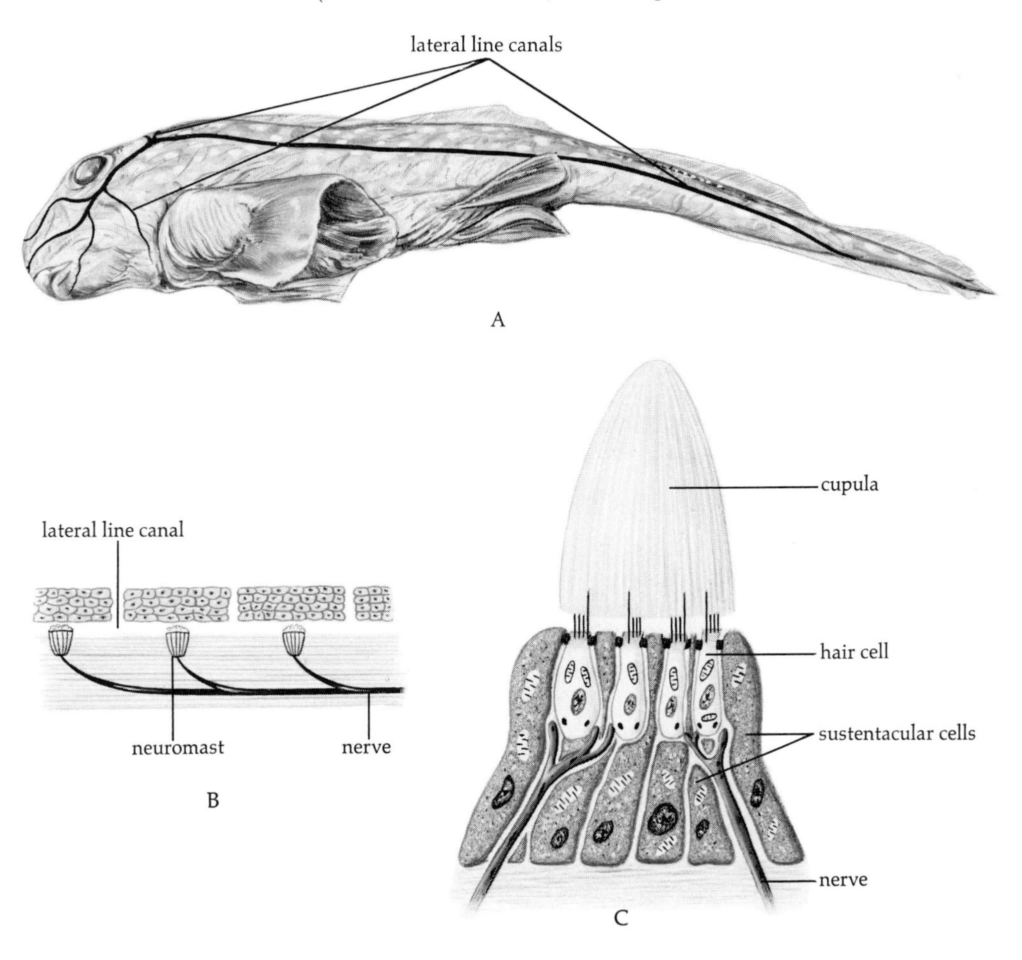

Fig. 10-14. (A) A lateral view of a holocephalan, the ratfish, in which the lateral line canals are clearly seen. (B) A low-power section along a lateral line canal, showing its relationship to the outside. (C) A high-power view of an individual neuromast organ.

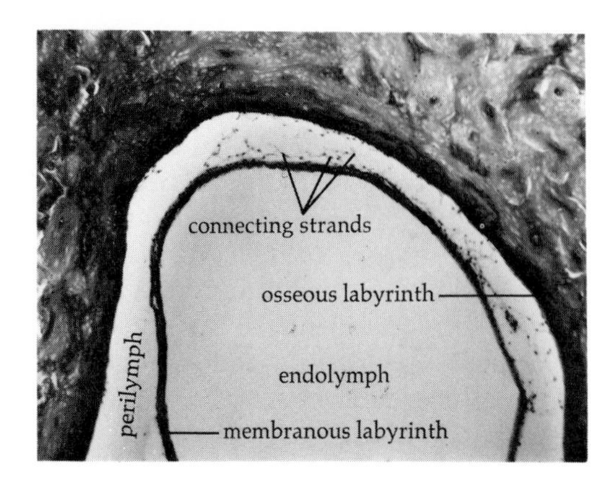

Fig. 10-15. A section through part of a semicircular canal. Note the compact bone which lines the perilymphatic space and forms the osseous labyrinth. Fine strands of connective and epithelial tissues traverse the perilymph and suspend the endolymph-filled membranous labyrinth in a sea of perilymph.

own body movements that is particularly adaptive since fishes lack muscle spindles.

In some fishes there are highly modified derivatives of the lateral line in which the hairs of the hair cells are lost. These ampullary organs apparently act as chemoreceptors, for they are quite sensitive to the chemical composition of the surrounding water.

In at least chondrichthyeans and a teleost group of weakly electric mormyrid fishes, highly modified lateral line organs act as electroreceptors. In the chondrichthyean head region there are many ampullae of Lorenzini, which are very sensitive to electrical fields. Their exact function is unclear, but they probably detect the muscle action potentials of other fish and perhaps the shark's own muscle action potentials.

In the mormyrids, the modifications of the lateral line are called tuberous organs. They are stimulated by the fish's own electrical potentials which are modified by nearby physical objects. It is thought that with this electric "radar" the mormyrid constantly scans for information about its environment. There is also some indication that they may be used in intraspecific communication.

THE INNER EAR

The inner ear, or membranous labyrinth, is a connective tissue structure in which there are areas of sensory epithelium containing hair cells. It is suspended in perilymphatic fluid within the spaces of the otic capsule; the hard, bony shell which in most forms encapsulates these spaces is called the osseous labyrinth. Connective tissue trabeculae run between the membranous labyrinth and the bone or cartilage of the otic capsule, and nerves run between the membranous labyrinth and the brain (Fig. 10-15).

Within the membranous labyrinth is another fluid, called endolymph. It has a high concentration of potassium ions and dissolved

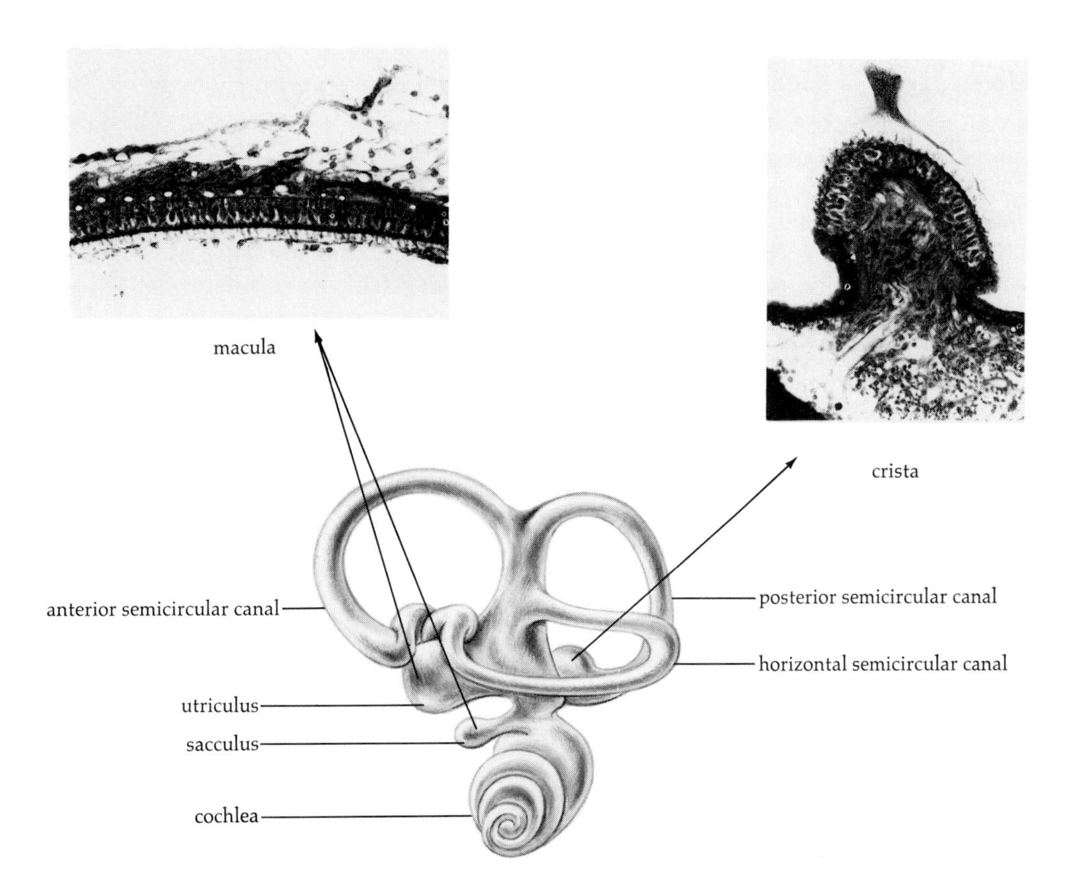

Fig. 10-16. Diagram of the membranous labyrinth of a mammal, with photomicrographs of a crista and macula. Except for the cochlea, the entire labyrinth is vestibular in function.

proteins and is, thus, more like an intracellular fluid than an extracellular fluid.

There are two sense organs in the membranous labyrinth. The vestibular apparatus is similar in all vertebrates except cyclostomes. The auditory apparatus varies considerably from group to group, with some interesting evolutionary trends. Primarily for this reason, the auditory apparatus will be discussed in a separate section and in considerably more detail than any other sense organ. This special treatment does not mean that the sense of hearing is more important or more complex than any other. It can be regarded however as a model from which basic principles as well as specific information can be understood.

The vestibular apparatus (*Fig. 10-16*)

The vestibular apparatus is made up of a large membranous sac, called the utriculus, connected with a smaller membranous sac, the sacculus. Coming off the utriculus are three membranous semicircular canals; one lies in each of the three planes of space.

The sacculus and utriculus each have one or more areas of sensory epithelium, the maculae. Associated with each macula is a gelatinous otolithic membrane filled with inorganic crystals in which the tips of the hair cells' cilia are embedded.

The portion of each semicircular canal adjacent to the utriculus is enlarged and is called the ampulla. Each ampulla contains an area of sensory epithelium, called a crista, whose hair cells' cilia are embedded in a gelatinous membrane (without crystals) called the cupula.

The sacculus and utriculus are connected through a membranous duct, the endolymphatic duct. It reaches the dorsal surface of the skin in chondrichthyeans, and in other vertebrates it terminates in a dilation, the endolymphatic sac, which lies just within the cranial cavity adjacent to the brain.

The sacculus and utriculus are organs of static equilibrium or balance. The otolithic membranes, which are heavy because of their inorganic crystals, bend the hairs on which they rest in a way that varies with the position of the head.

The cristae of the semicircular canals, on the other hand, respond to changes in acceleration, particularly angular acceleration. When an animal's velocity changes there is a brief period before the movement of the endolymph also changes; this causes a dragging effect on the cupulae, which bends the hairs of the hair cells.

In some fishes the sacculus and utriculus are confluent and have a variable number of maculae. In tetrapods, however, there is typically a single utricular macula and a single saccular macula.

Among cyclostomes, the hagfish has but a single semicircular canal in each ear, and the lamprey has two canals. Whether this is a primitive condition or one which has secondarily evolved in cyclostomes will probably not be known unless the semicircular canals of fossil ostracoderms can be examined.

Vertebrate Hearing

While the lateral line picks up near field sound in water, the auditory system picks up propagated sound in either water or air. In water, propagated sound differs from focal disturbances (near field sound) in that the vibrations are of smaller amplitudes but are carried much farther. The vibrations travel not because of gross movements of the water, but because of the elasticity and slightly compressible nature of water molecules. This also explains why the amplitude (strength) of the propagated sound wave diminishes very slowly over distance.

FISHES

All fishes tested can hear at least low frequencies of propagated sound waves. However, fishes are nearly the same density as the water in which they swim and many are essentially "transparent" to all but the strongest sound waves. In many teleosts, however, there are morphological modifications that facilitate more sensitive auditory reception.

The one organ in the teleost body with a significantly different density than water is the swim bladder, a dorsal gas-filled structure that

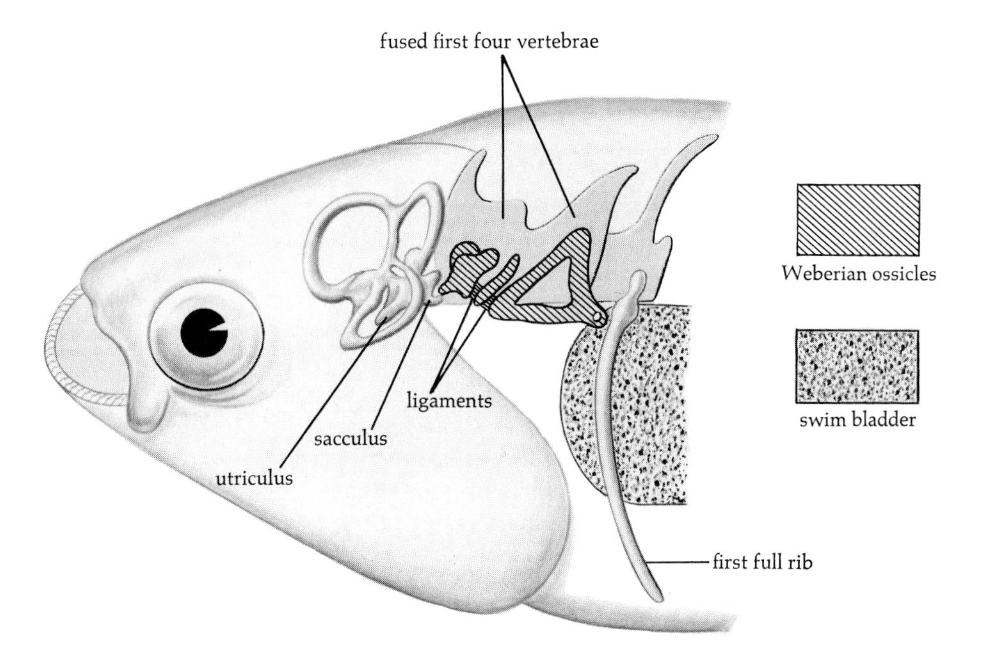

Fig. 10-17. Dorsolateral view of the anterior portion of an ostariophysian fish, showing the relationships of the Weberian ossicles to the swim bladder and inner ear.

functions as a hydrostatic organ. When gas is secreted into or absorbed from it, the teleost's specific gravity becomes equal to the specific gravity of the surrounding water, and the fish's level in the water is effortlessly maintained. This "air bubble," however, because of its vastly different density, is set into vibration by propagated sound waves. If there is a means of transferring these swim bladder vibrations to the sensory epithelium of the inner ear, hearing can occur.

In the Ostariophysi, a teleost group that includes catfish, carp, and goldfish, the lateral processes of the four anterior vertebrae are modified as tiny Weberian ossicles (Fig. 10-17). These run from the anterior part of the swim bladder to the perilymphatic spaces adjacent to the sensory epithelium of the sacculus. Propagated sound waves in the water set the swim bladder into vibration, and the vibratory energy is transferred from the swim bladder to the Weberian ossicles and then to the perilymphatic fluids; the movements of the perilymphatic fluid are transferred to the otolithic membrane of the sacculus, causing the macula of the sacculus to generate nerve impulses.

Other couplings of swim bladder and inner ear have evolved independently. There are forward extensions of the swim bladder in clupeid fish (herrings, sardines), and paired extensions that have been pinched off to form separate bubbles in the mormyrids. In both, sound waves passing the fish vibrate these gas-filled areas, and they in turn stimulate the nearby saccular macula.

There are small air bubbles dorsally in the complex gill chambers of the southeast Asian labyrinthine fish (named not for its inner ear, but for its gills). These bubbles, vibrated by propagated sound waves, stimulate the auditory sensory epithelium, which in this case is the utricular macula.

MIDDLE EAR

TETRAPODS (EXCEPT MAMMALS)

Larval amphibians (tadpoles) can also hear sound waves propagated in water. The lungs, which are not yet functional respiratory organs, are connected to the perilymphatic spaces by a small skeletal element called the bronchial columella (found only in tadpoles). There is an obvious similarity to the situation in some teleosts.

After metamorphosis, the amphibian's lungs are used for respiration, and for them and for all other tetrapods the problem of hearing is quite different. Airborne rather than water-borne sound waves must stimulate sensory epithelium which itself is in an aquatic medium. In order for the relatively weak aerial vibrations to be transmitted to the inner ear fluids, their force per unit area must be increased approximately twentyfold; this is called impedance matching. Without such impedance matching the sound energy is reflected off, rather than absorbed by, the fluid surfaces.

In tetrapods the middle ear is the transformer that accomplishes this impedence matching. With few exceptions, the middle ear includes a tympanic membrane and one or more middle ear ossicles traversing an air space to the inner ear within the otic capsule. The middle ear space, like the spiracle of sharks, is a derivative of the first pharyngeal pouch. Its outer boundary is the tympanic membrane composed of an inner layer of endoderm (from the pharyngeal pouch), a middle layer of connective tissue, and an outer layer of surface ectoderm. These layers are greatly thinned, making the tympanic membrane an extremely delicate structure. (Tympanic membrane is a better name for this structure than eardrum, since "drum" connotes tension. This membrane is quite loose so that it can be put into vibration by very weak sound waves.)

The embryonic connection between middle ear and pharynx is maintained as the Eustachian tube, through which pressure in the middle ear is equalized with that of the external environment.

In frogs, turtles, most lizards, crocodilians, and birds, there is one middle ear ossicle. It runs from the tympanic membrane to a "window," the fenestra ovale, leading to the perilymphatic spaces. This ossicle, called the columella, is a completely different structure than the bronchial columella of tadpole amphibians. It is almost always composed of two distinct but continuous portions: an outer extracolumella, frequently cartilaginous, which attaches to the tympanic membrane, and an inner, usually bony, columella proper, whose medial portion, the columellar footplate, fits into the fenestra ovale. The columella proper is homologous to the hyomandibular element of fishes.

Vibrations in the air set the tympanic membrane into motion, and this motion causes the extracolumella and columella to vibrate and to

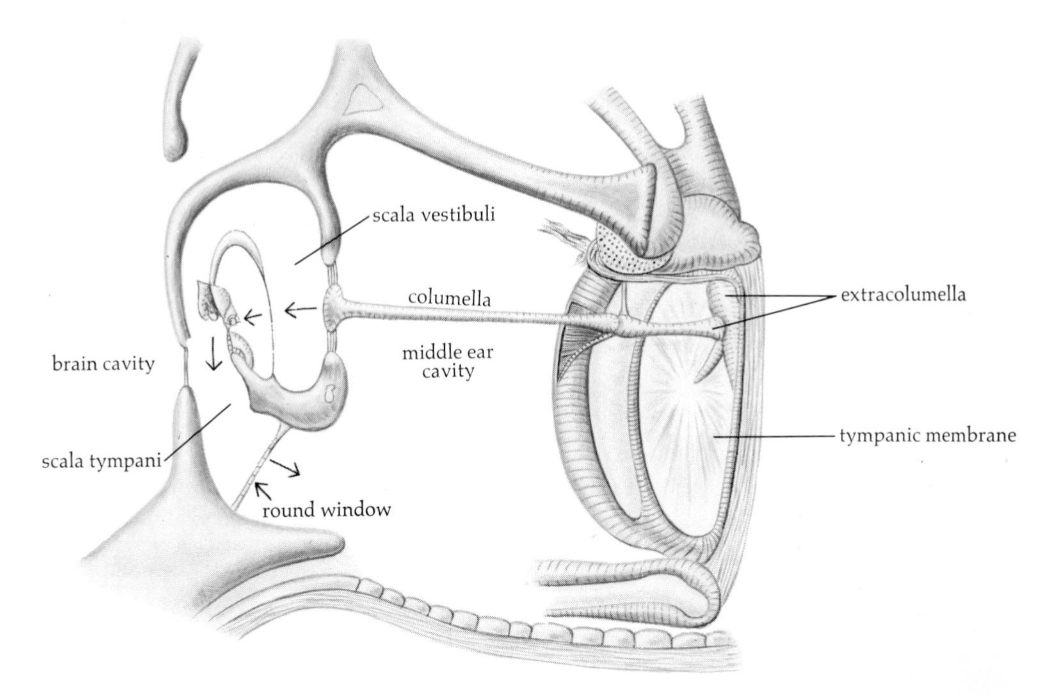

Fig. 10-18. Diagramatic view of the middle and inner ears of a lizard. After E. G. Wever (1965). Structure and Function of the Lizard Ear. J. Aud. Res. **5,** *331–371.*

carry these vibrations to the perilymphatic fluids of the inner ear (Fig. 10-18). Pressure is equalized in the inner ear at another membrane-covered window, the fenestra rotunda. The surface area of the tympanic membrane is at least twenty times greater than the surface area of the columellar footplate; in other words, the force striking the tympanic membrane, resolved onto the columellar footplate, increases the force per unit area by at least twentyfold. Thus aerial vibrations are transformed into vibrations of the inner ear fluids.

Some tetrapods (salamanders, snakes, and the limbless amphisbaenid lizards) have neither tympanic membrane nor middle ear cavity. In the adult salamander, the opercularis muscle runs from the suprascapular cartilage to the operculum; this operculum (not to be confused with that covering gill slits) is homologous to the basal part of the columella and fits into the fenestra ovale. Substrate vibrations are probably picked up through the forelimb and transmitted by way of this opercularis muscle to the operculum and then to the fluids of the inner ear. So far as is known salamanders are insensitive to airborne vibrations.

Snakes, on the other hand, lack a middle ear apparatus but have a good columella. They are sensitive to airborne as well as substrate vibrations, but the mechanism by which aerial vibrations are transformed to fluid vibrations is not known.

The rare amphisbaenids have neither tympanic membrane nor middle ear cavity, yet they are sensitive to airborne vibrations. In this case the extracolumella acts as a tympanic membrane. Greatly expanded, flattened, and thinned, the extracolumella lies just under the

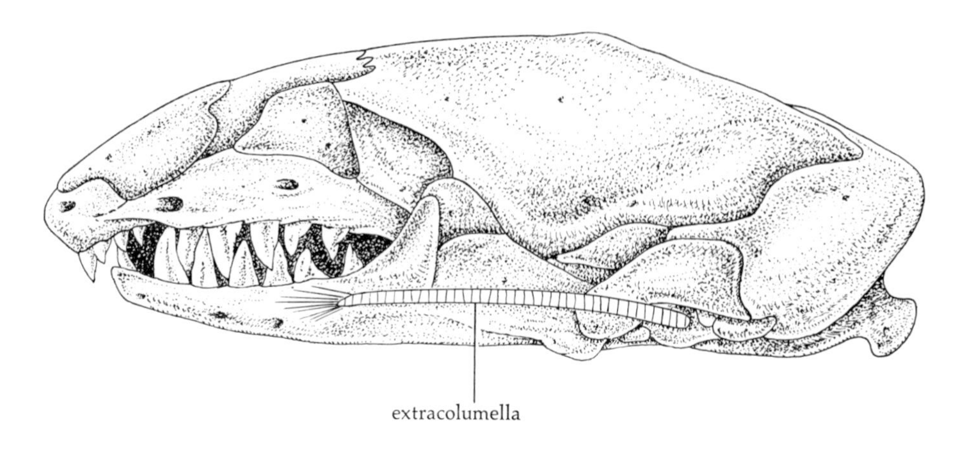

Fig. 10-19. Lateral view of the skull of an amphisbaenid, showing the elongated extracolumella extending anteriorly just lateral to the lower jaw bones. After C. Gans and E. G. Wever (1972). The Ear and Hearing in Amphisbaenia (Reptilia). J. Exp. Zool. **179,** 17–34.

skin of the head (Fig. 10-19). Vibrations striking the head set this modified extracolumella into vibration; these vibrations are transferred to the columella and then to the fluid spaces of the inner ear.

MAMMALS

In mammals a delicately suspended chain of three ossicles runs from the tympanic membrane to the inner ear. These are the malleus, incus, and stapes; the malleus and incus are homologous to the articular and quadrate bones of other tetrapods, and the stapes is homologous to the columella (Fig. 10-20).

The mammalian tympanic membrane is shaped like a low cone with its apex facing medially. Along its dorsal radius and embedded within its tissue is the manubrium of the malleus. Dorsally the malleus expands, forming an articulation with the incus. A small, slender, anterior process at the base of the manubrium supports the malleus in the middle ear. The incus has a process going posteriorly, from which a ligament attaches to the wall of the middle ear cavity. These two processes form the axis about which the ossicular system rocks when it is set in motion.

The shorter, articulating process of the incus parallels the manubrium of the malleus, makes a right angle turn, and then articulates with the stapes. The stapes is shaped like a stirrup; its two crura end in the oval footplate which is held into the oval window against the perilymphatic fluids of the inner ear.

Fig. 10-20. Diagramatic representation of the diversity of form of the mandibular arch and the hyomandibular, demonstrating their contributions to both the ear and the jaw joint in each of the gnathostome classes.

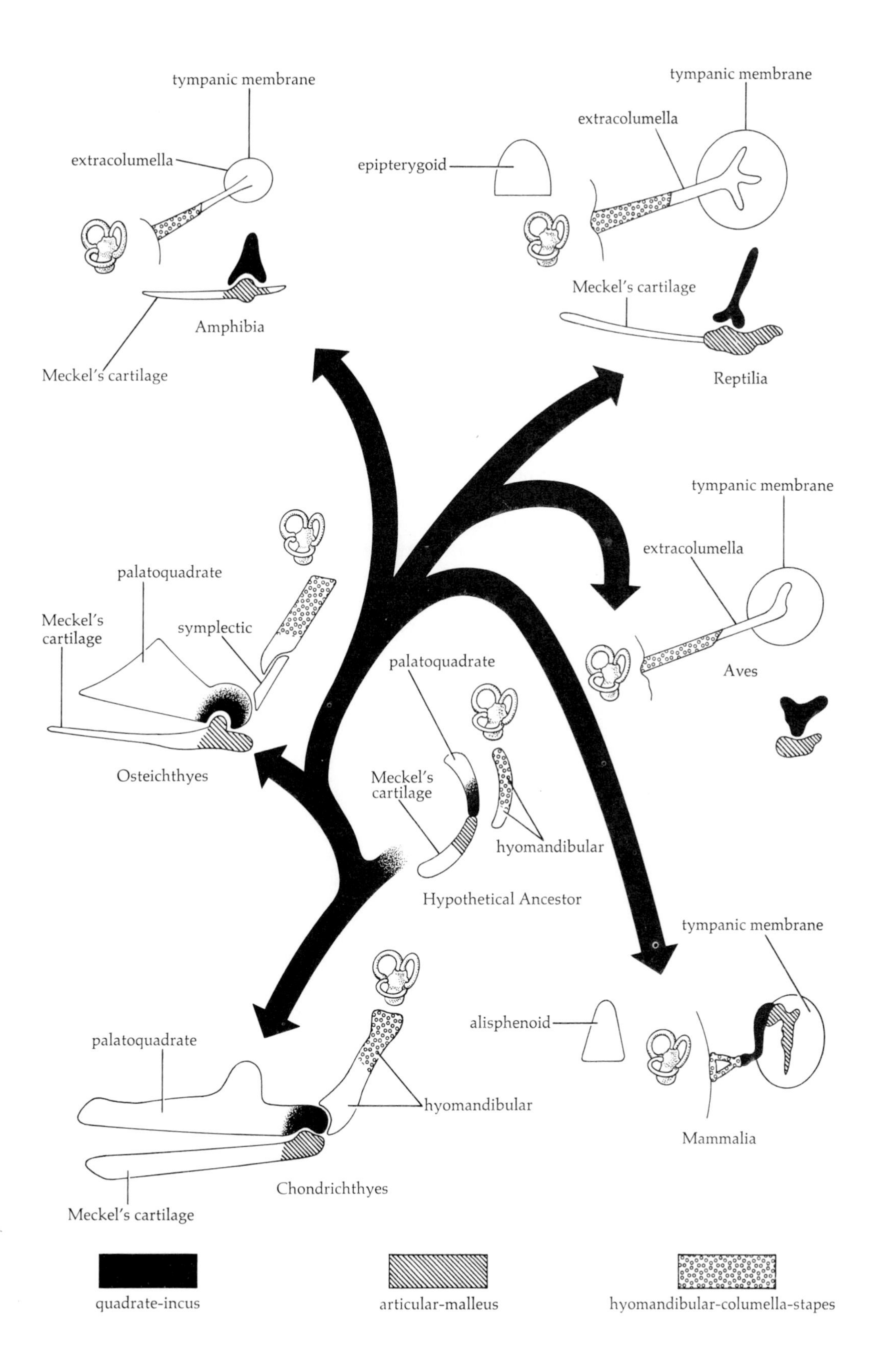
tympanic membrane
extracolumella
Amphibia
Meckel's cartilage
tympanic membrane
extracolumella
epipterygoid
Meckel's cartilage
Reptilia
tympanic membrane
extracolumella
Aves
palatoquadrate
Meckel's cartilage
symplectic
Osteichthyes
palatoquadrate
Meckel's cartilage
hyomandibular
Hypothetical Ancestor
tympanic membrane
alisphenoid
Mammalia
palatoquadrate
hyomandibular
Chondrichthyes
Meckel's cartilage
quadrate-incus
articular-malleus
hyomandibular-columella-stapes

The condensation, or positive pressure, portion of a sound wave moves the tympanic membrane medially, causing the ossicular system to swing on its axis: the manubrium and the articulating process of the incus move medially, and the articulating process of the incus pushes the stapes into the perilymphatic spaces. Then the rarefaction portion of the sound wave reverses the process.

The force per unit area at the footplate of the stapes is increased, first, as the ratio of tympanic membrane surface area to stapes footplate surface area. Furthermore, because the manubrium of the malleus is longer than the articulating process of the incus, force is also increased as a ratio of the two lever arms (Fig. 10-21).

The total increase in force from all causes is called the transformer ratio (expressed as a whole number). This transformer ratio varies in mammalian species from 20 to slightly over 100. It is always more than adequate to match the impedance of the air waves with the impedance of the inner ear fluids.

Attached to the auditory ossicles are two of the smallest skeletal muscles. The tensor tympani runs out from the medial wall of the middle ear cavity and inserts onto the manubrium of the malleus; when the tensor tympani contracts it stiffens the tympanic membrane. The even smaller stapedius muscle originates on the posterior wall of the middle ear and inserts on the neck of the stapes; the contraction of the stapedius stiffens the connection between stapes and fenestra ovale. These muscles contract reflexly in response to intense sound, protecting the inner ear. They also contract reflexly when an animal vocalizes and relax after vocalization, so that the ear is not overstimulated by the animal's own voice but is most sensitive and receptive to whatever follows.

AUDITORY INNER EAR

TETRAPODS (Fig. 10-22)

In amphibians there are two small diverticuli off the sacculus, each housing one or two maculae. In salamanders and anurans only the smaller diverticulum contains the macula of the amphibian papilla; it responds to sound. The larger diverticulum is the lagena, containing two maculae. At the distal end of the lagena is the lagenar macula, of unknown function. More proximally there is the basilar papilla macula, which responds to sound. Each of these three maculae has a delicate, gelatinous tectorial membrane in which the hairs of the hair cells are embedded.

In reptiles the lagena is larger; in birds it is larger still. As in amphibians, both reptiles and birds have a small lagenar macula distally, of unknown function. The larger basilar papilla varies widely in both size and structure among different reptiles and birds. However, the basilar papilla always contains hair cells, and it is the auditory transducing epithelium.

In mammals there is no lagenar macula. The lagena is called the cochlear duct, which is a long, coiled structure with one to five turns

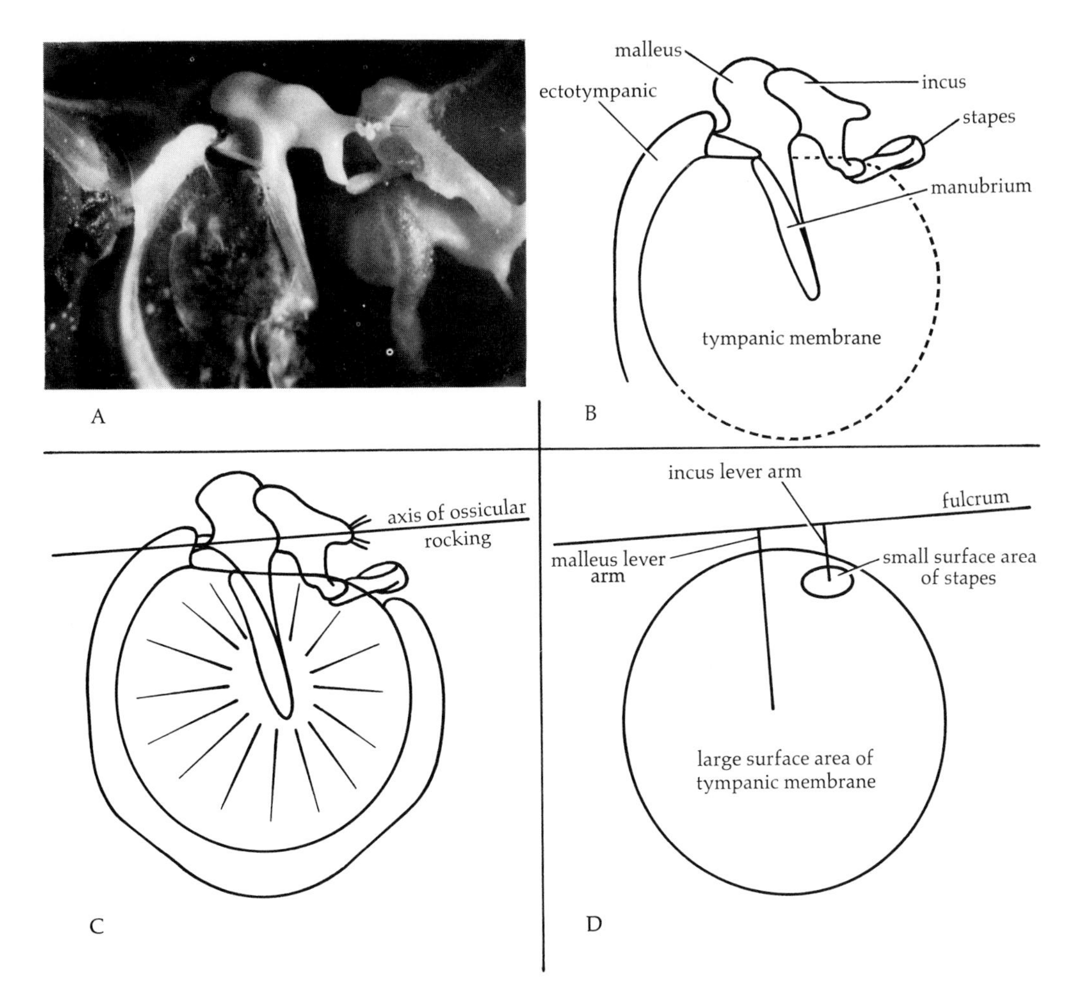

Fig. 10-21. Ventrolateral view of a mammalian tympanic membrane and auditory ossicles (A). The posterior half of the tympanic membrane is removed to show the entire incus and the stapes passing into the oval window. In B the parts are shown in outline and a dotted line completes the tympanic membrane. C and D picture the middle ear's functional attributes which increase the acoustical force in two ways: by collecting it on the large tympanic membrane and resolving it onto the small surface area of the stapedial footplate and by passing the vibrations through a lever system composed of the three auditory ossicles.

depending on species. It contains the auditory transducing epithelium, called the organ of Corti, which is a long structure both more complex and more stereotyped than the analogous structures of nonmammalians. Because the mammalian ear has been studied much more than that of any other vertebrate, it is better understood, although there are still many unanswered questions.

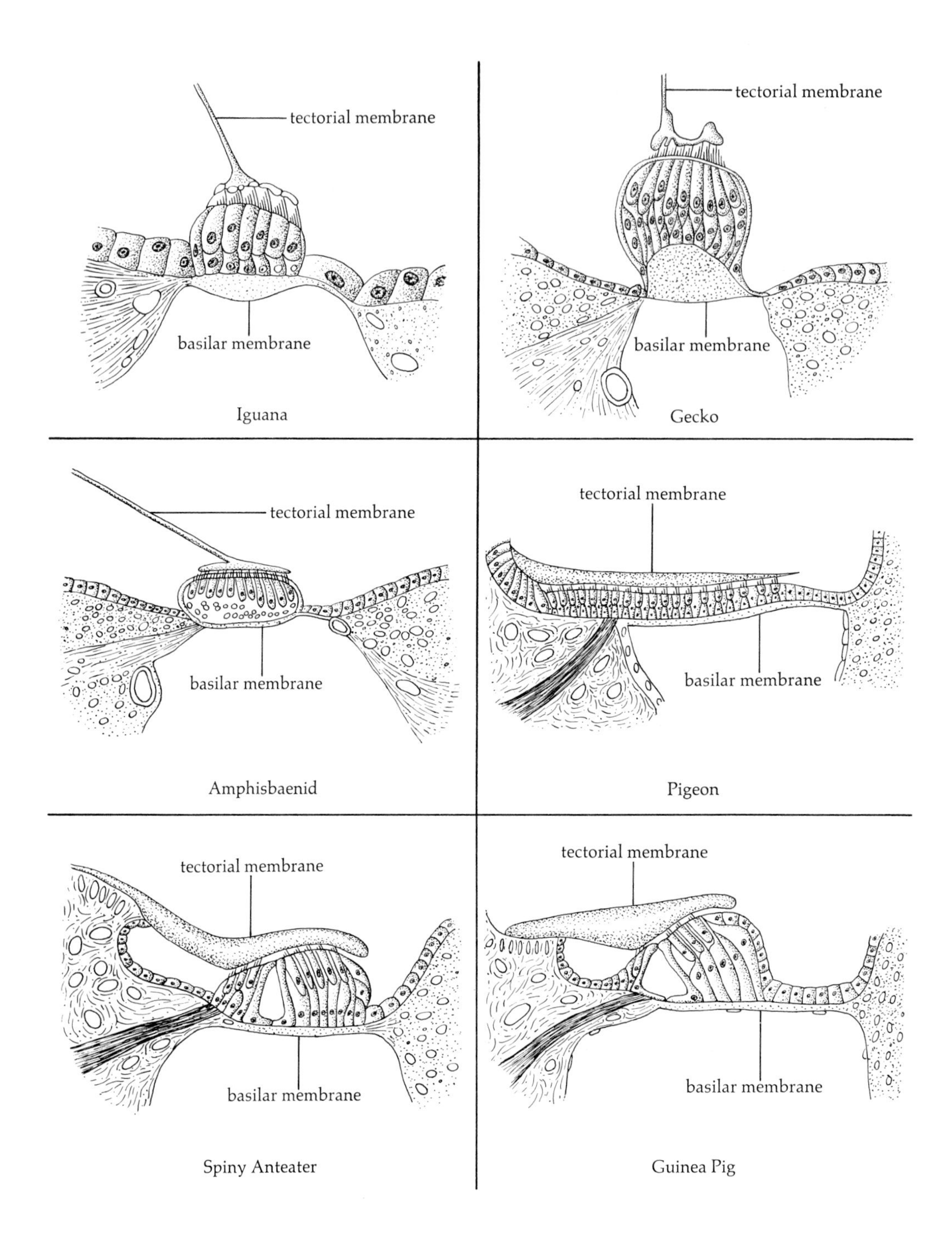
tectorial membrane
basilar membrane
Iguana
tectorial membrane
basilar membrane
Gecko
tectorial membrane
basilar membrane
Amphisbaenid
tectorial membrane
basilar membrane
Pigeon
tectorial membrane
basilar membrane
Spiny Anteater
tectorial membrane
basilar membrane
Guinea Pig

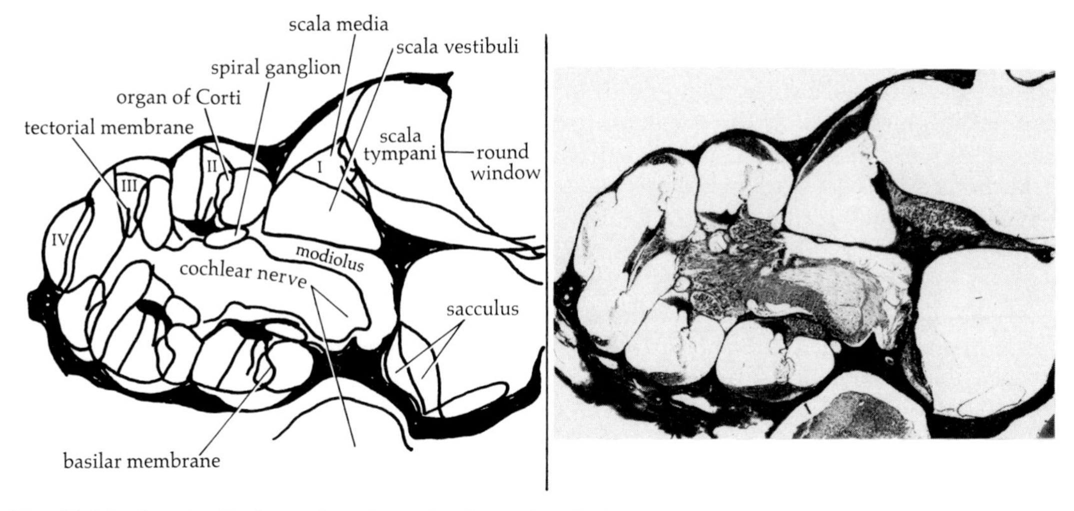

Fig. 10-23. Longitudinal section through the axis of the cochlea of a kangaroo rat. Roman numerals indicate turns of the cochlea.

DETAILS OF MAMMALIAN COCHLEAR DUCT (Figs. 10-23, 10-24)

General morphology

The cochlear duct turns about a central, bony core called the modiolus, within which lies the sensory ganglion (spiral ganglion) of the cochlear portion of the statoacoustic nerve (Fig. 10-23). The duct is divided lengthwise into three fluid-filled scalae. In the middle is the scala media, filled with endolymph; just above it is the scala vestibuli and just below it is the scala tympani. The latter two are each filled with perilymph and are continuous at the apex of the cochlea, at a small hole called the helicotrema.

The endolymph of the scala media remains separate from the perilymph of the other two scalae. The scala media is separated from the scala vestibuli by a thin, epithelial structure, Reissner's membrane, and from the scala tympani by the organ of Corti. This last structure,

Fig. 10-22. The diversity of the auditory epithelium of amniotes. The number and arrangement of hair cells and the shape of the tectorial membrane are particularly varied. Iguana and lizard after E. G. Wever (1967). The Tectorial Membrane of the Lizard Ear: Types of Structure. J. Morph. **122,** *307–320. Amphisbaenid after C. Gans and E. G. Wever (1972). The Ear and Hearing in Amphisbaenia (Reptilia).* J. Exp. Zool. **179,** 17–34. *Pigeon after T. Takasaka and C. A. Smith (1971). The Structure and Innervation of the Pigeon's Basilar Papilla.* J. Ultrastruct. Res. **35,** 20–65. *Spiny anteater and guinea pig after C. A. Smith and T. Takasaka (1971). Auditory Receptor Organs of Reptiles, Birds, and Mammals. In "Contributions to Sensory Physiology" (W. D. Neff, ed.), Vol. 5, pp. 129–178. Academic Press, New York.*

where the sensory epithelium is located, rests on a bony projection from the modiolus called the osseous spiral lamina, which is continuous with a fibrous basilar membrane. Just above the osseous spiral lamina, the inner wall of the cochlear duct projects out into the scala media as the spiral limbus, which supports the tectorial membrane. Vibrations enter the cochlear fluids by way of the footplate of the stapes, which lies in the fenestra ovale at the basal end of the scala vestibuli, and are dissipated at a membrane-covered window, the fenestra rotunda, at the base of the scala tympani.

At the side of the scala media furthest from the modiolus is the spiral ligament, which is firmly attached to the bony covering of the cochlea and which bears on its inner surface (bordering the scala media) a vascularized area, the stria vascularis. This is the only vascular supply to the organ of Corti; if there were direct vascularization of the hair cells themselves, the movements of the blood would cause a great deal of noise in the system.

Organ of Corti (*Fig. 10-24*)

To a great extent the functional characteristics of the organ of Corti are due to its elaborate morphology. This is easiest understood in cross section, but it must be borne in mind that the structure runs the length of the coiled cochlea. Thus what is seen in cross section as a single cell represents a row of cells extending from the basal to the apical end of the organ of Corti.

In cross section, then, from the modiolus outward, one sees resting on the osseous spiral lamina a single, inner sustentacular cell, supporting an inner hair cell. Next to the sustentacular cell is the inner pillar cell. The basilar membrane is continuous with the edge of the osseous spiral lamina and extends across to the spiral ligament. On the innermost part of the basilar membrane is the outer pillar cell, whose process reaches up and joins the upper process of the inner pillar cell, enclosing the fluid-filled tunnel of Corti. Pillar cells (both outer and inner) contain intracellular structural proteins in long fibers, giving them rigidity and helping them support the organ of Corti. Just external to the outer pillar cell are three rows of outer sustentacular cells, called Deiters' cells, supporting the outer hair cells. Next are the border cells: first tall cells of Hensen and then lower cells of Claudius which reach out to the spiral ligament.

The inner and outer hair cells differ in more than their arrangement into one or three rows. The pear-shaped, inner hair cell is supported on all sides except its apex by the inner sustentacular cell. The inner hair cell has a large, central nucleus and is rich in organelles, with many mitochondria and some rough endoplasmic reticulum. Extending out of the apical cuticular plate are three rows of stereocilia. The adult mammalian organ of Corti has no kinocilium (in inner or outer hair cells); the kinocilium develops in the embryo but is resorbed.

The outer hair cells are columnar rather than pear shaped, have a basally located nucleus, and are poor in organelles with but a few mitochondria apically and basally. The base of each of the outer hair cells rests on a cell of Deiters, which sends up a phalangeal process parallel

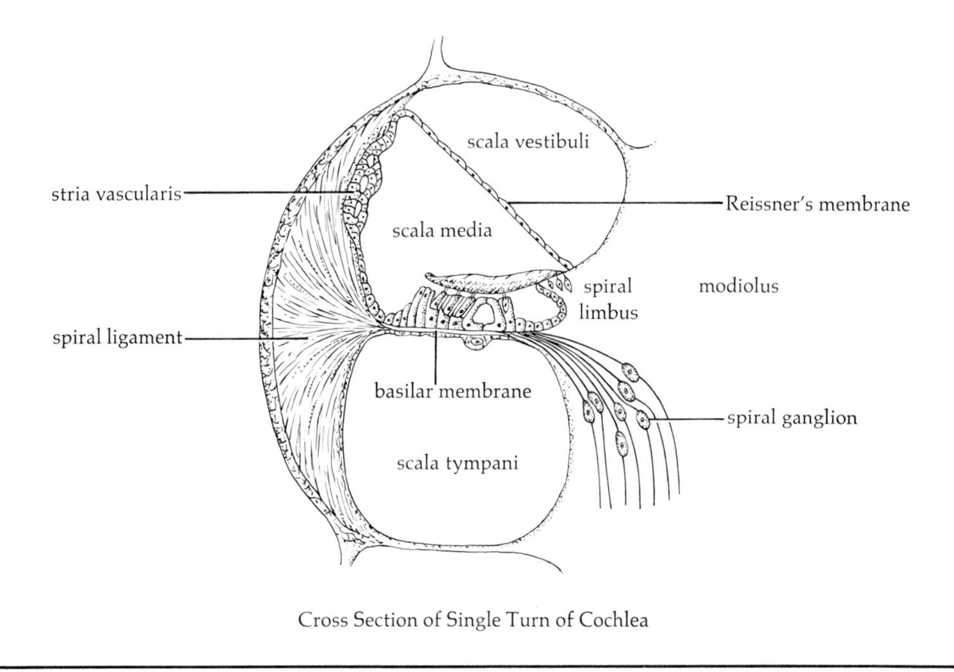

Cross Section of Single Turn of Cochlea

tectorial membrane
Hensen's cells
outer hair cells
inner hair cell
tunnel of Corti
Claudius' cells
inner sustentacular cell
inner pillar cell
nerve fibers
osseous spiral lamina
basilar membrane
Deiters' cells
outer pillar cell

Organ of Corti

Fig. 10-24. Detailed structure of the mammalian cochlea and organ of Corti.

to, but at a slight distance from, the outer hair cell. Thus there is an extracellular fluid space between the phalangeal process and the body of the hair cell; this is the space of Nuël. The phalangeal process, which contains structural filaments that give it rigidity, expands at its distal end and closely clasps the apex of the outer hair cell at the level of the cuticular plate.

This apical surface of the organ of Corti is a very dense, impermeable barrier, separating the extracellular fluids within the organ of Corti from the endolymph of the scala media beneath which the organ

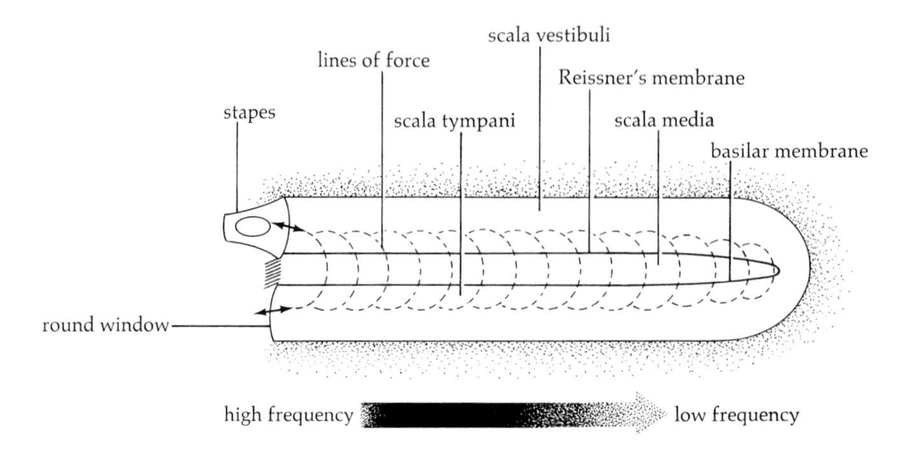

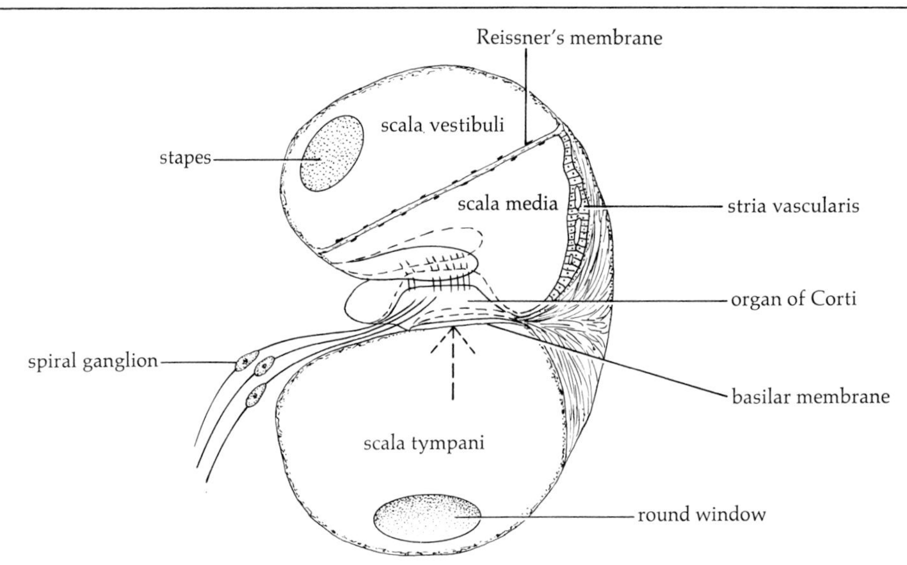

Fig. 10-25. Top: diagramatic view of a "straightened-out" cochlea, showing the portions of the cochlea responding to high and low frequencies. Bottom: cross section of a single turn of the cochlea, demonstrating how vibration of the basilar membrane causes a shearing action between the tectorial membrane and the organ of Corti, which bends the "hairs" of the hair cells.

of Corti lies. Similarly, these extracellular organ of Corti fluids are (probably) separated from the perilymph of the scala tympani by the basilar membrane. Coming out of the cuticular plate of each hair cell are three rows of stereocilia, arranged in a three-row "W" whose base faces the stria vascularis.

The kinetics of the basilar membrane and organ of Corti have been studied in some detail (Fig. 10-25). Since the cochlear duct and its components are essentially fluid-filled structures, vibrations brought in by the stapes do not have to pass the whole length of the cochlear duct before they are dissipated at the round window, but can pass through the basilar membrane, organ of Corti, and Reissner's membrane at any level, or at all levels, of the cochlear duct. When they pass through they stimulate these structures, which respond with vibrations. The vibrations in response to any sound are not uniform throughout the entire cochlear duct, but are greatest at one level. Experimental studies have shown that the basal turn, where the basilar membrane is narrowest, responds maximally to high frequencies and very weakly to low frequencies; the more apical turns of the cochlea respond maximally to low frequencies and not at all to high frequencies.

The vibrations of the basilar membrane and organ of Corti have a minute amplitude; the movement of the basilar membrane in response to a very intense tone is only 500 Angstroms (Å). However, these tiny vibrations push the entire organ of Corti toward the tectorial membrane which, being firmly attached to the spiral limbus, is not itself vibrated by the movements of the basilar membrane. This upward movement of the organ of Corti causes a shearing motion between the tectorial membrane and the hairs of the hair cells embedded in it. This bends the hairs, which is the final mechanical stimulus for the excitation of the hair cells.

Although it is not known precisely how this mechanical shearing action is transduced into nerve impulses, many characteristics of the phenomenon are known. Functionally, the cochlear duct at rest can be compared with a charged battery. By its active metabolism the stria vascularis maintains a resting bioelectric potential in the endolymph, called the endocochlear potential, with a voltage of about +90 millivolts (mV) relative to the surrounding perilymphatic spaces. The organ of Corti at rest maintains a potential of about −60 mV. Thus there is a potential charge of about 150 mV across the reticular lamina in the resting condition.

Upon acoustic stimulation there is an excitatory potential of the organ of Corti, called the cochlear microphonic. This is an alternating current potential whose frequency matches that of the stimulus and whose voltage varies directly with the intensity of the stimulus but rarely exceeds 2 mV. Although the source of this cochlear microphonic has not been proven, it is probably caused by a leakage of current across the reticular lamina owing to the bending of the hair cells' cilia. However, it is unlikely that the cochlear microphonic potential directly initiates the action potentials in the nerves, since the two phenomena can be disassociated; the application of certain drugs to the organ of Corti inhibits all action potentials without modifying the cochlear microphonic. Instead, because the synaptic connections between both inner and outer hair cells and their sensory neurons to the brain are morphologically typical synapses, it can be assumed that exciting action potentials in the neuron depends upon release of a (still unknown) chemical transmitter from the hair cell into the synapse.

The cell bodies of the sensory nerves are within the modiolus, in the structure known as the spiral ganglion. Their dendrites extend out through the osseous spiral lamina and, as bare nerve endings, traverse the fluid spaces to their terminal boutons on the hair cells. Those on inner hair cells are unbranched with a single bouton from a dendrite terminating on each inner hair cell; those on outer hair cells are branched with each dendrite sending synaptic endings to many outer hair cells. There is thus a possible analogy with the eye: a small bipolar cell in the retina receives stimuli either from many rods or from a single cone just as a spiral ganglion cell receives stimuli from many outer hair cells or from a single inner hair cell. If the situation is truly analogous, the outer hair cells would have very low thresholds but less specificity, and the inner hair cells would have higher thresholds but more specificity.

In addition to the sensory neurons leading from the hair cells with axons to the brain, some cells in the brain send axons to the hair cells, giving them efferent innervation and thus central control. Present experimental evidence indicates that this efferent mechanism acts as an inhibitory mechanism, reducing the sensitivity of the organ of Corti.

In summary, then, transduction of physical vibrations to nerve impulses in the organ of Corti involves traveling fluid waves in the perilymph and endolymph, which set up vibrations of the basilar membrane and organ of Corti, which in turn cause a shearing action that bends the hairs, which is the final mechanical stimulus to the hair cells. This mechanical stimulus is transduced to an electrical stimulus, the cochlear microphonic. It may be either the cochlear microphonic or the shearing force that causes the release of transmitter chemicals from the hair cells, exciting the dendrites of the sensory neurons leading to the brain.

SUGGESTED READING

Case, J. (1966). "Sensory Mechanisms." Macmillan, New York.

Eakin, R. M. (1970). A third eye. *Am. Sci.* **58,** 73–79.

Parsons, T. S. (1967). Evolution of the nasal structure in the lower tetrapods. *Am. Zool.* **7,** 397–413.

Prince, J. H. (1956). "Comparative Anatomy of the Eye." Thomas, Springfield, Illinois.

Spoendlin, H. H. (1966). "The Organization of the Cochlear Receptor." Karger, Basel.

Van Bergeijk, W. A. (1967). The evolution of vertebrate hearing. *In* "Contributions to Sensory Physiology" (W. D. Neff, ed.), Vol. 2, pp. 1–49. Academic Press, New York.

Webster, D. B. (1966). Ear structure and function in modern mammals. *Am. Zool.* **6,** 451–466.

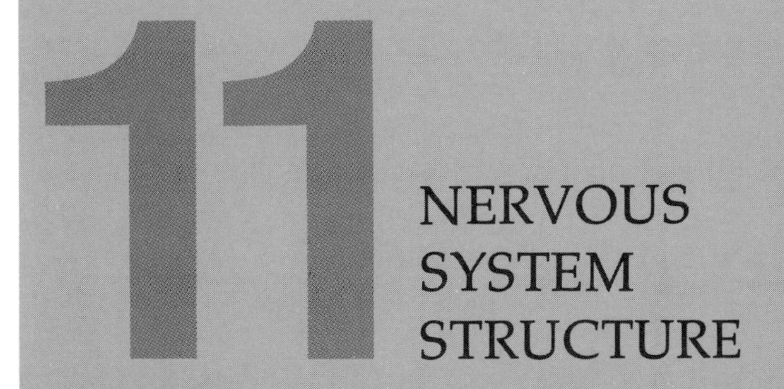

11 NERVOUS SYSTEM STRUCTURE

THE SUPRASEGMENTAL APPARATUS

DEVELOPMENT (Fig. 11-1)

Throughout its early stages the nervous system, like other organ systems, develops similarly in all vertebrates. Shortly after gastrulation a thickening called the neural plate forms in the dorsal surface ectoderm. Its lateral edges enlarge and become elevated above the rest of the surface ectoderm; these projections are known as the neural ridges. These two ridges grow medially and fuse with one another dorsally, forming a neural tube which even at this early stage is widened anteriorly as the presumptive brain and narrowed posteriorly as the presumptive spinal cord. During neural tube formation some ectodermal cells remain free between the tube and the surface ectoderm. These free ectodermal cells form the neural crest, which gives rise to the splanchnocranium, part of the neurocranium, the pigment cells, the primary sensory neurons, and the peripheral visceral motor neurons.

Individual cells of both the neural tube and the neural crest undergo rapid mitotic activity and then differentiate into neuroblasts, which are the precursors of neurons. No further mitotic activity is possible in the vertebrate nervous system; neuroblasts or neurons that die cannot be replaced. Neuroblasts migrate until they are close to where they will be located in the adult animal; then they differentiate into neurons. Axons grow from the neuroblasts' cell bodies to form tracts or nerves, and dendritic trees develop. The initiation of this growth signifies the transmutation of neuroblasts to neurons.

Not all cells of the neural crest or the neural tube differentiate into neurons, however. Some cells of the neural tube remain as ependymal cells lining the center of the hollow neural tube (brain + cord), which during the animal's life is filled with cerebrospinal fluid. Other cells of the neural crest and tube remain as Schwann cells (in the peripheral nervous system) or glial cells (in the central nervous system); Schwann cells and some glial cells protect the nerve fibers by their proximity or by the myelin they produce; the function of other glial cells is unknown.

During early differentiation of the neural tube, a groove, called the sulcus limitans, forms in the lateral walls of the central canal. Henceforth the dorsal and ventral portions of the central nervous system develop relatively independently. The dorsal portion, called the alar plate, differentiates into sensory structures, and the ventral portion, or basal plate, into motor structures.

Segmentation becomes apparent at an early stage. On each side of the body neural crest cells aggregate to a position opposite the center of each myotome. Dendrites grow from the neural crest cells toward the periphery, and axons grow into the dorsolateral aspect of the developing neural tube and make synaptic connections primarily in

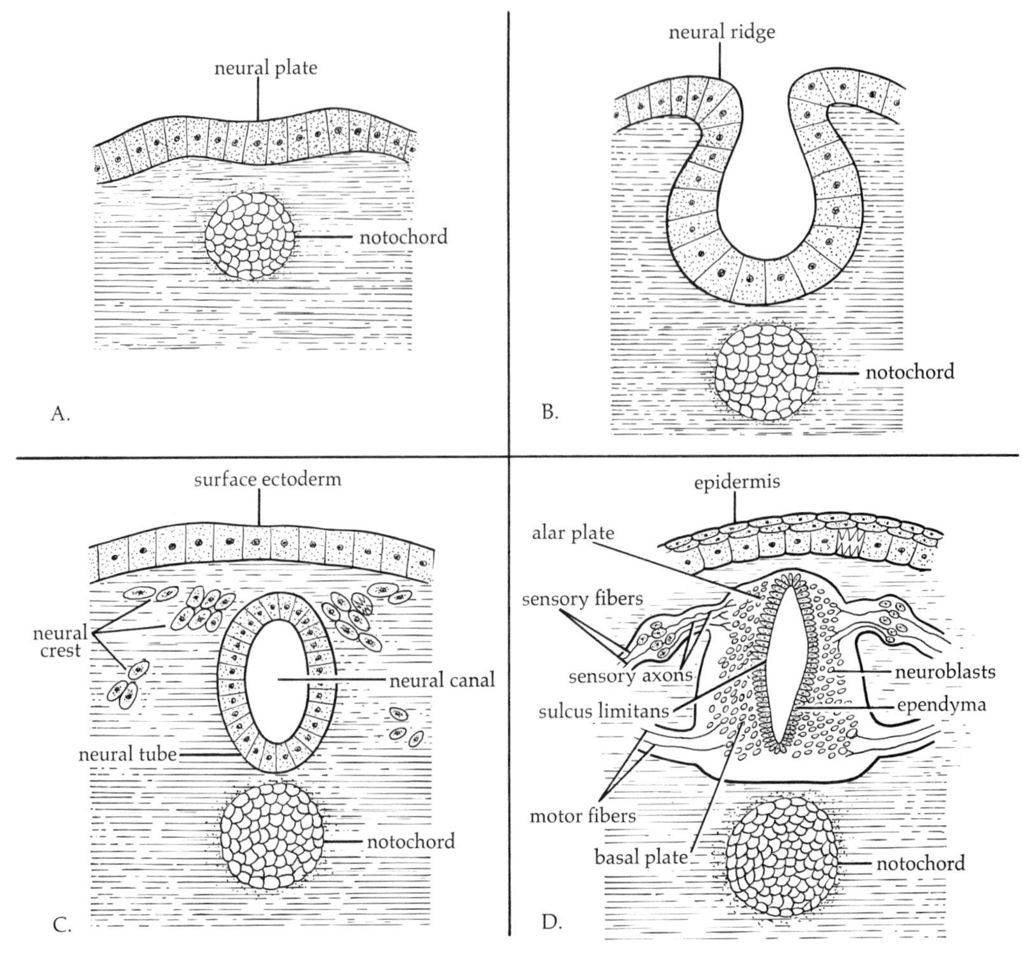

Fig. 11-1. Development of the spinal cord from early neural plate to the stage of neuroblast differentiation into neurons.

the dorsal portion (alar plate) of the developing central gray matter; these axons are the sensory fibers from peripheral structures. Concurrently neurons developing in the basal plate send their axons ventrolaterally out of the neural tube; these are the primary motor fibers innervating peripheral structures. Thus in each postcranial segment of the body there is a pair of spinal nerves, right and left, each containing both sensory and motor components. Their structure will be further described shortly. Each innervates a very specific portion of the body, which is confusingly called a dermatome, the same name as an embryonic precursor of the dermis.

Fig. 11-2. Lateral views of brain development in the pig. Abbreviations: dien, diencephalon; mes, mesencephalon; met, metencephalon; myel, myelencephalon; tel, telencephalon. After B. M. Patten (1927). "Embryology of the Pig." P. Blakiston's Son & Co., Philadelphia. Used with the permission of McGraw-Hill Book Company.

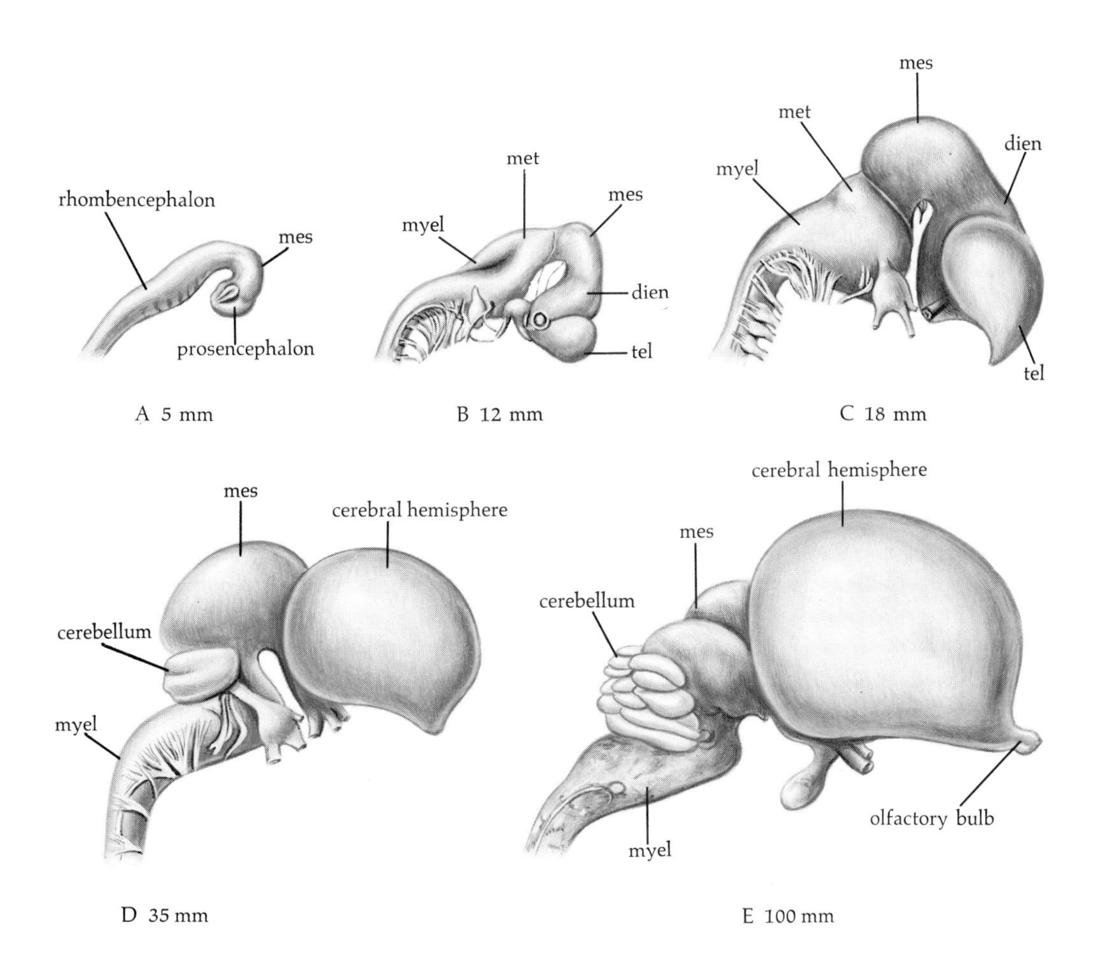
rhombencephalon
mes
prosencephalon
A 5 mm
met
myel
mes
dien
tel
B 12 mm
mes
met
myel
dien
tel
C 18 mm
mes
cerebral hemisphere
cerebellum
myel
D 35 mm
cerebral hemisphere
mes
cerebellum
olfactory bulb
myel
E 100 mm

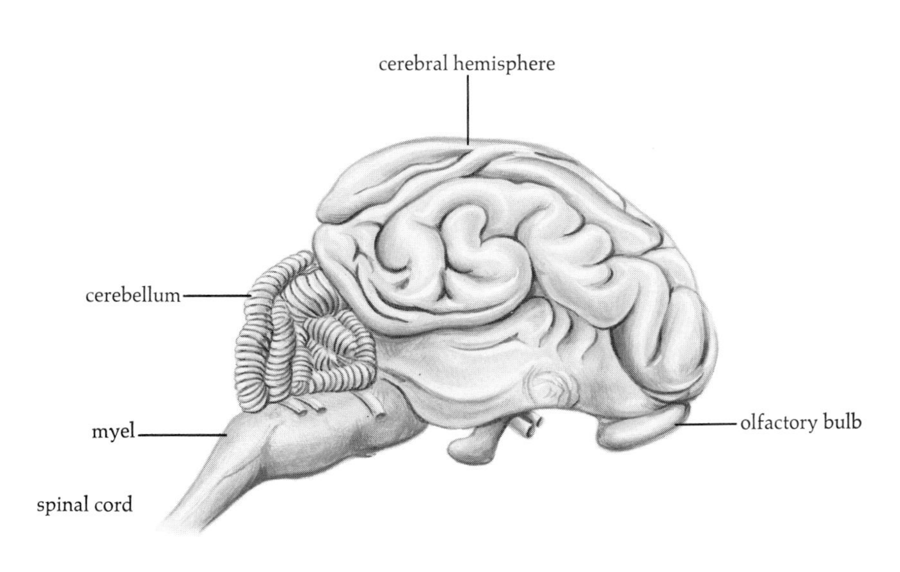
cerebral hemisphere
cerebellum
myel
olfactory bulb
spinal cord

F 200 mm

Segmentation in the developing brain is neither as regular nor as apparent as it is in the spinal cord. However, during early development (at least), segmentation can be determined in the basal portion, which is, therefore, known as the "segmental apparatus."

Soon the neural tube's anterior portion (presumptive brain) undergoes differential growth, developing three large thickenings: from anterior to posterior these are the prosencephalon or forebrain, the mesencephalon or midbrain, and the rhombencephalon or hindbrain (Fig. 11-2). Further differentiation produces subunits. The forebrain, or prosencephalon, becomes divided into the diencephalon and the more anterior telencephalon; the latter develops into large outgrowths, dorsally the cerebral hemispheres and anteriorly the olfactory bulb. The hindbrain, or rhombencephalon, differentiates into the myelencephalon caudally and, just anterior to this, into the metencephalon whose dorsal portion forms another outgrowth, the cerebellum. The diencephalon, mesencephalon, ventral metencephalon, and myelencephalon together comprise the brainstem. The two dorsal outgrowths, cerebrum and cerebellum, are the suprasegmental apparatus (Fig. 11-2).

The mammalian spinal cord develops its adult size and structure more rapidly than the surrounding vertebral column and skeletal muscles. Because of this the mammalian cord does not normally fill the neural canal and usually terminates at the level of the lumbar vertebrae. Moreover, as a result of this differential growth, the adult spinal nerves leave the vertebral column (through intervertebral foramina) at a point that is considerably caudal to their entrance into the cord, having travelled for some distance within the vertebral canal first. In fishes, amphibians, and reptiles, similar differential growth occurs in the cranial region; the brain does not fill the adult's cranial cavity, and the extra space is usually filled with fatty tissue. The spinal cord, however, extends the full length of the vertebral canal, indicating that the spinal cord and vertebral column grow at similar rates.

SPINAL CORD AND SPINAL NERVES

MAMMALS

In mammals the spinal cord extends from the foramen magnum to the lumbar region of the vertebral column. It is widest anteriorly and tapers to a fine tip in the lumbar region, but includes two enlargements: the cervical enlargement where nerves go to and from the forelimbs and a similar lumbosacral enlargement for the hindlimbs (Fig. 11-3). The entire spinal cord is covered by membranes called meninges. The outermost membrane is the tough, connective tissue dura mater, the middle is the weblike arachnoid layer, and the innermost is the pia mater, which is in intimate contact with the superficial white matter of the cord.

The cell bodies (gray matter) are gathered deeper in the spinal cord, appearing in cross section as a butterfly shape with the "central gray" surrounding the hollow canal. Each "wing" of the butterfly is organ-

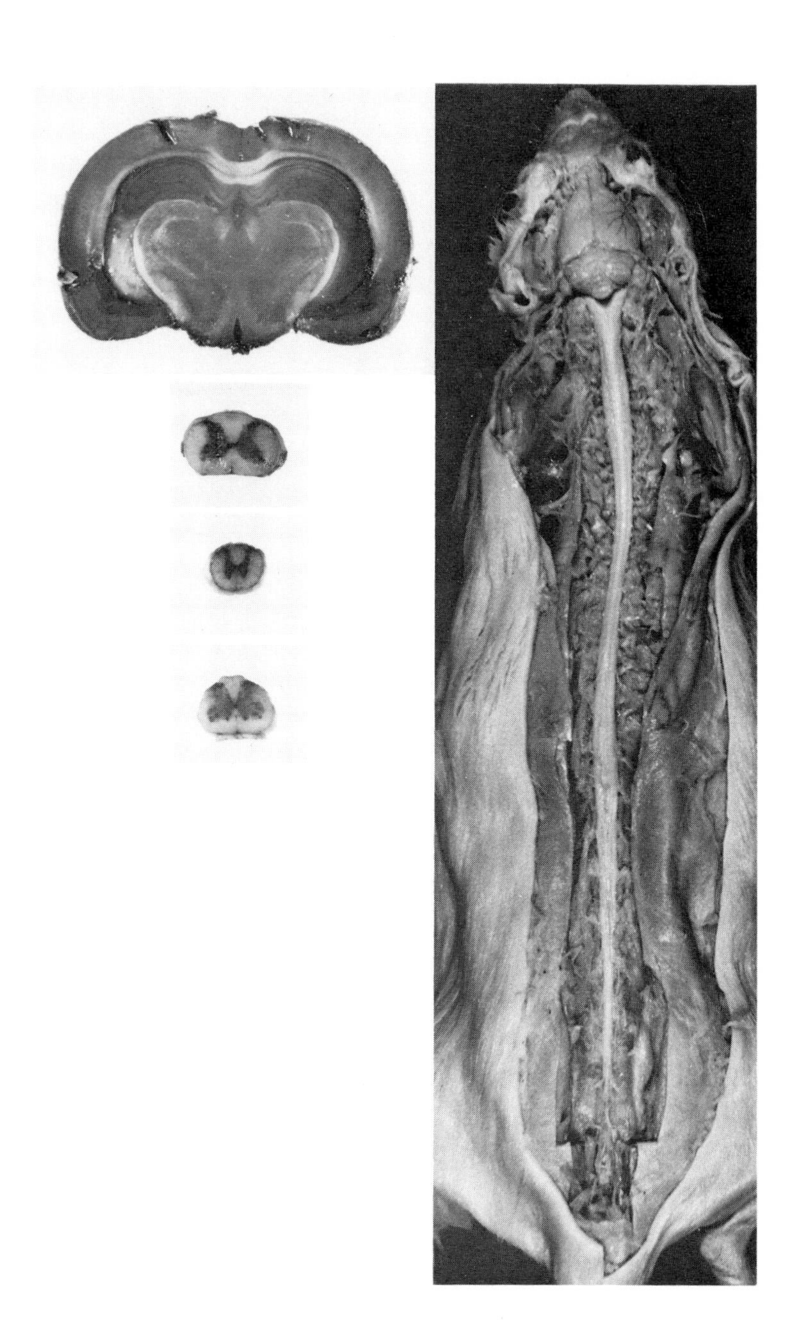

Fig. 11-3. The central nervous system of the rat. On the right is the dorsal view of a dissection, with the skull roof and neural arches and spines removed. On the left are cross sections through the cerebral hemispheres, cervical enlargement, thoracic cord, and lumbar enlargement; note the distribution of gray and white matter in different portions of the cord.

ized into four columns that run the length of the cord and are continued in the brain. Two columns are in the dorsal horn, the somatic sensory column dorsally, and the visceral sensory column ventrally, and two columns are in the ventral horn, the visceral motor column dorsally and the somatic motor column ventrally. In the somatic sensory

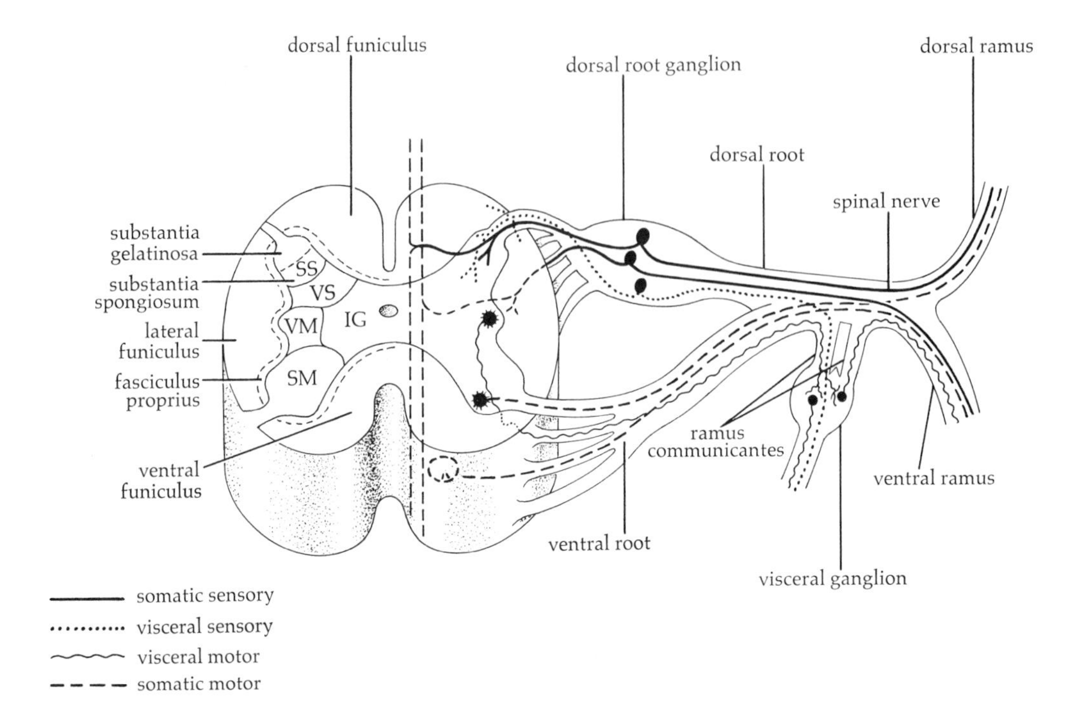

Fig. 11-4. The organization of the mammalian spinal cord and nerve as seen in a single segment. The organization of the fibers and cells of the cord are indicated on the left; the fibers are shown entering and leaving the cord on the right. (IG, intermediate column of cord; SM, somatic motor column; SS, somatic sensory column; VM, visceral motor column: VS, visceral sensory column.

column there are two subdivisions: the most dorsal part, comprised of very small cells, appears gelatinous in sections and is called the substantia gelatinosa; just ventral to this are scattered larger cells, forming what has been called the substantia spongiosa. The ventral horn is filled with large motor cells, many of whose axons leave through the ventral root. Between dorsal and ventral horns are several nuclear groups that may be classified together as the intermediate gray matter of the cord.

Fibers (white matter) lie superficially along the spinal cord throughout its length, traveling to and from various parts of the body; the white matter is proportionately greater in the anterior than in the posterior cord, since in the posterior spinal cord most of the white matter has left (or not yet joined) the cord. The white matter can be divided into the dorsal funiculus lying between the two dorsal horns, a similar ventral funiculus between the two ventral horns, and a lateral funiculus on each side. In the deepest white matter, next to the gray matter, are axons which run from one part of the spinal cord to another. This group of deep fibers is called the fasciculus proprius, and it is important for reflex connections.

Axons from primary sensory neurons in the dorsal root ganglia enter the dorsolateral portion of the cord; they make synaptic connections extensively in the substantia gelatinosa and spongiosa and few in the ventral horn. Branches from the primary sensory axons also enter the white matter.

The paired spinal nerves are formed by the dorsal root (containing the dorsal root ganglion and both visceral sensory and somatic sensory fibers) and the ventral root (containing visceral motor and somatic motor fibers); these come together just peripherally to the dorsal root ganglion and form the short spinal nerve proper, containing both sensory and motor elements. The nerve leaves the spinal cord through the intervertebral foramina and immediately bifurcates into a dorsal ramus to the epaxial musculature and overlying skin, and a ventral ramus to the hypaxial musculature and overlying skin (Fig. 11-4). Also leaving the spinal nerve ventrally are small rami communicantes that contain visceral motor and visceral sensory fibers innervating the viscera; these will be discussed with the visceral nervous system.

The ventral rami are generally larger and innervate larger areas than the dorsal rami, particularly at the level of the limbs. Here, the large ventral rami branch, intertwine, and rejoin to form complex networks called plexi: the brachial plexus for the forelimb and the lumbosacral plexus for the hindlimb (Fig. 11-5).

NONMAMMALS

Gnathostomes

The general organization of the spinal cord and spinal nerves is similar in all gnathostomes, although only in tetrapods are there distinct cervical and lumbosacral enlargements and corresponding brachial and lumbosacral plexi (Fig. 11-6).

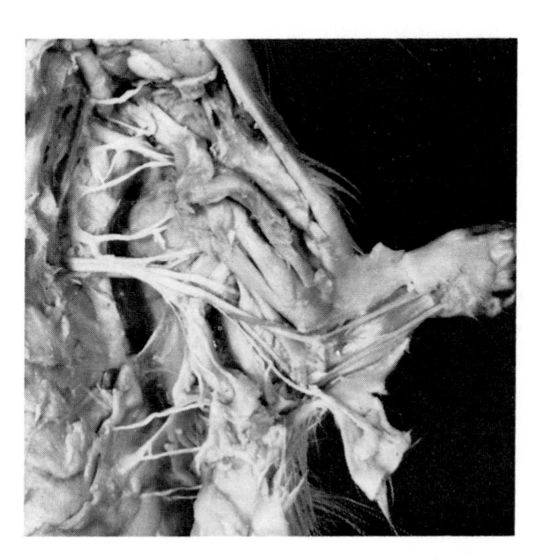

Fig. 11-5. Ventral view of the brachial plexus of a rat. Note the very large ventral rami sending large nerves out to the limb.

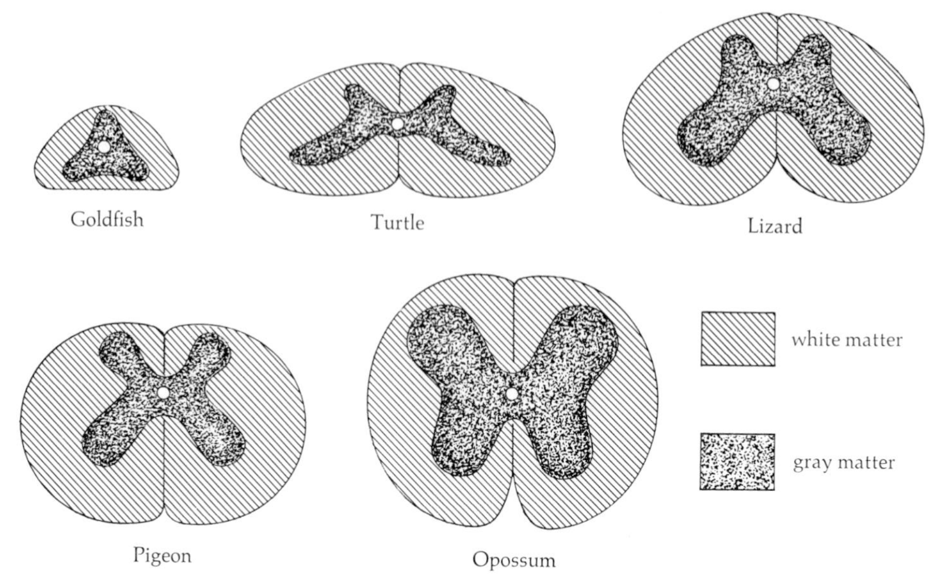

Fig. 11-6. Distribution of gray and white matter in the anterior spinal cords of diverse vertebrates.

In jawed fishes, however, there are visceral motor fibers in both the dorsal and ventral roots of the spinal nerves; this morphological condition is intermediate between tetrapods and agnathans.

Agnathans

The dorsal and ventral roots of agnathan spinal nerves alternate, with the dorsal roots emerging from the cord between myotomes and the ventral roots emerging in the center of the myotomes. The ventral roots contain only somatic motor fibers; the dorsal roots contain visceral motor, visceral sensory, and somatic sensory fibers. This is particularly important because of its similarity to the situation in the *cranial* nerves of all vertebrates, and it probably represents the primitive vertebrate condition. In the lamprey there is no connection between dorsal and ventral roots, and in the hagfish, only a slight connection.

The agnathan spinal cord has other peculiarities. There are no myelin sheaths, although glial cells are present. There are capillaries in the meninges but not in the substance of the cord as there are in other vertebrates. The spinal cord, flattened in shape, lies directly on the notochord; the cord has a small central canal and a thin layer of cells on each side of the canal. Only about four-fifths of the primary sensory neurons lie in dorsal root ganglia; the remaining ganglia are in the cord itself (Fig. 11-6).

CRANIAL NERVES

The cranial nerves of all vertebrates can be considered together in three groups (Fig. 11-7):

1. corresponding to the agnathan *ventral roots,* a group containing only or mostly somatic motor fibers and going to somatic striated musculature
2. corresponding to the agnathan *dorsal roots,* a group containing somatic sensory, visceral sensory, and visceral motor fibers
3. a group of *special sensory nerves,* carrying only sensory information from the specialized sense organs of the head

The cranial nerves are numbered I through X in anamniotes, and I through XII in amniotes. This numbering has no functional or phylogenetic significance, being merely the anterior to posterior order seen in the human; the nerves do not necessarily follow this order in all vertebrates, although in other respects they are remarkably similar.

SOMATIC MOTOR NERVES, CORRESPONDING TO VENTRAL ROOTS OF AGNATHAN SPINAL NERVES

nerve	*innervation*
III OCULOMOTOR IV TROCHLEAR VI ABDUCENS	EXTRINSIC OCULAR MUSCLES
XII HYPOGLOSSAL (AMNIOTES)	HYPOBRANCHIAL MUSCLES

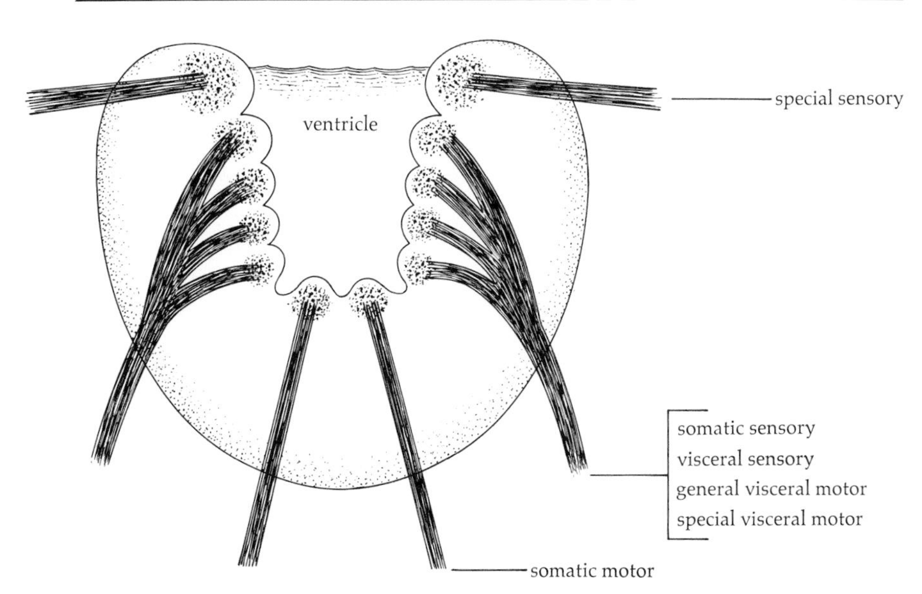

Fig. 11-7. The three functional types of cranial nerves are shown emanating from the myelencephalon of a fish. The four columns of the spinal cord continue, with minor modifications, into the brainstem. The visceral motor column divides into two parts: general visceral motor fibers innervate smooth muscles and glands and special visceral motor fibers innervate the striated branchiomeric muscles. A discontinuous special sensory column is located dorsal to the other four columns.

All these except the trochlear nerve leave the brain ventrally; the trochlear leaves the brain dorsally just in front of the cerebellum.

In anamniotes there are several occipital nerves homologous to the hypoglossal nerve of amniotes that exit from the anterior spinal cord rather than from the brain.

SOMATIC SENSORY, VISCERAL SENSORY, AND VISCERAL MOTOR NERVES, CORRESPONDING TO AGNATHAN DORSAL ROOTS (Fig. 11-8)

	nerve	*innervation*
0	NERVUS TERMINALIS	ANTERIOR NASAL EPITHELIUM (SENSORY ONLY)
V	TRIGEMINAL	MUSCLES, SKIN, ETC., IN FIELD OF MANDIBULAR ARCH AND DERIVATIVES
VII	FACIAL	MUSCLES, SKIN, ETC., IN FIELD OF HYOID ARCH AND DERIVATIVES
IX	GLOSSOPHARYNGEAL	MUSCLES, SKIN, ETC., IN FIELD OF GLOSSOPHARYNGEAL ARCH AND DERIVATIVES
X	VAGUS	MUSCLES, SKIN, ETC., IN FIELD OF ARCHES 4,5,6 AND DERIVATIVES; ALSO A LARGE VISCERAL PORTION TO MOST BODY GLANDS
XI	SPINAL ACCESSORY (AMNIOTES)	SOME MUSCLE DERIVATIVES OF LAST THREE PHARYNGEAL POUCHES

The nervus terminalis is numbered "0" because it was not known when the numbering system was adopted. It is found persistently in vertebrates, however, and consists of a small bundle of sensory fibers from the snout region. Since living cyclostomes and ancient ostracoderms have more gill slits than do living jawed fishes, it is likely that there were premandibular pharyngeal pouches which have been lost during the evolution of jaws; the nervus terminalis probably innervated one of these in ancestral fishes.

The spinal accessory is a discrete nerve only in amniotes. It is derived from part of the vagus plus anterior spinal dorsal root elements.

For a better understanding of these mixed nerves, consider a generalized fish gill slit. The inner (endodermal) part is innervated by visceral sensory fibers, and the outer (ectodermal) part and surrounding skin by somatic sensory fibers; the branchiomeric muscles of the gills are striated muscles derived from visceral mesoderm and, therefore, are innervated by visceral motor fibers. Because striated muscles have

Fig. 11-8. Diagrams illustrating the distribution of the four major branchiomeric nerves in jawed fishes. Numbers indicate gill slits. After J. S. Kingsley (1917). "Outlines of Comparative Anatomy of Vertebrates." P. Blakiston's Son & Co., Philadelphia.

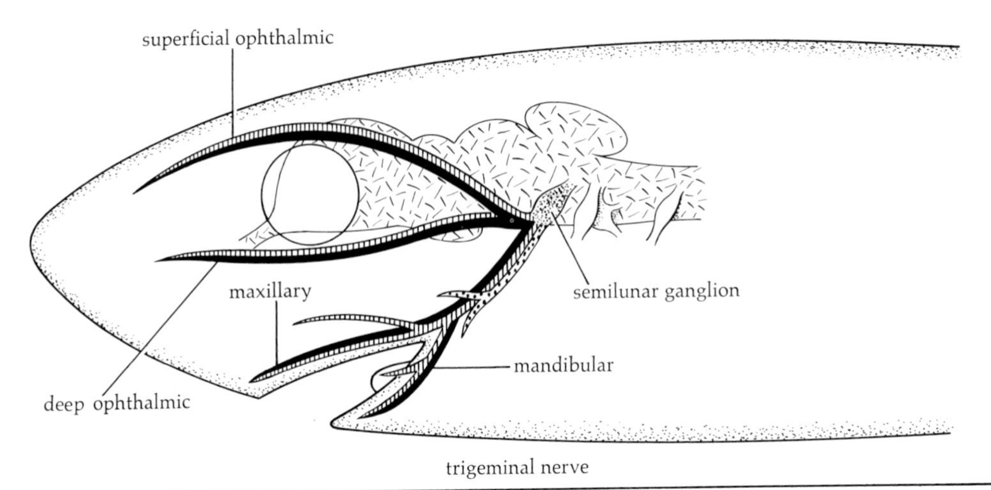

trigeminal nerve

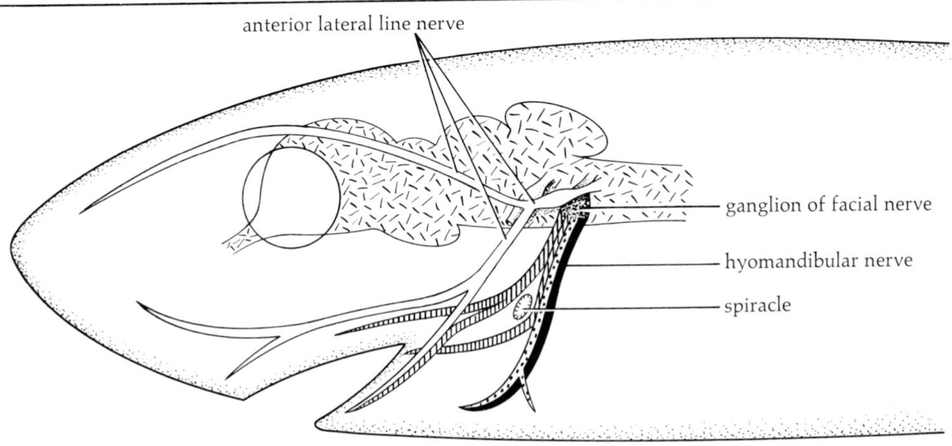

facial nerve

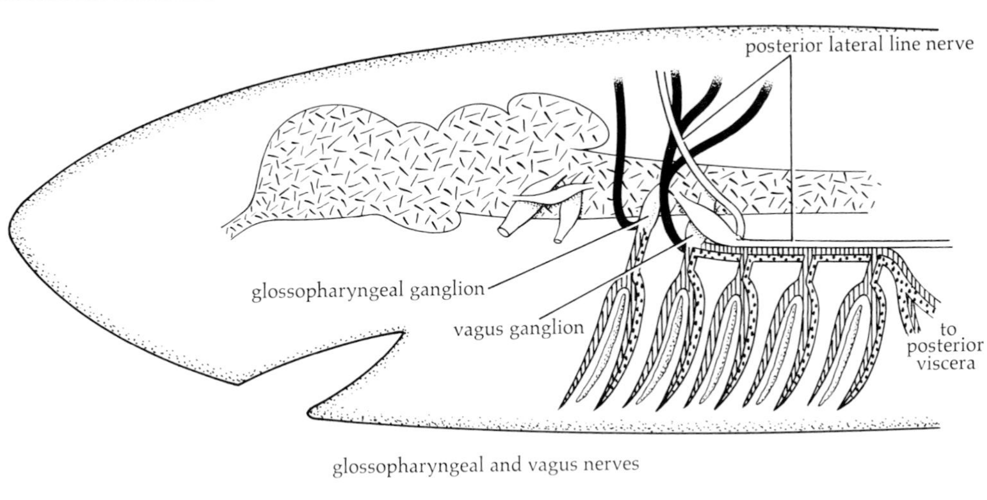

glossopharyngeal and vagus nerves

lateral line | visceral motor | visceral sensory and taste | somatic sensory

more innervation from larger neurons than do smooth muscles, these visceral motor fibers are specialized and are called the *special visceral motor fibers*. There are also smaller, more diffuse fibers called *general visceral motor fibers* innervating the smooth muscles of blood vessels and glands in the same region.

Because of the extent to which the mandibular and hyoid gill slits are modified in the formation of jaws, their nerves, the trigeminal and facial, are more modified than any others in fishes. As gill slits became modified into diverse derivatives during tetrapod evolution, even greater modifications occurred in these nerves. The trend toward increased cephalization resulted in the contribution of both endoderm and ectoderm to the evolution of special sense organs (lateral line, taste), with still further modifications of the nerves. The result of these modifications is these mixed nerves, with somatic sensory, visceral sensory, visceral motor (both general and special), and even special sensory fibers.

Somatic sensory components, V

With the loss of premandibular gill slits and arches, the entire "facial" region receives somatic sensory innervation from the trigeminal nerve, one part of which represents the somatic sensory portion of one or more premandibular nerves (not including the nervus terminalis). The trigeminal has three branches: the ophthalmic, from the upper part of the face; the maxillary, from the upper jaw; and the mandibular, from the lower jaw (Fig. 11-8). All the cutaneous somatic sensory fibers converge near the brainstem where their cell bodies lie in the large semilunar ganglion. Their axons extend into large somatic sensory nuclei in the brainstem and upper cord.

Somatic sensory components, VII, IX, X

The somatic sensory portions of the facial, glossopharyngeal, and vagus are much smaller than that of the trigeminal. In fishes only the skin immediately around the external gill slits receives sensory innervation from these three nerves. In tetrapods, the loss of gill slits results in further reduction of the somatic sensory fields of these nerves; in mammals, the only somatic sensory innervation is from part of the external ear.

Visceral sensory components

Visceral sensory fibers from the organs innervated by the general visceral motor fibers of VII, IX, and X accompany the general visceral motor fibers of these nerves to and from the same organs. There is no visceral sensory component in V.

General visceral motor components

Glands, blood vessels, and smooth muscles of the head and most of the trunk receive general visceral motor fibers from cranial nerves VII,

IX, and X (also some from the primarily somatic motor nerve III, none from V). These represent a substantial portion of the autonomic nervous system to be described below.

Special visceral motor components

The motor axons to the branchiomeric muscles make up the special visceral motor fibers. They travel in nerves V, VII, IX, X, and, in amniotes only, XI; they innervate all the muscles of each arch derivative, as described in Chapter 7.

Special sensory components

Although not true "special sensory nerves," three of these mixed nerves contain fibers from the taste and lateral line senses; these are VII, IX, and X. Taste fibers from the anterior portion of the tongue (plus the external taste buds of some scaleless teleosts such as catfish) are carried in the facial nerve, those from the posterior tongue and the oral cavity in the glossopharyngeal, and those from the pharynx and epiglottis in the vagus.

The lateral line nerves of fishes and larval amphibians are actually quite separate from the other cranial nerves. The lateral line organs of the head are innervated by the anterior lateral line nerve, which enters the brain next to the facial nerve and is generally considered to be a part of it. The lateral line organ of the trunk is innervated by the posterior lateral line nerve, which enters the brain with the glossopharyngeal and vagus nerves and is thus often regarded as a part of one or the other of these nerves. Both lateral line nerves have their own ganglia, however, separate from the main sensory ganglia of the nerves they accompany, and should be considered to be separate (unnumbered) cranial nerves characteristic of fishes and larval amphibians.

SPECIAL SENSORY NERVES

nerve		*innervation*
I	OLFACTORY	OLFACTORY EPITHELIUM; USUALLY AS SEPARATE FASCICLES
	(VOMERONASAL)	(VOMERONASAL EPITHELIUM)
II	OPTIC	ACTUALLY A BRAIN TRACT, CARRYING AXONS OF RETINAL GANGLION CELLS
VIII	STATOACOUSTIC	VESTIBULAR AND AUDITORY PORTIONS OF INNER EAR

These nerves have been discussed in Chapter 10; their central connections will be discussed in Chapter 12.

VISCERAL NERVOUS SYSTEM

The visceral nervous system is not a distinct system but an artificial classification including some parts of the nervous system already discussed. It comprises all the nuclei, ganglia, tracts, and nerves—motor and sensory—that innervate the viscera or its derivatives; thus, the visceral nervous system involves the innervation of digestive, respiratory, urinary, and reproductive viscera and also of blood vessels, heart, and integumentary glands.

The motor components of the visceral nervous system, carrying "commands" from the central nervous system to viscera and glands, comprise the autonomic nervous system. The autonomic system is structurally and functionally divided into antagonistic portions, the sympathetic and the parasympathetic, which provide dual innervation to most glands and viscera.

In general, excitation of the sympathetic system prepares an animal for an emergency situation—for "flight or fight"—by increasing heart rate, dilating pupils, erecting mammalian hairs and by inhibiting the animal's vegetative organs such as digestive glands, alimentary canal, and urinary bladder. Excitation of the parasympathetic system causes the opposite effects, facilitating digestion and generally slowing down somatic activities. The antagonistic actions of these two parts of the autonomic nervous system are mediated by different neural transmitting humors, adrenaline (or noradrenaline) for the sympathetic and acetylcholine for the parasympathetic.

The sensory fibers of the visceral nervous system travel with the motor (autonomic) fibers, carrying information toward the central nervous system. Their cell bodies lie in the dorsal root or cranial ganglia; axons from these cell bodies extend into the visceral sensory gray columns.

MAMMALIAN AUTONOMIC SYSTEM (Fig. 11-9)

There are a few general principles of the autonomic nervous system that make it more understandable. By definition it includes only visceral motor elements (no sensory), and of those elements, only general visceral motor (no special). From the central nervous system to the effector organ there is always a two-neuron chain. The first, called the

Fig. 11-9. Diagram of the anatomical relationships of the autonomic nervous system of the rat, with only the sympathetic portion shown on the right and only the parasympathetic on the left. The connections via gray rami communicantes to the vasculature and glands of the skin are omitted. Roman numerals indicate the oculomotor, facial, glossopharyngeal, and vagus cranial nerves; C1, T1, L1, S1, and Ca1 indicate the first cervical, thoracic, lumbar, sacral, and caudal spinal nerves, respectively.

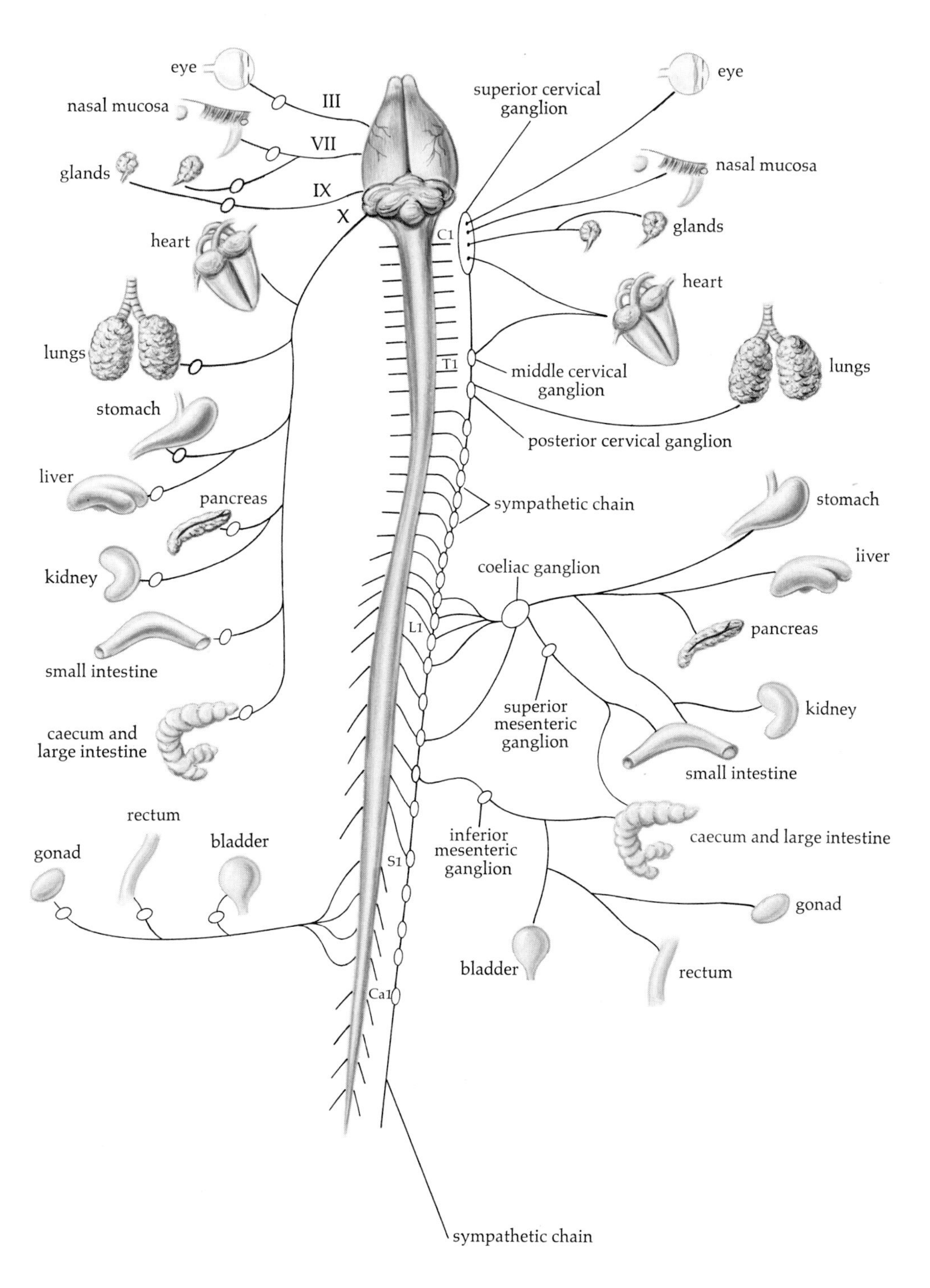
Parasympathetic
Sympathetic
eye
nasal mucosa
glands
III
VII
IX
X
heart
lungs
stomach
liver
pancreas
kidney
small intestine
caecum and
large intestine
rectum
gonad
bladder
superior cervical
ganglion
eye
nasal mucosa
glands
C1
heart
T1
middle cervical
ganglion
lungs
posterior cervical ganglion
sympathetic chain
stomach
liver
coeliac ganglion
pancreas
L1
superior
mesenteric
ganglion
kidney
small intestine
inferior
mesenteric
ganglion
caecum and large intestine
S1
gonad
bladder
rectum
Ca1
sympathetic chain

preganglionic neuron, lies in the central nervous system and sends its axon to a peripheral autonomic ganglion. The second, or postganglionic neuron, lies in the autonomic ganglion with its axon extending to the effector organ.

For the sympathetic system, the preganglionic cell bodies lie in the lateral horn of the thoracic and lumbar portions of the cord; for the parasympathetic system, the preganglionic cell bodies lie in the brain and in the lateral horn of the sacral region. Furthermore, in the sympathetic system the preganglionic axon is short and the postganglionic long, while in the parasympathetic system the opposite is true.

Sympathetic system

Preganglionic neurons of the sympathetic system send axons through the ventral root of the spinal nerve. These terminate on second-order (postganglionic) neurons located in peripheral sympathetic ganglia, either ventrolateral to the vertebral column or near large blood vessels in the dorsal mesentery (Fig. 11-9). Of the sympathetic ganglia in the dorsal mesentery, the largest (coeliac and superior mesenteric) contain most of the postganglionic cell bodies of the axons innervating the major thoracic and abdominal viscera. The remaining sympathetic ganglia form a chain running throughout the lumbar and thoracic regions and include three cervical sympathetic ganglia, the posterior, the middle, and the superior. From some of the postganglionic cells in the sympathetic chain, small, lightly myelinated, or unmyelinated axons rejoin the spinal nerves (through fine gray rami communicantes) and travel in the dorsal and ventral rami to innervate peripheral blood vessels, smooth muscles, and skin glands (Fig. 11-4).

Parasympathetic system

Preganglionic parasympathetic fibers leave the brain with cranial nerves III, VII, IX, and especially X. In the oculomotor nerve these preganglionic parasympathetic fibers synapse with postganglionic cells in the ciliary ganglion just outside the eyeball; postganglionic axons innervate smooth muscles that control pupillary constriction in the iris and accommodation in the ciliary body. The preganglionic parasympathetic fibers of the facial and glossopharyngeal nerves synapse with postganglionic cells in small cranial ganglia; their postganglionic axons primarily control the secretions of the salivary and lacrimal glands.

Most of the parasympathetic fibers travel with the vagus through the cervical, thoracic, and abdominal regions, and branch into all but the most caudal viscera. These branches travel to postganglionic cell bodies, located in small intramural ganglia in the walls of the viscera; from these ganglia very short postganglionic fibers innervate the organs.

A small parasympathetic outflow goes from the sacral region of the cord to the lower alimentary canal and genital organs; again the postganglionic cells are in intramural ganglia.

NONMAMMALIAN AUTONOMIC SYSTEM

The visceral nervous systems of reptiles and birds are similar to that of mammals, but in anamniotes there are differences, particularly in the autonomic portion. In teleosts and amphibians, preganglionic sympathetic fibers emerge from the trunk spinal cord and pass to sympathetic ganglia lying close to the spinal nerves (as in mammals), but the chain of sympathetic ganglia extends all the way up to the cranial region. Postganglionic sympathetic fibers extend to the viscera or rejoin the spinal nerve and pass to the skin; pigment expansion or contraction in fish chromatophores is largely controlled by the autonomic nervous system. The vagus nerve contains most of the visceral efferents of the parasympathetic system, and this nerve extends to most of the viscera. In amphibians most organs receive antagonistic visceral efferent innervation (both sympathetic and parasympathetic); however, in fishes many organs receive only one of these innervations.

In elasmobranchs, autonomic ganglia lie near the spinal cord, one per segment, and send their postganglionic fibers to the organs to be innervated. The vagus also sends branches to most viscera. However it is difficult to distinguish either anatomically or physiologically between the sympathetic and parasympathetic portions.

In cyclostomes, the visceral nervous system is even more diffuse. The vagus innervates most of the alimentary canal, but nowhere are there discrete sympathetic or parasympathetic ganglia. Visceral efferents run through the dorsal root of the spinal nerve and go to blood vessels and glands, but they have no second neuron interposed along their route. Therefore it would seem that a definitive autonomic nervous system evolved sometime after the cyclostomes branched away from the other lines of vertebrate evolution.

STRUCTURE OF THE BRAIN (Fig. 11-10, see pages following p. 177)

The spinal cord's functional gray columns continue through the brainstem but become discontinuous as additional fiber tracts appear in the white matter and neural groups (nuclei) are segregated in the gray matter.

Myelencephalon

Within the myelencephalon the central canal widens into the fourth ventricle, which is the most posterior of the brain's large cerebrospinal spaces. The alar plate forms the ventricle's lateral walls; its roof is an ependymal layer, the tela choroidea, containing a vascular network called the choroid plexus of the fourth ventricle.

The extension of the somatic motor column into the myelencephalon includes the nucleus of the hypoglossal nerve and the nucleus of the abducens nerve (Fig. 11-11).

The visceral motor column is broken into two portions. Just lateral and slightly dorsal to the somatic motor column are the three nuclei of

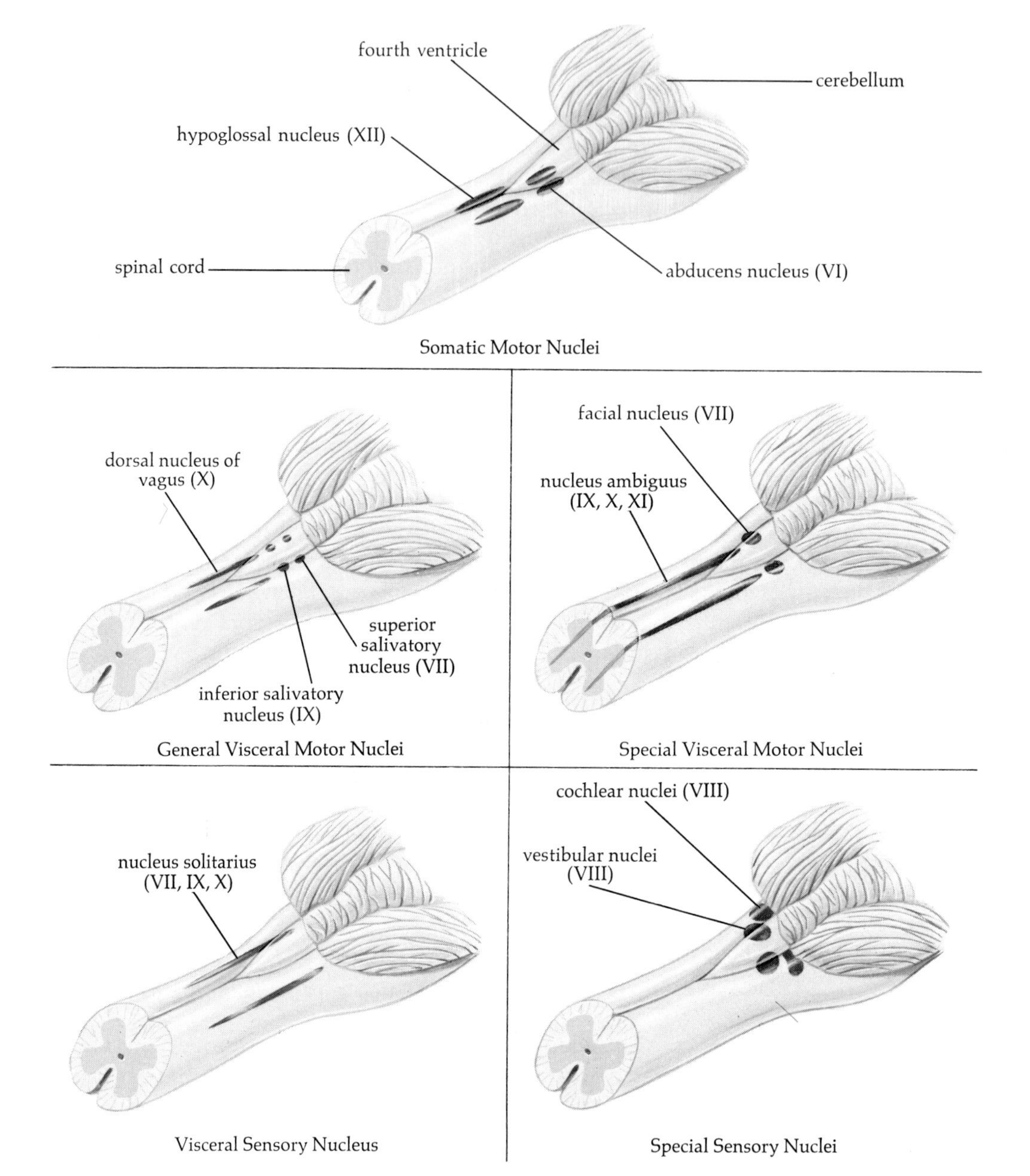

Fig. 11-11. Diagrams showing the positions of the central nuclei of the cranial nerves emanating from the mammalian myelencephalon (VI, VII, VIII, IX, X, XI, XII).

origin of the (parasympathetic) general visceral motor fibers: the dorsal nucleus of the vagus, giving rise to all the preganglionic parasympathetic fibers of the vagus nerve; the inferior salivatory nucleus, whose fibers join with and form part of the glossopharyngeal nerve to innervate the parotid gland; and the superior salivatory nucleus, whose fibers form part of the facial nerve and innervate the submaxillary, sublingual, and lacrimal glands (Fig. 11-11).

Ventral to these general visceral motor nuclei are the special visceral motor nuclei whose fibers go directly to the striated branchiomeric musculature. The most posterior of these is a long nucleus, the nucleus ambiguus, which gives rise to special visceral motor fibers of the glossopharyngeal, vagus, and (in amniotes) spinal accessory nerves. Further anterior is the facial nucleus giving rise to motor fibers. In mammals these fibers innervate the muscles of facial expression, the posterior digastric, and the stapedius; in nonmammals they innervate the muscles derived from the hyoid arch.

The nuclei of these special visceral motor nerves develop in the dorsal part of the myelencephalon, near the fourth ventricle's ventrolateral wall, and then migrate to a ventrolateral position in the myelencephalon. Their route of migration is shown by their axons which travel dorsomedially from the nuclei and then make a "U turn" and exit dorsal and lateral to the nuclei. The "U turn" is called the genu of the nerve (Fig. 11-12).

The visceral sensory column has one nucleus in the myelencephalon, the nucleus solitarius, which lies just dorsal and lateral to the parasympathetic nuclei. General sensory and taste fibers enter the nucleus solitarius from three cranial ganglia lying just outside the brain: the nodose ganglion of the vagus nerve, the petrous ganglion of the glossopharyngeal, and the geniculate ganglion of the facial. Within the nucleus the fibers of taste and general sense are separated.

Somatic sensory nerve fibers in the myelencephalon are represented only by a few fibers of the facial, glossopharyngeal, and vagus nerves. These fibers terminate in trigeminal sensory nuclei, which will be described below with the metencephalon.

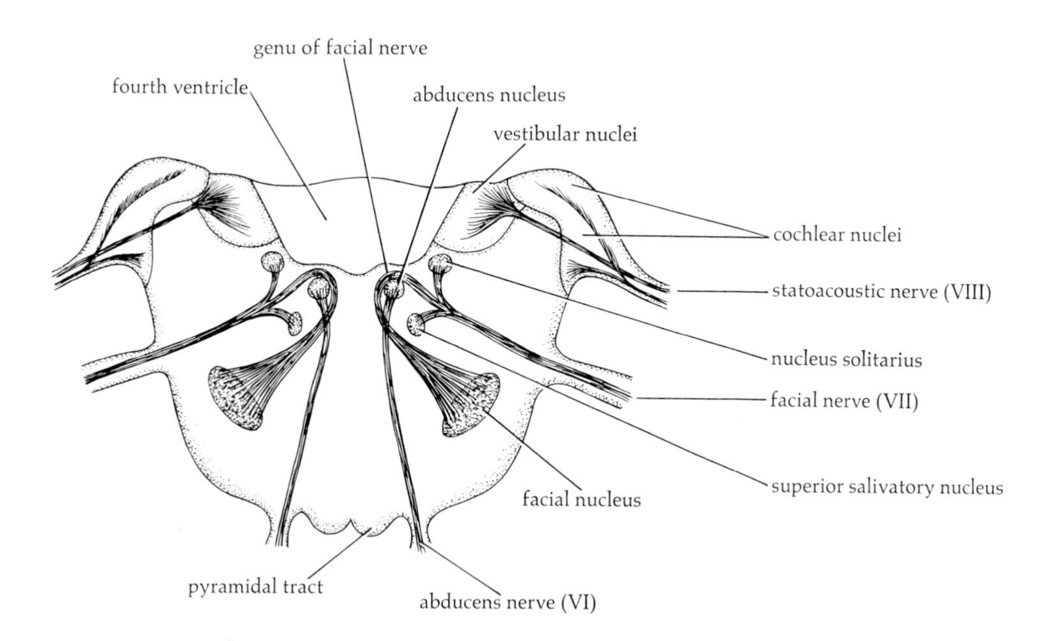

Fig. 11-12. Cross section through a mammalian upper myelencephalon, just behind the cerebellum, showing the relationships of the abducens, facial, and statoacoustic cranial nerves.

Throughout the brain there is a discontinuous fifth gray column, dorsal to the somatic sensory column, which receives primary neurons of special sense. This special sensory column is present in the anterior myelencephalon where it receives primary axons from the vestibular and acoustic branches of the statoacoustic nerve (VIII). The sensory ganglion of the vestibular portion lies in the otic capsule; its axons enter the brainstem dorsally and travel to the superior, inferior, medial, and lateral vestibular nuclei which are located in the lateral walls of the fourth ventricle. The acoustic portion's sensory ganglion is the spiral ganglion within the modiolus; its axons terminate in the dorsal and ventral cochlear nuclei, located dorsolateral to the vestibular nuclei. In nonmammalian vertebrates, these are called the nucleus angularis and the nucleus magnocellularis, respectively, but they maintain approximately the same positions.

The myelencephalon has several other important features. All amniotes have two prominent pairs of dorsal nuclei forming the posterior walls of the fourth ventricle, i.e., on each side, the nucleus gracilis medially and the nucleus cuneatus laterally; both pairs of nuclei are sensory in function. Prominent in birds and mammals but small or absent in other vertebrates is the inferior olivary nucleus, which lies about midway in the myelencephalon near its ventral aspect. The inferior olive and cerebellum are functionally related, and in species where one is large the other is large also. In some mammals the inferior olive is so large that it forms a prominent bulge ("the olive") on the ventral surface. In mammals only, paired pyramidal tracts lie prominently on each side of the midventral line.

Throughout the core of the brainstem (including myelencephalon) all vertebrates have a loose, diffuse plexus or network of generally large neurons and fibers, both sensory and motor; this is called the reticular formation.

While this description of the myelencephalon applies basically to most vertebrate groups, teleosts show some startling variations. In those with taste buds over the body surface, such as catfish, the visceral sensory nuclei for taste are often greatly enlarged, and the facial or vagal nuclei (or both) form lobes bulging out from the fourth ventricle; these are sometimes larger than the cerebellum itself (Fig. 11-13). In other teleosts, such as the mormyrids, the central nuclei of the lateral line system (closely associated with the vestibular nuclei) are greatly enlarged and form similar lobes. In other vertebrates, both the visceral sensory and the lateral line nuclei are relatively small or absent.

Metencephalon (except cerebellum)

The metencephalon, a comparatively short brain segment, includes the anterior portion of the fourth ventricle. In mammals the metencephalon contains the pons, a prominent ventral structure involved in interconnections between the cerebral hemispheres and cerebellum. This structure is largest in primates, in which it contains the decussating fibers from massive nuclear groups. (To decussate is to cross the midline; *pons* is Latin for bridge). The decussated fibers leave the pons

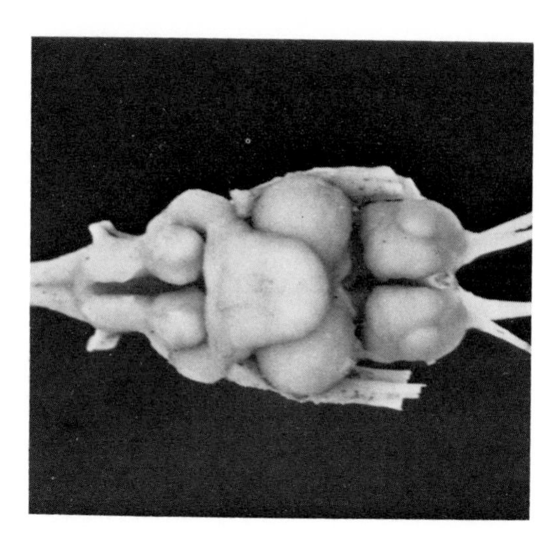

Fig. 11-13. Dorsal view of the brain of a catfish. Note the large facial and vagal lobes behind the cerebellum. These lobes are nerve centers for taste; catfish have taste buds over their entire body, which are innervated by the facial and vagus nerves.

as the brachium pontis, which forms the ventral and lateral walls of the metencephalon and passes into the cerebellum. While it is only in mammals that the pons appears as a large, grossly visible structure, there are similar nuclear and fiber groups in birds and possibly in other vertebrates as well.

The continuation of the gray columns into the metencephalon is influenced by the only cranial nerve entering this segment, the trigeminal (V). Since in mammals this nerve has only somatic sensory and visceral motor components, the other three columns, somatic motor, visceral sensory, and special sensory, do not appear in the mammalian metencephalon.

The special visceral motor column is represented by the masticator nucleus, which is the motor nucleus of the trigeminal nerve. In all vertebrates the masticator nucleus sends axons to the striated branchiomeric muscles associated with the mandibular arch. The largest of these are the muscles which close the jaws: the adductor mandibulae (in sharks), the levator mandibulae (in amphibians), and the masseter and temporalis (in amniotes). Other derivatives of the mandibular arch muscles are also innervated by the motor nucleus of V, for example, in mammals, the mylohyoid, the anterior belly of the digastricus, and the tensor tympani.

There is a large somatic sensory input from the head to the metencephalon by way of the trigeminal and a correspondingly large somatic sensory column (Fig. 11-14). The skin of the head is quite sensitive to tactile stimulation; its many sensory fibers have cell bodies in the large semilunar ganglion, from which axons terminate in two sensory nuclei of the trigeminal nerve. The smaller of the two nuclei, at the level of

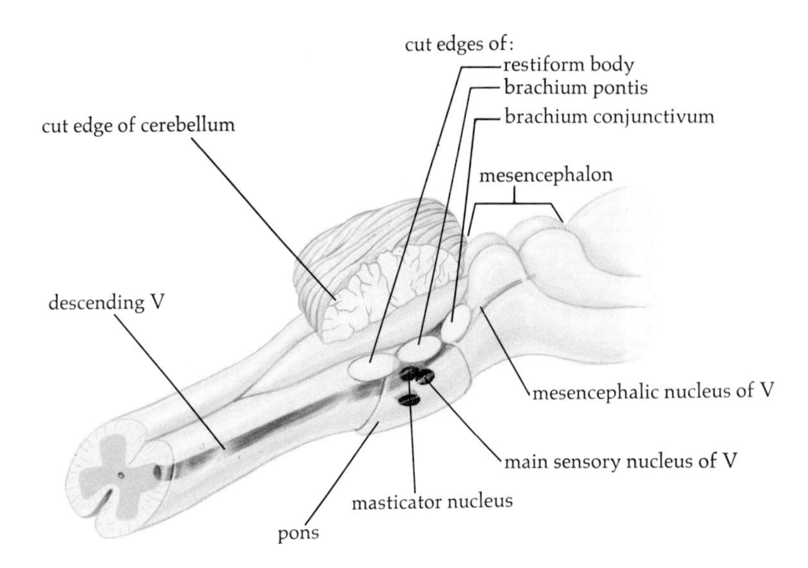

Fig. 11-14. Schematic view of the posterior part of a mammalian brain to show the extent of the trigeminal nerve nuclei. The right half of the cerebellum is removed, showing the cut edges of the cerebellar peduncles.

entrance of the trigeminal nerve, is called its main sensory nucleus. Much larger is the descending, or spinal, nucleus of the trigeminal nerve (descending V for short); the descending nucleus extends from the metencephalon caudally through the entire myelencephalon and into the anterior part of the spinal cord and is then continuous with the dorsal horn of the cord. Axons from the semilunar ganglion enter at the level of the anterior part of the nucleus and send a superficial tract along it, called the tract of descending V; this becomes smaller caudally as the fibers terminate in the cells of the nucleus.

Entering with the trigeminal are the proprioceptive fibers from muscles and the sensory fibers from the teeth and gums; these fibers then pass anteriorly on each side of the ventricular system up to their cell bodies in the mesencephalon. These cell bodies compose the mesencephalic nucleus of V, an unusual, perhaps unique, nucleus since it contains the *primary* sensory cell bodies which are elsewhere found only in peripheral ganglia.

The metencephalon also includes a continuation of the reticular formation, as described for the myelencephalon.

The most prominent feature of the metencephalon is its dorsal outgrowth, the cerebellum, part of the suprasegmental structure of the brain. It will be discussed more fully in a later section. In mammals, cerebellar input and output occur by way of three paired tracts, called peduncles, which arch above the fourth ventricle (Fig. 11-14). From posterior to anterior these are the posterior cerebellar peduncle (restiform body), the middle cerebellar peduncle (brachium pontis), and the anterior cerebellar peduncle (brachium conjunctivum).

In nonmammalian vertebrates, large tracts, not distinctly separated from one another, pass from metencephalon to cerebellum and back.

Mesencephalon

As the fourth ventricle enters the midbrain it becomes a narrow canal. In mammals it passes through this section of the brain as the aqueduct of Sylvius. In other vertebrates the ventricle widens again into large lateral extensions which lie just deep to optic centers and are, therefore, called the right and left optic ventricles.

In the midbrain the alar plate is called the tectum, and the basal plate, the tegmentum. The mammalian tectum is made up of two pairs of bulges that are collectively called the corpora quadrigemina. The anterior pair is the superior colliculi (visual centers), and the posterior pair is the inferior colliculi (auditory centers). This entire alar plate is an extension of the special sensory column (Fig. 11-15).

In nonmammalian vertebrates only the visual centers, corresponding to the mammalian superior colliculi, come to the surface of the brain; thus nonmammalians have a corpora bigemina, otherwise called the optic tectum or optic lobes. A nonmammalian auditory center, comparable in function to the mammalian inferior colliculus, lies just deep to the optic ventricle as a gross swelling, the torus semicircularis, which also contains somatic sensory centers (Fig. 11-15).

The somatic sensory column is very limited, with only a few cells in the deep part of the tectum. The visceral sensory column does not extend past the nucleus solitarius in the anterior myelencephalon.

The visceral motor column has only one nucleus, the parasympathetic nucleus of Edinger-Westphal, in the anterior midbrain; this nucleus lies in the dorsal part of the tegmentum near the midline and sends parasympathetic fibers with the oculomotor nerve to the ciliary ganglion where information is relayed to the iris and ciliary body.

The midbrain somatic motor column has two nuclei: posteriorly the trochlear nucleus, giving rise to the trochlear nerve; anteriorly the oculomotor nucleus, giving rise to the oculomotor nerve. The root from the trochlear nerve is unusual in two respects: it is the only cranial nerve that exits the brain dorsally (between tectum and cerebellum); and it is the only somatic motor nerve that decussates completely after leaving the nucleus (Fig. 11-16). It innervates the superior oblique muscle on the opposite side of the body.

The ventrolateral walls of the mammalian midbrain have conspicuous bulges formed by the cerebral peduncles. Between the peduncles is a recessed space, the interpeduncular fossa. The peduncles carry information from the mammalian cerebral cortex to other parts of the brain and to the spinal cord. They are peculiar to mammals, however, and the ventrolateral aspects of the nonmammalian midbrain are relatively smooth.

In all vertebrates the reticular formation extends through the midbrain. Another characteristic structure deep in the tegmentum is the red nucleus, which is very large and prominent in mammals and less

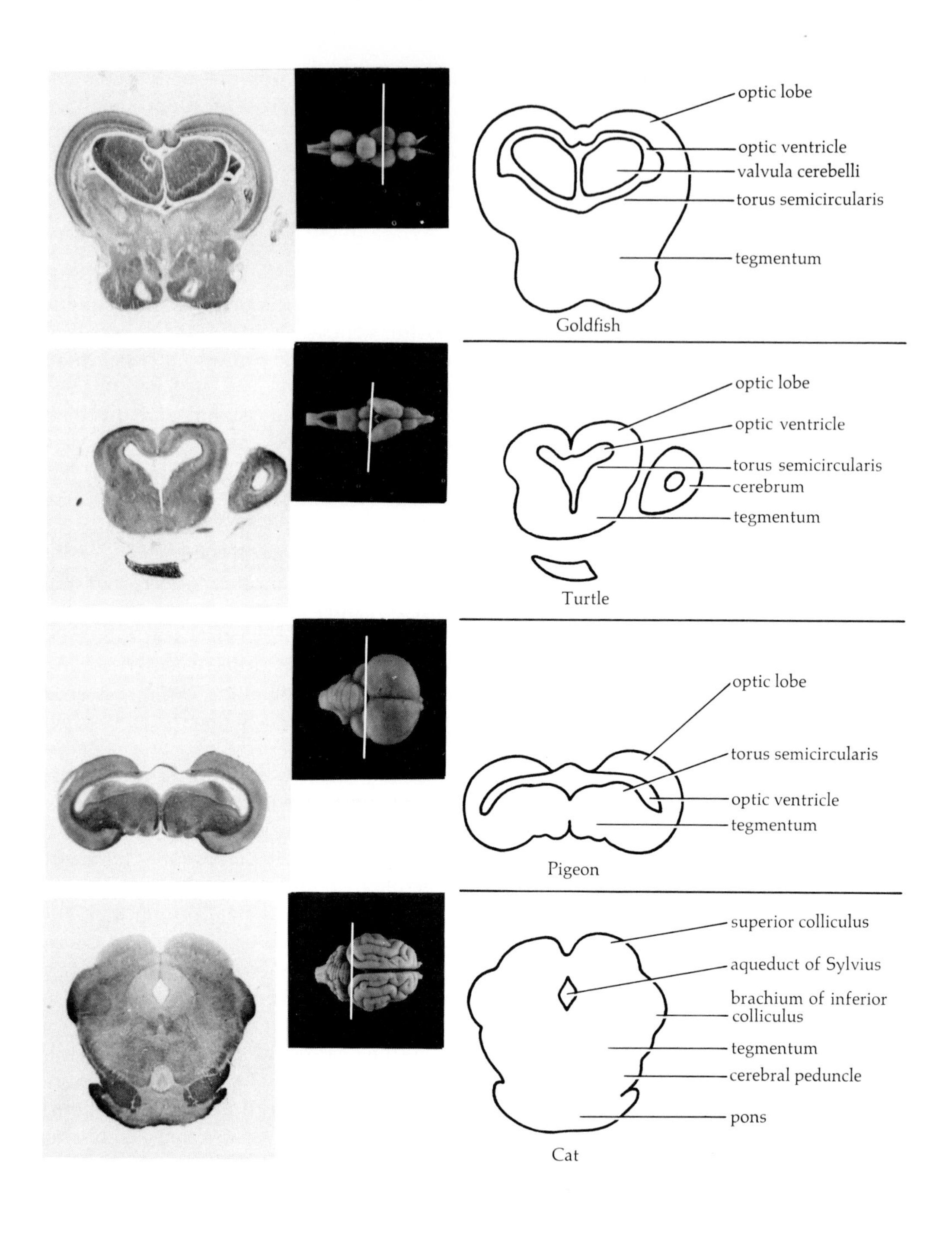

Fig. 11-15. On the left is a transverse section through the mesencephalon in each of four vertebrates. The position of the section is indicated by a white line on the dorsal view of each brain. Cerebellum and cerebrum are omitted from the sections of the pigeon and the cat.

prominent in nonmammals. The red nucleus plays an important role in cerebellar functions and in sending motor impulses to the spinal cord.

Diencephalon

The third ventricle runs through the diencephalon as a narrow, deep slit in mammals and as a wider space in other vertebrates. It is covered by a tela choroidea, with a choroid plexus similar to that of the fourth ventricle.

Although details are complex, the general functional organization of the diencephalon is relatively simple. It is divided dorsally to ventrally into epithalamus, dorsal thalamus, ventral thalamus, and hypothalamus, and it has only one entering nerve, the optic nerve (Fig. 11-17).

The most dorsal portion is called the epithalamus; it is concerned (although not critically) with the special sense of olfaction, and it contains the habenular nuclei.

Ventral to the epithalamus is the dorsal thalamus. In amniotes this is the largest portion of the diencephalon; in anamniotes, on the other hand, the third ventricle is considerably larger and the dorsal thalamus considerably smaller than in amniotes. The dorsal thalamus is primarily somatic sensory in nature, but it also contains special sense centers for vision and audition. Indeed in mammals (and perhaps all vertebrates) all sensory information except that of olfaction must synapse in a nuclear component of the dorsal thalamus before reaching the cerebral cortex.

Just ventral to the dorsal thalamus is the ventral thalamus (sometimes called the subthalamus), which is primarily concerned with somatic motor functions. No somatic motor nerves emanate from the ventral thalamus, but it sends information to motor nuclei.

Most ventral is the hypothalamus, the largest part of the diencephalon in anamniotes and complex in structure in all vertebrates. It

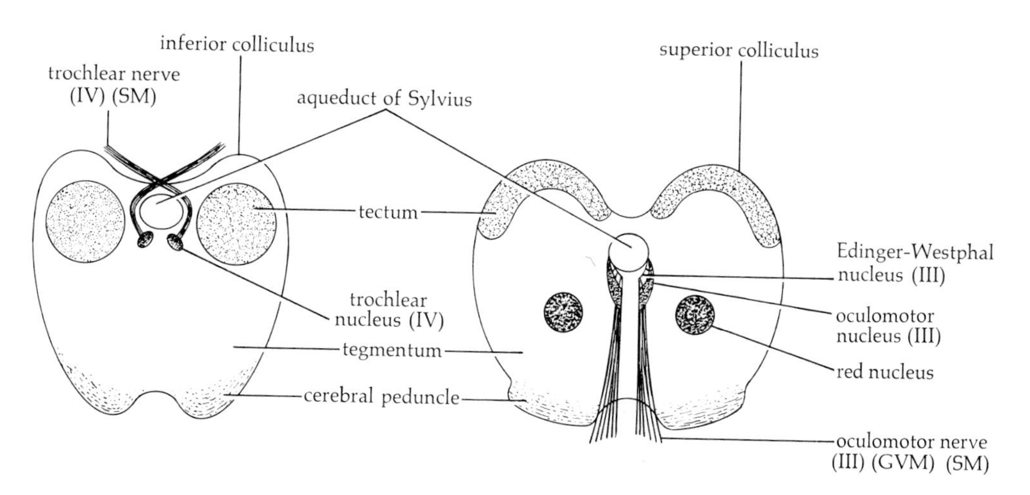

Fig. 11-16. Cross sections through a mammalian midbrain. On the left the section is further posterior than on the right. (SM, somatic motor; GVM, general visceral motor).

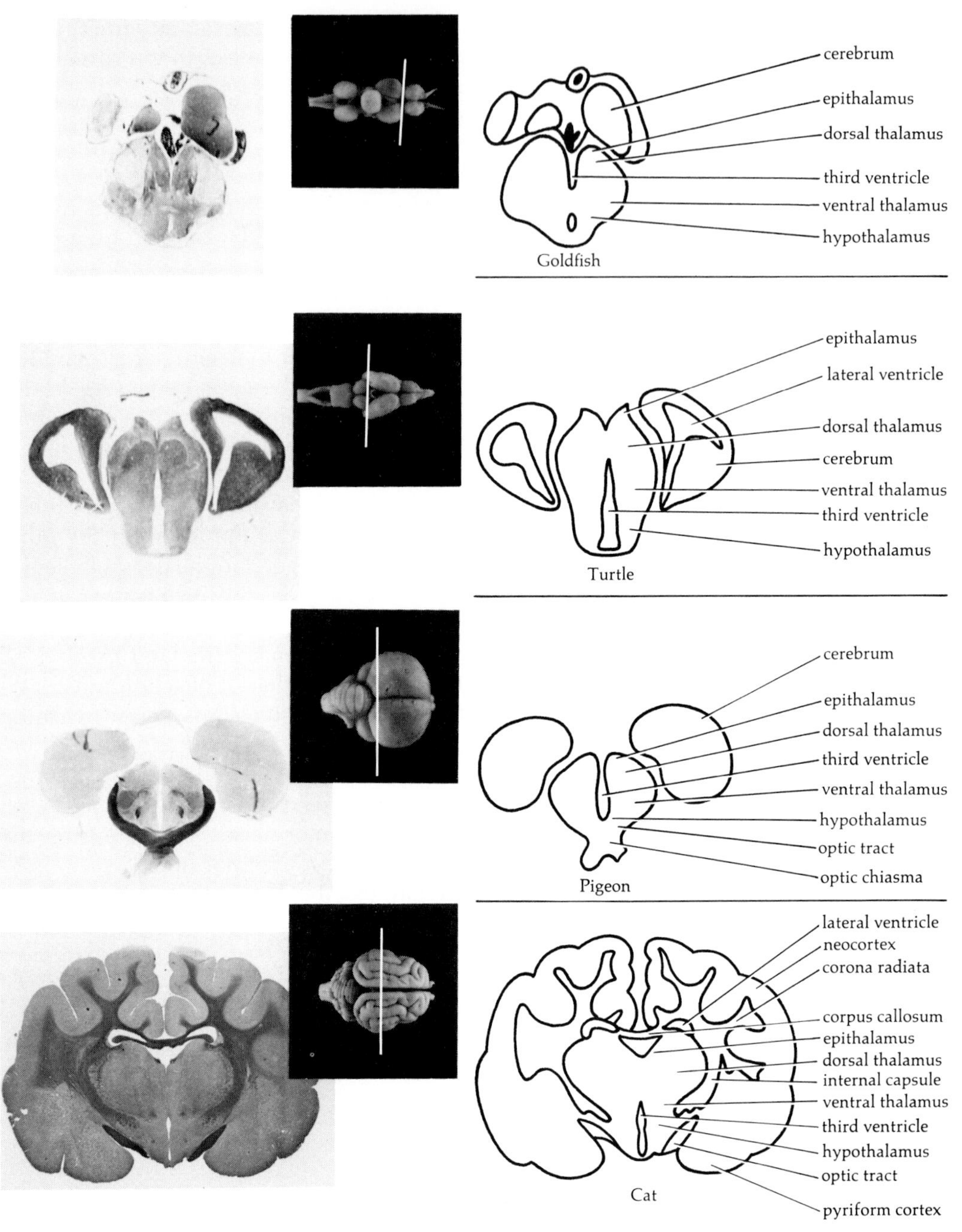

Fig. 11-17. On the left is a transverse section through the diencephalon in each of four vertebrates. The position of the section is indicated by a white line on the dorsal view of each brain.

is intimately concerned with most sensory and motor visceral functions. It mediates, among other things, the central nervous control of digestion, maintenance of osmotic independence, and sexual activities. As will be discussed in the chapter on endocrines, the hypothalamus also contributes to control of the pituitary gland.

The right and left optic nerves meet and cross at the optic chiasma, which is a gross structure in the anterior ventral portion of the diencephalon. In almost all nonmammals there is a complete decussation, with all optic nerve fibers from one side crossing to the other; in most mammals there is only a partial decussation. In all vertebrates the axons forming the optic nerve and optic chiasma continue as the optic tract, which travels on each side of the brain up to diencephalic and mesencephalic visual centers.

THE SUPRASEGMENTAL APPARATUS

The cerebellum

In the brain's early embryology paired outgrowths develop from the metencephalic alar plates on each side of the fourth ventricle. These outgrowths undergo rapid mitotic division, enlarge, and fuse above the ventricle forming the cerebellum: the ventrolateral portions form auricular lobes of the cerebellum (mammalian flocculonodular lobes), and the central portion forms the body of the cerebellum or corpus cerebelli. In mammals, and to a lesser extent in birds, enlargements form between the corpus cerebelli and the flocculonodular lobe; these enlargements are called the cerebellar hemispheres.

Although the cerebellum varies in relative size in different taxa, its basic cellular structure remains remarkably constant. There is superficial gray matter, the cerebellar cortex, overlying deeper white matter which contains the fibers going to and from the cortex. In the deeper portions there are some cerebellar nuclei. Fibers leaving the cerebellar cortex synapse in the cerebellar nuclei; from these nuclei axons leave the cerebellum and travel to other parts of the brain.

The cerebellar cortex is composed of three layers (Fig. 11-18). In the deepest layer, next to the white matter, are many small, closely packed

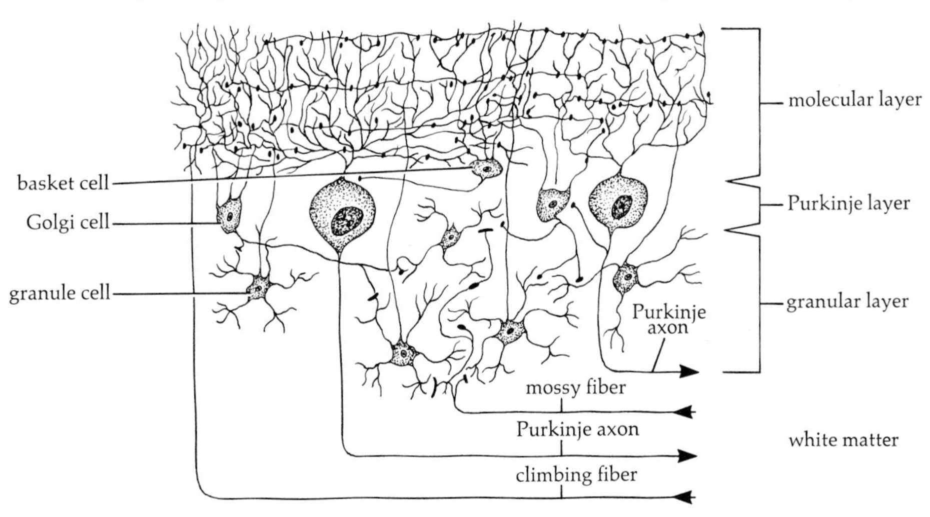

Fig. 11-18. A simplified diagram of the cerebellar cortex.

granule cells; therefore, this is called the granular layer. The most superficial layer is composed primarily of a neuropil of dendrites with few cells; this is the molecular layer. Between the granular and molecular layers are large cells called Purkinje cells that are unique to the cerebellum. In most vertebrates these are arranged in a layer which is only one cell thick. Purkinje cell axons run down through the white matter and synapse with the cerebellar nuclei, while their huge arborizing dendritic trees form much of the neuropil of the molecular layer. The rest of this neuropil contains the dendrites of the basket cells of the granular and molecular layers and the Golgi cells (infrequent large cells of the granular layer).

Fibers coming to the cerebellum travel through the white matter and then enter the cerebellar cortex as either climbing or mossy fibers. Climbing fibers make complex synaptic connections along the dendritic trees of the Purkinje cells. Mossy fibers, which are slightly smaller, enter only into the deep granular layer where they make complex synaptic connections on the dendrites of the granule cells. The axons of the granule cells extend into the molecular layer, where they travel parallel to the surface of the cortex and make synaptic connections with Purkinje cells.

This complex cerebellar circuitry smooths out movements, making them graceful and coordinated; the flocculonodular lobes are primarily concerned with balance and equilibrium, the corpus cerebelli with locomotor movements, and the cerebellar hemispheres, when present, with the fine use of smaller musculoskeletal components. An animal whose cerebellum is missing or nonfunctional moves jerkily, without fine coordination.

Although the flocculonodular lobes vary little in relative size among different taxa (equilibrium and balance being important functions in all animals), it is not surprising that the cerebellar hemispheres are largest in animals which are capable of very fine movement and smallest in those animals which are not, or that the corpus cerebelli is largest in very actively moving animals and smallest in sluggish ones (Fig. 11-19).

In snakes and salamanders the cerebellum is little more than a ridge above the fourth ventricle; it is much larger in other reptiles and amphibians, such as lizards, turtles, frogs, and toads. In teleost and chondrichthyean fishes the cerebellum is relatively large, particularly the corpora cerebelli. In most teleosts the anterior corpus cerebelli forms a large structure which extends into the optic ventricles; this portion is called the valvula. In some teleosts this anterior growth is so large that it bulges out of the optic ventricles and covers the superficial part of the brain. In the weakly electric mormyrid fishes, for instance, it obscures the brain's entire dorsal surface, and it is suspected that the cerebellum mediates the fine control of their electroreceptor and electroeffector organs.

The cerebrum

The primordial telencephalon undergoes rapid and lengthy growth, developing into the cerebral hemispheres (containing lateral ventricles) and, anteriorly, into the olfactory bulbs, which receive the

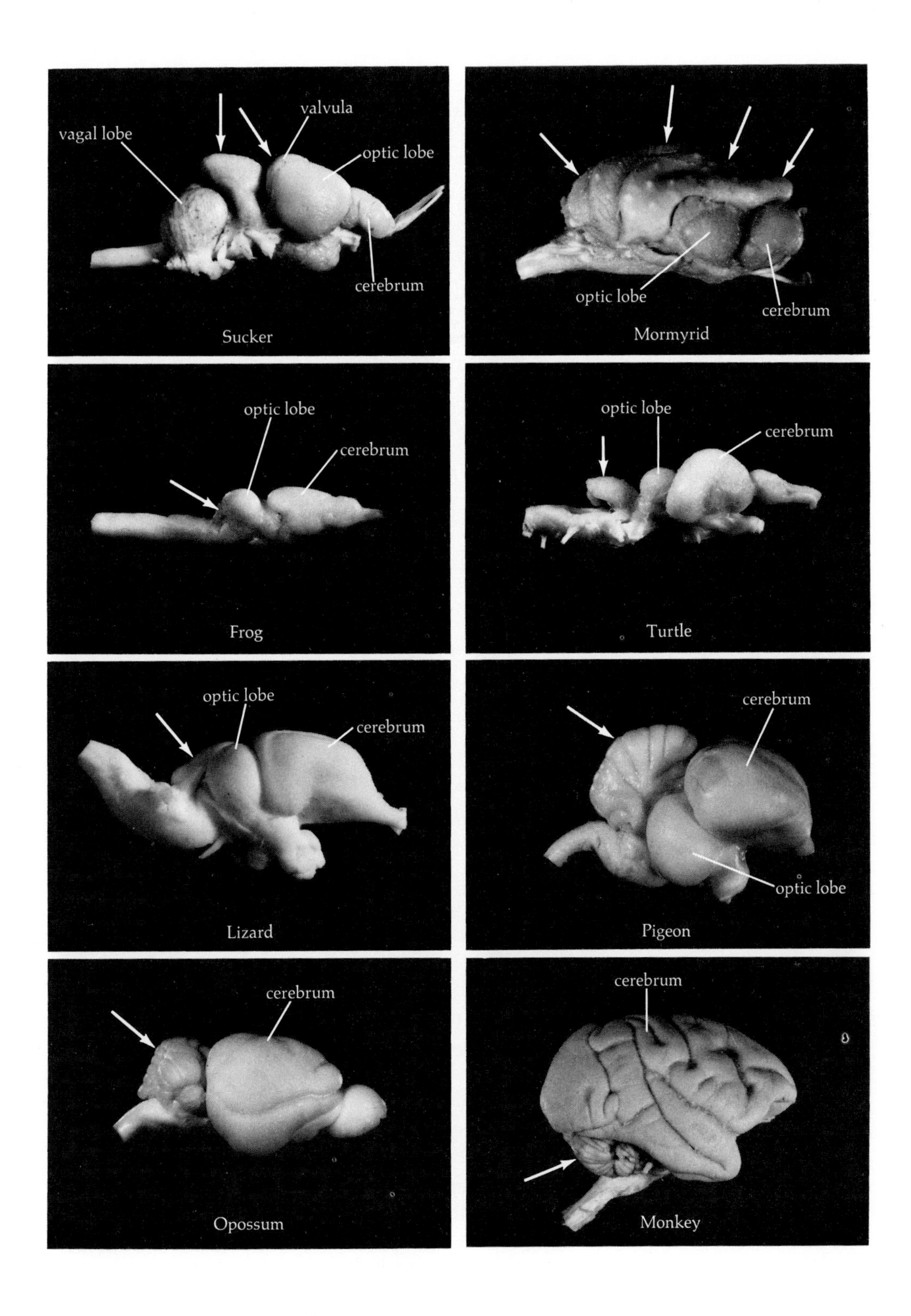

Fig. 11-19. Lateral views of the brains of eight vertebrates, with arrows indicating their cerebella.

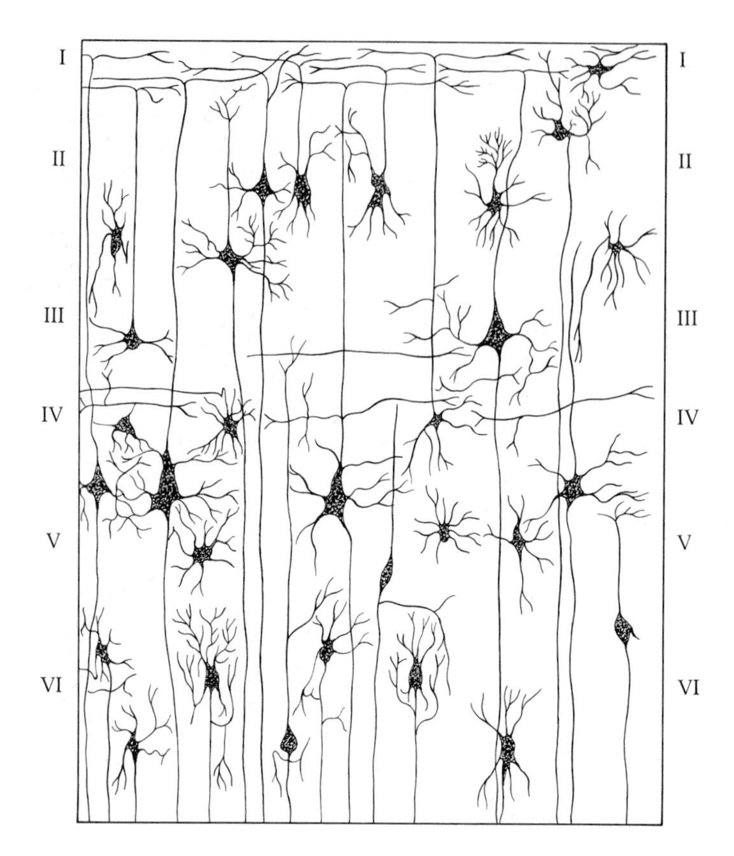

Fig. 11-20. Neuron diversity and organization in the mammalian neocortex. Only a small percentage of the neurons present are shown here. Roman numerals indicate layers of the neocortex, with layer I being the most superficial.

olfactory nerves. The development of the rest of the cerebral hemispheres varies greatly in different classes.

MAMMALS

The dorsolateral telencephalon just behind the olfactory bulb differentiates into the mass of the cerebral hemisphere. This portion undergoes great expansion and continues its growth far longer than the rest of the nervous system, forming, in the adult mammal, the typical six-layered neocortex (Fig. 11-20). In many mammals it grows so far anteriorly that it overlies the olfactory bulbs and so far posteriorly that it overlies the diencephalon, the mesencephalon, and often even the metencephalon and the myelencephalon. It also forms the most lateral and most dorsal portions of the brain. In large mammals it becomes convoluted, forming distinct grooves (sulci) and bulges (gyri), and is called a gyrencephalic cortex. In smaller mammals, such as rats and mice, it remains smooth and is called a lissencephalic cortex.

There are three major fiber connections of the neocortex.

1. From the dorsal thalamus massive numbers of fibers enter a large tract called the internal capsule which leads up to the neocortex. As it

approaches the neocortex the tract spreads out into the corona radiata, sending fibers to all portions of the neocortex and forming thus the white matter underlying the six-layered neocortex. Among these ascending fibers of the internal capsule are some fibers travelling in the other direction, i.e., from neocortex to dorsal thalamus.

2. Many of the fibers leaving the neocortex pass through the corona radiata and internal capsule but bypass the dorsal thalamus; these fibers form the cerebral peduncles which carry neocortical information to lower brain and spinal cord structures.

3. Some fibers connect the right and left neocortex, forming a commissural system. (A commissure is any tract connecting identical right and left structures of the brain or spinal cord, whereas a decussation connects different right and left structures.) In placental mammals, the commissure of the neocortex is the corpus callosum. In marsupials, neocortical structures as well as more basal structures of the cerebral hemispheres are joined by an independently evolved neocortical commissure derived from fibers of the corpus callosum plus those of the anterior commissure (phylogenetically older and also found in placentals).

The dorsomedial edge of the developing cerebral hemisphere does not expand as much as does the dorsolateral; during neocortical growth it becomes folded under. It forms the hippocampus, which lies deep to the neocortex (Fig. 11-21) and has fiber connections to the neocortex, hypothalamus, and (indirectly) olfactory bulbs.

The ventrolateral portion of the developing cerebral hemisphere is called the piriform cortex (Fig. 11-21). It too has a shorter and less spectacular growth than the neocortex. It becomes folded under and lies only ventrally in the adult cerebral hemisphere, with connections primarily to the olfactory bulbs.

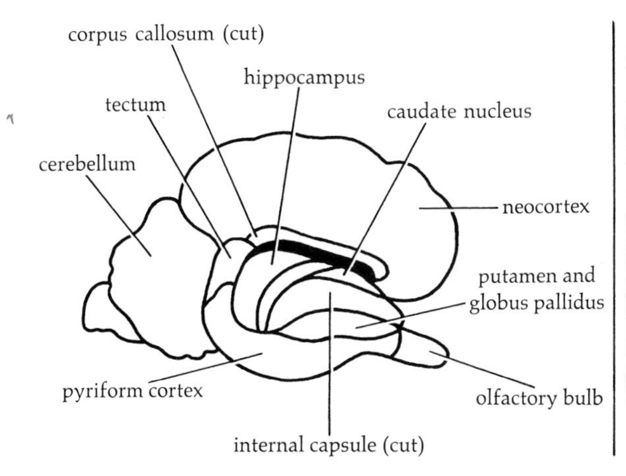

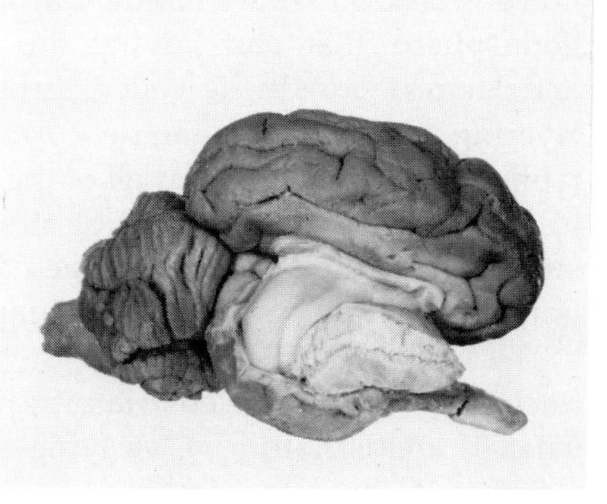

Fig. 11-21. A dissection of the sheep brain with the right neocortex and underlying white matter removed. The relationships of the basal ganglia, internal capsule, and hippocampus can be seen. Note also the extent of the neocortex and the white matter in the telencephalon.

Developing from the ventral and ventromedial portion of the cerebral hemisphere are two groups of structures. The septal nuclei develop in and remain in the anterior part of the ventromedial aspect. The basal ganglia grow dorsally into deeper portions of the cerebral hemisphere where the internal capsule, en route from dorsal thalamus to neocortex, pierces them. This produces the striated effect of fibers intermixed with a nuclear group and gives the basal ganglia their other name, the corpus striatum. This structure contains several important nuclei, such as the caudate nucleus medially and the globus pallidus and putamen ventrolaterally (Fig. 11-21). The main input to the basal ganglia is from the dorsal thalamus, and the output is to the ventral thalamus and the midbrain reticular formation.

There are two relatively small commissures involved in the deeper portions of the cerebral hemispheres. Right and left hippocampi are connected by the hippocampal commissure which runs with the corpus callosum in placental mammals and independently in marsupials. Olfactory structures are interconnected in both placentals and marsupials by the anterior commissure; this commissure is much larger in marsupials, in which it is joined by the major neocortical commissure.

BIRDS

The bird's cerebral hemisphere has no neocortex, no convolutions, and only small olfactory bulbs. There are three structures of probable homology with mammalian structures: a very small hippocampus dorsomedially, an equally small septal region in the medial wall ventrally, and a small strip of piriform cortex along the lateral wall of the hemisphere.

Dorsolaterally, between the hippocampal and piriform areas, there is a very thin strip of corticoid tissue, not six-layered, which may or may not be related to the mammalian neocortex. The bulk of the cerebral hemisphere, that part having the greatest developmental expansion and longest growth, is called striatal because it lacks the definitive layering of the mammalian neocortex (Fig. 11-22). It is not definitely established whether some or all of the bird's striatal structures are related to the mammalian striatum, or if some of the more superficial avian striatal structures, which are actually large telencephalic nuclei, are homologically related to mammalian cortical structures.

The ventromedial portion of the hemisphere is made up of a large nuclear group, the paleostriatum; just lateral to this is the somewhat smaller archistriatum. Overlying the posterior part of the paleostriatum is the ectostriatum, and, overlying this, the very large neostriatum which also extends considerably anteriorly. Most of the complex cellular portions of the cerebral hemisphere are immediately dorsal to the neostriatum in the hyperstriatum. The hyperstriatum is divided into the pars ventrali, the pars dorsali, and, most dorsally and medially, into the hyperstriatum accessorium. In most birds the accessorium forms the wulst—a distinct, grossly visible ridge or bulge along the dorsomedial portion of the hemisphere.

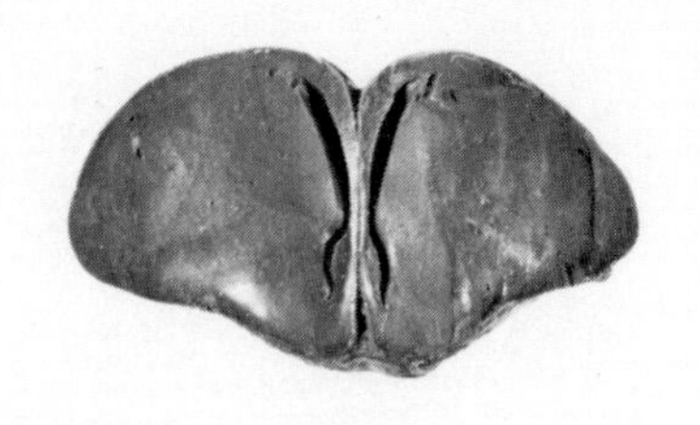

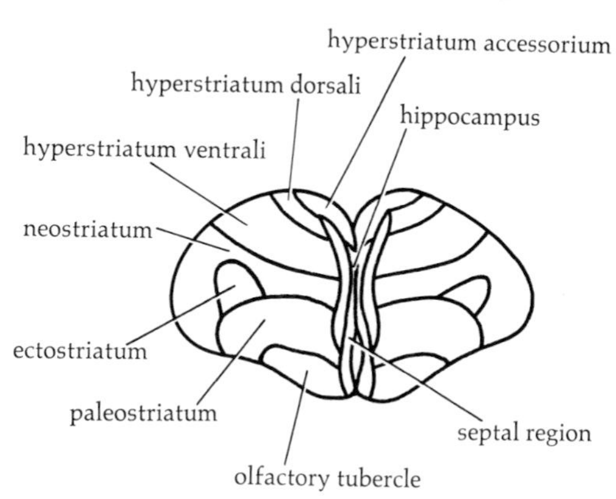

Fig. 11-22. Transverse section through the telencephalon of the pigeon.

A sulcus called the vallecula divides the wulst from the more lateral parts of the hemisphere.

The fiber connections and functions of the avian cerebrum will be discussed in Chapter 12.

REPTILES

The less extensive reptilian cerebrum has a shorter growth period than the cerebrum of birds or mammals and does not fill the cranial cavity. Behind relatively large olfactory bulbs the major portion of the cerebral hemisphere is made up of a dorsomedial hippocampus, a lateral piriform, and a small amount of general cortex between the two; there is no six-layered neocortex. Along the ventromedial wall is a septal region, but the bulk of the hemisphere is made up of striatal structures, smaller and less differentiated than those of birds but having the same general positions. Although some reptilian striatal structures are probably phylogenetically related to structures in birds or mammals, the details are still unknown (Fig. 11-23).

AMPHIBIANS, CHONDRICHTHYEANS, SARCOPTERYGIANS, AND CYCLOSTOMES

In all of these groups there are anterior olfactory bulbs and, behind them, modest cerebral hemispheres which are usually smaller than the midbrain. Most of the cells of each hemisphere lie in deep portions,

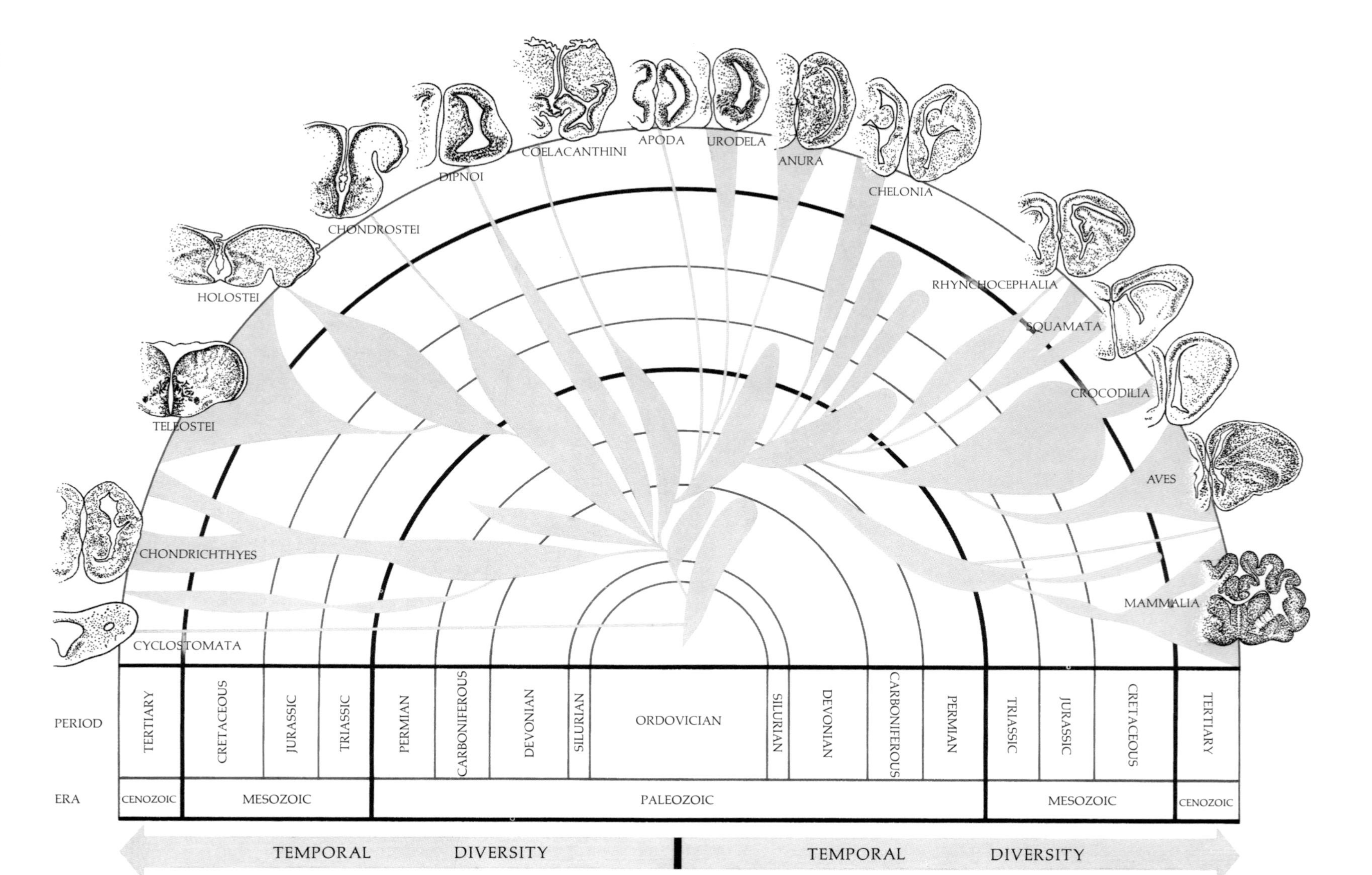

TELEOSTEI
HOLOSTEI
CHONDROSTEI
DIPNOI
COELACANTHINI
APODA
URODELA
ANURA
CHELONIA
RHYNCHOCEPHALIA
SQUAMATA
CROCODILIA
AVES
MAMMALIA
CHONDRICHTHYES
CYCLOSTOMATA
PERIOD
TERTIARY
CRETACEOUS
JURASSIC
TRIASSIC
PERMIAN
CARBONIFEROUS
DEVONIAN
SILURIAN
ORDOVICIAN
SILURIAN
DEVONIAN
CARBONIFEROUS
PERMIAN
TRIASSIC
JURASSIC
CRETACEOUS
TERTIARY
ERA
CENOZOIC
MESOZOIC
PALEOZOIC
MESOZOIC
CENOZOIC
TEMPORAL
DIVERSITY
TEMPORAL
DIVERSITY

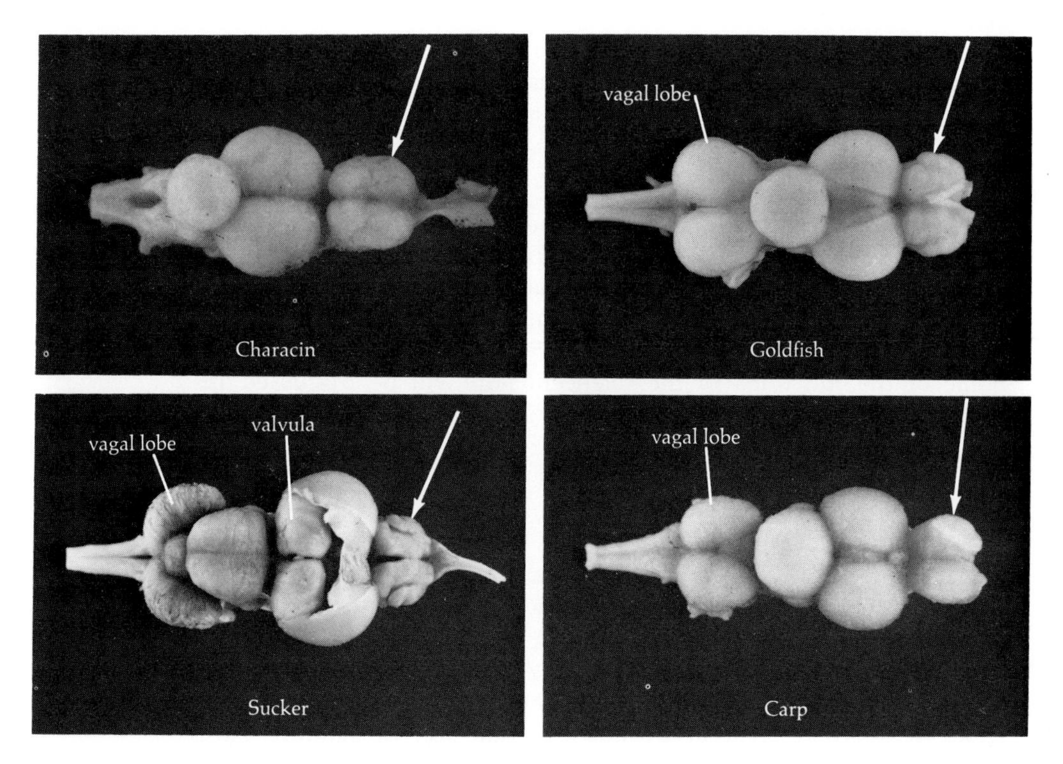

Fig. 11-24. Dorsal views of the brains of four teleosts. Note the variations in the form of the cerebrum (arrows). Also note the prominent valvula cerebelli in the sucker and the large vagal lobes in three of the four brains.

near the ependymal lining of the lateral ventricle (Fig. 11-23). More superficially, there is a dendritic neuropil. Only some parts of the hemisphere have been named: the dorsal portion is the hippocampus, and the lateral is the piriform; ventrolaterally there is a striatal region, and ventromedially a septal region. However, despite the use of these names, homologies with amniote structures are questionable.

ACTINOPTERYGIANS

Only the olfactory bulb portion develops as it does in other vertebrates. Behind the olfactory bulb, the telencephalic tissue everts, that is, its dorsal portion folds laterally and pulls a thin tela choroidea along with it; it then folds ventrally again so that the hemispheric wall is folded over upon itself laterally in each hemisphere, producing a solid mass of cerebral hemisphere with a very thin tela choroidea stretched over all (Fig. 11-23). Although named the striatum, or some-

Fig. 11-23. Spatial diversity of the vertebrate telencephalon as seen in transverse sections through the middle of the cerebral hemisphere. Stippling indicates the concentration of nerve cells. Homologies are not indicated because so many are still uncertain.

times epistriatum, one would suspect from its developmental pattern that portions of this hemisphere (perhaps major portions) are homologically related to cortical rather than striatal structures of other vertebrates (Fig. 11-24).

This great diversity of telencephalic structures from class to class has probably helped perpetuate the myth of mammals as the "highest" class. Recent and current studies on the fiber connections and their functions in nonmammals are indicating a good deal more complexity and "sophistication" in the brains of other groups than had been suspected.

Some of the functions of the telencephalon will be discussed in Chapter 12.

SUGGESTED READING

Gardner, E. 1968. "Fundamentals of Neurology." Saunders, Philadelphia, Pennsylvania (A good general account of the structure of the human brain.)

Nauta, W. J. H., and Karten, H. J. (1970). A general profile of the vertebrate brain, with sidelights on the ancestry of cerebral cortex. *In* "The Neurosciences: Second Study Program" (F. O. Schmitt, ed.), pp. 7–26. Rockefeller Univ. Press, New York. (An excellent brief account.)

Papez, J. W. (1929). "Comparative Neurology." T. Y. Crowell Co. Publ., New York. (Gross structure excellent but fiber connections are often wrong as demonstrated by recent experimental studies.)

Petras, J. M., and Noback, C. R. (1969). Comparative and evolutionary aspects of the vertebrate central nervous system. *Ann. N. Y. Acad. Sci.* **167,** 1–513. (The only volume now in existence dealing with this subject and based on modern experimental findings.)

Zeman, W. J., and Innes, R. M. (1963). "Craigie's Neuroanatomy of the Rat." Academic Press, New York. (The best account of a nonhuman mammalian brain.)

12 NERVOUS PATHWAYS

Thus far we have discussed major elements of the nervous system. These elements are connected by complex fiber pathways, including many feedback loops, and organized into sensory, motor, and integrative systems. Particularly in nonmammals much of this organization is not fully understood, and this chapter will necessarily emphasize mammals.

PATHWAYS OF THE SOMATIC SENSORY SYSTEM

MAMMALS

The somatic sensory fiber systems can be divided functionally and morphologically into three groups. The largest, most heavily myelinated fibers carry information of fine touch and proprioception and are called epicritic fibers. Slightly smaller fibers with considerable myelin carry information of pressure and gross touch and are called intermediate fibers. There is now some dispute as to whether the intermediate fibers form a separate pathway or are mixed with the protopathic fibers. These smallest fibers have little or no myelin, and carry information of pain, heat, and cold. When these three kinds of fibers carry information from the body they enter the spinal cord and follow distinct pathways to the brain, called the epicritic, intermediate (perhaps not fully separate), or protopathic systems; when they carry information from the head they enter the brain in the trigeminal nerve (V) and, to a very slight extent, in VII, IX, and X.

All these pathways share certain characteristics. There is at least a three-neuron chain. The cell bodies of the second and third neurons are within the central nervous system (CNS). With rare exception, the cell body of the first (most peripheral) neuron is outside the CNS—in a dorsal root ganglion of the spinal cord or in the semilunar ganglion of the trigeminal nerve. Upon entering the CNS, the sensory fiber bifurcates into an ascending and a descending branch; the latter usually makes only reflex connections. The axon of the second-order neuron decussates. The third-order neuron lies within a thalamic nucleus and projects into the neocortex.

Epicritic system (fine touch)

The cell bodies of the primary epicritic fibers (Fig. 12-1) lie in dorsal root ganglia. Their axons enter the dorsal root, bifurcate, and send collaterals into the gray matter and for short distances up and down the spinal cord where they make reflex connections. The main part of the

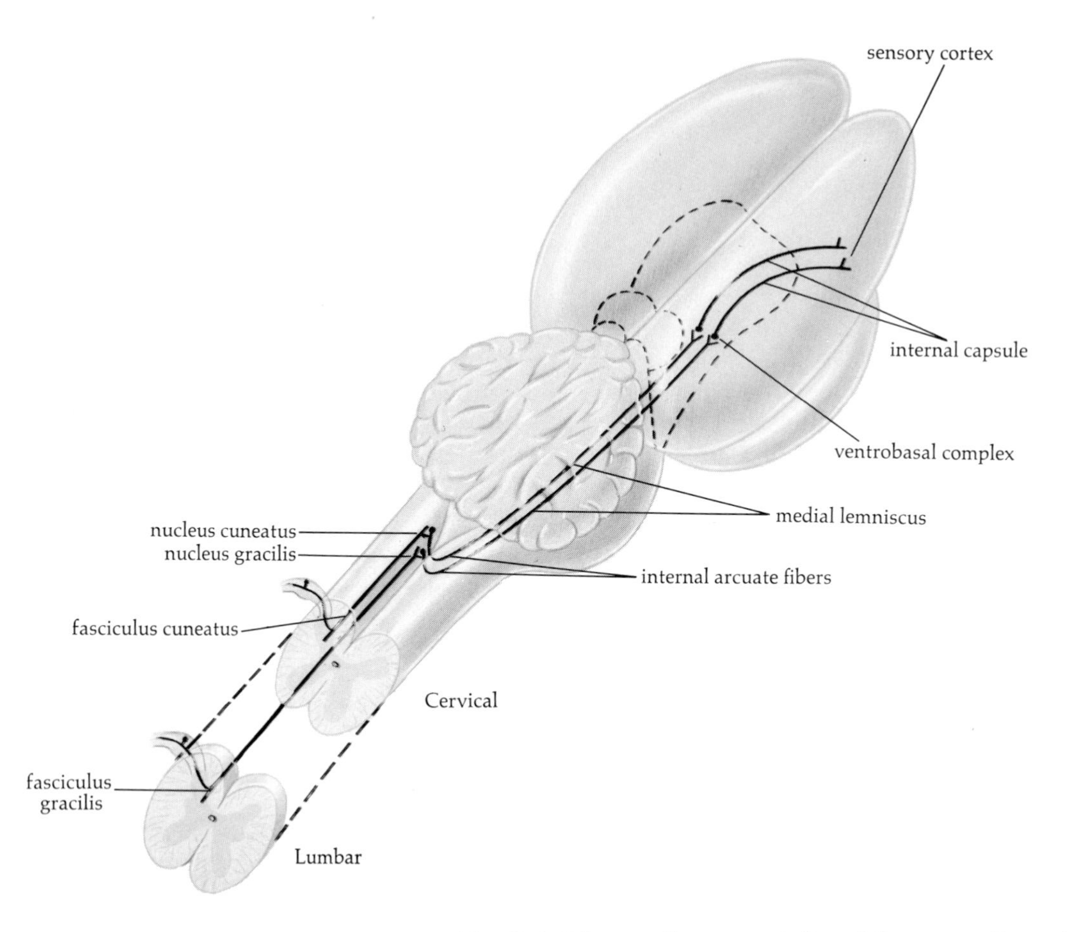

Fig. 12-1. Diagramatic representation of the mammalian epicritic system, representing large tracts by single neurons.

axon, however, enters directly into the dorsal funiculus and ascends within it to the brainstem. Epicritic fibers from the lumbar, sacral, and caudal regions enter the medial part of the dorsal funiculus and constitute the fasciculus gracilis; epicritic fibers from the thoracic and cervical regions enter the lateral part of the dorsal funiculus and constitute the fasciculus cuneatus. These two fasciculi terminate on second-order neurons of the dorsal column nuclei—either the nucleus gracilis or the nucleus cuneatus—both of which are located at the caudal end of the fourth ventricle.

The axons of these second-order neurons are called the internal arcuate fibers. They swing ventrally and medially, cross the ventral midline, turn immediately anteriorly, and thus form a large ascending tract. This tract is the medial lemniscus, which continues through the hindbrain and midbrain and terminates in the ventrobasal complex of the dorsal thalamus (or, as it is also called, the nucleus ventralis posterior pars lateralis of the dorsal thalamus).

The tertiary sensory neurons in the nucleus ventralis posterior pars lateralis send their axons by way of the internal capsule and corona radiata up to the sensory portion of the neocortex, which lies in the frontal lobe or, in some mammals, slightly more caudally in the parietal lobe.

Intermediate system (pressure, gross touch)

Intermediate fibers (Fig. 12-2) also have cell bodies in dorsal root ganglia and enter the dorsal roots of spinal nerves; in the dorsal spinal cord they too send ascending and descending collaterals which terminate in the gray matter. The main axon ascends several segments in the ventral part of the dorsal funiculus and then synapses in the deeper portion of the dorsal horn.

The second-order fibers arising from the dorsal horn cells immediately decussate in the ventral white commissure of the spinal cord and then turn anteriorly in the ventral part of the lateral funiculus, forming the ventral spinothalamic tract. This tract ascends the spinal cord to the medulla oblongata where it joins the medial lemniscus to travel forward and terminate in the ventrobasal complex of the dorsal thalamus.

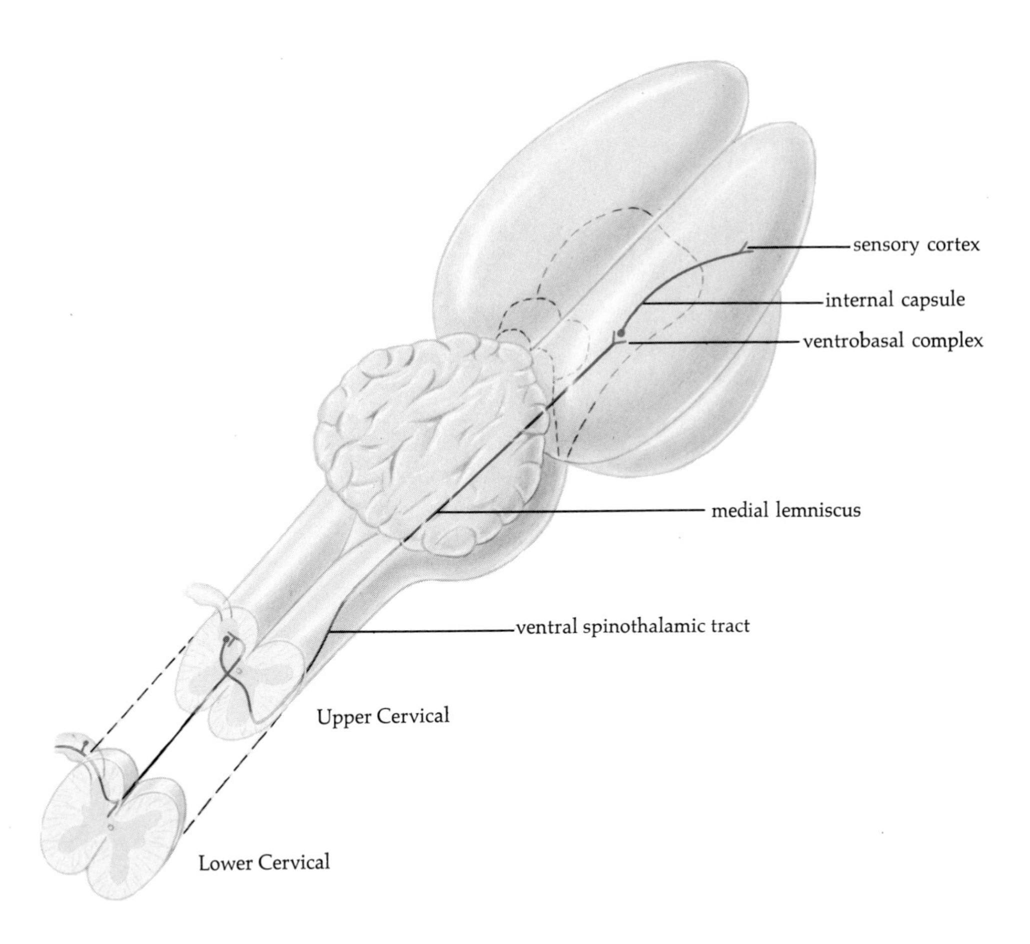

Fig. 12-2. Diagramatic representation of the mammalian intermediate sensory system as shown by single neurons at each level beginning with the lower cervical spinal cord.

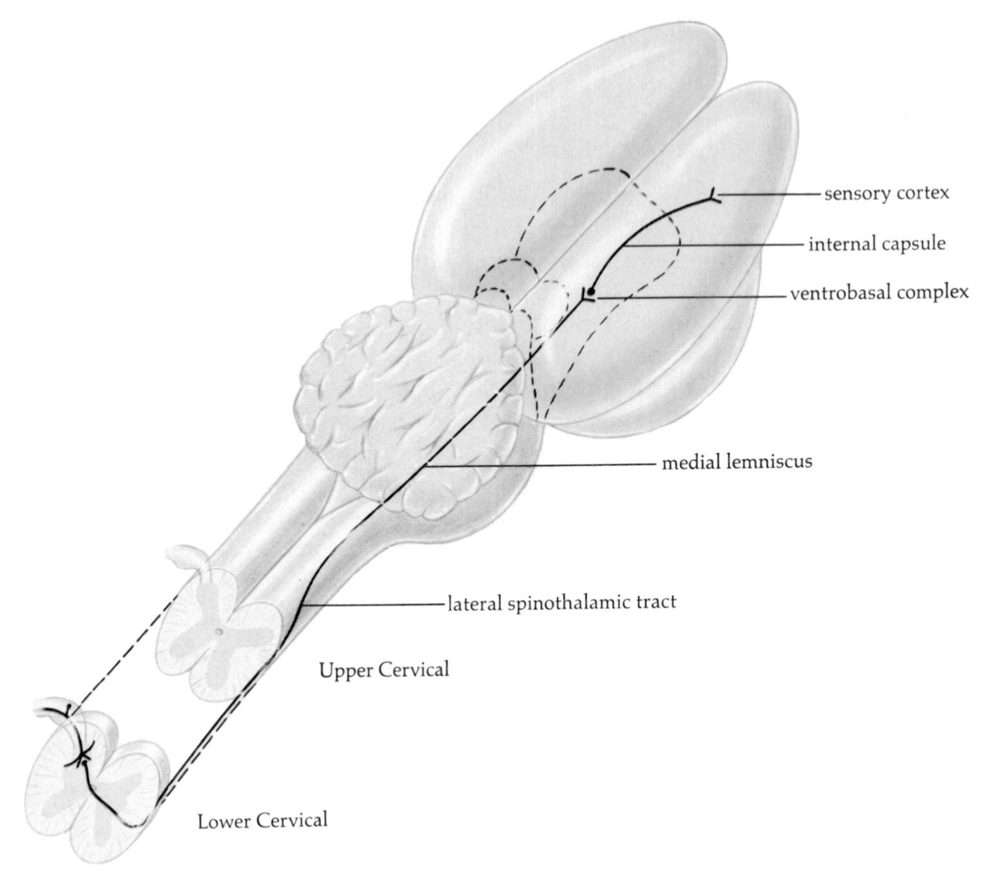

Fig. 12-3. Diagramatic representation of the mammalian protopathic sensory system as shown by single neurons at each level beginning with the lower cervical spinal cord.

The projection from the thalamus to the somatic sensory cortex is the same as that for the epicritic system.

Protopathic system (pain, heat, cold)

The often unmyelinated protopathic fibers (Fig. 12-3) arise from small cell bodies in the dorsal root ganglia. These fibers enter the dorsolateral fasciculus of the spinal cord, send short ascending and descending fibers for one or two segments, and then make synaptic connections in the most dorsal part of the dorsal horn, the substantia gelatinosa.

From the substantia gelatinesa, second-order fibers decussate in the ventral commissure of the cord and pass anteriorly in the lateral funiculus as the lateral spinothalamic tract. Like the ventral spinothalamic tract, with which it is confluent, the lateral spinothalamic tract joins the medial lemniscus in the brainstem. From the posterior brainstem the axons travel up to and synapse in the ventrobasal nuclei of the dorsal thalamus. Third-order neurons travel from the dorsal thalamus through the internal capsule and corona radiata to the somatic sensory cortex.

Trigeminal input (*Fig. 12-4*)

Most somatic sensory information from the head is carried by the trigeminal nerve (V). The primary neurons have their cell bodies in the large semilunar ganglion; their axons enter the brainstem and terminate in the main nucleus and the descending nucleus of the trigeminal nerve. A different arrangement exists for fibers carrying proprioceptive and sensory information from the teeth and gums. These fibers enter directly into the brain; the primary cell bodies are in the mesencephalic nucleus of V, and their axons terminate in the motor nucleus of the trigeminal nerve. Secondary fibers from the main nucleus of V are another group of internal arcuate fibers, which parallel the course of those from the dorsal column nuclei and then join them in the medial lemniscus to ascend to the thalamus. (Since these fibers remain somewhat distinct they are sometimes named the trigeminal lemniscus.) In the thalamus these fibers terminate in the medial, rather than lateral, part of the ventrobasal complex (also called the nucleus ventralis posterior pars medialis of the dorsal thalamus). Tertiary

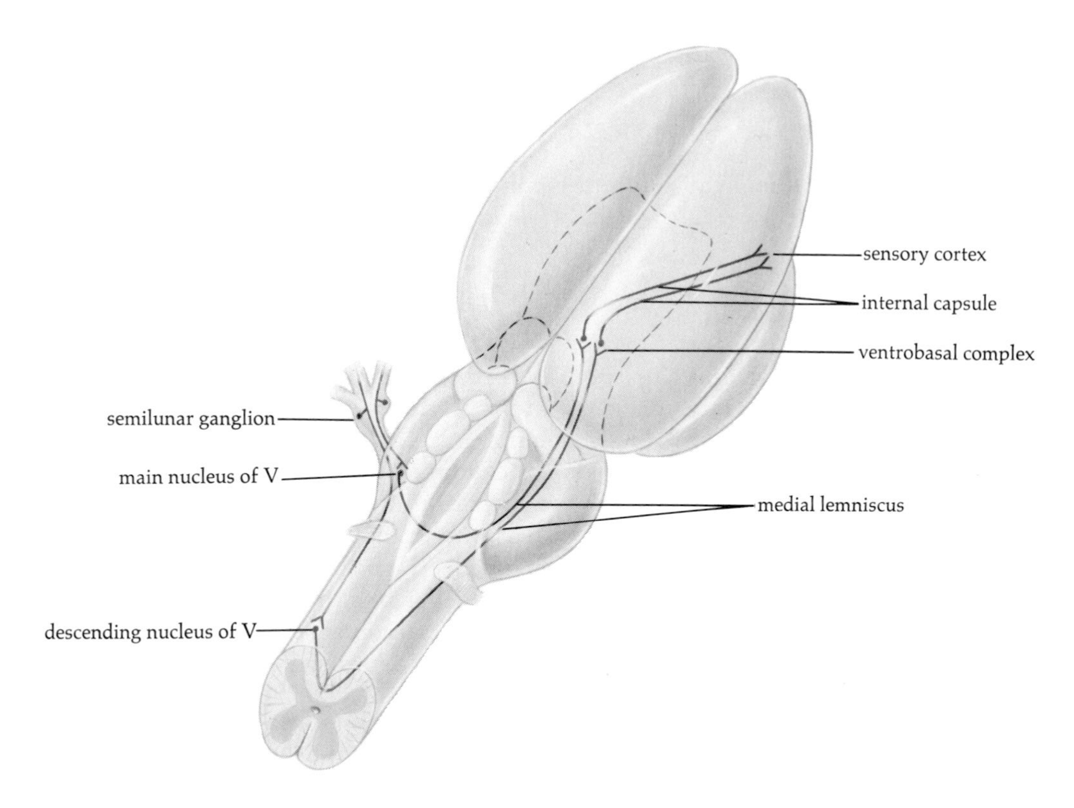

Fig. 12-4. Diagramatic representation of the mammalian trigeminal sensory system from the semilunar ganglion to the sensory cortex. The cerebellum has been removed in order to show the nuclei and pathways with greater clarity.

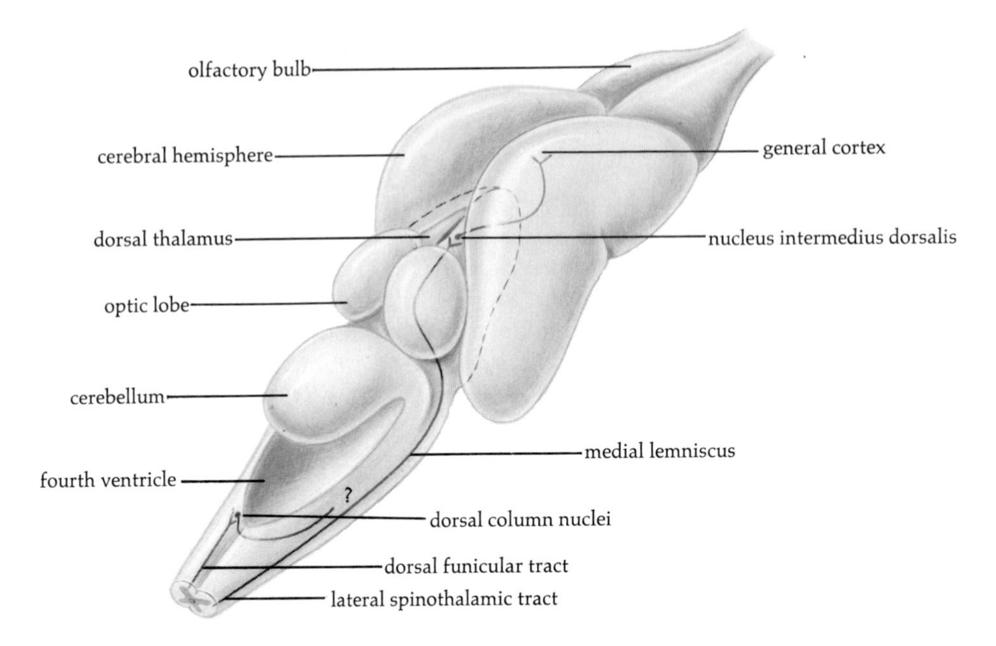

Fig. 12-5. Drawing of a turtle brain showing as much as is known of the somatic sensory pathway to the cerebrum. The question mark indicates that the projection of dorsal column nuclei is still to be determined.

fibers follow the internal capsule and corona radiata to the somatic sensory neocortex.

NONMAMMALIAN VERTEBRATES

Much less is known about the central connections of the nonmammalian somatic sensory systems. In frogs, lizards, alligators, snakes, and turtles, there are dorsal column tracts running in the dorsal funiculus and terminating in dorsal column nuclei. From the dorsal column nuclei fibers decussate and ascend to unknown terminations. Somatic sensory system fibers also ascend in the lateral funiculus of the spinal cord; these terminate in deep portions of the tectum, including specific parts of the torus semicircularis. In reptiles, but not in amphibians, some lateral funicular fibers continue as the medial lemniscus up into the dorsal thalamus. They terminate in the nucleus intermedius dorsalis, which is possibly homologous to the ventrobasal nuclei of mammals. In turtles and perhaps other reptiles fibers project from the nucleus intermedius dorsalis to the general cortex (Fig. 12-5).

Very little is known about somatic sensory projections in fishes, although one unusual situation has been discovered in sharks: fibers leaving the thalamus decussate just above the optic chiasma, projecting to the contralateral telencephalon.

Special Sensory Systems

AUDITORY PATHWAYS

MAMMALS (Fig. 12-6)

From cell bodies in the spiral ganglion (within the cochlea) axons enter the dorsolateral portion of the anterior medulla, bifurcate, and terminate in the dorsal and ventral cochlear nuclei. Secondary fibers leave the cochlear nuclei via the trapezoid body and travel ventrally and then medially along the brainstem's surface. Most secondary fibers terminate in a ventral nuclear group collectively called the superior olivary complex: in the ipsilateral, or, after decussating, in the contralateral superior olivary complex. Other fibers bypass the superior olivary complex but after decussating turn anteriorly and join its fibers.

Past the superior olivary complex these fibers together form the lateral lemniscus, which terminates massively in the inferior colliculus. From the inferior colliculus a distinct tract, the brachium of the inferior colliculus, leads to the medial geniculate body. This nucleus forms a swelling or bulge in the posterior lateral portion of the thalamus, and here auditory information synapses. Then auditory information travels by way of the internal capsule and corona radiata to the temporal lobe of the neocortex.

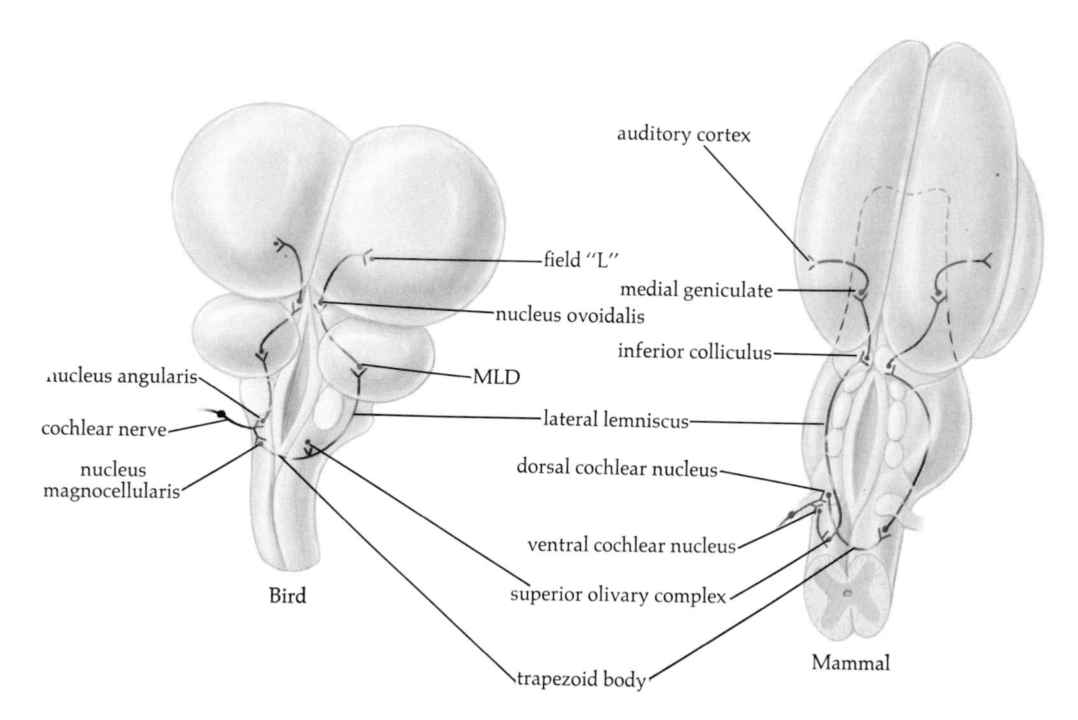

Fig. 12-6. The central auditory pathways in a bird and a mammal. The cerebellum of each has been removed for clarity.

NONMAMMALS

Most studies of nonmammalian central auditory pathways have been done on birds (Fig. 12-6). As in mammals, the primary fibers project to the cochlear nuclei: the bird's dorsal cochlear nucleus is called the nucleus angularis, and its ventral, the nucleus magnocellularis. Some fibers from these nuclei project to a relatively small superior olivary complex, but most decussate in the trapezoid body and travel anteriorly in the lateral lemniscus to a deep part of the tectum. There they terminate in the torus semicircularis, in a specific nucleus called the nucleus mesencephalicus lateralis pars dorsalis, otherwise known as the MLD. On the basis of neuron morphology, fiber connections, and embryology, the bird's MLD appears to be homologous to the mammal's inferior colliculus. From the MLD axons project to the nucleus ovoidalis, located medially in the dorsal thalamus and apparently analogous (rather than homologous) to the mammalian medial geniculate body. The nucleus ovoidalis in turn projects to the telencephalon, where the fibers terminate in a very specific portion of the neostriatum called field L.

The central auditory pathways of reptiles and amphibians are similar to that of birds through the midbrain, but there are no known auditory projections beyond that point. In fishes no one has yet been able to distinguish auditory from vestibular fibers, and virtually nothing is known of their auditory pathways.

VESTIBULAR PATHWAYS

Most primary axons from the vestibular ganglion project to a group of vestibular nuclei in the dorsal medulla; in mammals and probably all vertebrates some primary vestibular axons project directly to the cerebellum. There is some physiological evidence of secondary projections from the vestibular nuclei to the thalamus and then the cortex, but they have not been traced anatomically.

The vestibular system plays an important role in cranial and cervical reflexes, and these have been traced. Vestibular nuclei axons enter a tract, the medial longitudinal fasciculus, which runs along the brain's long axis just ventral to the ventricle, throughout the brainstem and into the cervical part of the spinal cord. The vestibular fibers of this tract make several synaptic connections: with the motor nuclei of the abducens, trochlear, and oculomotor cranial nerves; just anterior to this with the pretectal nuclei which also receive visual information; and caudally with the cervical portion of the cord (Fig. 12-7). Through the medial longitudinal fasciculus are mediated reflex responses elicited by vestibular and visual stimuli. These responses usually involve the neck and the extrinsic ocular muscles, for instance, the following movements of head and eye.

There is a distinct vestibulospinal tract descending from the lateral vestibular nucleus to the motor portion of the spinal cord; its function is uncertain.

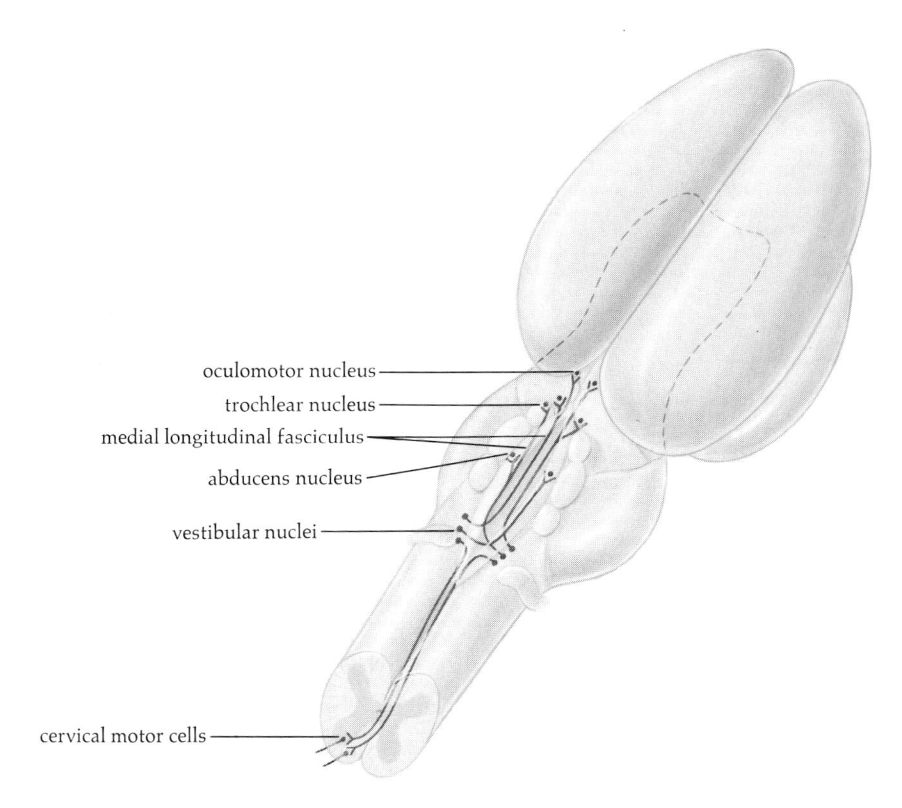

Fig. 12-7. The medial longitudinal fasciculus connects the vestibular nuclei with both the motor nuclei of the extrinsic ocular muscles and the nerves to the cervical muscles. In this way head and eye movements can be coordinated.

In fishes, the region of the brainstem containing the vestibular nuclei also contains lateral line nuclei which are so large that in some they form prominent lateral line lobes. This area no doubt also contains the central auditory nuclei.

VISUAL PATHWAYS

Both peripherally and centrally the visual system is the best understood of the special senses. Unlike most sensory systems the peripheral portion has primary neurons (rods and cones), secondary neurons (bipolar cells), and tertiary neurons (ganglion cells). Third-order fibers travel from the ganglion cells to the brain proper, as the so-called optic nerve which is actually a "misplaced" tract of the brain.

MAMMALS

In cetaceans, some bats, and possibly other mammals, the optic nerves decussate completely at the optic chiasma on the ventral surface of the hypothalamus. In most mammals, however, there is only a partial

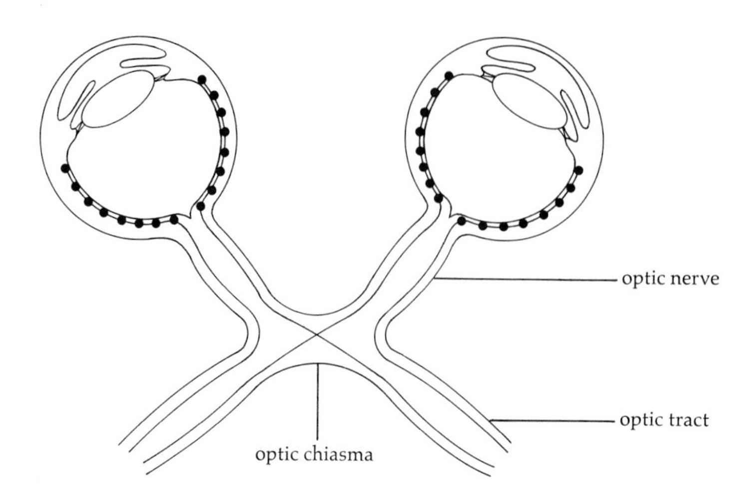

Fig. 12-8. In most mammals there is only a partial decussation of optic nerve fibers at the optic chiasma. Nerve fibers from the lateral part of the retina stay on the same side, whereas those from the medial side of the retina cross to the opposite optic tract.

decussation: fibers from lateral portions of the retina stay on the same side, whereas those from medial portions cross to the opposite side (Fig. 12-8).

Past the optic chiasma the same fibers are called optic tracts. These travel dorsally and posteriorly in either of two paths. Some fibers terminate in a nucleus of the dorsal thalamus, called the lateral geniculate body; from there fibers project by way of the internal capsule and corona radiata to the visual cortex at the occipital pole of the neocortex. Other fibers bypass the lateral geniculate and terminate on cells in the superior colliculus, whose fibers project to a posterior group of dorsal thalamic nuclei, the pulvinar; pulvinar neurons project by way of the internal capsule and corona radiata to visual cortical areas (Fig. 12-9).

BIRDS

Like mammals, birds have a double projection to the telencephalon (Fig. 12-9). The optic nerves completely decussate in the optic chiasma and continue as the optic tracts. Some axons terminate in the lateral geniculate body; fibers project from there to the dorsomedial part of the telencephalon, terminating in the hyperstriatum dorsali or, just dorsal to it, in the hyperstriatum accessorium (wulst). Most axons in the optic tract project to the very large optic lobes, homologous to the mammalian superior colliculi. Fibers from the optic lobes pass to a prominent nucleus of the dorsal thalamus, the nucleus rotundus, which relays visual information to the ectostriatum in the telencephalon. Despite the similarities with mammals, however, it is questionable whether the telencephalic structures involved in these visual pathways are homologous. In fact, homological relationships between the avian and mammalian lateral geniculate bodies, and between the nucleus rotundus and the pulvinar, are still uncertain.

REPTILES, AMPHIBIANS, AND FISHES

In the chondricthyeans, teleosts, amphibians, and most of the reptiles so far examined, the optic nerves decussate completely and send optic tracts to the so-called lateral geniculate bodies (homologies uncertain) and to the optic lobes of the tectum, thus repeating the pattern of a twofold visual input to diencephalon and mesencephalon.

In reptiles the lateral geniculate projects to the general cortex, and the tectum projects to the nucleus rotundus which in turn projects to a portion of the striatum called the ventricular ridge (all homologies uncertain). Some tectothalamic connections have been found in amphibians and teleosts, but it is not known whether this is a general pattern. Projections to the telencephalon in amphibians and fishes are unknown.

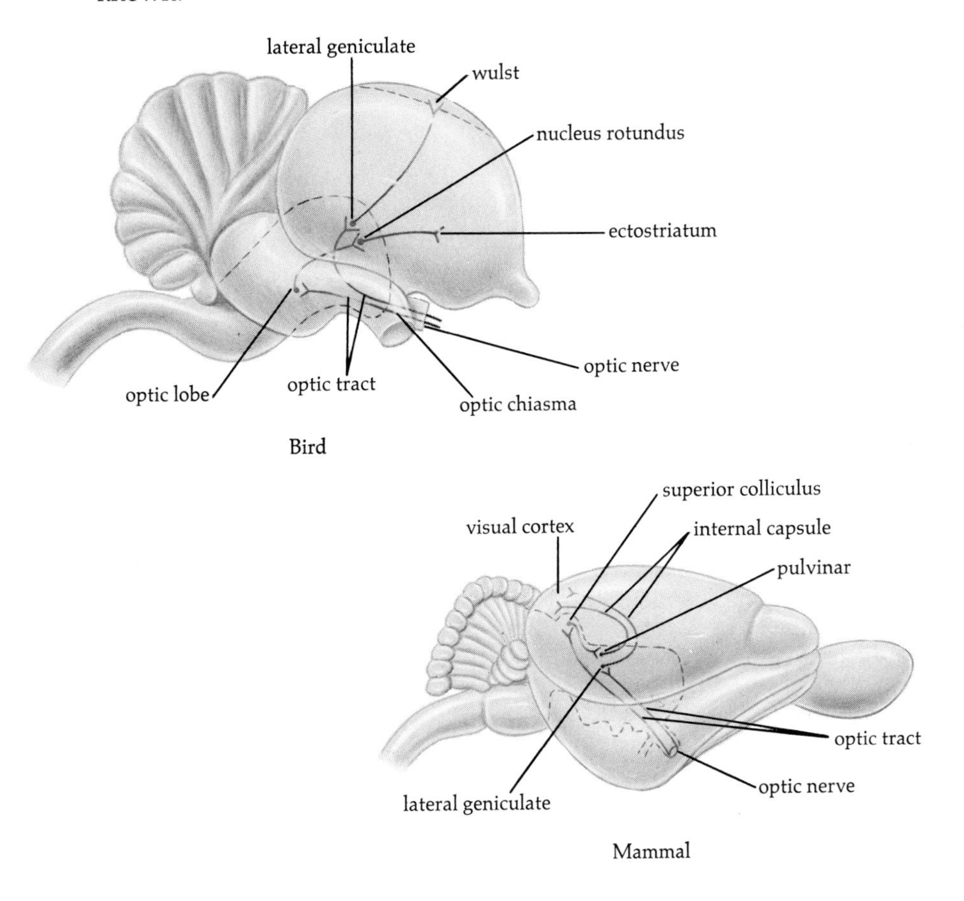

Fig. 12-9. Lateral views of the major visual pathways of a bird and a mammal. Note that each has both a retino–thalamo–telencephalic pathway and a retino–tecto–thalamo–telencephalic pathway.

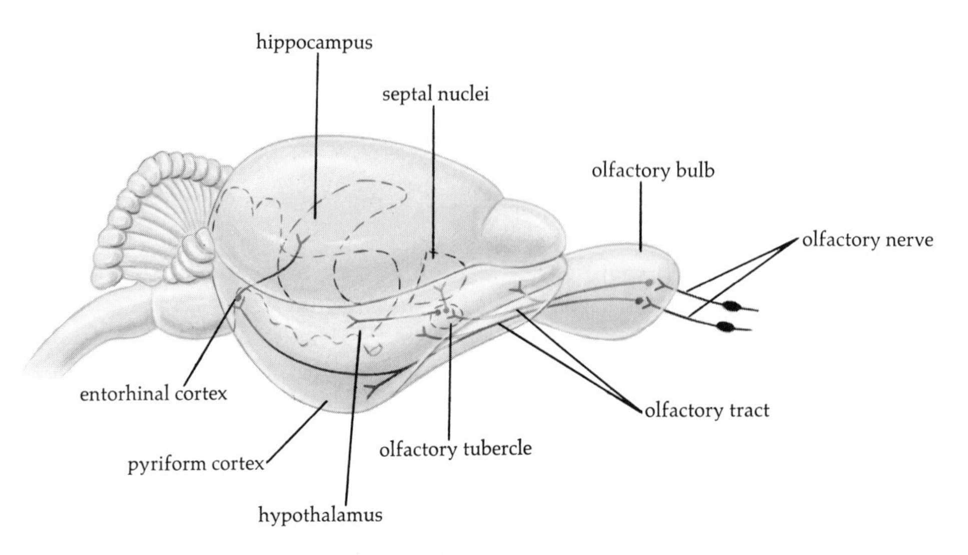

Fig. 12-10. A lateral view of a mammalian brain showing olfactory connections.

OLFACTORY PATHWAYS

MAMMALS

Axons from mammalian primary olfactory neurons terminate amid the complex neuropil of the olfactory bulbs. From there, axons project to several ventral structures of the telencephalon: the olfactory tubercle, gray matter which lies along the lateral olfactory tracts; the piriform cortex; and the entorhinal cortex, which lies just behind the piriform and which is structurally intermediate between piriform and neocortex.

From these olfactory centers axons run to the hypothalamus, habenulae, septal nuclei, and hippocampus (Fig. 12-10). Some axons from these structures project to the neocortex, but no part of the neocortex is considered in the direct olfactory pathway.

NONMAMMALIAN VERTEBRATES

The entire nonmammalian telencephalon, including its cerebral hemispheres, was long believed to have evolved from olfactory structures and to receive abundant olfactory input. However, recent experimental anatomical and physiological studies have indicated a much smaller olfactory input to the telencephalon than expected. There is only limited projection of olfactory information posterior to the olfactory bulbs.

In lizards, the olfactory tract projects from the olfactory bulb to a ventral, anterior olfactory nucleus and to the piriform cortex (Fig. 12-11). Some fibers then cross to the contralateral side of the cerebrum and terminate along a narrow, ventromedial surface of the telencephalon. When a vomeronasal organ is present it projects to a paraolfactory bulb, just next to the olfactory bulb; from the paraolfactory bulb fibers project primarily to the nucleus sphericus of the striatal complex (Fig. 12-11). In all nonmammalians, tertiary projections of olfactory information must still be determined.

In sharks the entire telencephalon was thought to be an olfactory structure. However, destruction of the olfactory bulbs demonstrates projections only to a relatively small strip of gray matter along the anterolateral and posteroventral portions of the telencephalic hemisphere—basically that region regarded as piriform cortex.

It is thus apparent that the nonmammalian telencephalon is not overwhelmingly olfactory as had previously been thought.

Motor Systems

CORTICOSPINAL SYSTEM

In the frontal lobe of the neocortex of all mammals studied there are very large cells, called Betz cells, whose axons, along with those of smaller motor cortical cells, project into the spinal cord to form the corticospinal pathway. This large tract goes through the corona radiata and the lateral portion of the internal capsule, past the thalamus, to the ventrolateral surface of the mesencephalon. In the mesencephalon it enters the large cerebral peduncles which pass through the pons. Caudal to the pons on each side of the ventral midline of the medulla, the corticospinal tracts reappear; from here on they are called the pyramidal tracts (Fig. 12-12). Where the medulla joins the spinal cord the pyramidal tracts cross one another in a complete or nearly complete decussation, and then these same axons travel down the white funiculi of the spinal cord to terminate finally in either the dorsal or ventral horn of the spinal cord. As this tract courses through the brain some axons leave it for terminations in various brainstem sensory or motor nuclei; these fibers constitute the corticobulbar system.

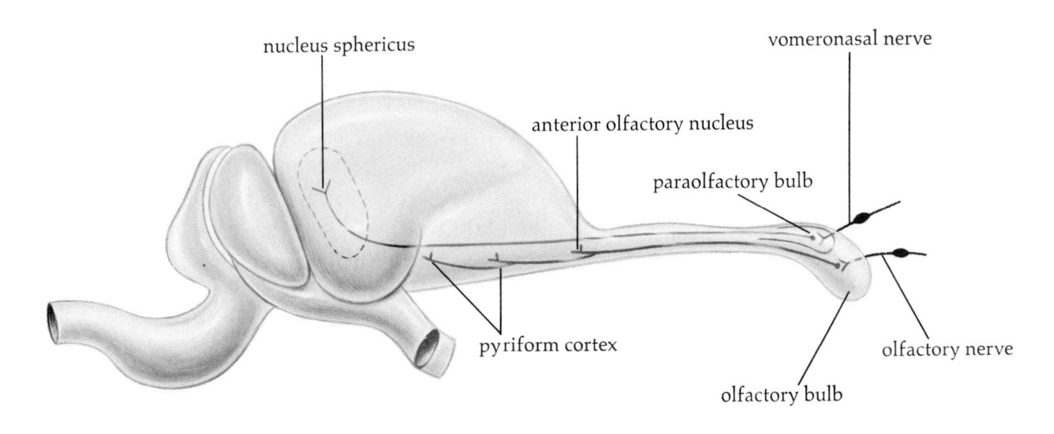

Fig. 12-11. Lateral view of a lizard brain showing the separate olfactory and vomeronasal projections to the telencephalon.

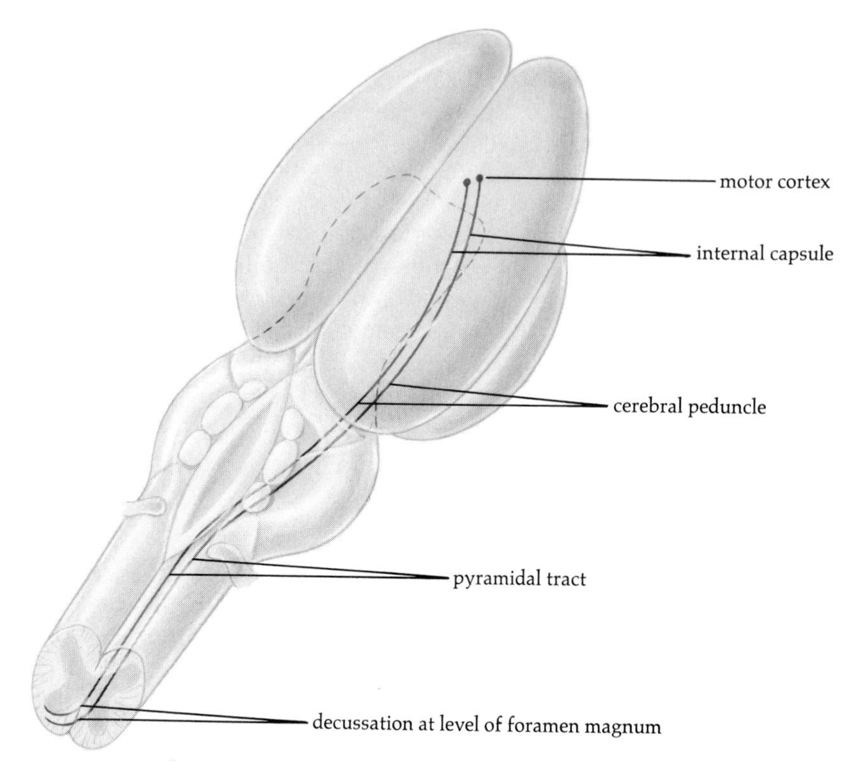

Fig. 12-12. The course of the neocortical Betz cells' axons from the motor cortex throughout the brain. (Cerebellum is removed.)

There is considerable variation in the spinal portion of the corticospinal tracts of different mammals (Fig. 12-13). In the rat, the tree shrew, and marsupials, corticospinal fibers run in the ventral portion of the dorsal funiculus. In carnivores and primates most corticospinal fibers travel in the lateral funiculus, with a few in the dorsal portion of the ventral funiculus. In the armadillo and insectivores a few corticospinal fibers run in the lateral funiculus, with most in the ventral funiculus. These fibers extend to the thoracic region in most mammals, to the cervical region in a few mammals, and to the end of the spinal cord in monkeys, apes, and humans, terminating in all cases on the gray matter of the spinal cord throughout their course.

The corticospinal fibers synapse on different spinal cord cells in different mammalian groups, with some interesting functional correlations. As previously described, axons to skeletal muscles arise from the large motor cells in the most ventral portion of the ventral horns; the corticospinal tract terminates on these cells only in monkeys, apes, humans, the raccoon, and the kinkajou (of those tested)—the same mammals with the greatest degree of fine digital control. In other mammals, corticospinal fibers terminate in more dorsal portions of the gray columns, and from there short axons travel to the ventral horn cells. Thus there is another synaptic junction and an extra neuron in the motor route.

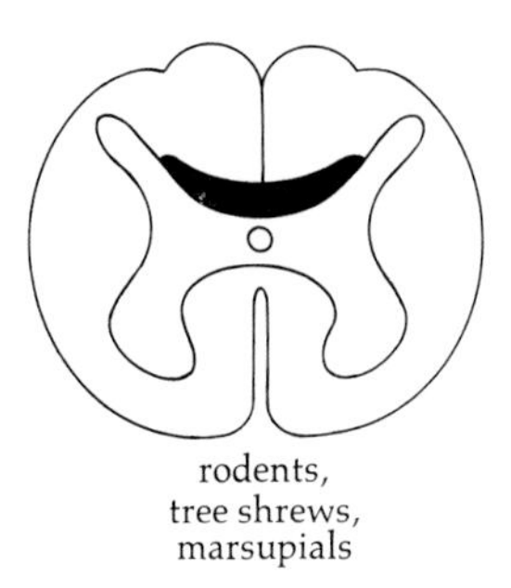

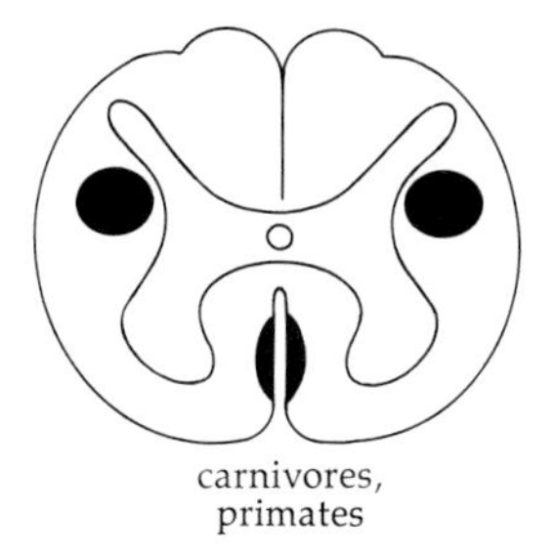

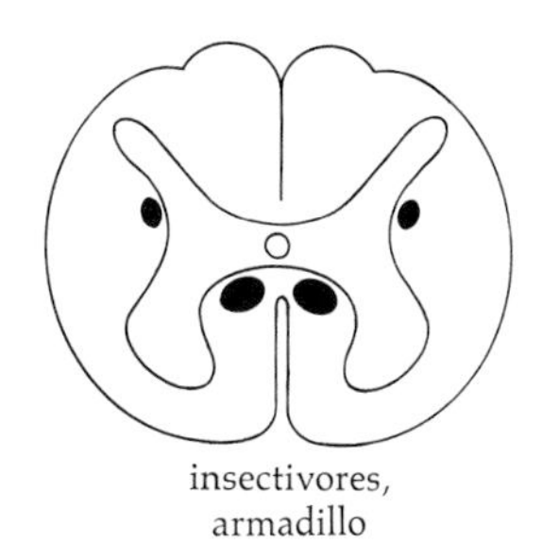

Fig. 12-13. Corticospinal fibers travel in different funiculi of the spinal cord in different mammals. The blackened areas in these cross sections of the cervical spinal cord indicate the positions of the corticospinal tracts.

The corticospinal system is called the pyramidal system. In humans at least it is the "conscious motor system"—that which has major control over voluntary movements. It had been thought that in nonmammalians more neurons and synapses were involved in motor outflow from brain to spinal cord and that only in mammals was there such a direct, one-neuron route. However it is now known that in at least some birds the region of the anterior wulst, the hyperstriatum accessorium, sends fibers through the brainstem to the cervical cord and possibly further. Studies on other nonmammalians are insufficient to show whether or not they too have corticospinal tracts, or what their extent might be.

EXTRAPYRAMIDAL MOTOR SYSTEM

Broadly speaking, the extrapyramidal motor system excludes the fibers of the corticospinal system but includes all other pathways terminating on motor nuclei. In most vertebrates, it is the largest and most complex motor system, and in many vertebrates it may be the only motor system.

MAMMALS

Basal ganglia

In mammals the nuclei of the largest extrapyramidal group do not project directly to motor nerves. This group is collectively called the basal ganglia, or corpus striatum (Fig. 12-14). The basal ganglia are located in the telencephalon and are traversed by internal capsule fibers en route from the dorsal thalamus to the neocortex, particularly to the frontal lobe of the neocortex; the basal ganglia have complex relationships with both these areas.

The caudate–putamen, which can be regarded either as two separate nuclei or one partially discontinuous nucleus of the basal ganglia,

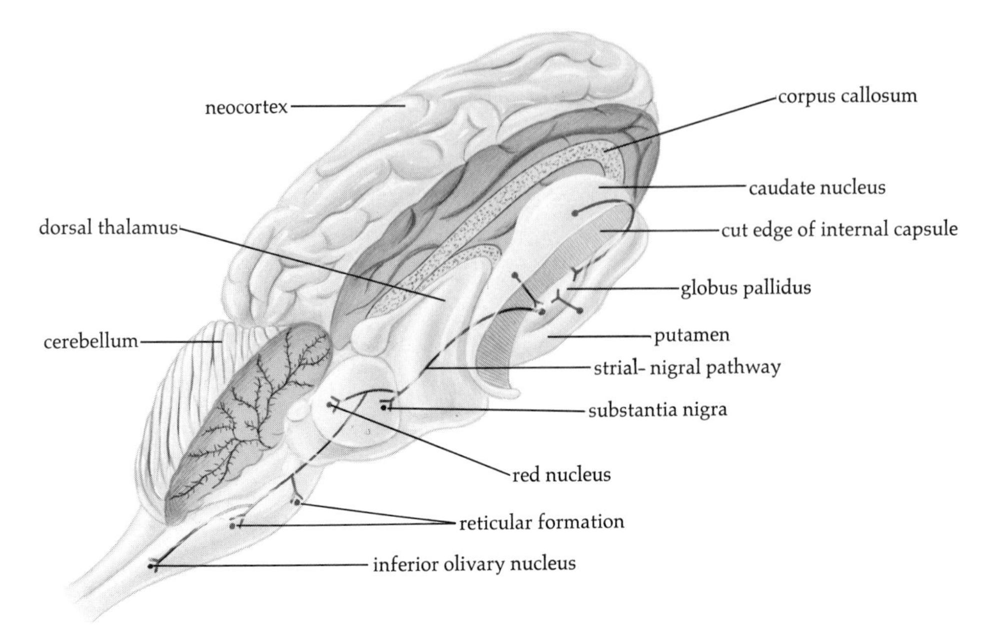

Fig. 12-14. The basal ganglia and their major fiber projections are shown in a sheep brain. The right half of the cerebellum and the right cerebral cortex and underlying white matter have been removed.

sends major projections directly to another nucleus of the group, the globus pallidus. Only a few fibers go directly from the caudate–putamen to extrapyramidal nuclei elsewhere in the brain.

Fibers leave the globus pallidus in the strionigral pathway. Many fibers terminate in a nucleus called the substantia nigra, which lies just dorsal to the cerebral peduncles on the ventral surface of the midbrain; some terminate in the red nucleus of the tegmentum, some in the inferior olivary nucleus, and others in the reticular formation of the hindbrain.

Finally the extrapyramidal system projects to the spinal cord. The rubrospinal tract leaves the red nucleus, and the reticulospinal tract leaves the reticular formation (Fig. 12-15). Both tracts decussate, and then most fibers terminate in the deep portions of the dorsal horn on the opposite side of the spinal cord; some fibers terminate in the more dorsal portions of the ventral horn.

The functional characteristics of this system are not thoroughly understood, but it is known that if these nuclei malfunction there is involuntary shaking, particularly in the extremities. In humans and probably other mammals this can be controlled to a degree by volitional corticospinal activity (the pyramidal system).

Tectospinal tract (Fig. 12-15)

Another prominent component of the extrapyramidal system is the tectospinal tract, which originates deep in the superior colliculi, decussates in the midbrain, and travels various distances down the spinal cord, usually terminating in the deep portion of the dorsal horn of the cervical region of the spinal cord. This tract is probably important for reflex functioning.

Vestibulospinal tract (*Fig. 12-15*)

The third and final major extrapyramidal component in mammals is the vestibulospinal tract. It originates from large cells in the lateral vestibular nucleus, extends usually the entire length of the spinal cord, and terminates in the deep portion of the dorsal horn of the cord.

NONMAMMALIAN EXTRAPYRAMIDAL SYSTEMS

There are rubrospinal, reticulospinal, and vestibulospinal (but not tectospinal) tracts in reptiles and probably in birds, which are morphologically similar to these tracts in mammals. Pathways beyond the thalamus are unclear. Behavioral studies involving stimulation of deep telencephalic portions suggest that the striatum is important in reptilian motor control, but precise pathways are undetermined.

Amphibians are apparently similar, except that at least frogs have a prominent, crossed tectospinal pathway.

In teleosts and cyclostomes the most prominent portions of the motor system are certain large, or giant, cells in the brainstem. In teleosts these are the Mauthner cells, located in the reticular formation near the entrance of the lateral line and eighth cranial nerves. In cyclostomes there are similar Mauthner cells in the hindbrain, and Muller cells in the midbrain. The axons from these giant cells are so large that they can be seen almost unaided, and so long that they extend the entire length of the spinal cord, making synaptic connections with the dendrites of many of its cells. On the basis of experimental

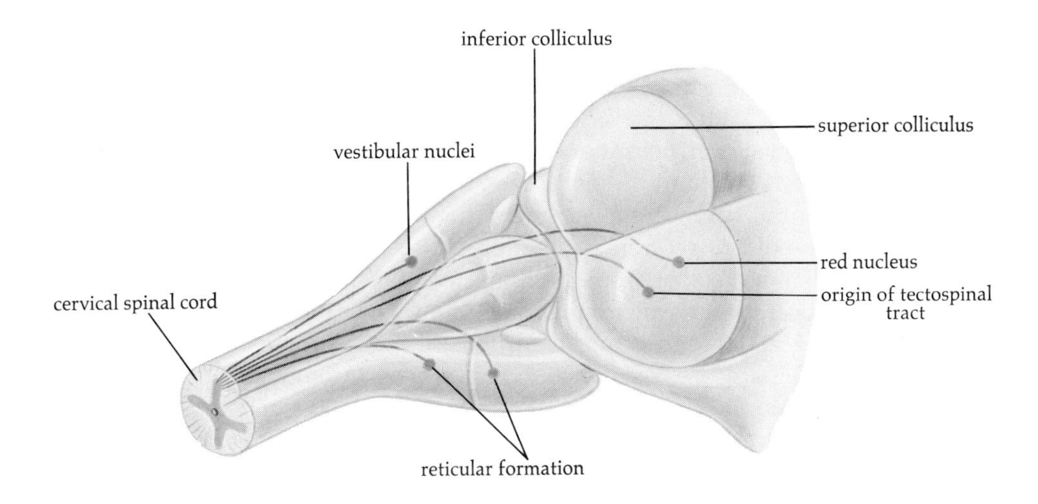

Fig. 12-15. Various brainstem nuclei give rise to fibers that carry extrapyramidal motor information to the spinal cord.

evidence it is these giant cells of the brainstem that primarily control the rhythmic contractions of lateral undulations, in which the myotomes contract first on one side and then the other. These giant cells are not found in Chondrichthyes, however, whose similar lateral undulations are apparently controlled by smaller axons.

CEREBELLAR PATHWAYS

Although the cerebellum does not itself send fibers to spinal or cranial motor nerves, it is nevertheless important in synergizing, or coordinating, all motor activity. Understanding its pathways is vital to understanding movement.

MAMMALS (Fig. 12-16)

The cerebellum receives information from spinal cord, brainstem, and neocortex. All pathways enter the cerebellum through one or more of its three pairs of peduncles, and the axons extend into the cerebellar cortex as either climbing or mossy fibers.

Mammals have two spinocerebellar pathways that are quite distinct from one another. In the ventral spinocerebellar system, first-order neurons come from the periphery to the large cell bodies of the dorsal root ganglia. From the ganglia, fibers enter the spinal cord, ascend a short distance in the dorsal funiculus, and then synapse in the nucleus proprius of the dorsal horn. Second-order fibers ascend in the most lateral portion of the lateral funiculus on the ipsilateral side; this is the ventral spinocerebellar tract. This tract travels to the brainstem, turns dorsally, and enters the anterior cerebellar peduncle (brachium conjunctivum) in which the second-order fibers enter the cerebellar cortex.

The dorsal spinocerebellar system is similar, except that its primary fibers ascend a considerable distance in the lateral funiculus before synapsing. Fibers from caudal, sacral, and lumbar portions of the cord ascend to the level of the thorax and synapse in Clark's column, a group of large, dorsal horn cells found only in the thoracic and cervical regions. From Clark's column, fibers ascend on the ipsilateral side just dorsal to the ventral spinocerebellar tract. This tract enters the brainstem, joins the posterior cerebellar peduncle (restiform body), and enters the cerebellum. Axons from the thoracic and cervical regions of the dorsal spinocerebellar system ascend all the way to the brainstem and synapse just lateral to the cuneate nucleus in the external cuneate. From the external cuneate, secondary fibers project through the restiform body up into the cerebellum.

Several afferent cerebellar pathways involve the brainstem rather than the spinal cord. The vestibulocerebellar tract runs from the vestibular nuclei in the brainstem through the restiform body to terminate in the cerebellar cortex, primarily, if not exclusively, in its flocculonodular lobe. The olivocerebellar tract runs from the inferior

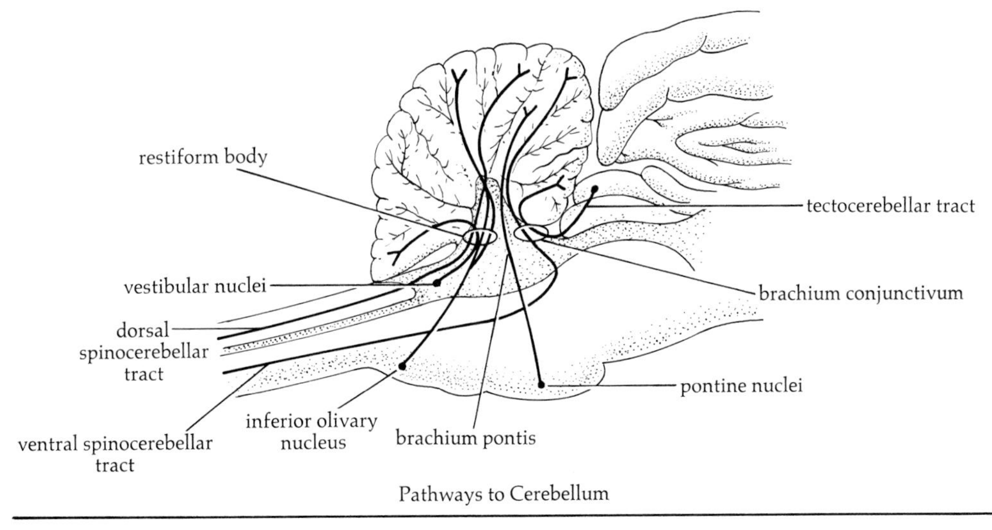

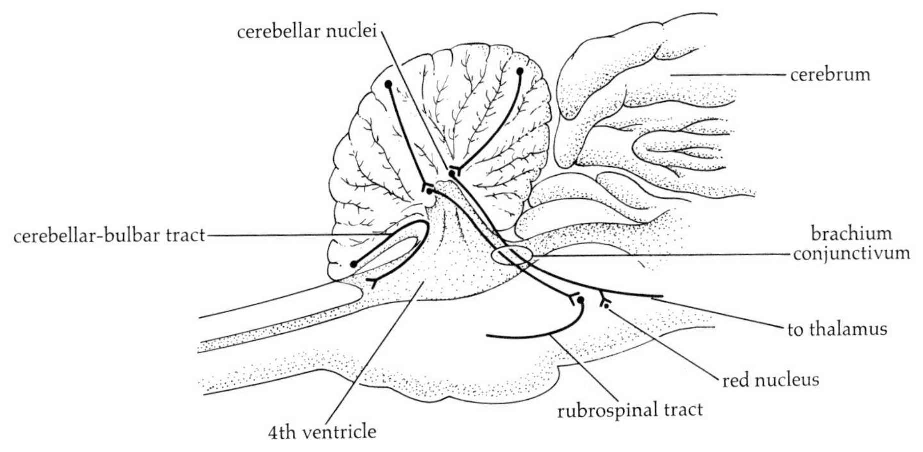

Pathways from Cerebellum

Fig. 12-16. The pathways to and from the mammalian cerebellum as projected onto a sagittal view of the sheep brain. The ventral part of the midline cerebellum (vermis) is removed to show the cerebellar peduncles.

olivary nucleus near the ventral surface of the brainstem just posterior to the trapezoid body; the tracts decussate and enter the cerebellum via the contralateral posterior peduncle. The largest input to the cerebellum comes through the middle cerebellar peduncle (brachium pontis). This peduncle is composed of many heavily myelinated fibers from the pontine nuclei, which decussate within the pons and enter the cerebellar hemisphere on the opposite side. Finally there are relatively small tectocerebellar tracts, which come from deep tectal nuclei and pass directly up through the anterior cerebellar peduncle (brachium conjunctivum) into the cerebellar cortex.

The output from the cerebellar cortex is by way of Purkinje cell axons. A few of these axons go directly to the vestibular nuclei and the reticular formation, forming the small cerebellar–bulbar tract which

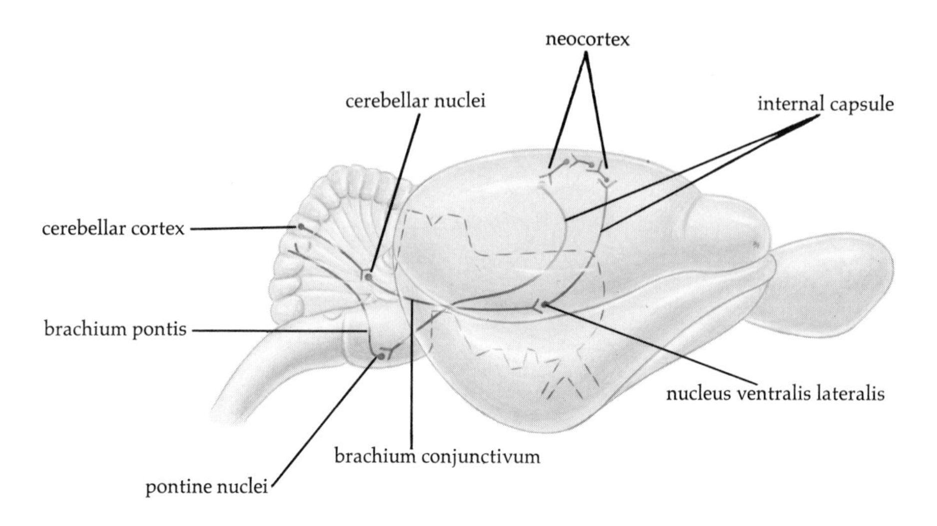

Fig. 12-17. Feedback circuitry between the mammalian cerebellar cortex and neocortex plays an important role in the central control of motor activity.

runs through the posterior cerebellar peduncle to the brainstem. Most fibers from Purkinje cells, however, synapse first in deep cerebellar nuclei; fibers from there leave the cerebellum as the anterior cerebellar peduncle (brachium conjunctivum). These fibers decussate in the midbrain and then form a large capsule around the red nucleus; many fibers terminate in the red nucleus, particularly on its larger cells, whose fibers in turn form the rubrospinal tract providing direct cerebellar influence on spinal motor neurons.

Equally many cerebellar fibers bypass the red nucleus after encapsulating it and continue anteriorly to terminate in the dorsal thalamus, primarily in its nucleus ventralis lateralis. Some of the projection from this nucleus (and other thalamic nuclei) is to the basal ganglia. Its major projection, however, is through the internal capsule and corona radiata to broad terminations in the neocortex.

Then from all over the neocortex (not just its motor cortex) fibers pass through the corona radiata and internal capsule back to the cerebral peduncles. Some of these fibers are the direct corticospinal fibers of the pyramidal system which will form the pyramidal tracts in the medulla. Most of these fibers, however, terminate in the pontine nuclei, which in turn project to the contralateral cerebellar cortex by way of the brachium pontis.

Thus there is strong feedback circuitry between cerebellum and neocortex, involving brachium conjunctivum, nucleus ventralis lateralis, neocortex, cerebral peduncles, pontine nuclei, and brachium pontis (Fig. 12-17). Through this feedback system the cerebellum controls all motor activity, both pyramidal and extrapyramidal, to fulfill its synergistic function.

NONMAMMALS

Little is known about cerebellar fiber connections in nonmammalian vertebrates. The few published experimental studies indicate that they are complex and that the older literature, which was based on studies of normal nondegenerating fibers, was frequently in error. In nonmammals as well as mammals lesion studies have proven that the cerebellum assists in equilibrium and in synergistic muscular activity.

FINAL COMMON PATHWAY

Multiple excitatory and inhibitory impulses from the complex circuitry of brain and spinal cord finally must be reduced to one specific signal, which is sent out a cranial motor nerve or the ventral root of a spinal nerve. This cranial motor nerve or ventral root is thus called the final common pathway, for it is here that all the influences from sensory input, central cranial activity, and various motor systems are brought to bear on an effector organ to cause the animal's coordinated behavior (Fig. 12-18).

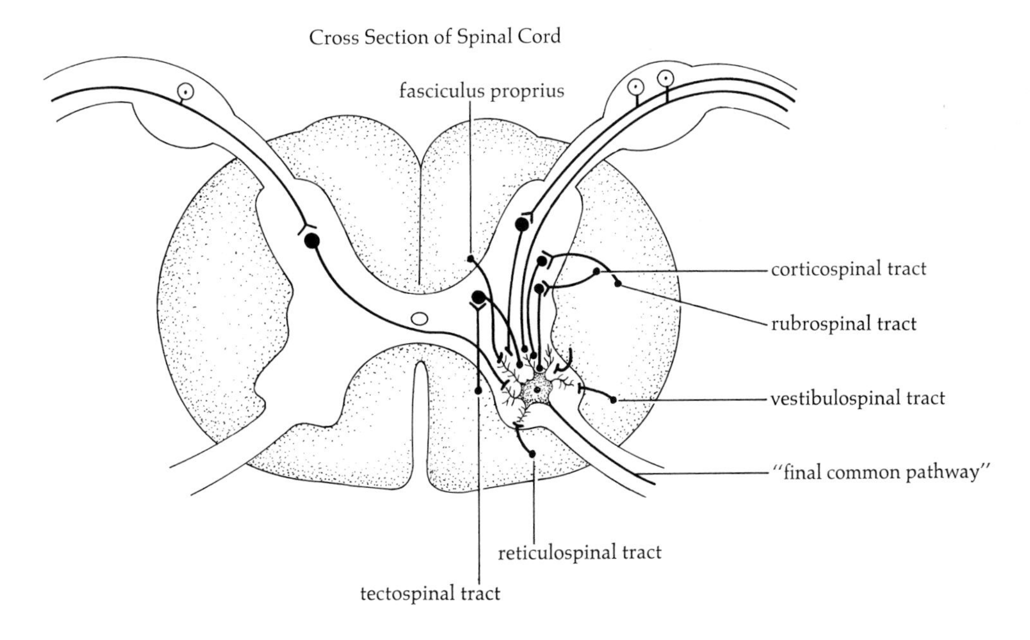

Fig. 12-18. A diagram illustrating that each motor neuron to the periphery is controlled, through a balance of excitation and inhibition, by a great many other neurons coming from many parts of the central and peripheral nervous system. In this way the "final common pathway" conveys coordinated motor information to the periphery.

THE MAMMALIAN "LIMBIC LOBE"

Several intimately connected nuclear groups and fiber pathways within the forebrain constitute the limbic lobe. These include the hippocampus, mammillary bodies, anterior dorsal thalamic nuclei, and cingulate cortex—the last being the region of cortex just above the corpus callosum whose morphology is intermediate between neocortex and the simpler hippocampus and piriform cortex. These structures were originally thought to be primarily olfactory in function but now have been demonstrated to be involved in emotional responses, general arousal activity, and the laying down of memories.

The major efferent path from the hippocampus is the fornix, whose fibers go anteriorly, ventrally, and then posteriorly and terminate massively in the mammillary body and surrounding hypothalamus. From the mammillary body the prominent mammillothalamic tract projects to the anterior nuclei of the dorsal thalamus; these nuclei in turn project through the internal capsule to the cingulate cortex. A large association tract, the cingulum, goes from the cingulate cortex caudally and then ventrally, terminating in the entorhinal cortex. From the entorhinal cortex fibers travel back to the hippocampus, completing the circle of tracts and nuclei which form the limbic lobe (Fig. 12-19).

Olfactory information enters the limbic lobe through the entorhinal cortex, and other sensory information enters through thalamic projections to the anterior thalamic nuclei. Nonspecific ascending reticular pathways from the brainstem enter via the thalamus. Output from the limbic lobe is primarily to the hypothalamus and reticular formation.

Clues to various types of behavior are to be found in the morphology and physiology of the limbic lobe. Its structures are not unique to mammals, and indeed most of its fiber pathways are present in non-

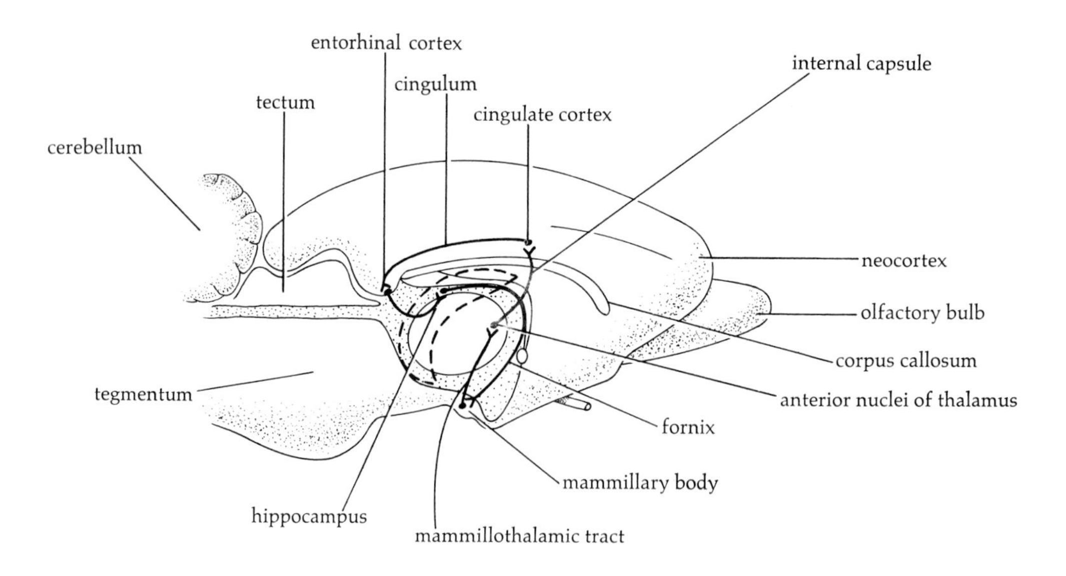

Fig. 12-19. Limbic lobe circuitry projected onto the medial surface of a rat's forebrain. The laterally placed hippocampus is indicated by the dashed line.

mammalian vertebrates. It may represent a very ancient portion of the vertebrate brain which plays a vital role in the general arousal of the nervous system necessary for such diverse functions as emotional responses and memory formation.

MAMMALIAN NEOCORTEX

The neocortex, whose distinctive six layers are unique to mammals, is a complex structure (as what part of the nervous system is not!). It can be more easily studied and its organization more clearly understood if we approach smaller parts of it. To this end it has been divided in several ways (Fig. 12-20).

Most readily, it can be considered as four distinct lobes: frontal, parietal, occipital, and temporal, named for the bones beneath which they lie.

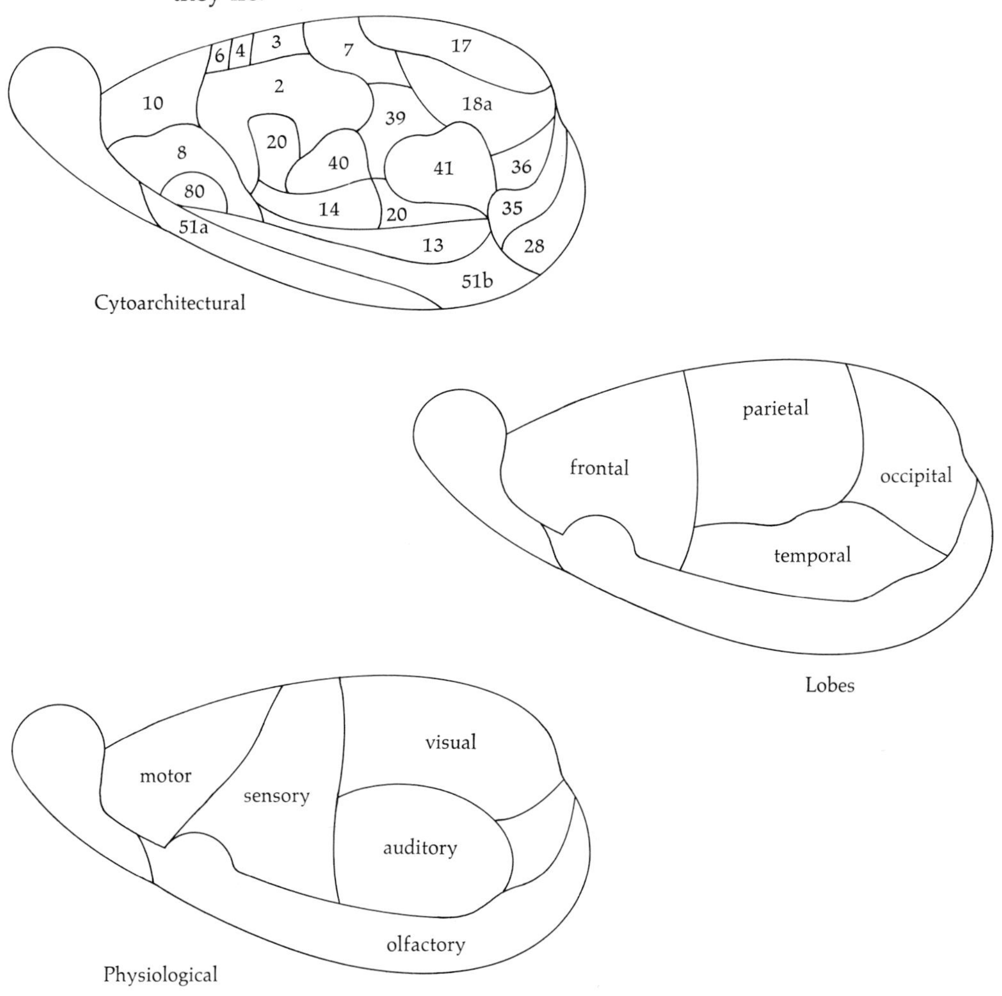

Fig. 12-20. Lateral views of the rat cerebrum divided into (1) areas determined by the cytology of the cortex, (2) lobes named for the bones beneath which they lie, and (3) areas determined by electrophysiological responses.

It can also be divided on the basis of slight cytoarchitectural differences in the neuropil. Thus the human neocortex can be divided into 52 areas which differ from each other slightly in the architecture of their layers. Cytoarchitectural studies on some other mammals (rhesus monkey, cat, laboratory rat) have demonstrated similar cytoarchitectural areas in similar topographic areas.

The neocortex can also be divided on physiological grounds into areas of response; for example, those which respond to auditory stimulation, to visual stimulation, and to somatic sensory stimulation. When these studies are combined with cytoarchitectural studies it is found that congruent areas are being determined; for example, all of areas 17, 18, and 19 are visual cortex and all of area 41 is the primary auditory cortex.

Finally, and perhaps most meaningfully, the neocortex can be divided on the basis of incoming thalamic projections and outgoing motor projections. Again, a close relationship is found between the anatomy of fiber projections, the physiology of responses, and the cytoarchitecture of the areas.

The percentage of the neocortex taken up by sensory projections or motor output is large in the lissencephalic cortex of small mammals, but quite small in the gyrencephalic cortex of a large mammal. Between specific cortical projection areas and motor output areas are the so-called association or silent areas—by no means silent physiologically, since they have constant neural activity except when under deep anesthesia. In general, the larger the mammal, the larger is its association cortex, and the larger the association cortex, the more capable the mammal is of complex learning and of behavioral plasticity.

It is evident that the cerebral cortex is still at best only a partially understood structure, both anatomically and physiologically.

SUGGESTED READING

Note: Most of the reliable data for nonmammalian vertebrates are reported in a broad range of detailed papers published in recent years in *Journal of Comparative Neurology, Brain Research,* and *Brain, Behavior and Evolution.* They have not yet been brought together into one or more reference or textbooks.

Gardner, E. (1968). "Fundamentals of Neurology." W. B. Saunders, Philadelphia.

Petras, J. M., and Noback, C. R., eds. (1969). Comparative and evolutionary aspects of the vertebrate central nervous system. *Ann. N. Y. Acad. Sci.* **167,** 1–513.

Polyak, S. (1958). "The Vertebrate Visual System." Univ. of Chicago Press, Chicago.

Whitfield, I. C. (1967). "The Auditory Pathway." Williams & Wilkins, Baltimore, Maryland.

13 ENDOCRINES

PITUITARY GLAND OR HYPOPHYSIS

STRUCTURE

- NEUROHYPOPHYSIS
- ADENOHYPOPHYSIS
  - PARS TUBERALIS
  - PARS DISTALIS
  - PARS INTERMEDIA

HORMONES OF THE PITUITARY

- NEUROHYPOPHYSIS
- ADENOHYPOPHYSIS
  - PARS DISTALIS
  - PARS INTERMEDIA

THYROID GLAND

STRUCTURE

HORMONES OF THE THYROID

- FOLLICULAR HORMONES
- PARAFOLLICULAR HORMONE: CALCITONIN

PARATHYROID GLANDS

PANCREATIC ISLETS OF LANGERHANS

HORMONES OF THE ISLETS

ADRENAL GLANDS AND TISSUES

- CHROMAFFIN TISSUES
- STEROIDOGENIC TISSUES

GASTROINTESTINAL HORMONES

MINOR AND LESS WELL UNDERSTOOD ENDOCRINES
PINEAL
THYMUS
UROPHYSIS
ANGIOTENSIN

REPRODUCTIVE HORMONES

SUGGESTED READING

The glands discussed so far are exocrine glands, the secretions of which pass through ducts to an external or internal body surface. An endocrine gland, by contrast, is ductless; its secretion, called a hormone, either passes from the synthesizing cell directly into surrounding capillaries or is first stored and later released into the capillaries. Once in the circulatory system, a hormone is carried throughout the body. Each hormone has a very specific effect, but only on particular "target organs." (Other substances in the circulatory system, such as carbon dioxide and urea, also cause specific responses in certain organs; they are not hormones but are primarily the excretory products of cell metabolism.)

Anatomically speaking the endocrine glands do not form an organ system. They lie in diverse parts of the body, with no distinct interconnections. Moreover, although some are discrete organs (e.g., thyroid, adrenals), others are simply small parts of organs (e.g., islets of Langerhans within the pancreas, or endocrine cells within the alimentary canal). Endocrine tissues arise from any of the three germ layers. Their hormones vary chemically according to their embryonic origin: protein, polypeptide, or amino acid hormones are secreted by endocrine glands of endodermal or ectodermal epithelial origin; steroids, by those of mesodermal origin.

However, although not anatomically an organ system, endocrines do function as a system in controlling and regulating diverse portions of the body. In this they do not work alone. The pituitary, an extreme case, is not only controlled by the hypothalamus but some of its secretions are actually synthesized in the hypothalamus. Many other endocrines are influenced by autonomic innervation.

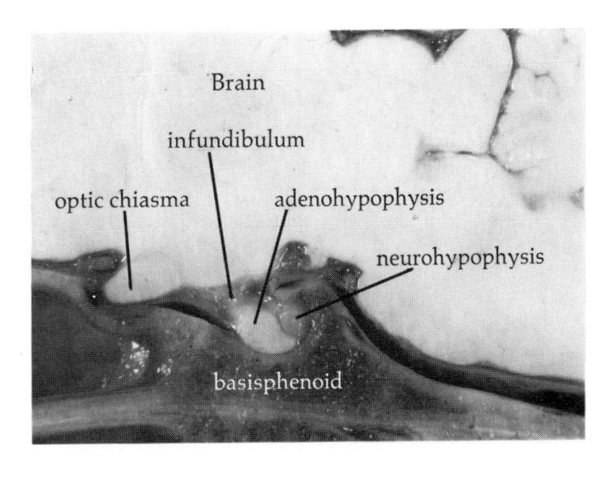

Fig. 13-1. A vertical section, just off the midline, through the basisphenoid and pituitary of a cat.

PITUITARY GLAND OR HYPOPHYSIS

The pituitary gland, which is located just below and attached to the hypothalamus (Fig. 13-1), is structurally and functionally the most complex of endocrine glands. It releases at least nine protein and polypeptide hormones which influence every portion of the body. Although frequently called the master gland, the pituitary itself is regulated, through feedback mechanisms, by the nervous system and by other endocrines. It would more properly be thought of as a coordinating organ.

STRUCTURE

The pituitary has two major components, which develop from separate tissues and grow together (Fig. 13-2). The neurohypophysis is formed by a ventral growth from the hypothalamus; the adenohypophysis, by

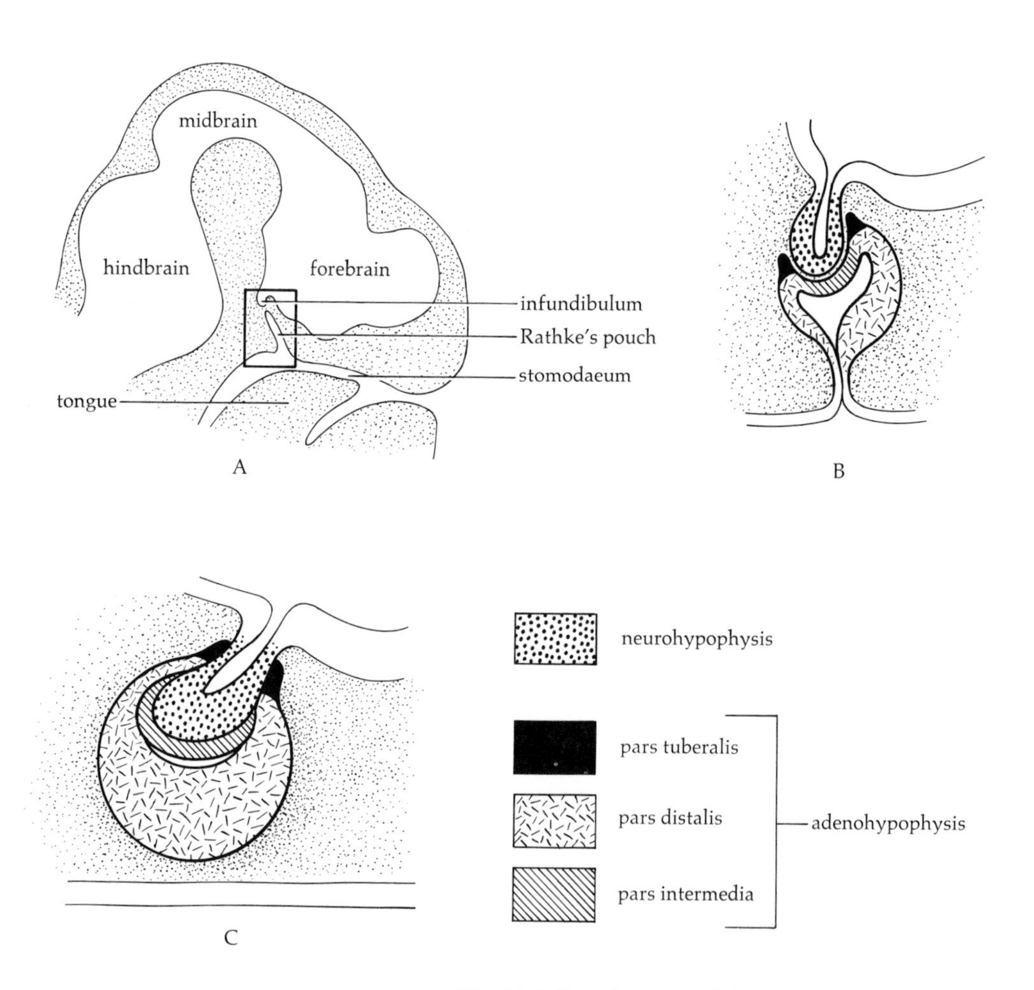

Fig. 13-2. Development of the pituitary as seen in sagittal sections. A, sagittal section through entire head; B and C, higher magnifications of later stages in pituitary development.

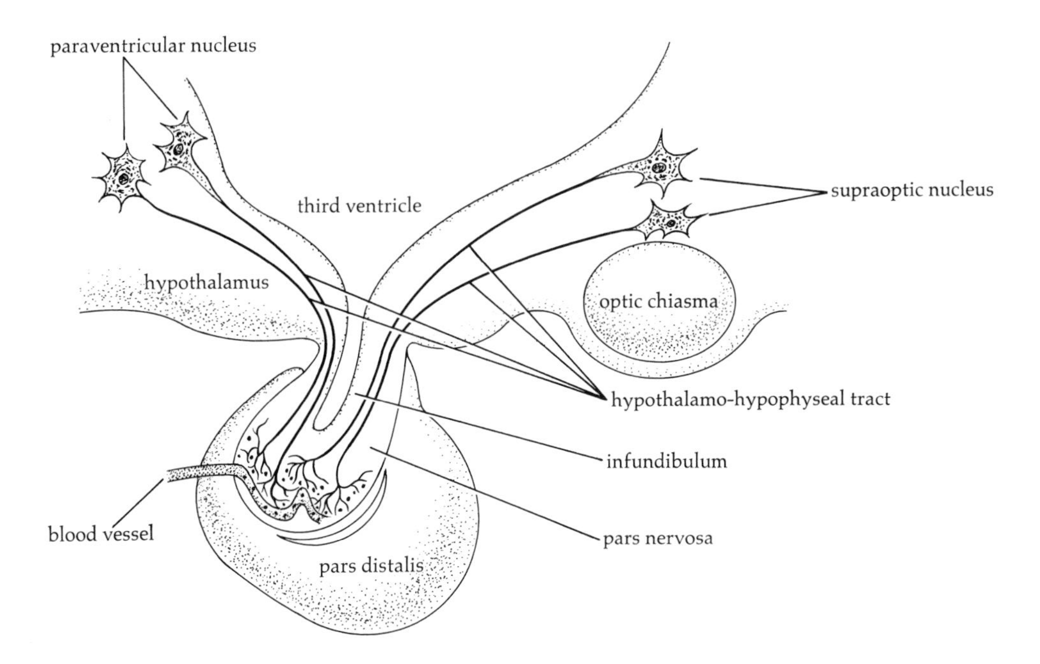

Fig. 13-3. Diagramatic representation of the structural and functional relationships of the hypothalamus and the pituitary's pars nervosa.

a dorsal evagination from the roof of the mouth. This evagination, called Rathke's pouch, grows up anterior to the neurohypophysis, adheres to it, and then loses its contact with the oral cavity.

Neurohypophysis

The neurohypophysis includes the pars nervosa and the infundibulum (the stalk connecting pars nervosa and hypothalamus). The cells of the pars nervosa are glial elements called pituicytes; they neither synthesize nor secrete hormones. Hormones are synthesized in the cell bodies of the supraoptic and paraventricular nuclei of the hypothalamus. The hormones pass down the axons of these hypothalamic cells, which run through the infundibulum in the hypothalamohypophyseal tract. They are then released into the highly vascularized space around the pituicytes, passing into the circulatory system (Fig. 13-3).

In jawed fishes, except sarcopterygians, the part of the hypothalamus just behind the infundibulum is a highly folded and vascularized structure of unknown function called the saccus vasculosus. Although frequently described as part of the piscine pituitary, it is truly part of the hypothalamus.

Adenohypophysis

PARS TUBERALIS

In tetrapods, but not fishes, a small portion of the adenohypophysis wraps around the infundibulum and is called the pars tuberalis. It receives autonomic innervation and is highly vascularized; its large cuboidal epithelial cells contain granules. It apparently produces no hormones.

PARS DISTALIS

This largest and most prominent part of the adenohypophysis is composed of irregular cords of polyhedral epithelial cells separated by vascular sinusoids. There are three types of cells: acidophils have granules which stain prominently with acid stains; basophils have granules which stain with basic stains; and chromophobes have granules which stain with neither (Fig. 13-4).

In tetrapods the pars distalis has no innervation, but is connected with the median eminence (or tuber cinereum) of the hypothalamus by a unique portal vascular system. From a capillary bed in the median eminence venules lead down through the infundibulum to the pars distalis. These venules form the sinusoids in tetrapods. Specific hormone-releasing factors synthesized by cells in the median eminence are carried by this hypothalamohypophyseal portal system to the pars distalis. There they regulate its flow of hormones (Fig. 13-5). A smaller portal system is present in chondrosteans, holosteans, and elasmobranchs. In teleosts, however, there is nervous innervation. The hormone-releasing factors from the cells of the median eminence travel down the cells' axons through the infundibulum and are released in the pars distalis, which in teleosts is densely innervated. In cyclostomes neither a distinct portal system nor nervous innervation of the pars distalis has been demonstrated. The pars distalis of cyclostomes, teleosts, and chondrichthyeans is divided into rostral and proximal portions, frequently separated by connective tissue septae and having different staining characteristics. In chondrichthyeans there is also a separate ventral lobe of unknown significance (Fig. 13-6).

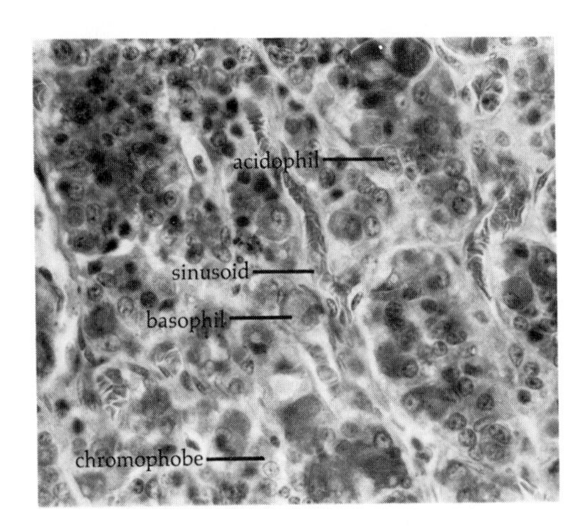

Fig. 13-4. Microscopic section of the pars distalis of the pituitary.

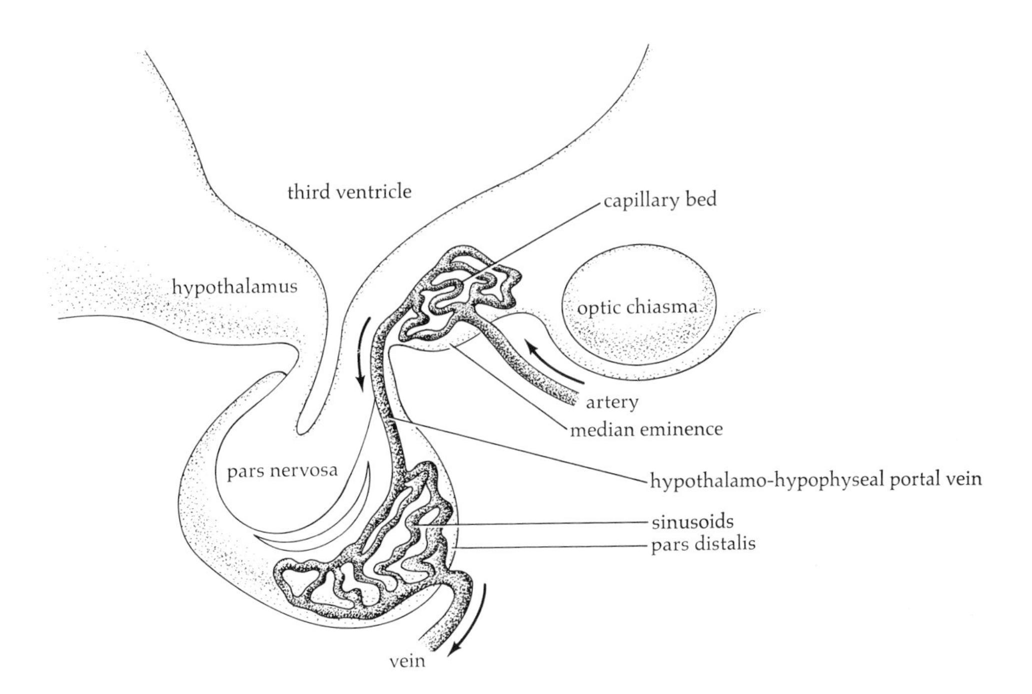

Fig. 13-5. Diagramatic representation of the structural and functional relationships of the hypothalamus and the pituitary's pars distalis.

PARS INTERMEDIA

That portion of Rathke's pouch which makes contact with the neurohypophysis is called the pars intermedia. It is separated from the pars distalis by a small extracellular space, or by an area of mixed pars distalis and pars intermedia cells. The pars intermedia has distinct polyhedral cells with basophilic granules. In at least frogs, elasmobranchs, and the bowfin, it receives autonomic innervation, with terminals directly on the gland cells. There is no pars intermedia in birds.

HORMONES OF THE PITUITARY

The pituitary hormones have been most extensively studied in mammals; nonmammals have the same hormones, with but minor differences in amino acid sequences.

Neurohypophysis

Two octapeptide hormones produced in the hypothalamus are stored in and finally released by the neurohypophysis. Both affect the adenohypophysis as well as other target organs.

Oxytocin helps regulate the release of various hormones of the adenohypophysis. In mammals it also causes contraction of the uterine

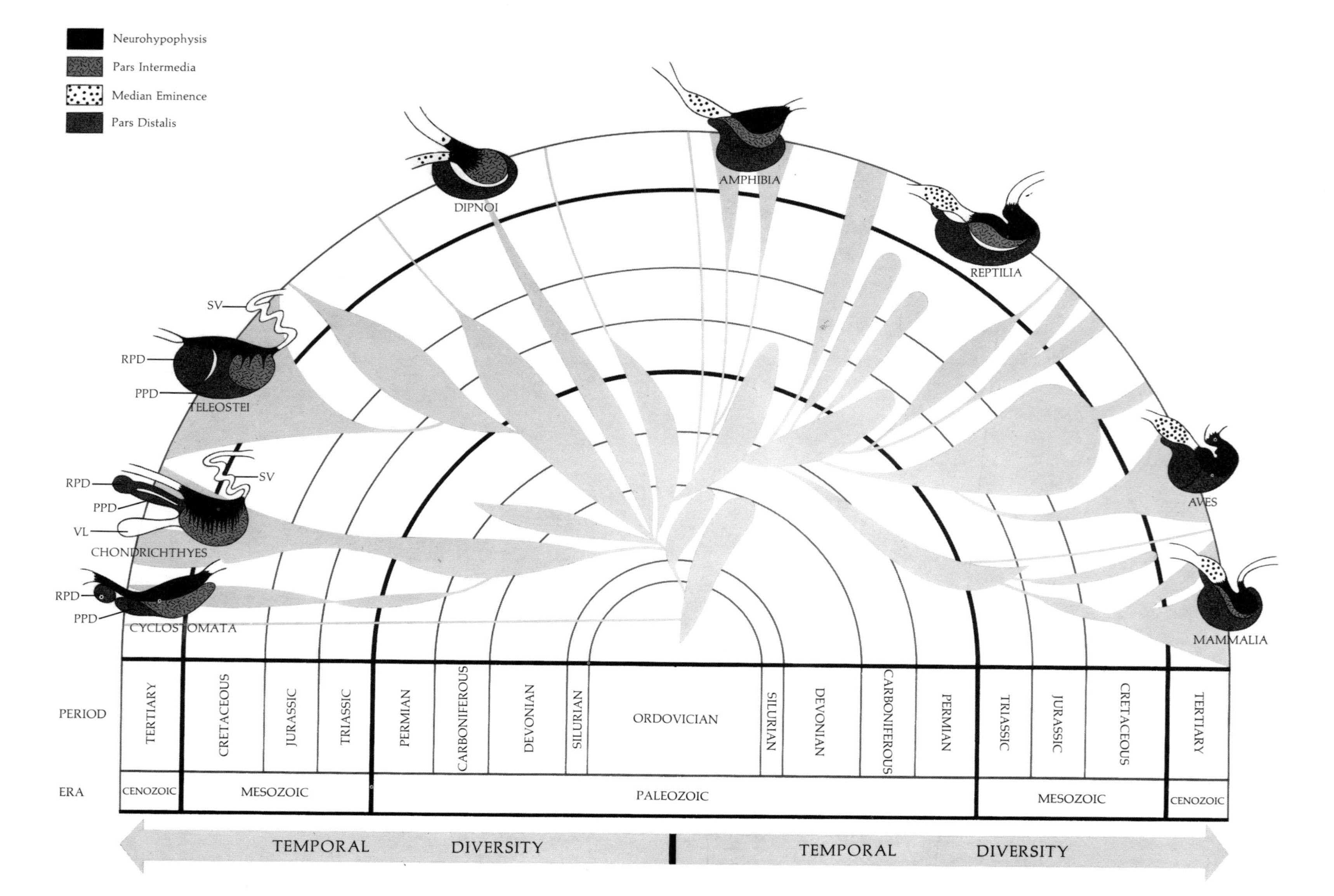
Neurohypophysis
Pars Intermedia
Median Eminence
Pars Distalis
DIPNOI
AMPHIBIA
REPTILIA
SV
RPD
PPD
TELEOSTEI
SV
RPD
PPD
VL
CHONDRICHTHYES
RPD
PPD
CYCLOSTOMATA
AVES
MAMMALIA
PERIOD
TERTIARY
CRETACEOUS
JURASSIC
TRIASSIC
PERMIAN
CARBONIFEROUS
DEVONIAN
SILURIAN
ORDOVICIAN
SILURIAN
DEVONIAN
CARBONIFEROUS
PERMIAN
TRIASSIC
JURASSIC
CRETACEOUS
TERTIARY
ERA
CENOZOIC
MESOZOIC
PALEOZOIC
MESOZOIC
CENOZOIC
TEMPORAL DIVERSITY
TEMPORAL DIVERSITY

muscles and of the myoepithelial cells surrounding the mammary gland alveoli (causing milk to be "let down," i.e., stored prior to release by suckling).

Vasopressin, the antidiuretic hormone, helps regulate the release of pars distalis hormones. It also has two other important functions: to increase water resorption in the kidney, which helps keep the osmotic concentration of the internal fluids at a constant level, and to stimulate contraction of the smooth muscles of blood vessels, which increases blood pressure. In birds, oddly enough, vasopressin causes the smooth muscle of blood vessels to relax.

Adenohypophysis

PARS DISTALIS

Six known hormones are produced by the pars distalis. Research has shown which cells produce each and has strongly suggested that no cell produces more than one.

The growth hormone, somatotropin, produced by acidophilic cells, is a branched protein with about 200 amino acids. It is necessary for the normal growth and development of skeletal and muscular tissues.

The lactogenic hormone, prolactin, also produced by acidophilic cells, is a protein with a molecular weight of about 25,000. In mammals it promotes the synthesis of milk, and in birds the production of a proteinaceous secretion of the crop called crop milk. In all vertebrates it affects sexual behavior and the maintenance of the postovulatory structures of the ovary.

The adrenocorticotropic hormone, ACTH, is a 39-amino acid polypeptide produced by chromophobe cells. It stimulates the activity and secretion of the adrenal cortex (interrenal tissue in anamniotes) and also in most vertebrates affects the dispersion of melanin within melanocytes.

The luteinizing hormone, LH, and the follicle stimulating hormone, FSH, both affect the reproductive organs. They are produced by slightly different basophilic cells, and are both small glycoproteins. In females, FSH activates the growth of the ovarian follicle, and LH stimulates the interstitial cells of the ovary and the development of the corpus luteum. In males, FSH stimulates activity of the seminiferous tubules of the testis, and LH stimulates the activity of its interstitial cells.

The thyroid-stimulating hormone, TSH, is produced by basophilic cells and is a glycoprotein whose molecular weight is approximately 10,000. Its primary function, as its name implies, is to stimulate the secretion of thyroid hormone.

Fig. 13-6. Spatial diversity of parts of the vertebrate pituitary as seen in schematic sagittal sections. PPD, proximal pars distalis; RPD, rostral pars distalis; VL, ventral lobe; SV, saccus vasculosus. Drawings adapted from C. D. Turner and J. T. Bagnaria (1971). "General Endocrinology." W. B. Saunders, Philadelphia.

The production and release of these six pars distalis hormones are promoted by specific releasing factors which are synthesized and secreted by the median eminence of the hypothalamus into the hypothalamohypophyseal portal system. Pars distalis activity is also directly influenced by feedback from other endocrines.

PARS INTERMEDIA

The pars intermedia produces only the melanin-stimulating hormone, MSH, a polypeptide with 13 to 18 amino acids. Its action is most pronounced in anamniotes, where it stimulates the synthesis and dispersion of melanin within melanocytes, and the contraction of the purine particles in iridophores.

THYROID GLAND

STRUCTURE

The thyroid is the chief endocrine gland that regulates metabolic rate. In all vertebrates it is composed of spherical or oval follicles within a highly vascularized, loose connective tissue. The follicles contain a viscous, semigel colloid. The walls of the follicle are composed of a single layer of secretory epithelial cells, resting on a fine, connective tissue basement membrane which forms the outer wrapping of the follicle (Fig. 13-7). Against this basement membrane is a highly folded plasma membrane. On their apical surfaces the secretory epithelial cells have microvilli and occasionally flagellae or cilia which extend into the colloid (Fig. 13-8). Between follicular cells there are tight junctions, with no intercellular space. The cytoplasm of the follicular cells contains complex ergastoplasmic membranes and large cisternae containing follicular fluid. In addition to the follicular cells there are occasional parafollicular glandular cells, which stain more lightly than do follicular cells and contain granules rather than cisternae (Fig. 13-8).

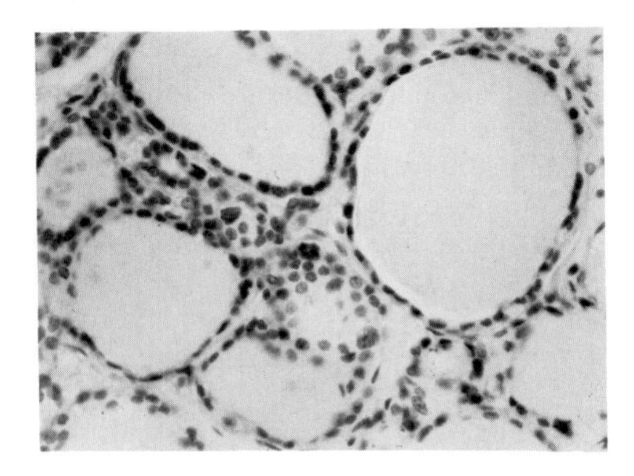

Fig. 13-7. Microscopic section of a mammalian thyroid gland, showing follicle morphology.

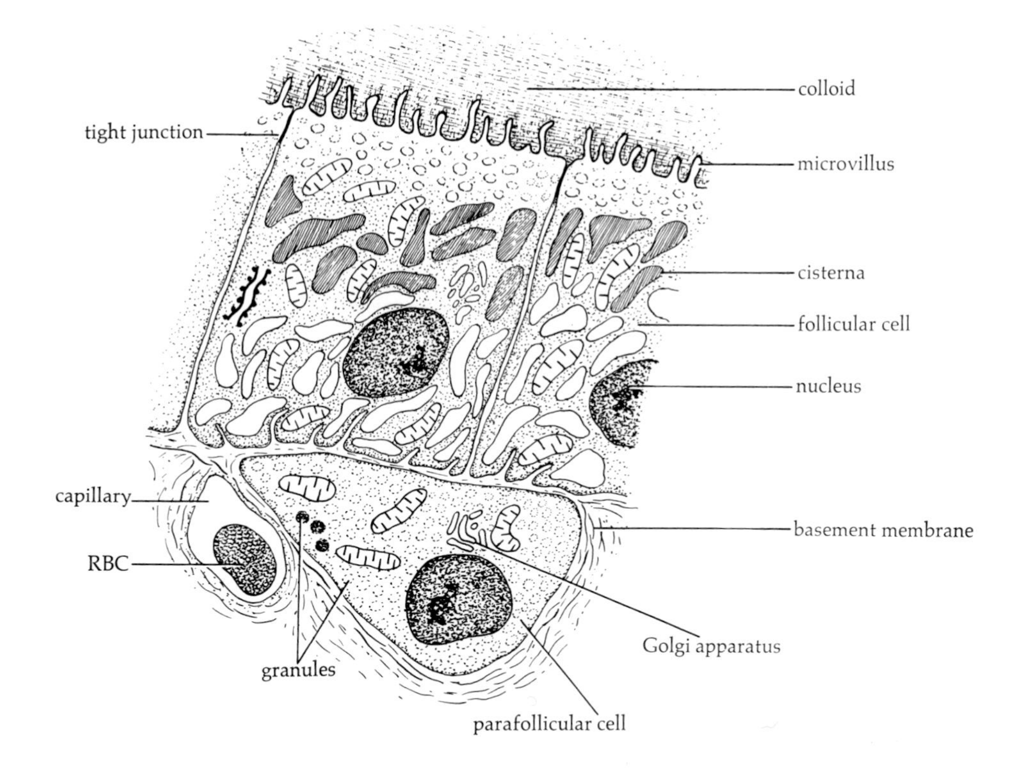

Fig. 13-8. Diagram of a small portion of a thyroid follicle, showing ultrastructural features of both follicular and parafollicular cells. RBC, red blood cell.

In all vertebrates the thyroid develops as a midline structure, evaginating from the floor of the pharynx. Throughout early development it retains contact with the pharynx by a hollow or solid column of epithelial cells. By the adult stage this contact has been lost and the thyroid is a true endocrine gland, vascularized and ductless.

In the lamprey we can see what the phylogeny of the thyroid may have been. In the ammocoete larva it is called the subpharyngeal gland and is continuous with the pharynx via a duct; the gland lies in a groove called the endostyle which is also continuous with the pharynx. At metamorphosis the floor of the pharynx is closed off, and the thyroid tissue becomes endocrine. It is likely, therefore, that the vertebrate thyroid gland evolved from an exocrine gland very early in vertebrate evolution, or perhaps even in "prevertebrate" chordates.

Despite similarities of development and microscopic structure, the thyroid's gross structure varies greatly among vertebrates (Fig. 13-9). Among bony fishes discrete thyroid glands are found in only a few teleosts; other bony fishes and adult lampreys have thyroid follicles dispersed along the ventral aorta below the pharynx. In cartilaginous fishes compact thyroid follicles encapsulated by a loose connective tissue lie among the hypobranchial musculature in the anterior pharynx, deep to its epithelium. Such isolated thyroid follicles among head and neck tissues are not unusual even in vertebrates that have

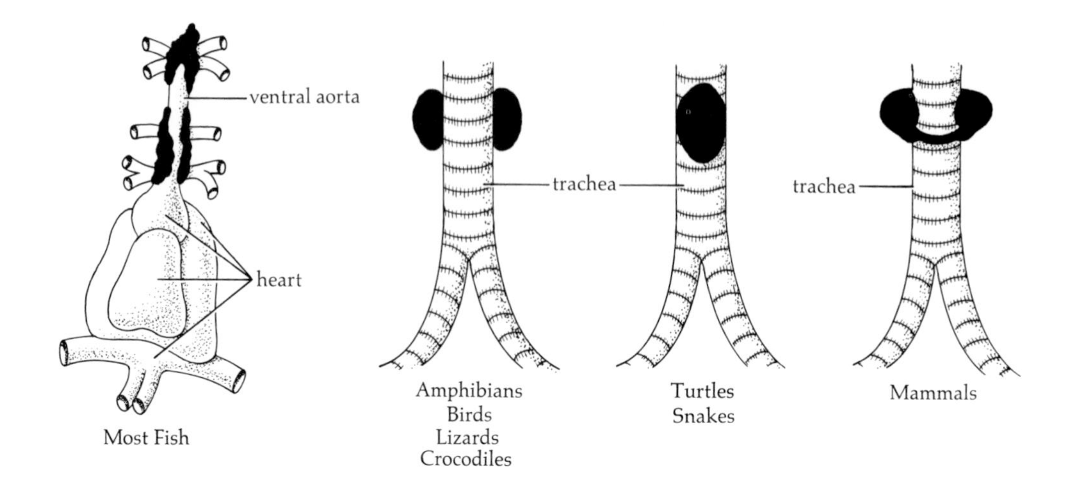

Fig. 13-9. Diagramatic representations of the distribution of thyroid tissue (in black) in diverse groups of vertebrates.

discrete thyroid glands; in mammals, for instance, they frequently occur at the base of the tongue. These isolated particles are "leftovers" from the migrations of the thyroid tissue anlagen.

In amphibians, birds, and some reptiles, the evagination from the floor of the pharynx divides, forming paired thyroids on either side of the trachea. In snakes and turtles, however, there is a single, midline thyroid. In adult mammals a bilobed thyroid lies ventrolaterally on either side of the larynx; there is usually a thin strand of thyroid tissue ventrally crossing the larynx and connecting the two lobes.

HORMONES OF THE THYROID

The follicular and parafollicular cells of the thyroid synthesize and secrete different hormones.

Follicular hormones

The follicular cells concentrate the body's iodine and utilize it in synthesizing two thyroid hormones. The two hormones, triiodothyronine and tetraiodothyronine (thyroxine), are iodinated amino acids having similar effects on the body; triiodothyronine is produced in smaller amounts but is considerably more potent than tetraiodothyronine.

The final synthesis of these hormones occurs not within the follicular cells but within the colloid contained in the follicle. A major component of the colloid is a large protein, called thyroglobulin, secreted by the follicle cells. In the colloid, iodine becomes attached to amino acids of the thyroglobulin. The resulting product splits off as one of the thyroid hormones (which one depends upon whether three or four iodine molecules are involved). The completed hormone must pass back through the thyroid cells to reach the capillaries outside the follicle.

The production and release of thyroid hormone are stimulated by TSH, one of the hormones of the pars distalis of the pituitary. In turn, the production and release of TSH are accelerated by releasing factors from the hypothalamus; both production and release of TSH are also limited by the secretion of thyroid hormone and are stimulated by its absence. This is an example of the intricate feedback mechanisms of the endocrine and nervous systems.

The primary physiological effect of thyroid hormone is to increase the oxidative respiration of cells and therefore to increase the animal's metabolic rate—particularly important in the homeothermic mechanisms of mammals and birds. Thyroid hormone also works in conjunction with the growth hormone from the pituitary. In amphibians it initiates metamorphosis; if the gland is removed from amphibian tadpoles they will not metamorphose. In neotenic amphibians such as *Necturus* (mud puppy) the thyroid, although present, is very small.

Parafollicular hormone: calcitonin

The parafollicular cells of the thyroid gland produce quite a different hormone, called calcitonin or thyrocalcitonin, which is a polypeptide with 32 amino acids. This hormone lowers the level of calcium in the blood by causing more calcium to be deposited in bones; less calcium is, therefore, excreted with urine. Calcitonin can also cause calcium to be deposited in the kidney.

Current evidence indicates that parafollicular cells develop differently than do follicular cells, being derivatives of the ultimobranchial bodies. These ultimobranchial bodies originate from the corners of the posterior pharyngeal pouches in all vertebrates. They give rise to parafollicular cells of the thyroid and parathyroid only in tetrapods, remaining as ultimobranchial bodies in fish. Apparently, however, the ultimobranchial bodies themselves secrete both calcitonin and parathyroid hormone in adult fish.

PARATHYROID GLANDS

The parathyroids occur as discrete glands only in adult amphibians and amniotes. They are composed of irregular epithelial cords of polyhedral cells, called chief cells. Like all endocrine glands they are highly vascularized (Fig. 13-10). Although endodermal derivatives of pharyngeal pouches two, three, and four, they may migrate considerable distances: they are usually found in close association with the thyroid gland, often being embedded in its tissues near the cervical and upper thoracic ventral blood vessels.

The hormone of the parathyroid is a polypeptide of 75 amino acids. Its rate of secretion is controlled by the level of calcium and phosphate in the blood. Its target organs are bone, where it mobilizes calcium by causing bone resorption, and kidney, where it causes both the resorp-

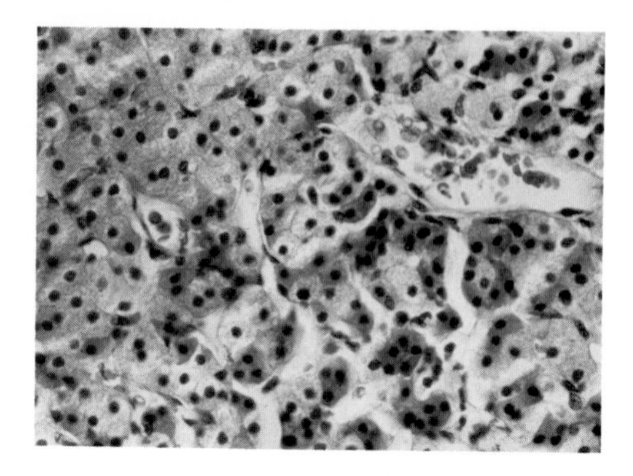

Fig. 13-10. Microscopic structure of the mammalian parathyroid gland, showing the irregular cords of the chief cells. The dark cells, called oxyphile cells, are few in number and are of unknown function.

tion of calcium from urine and the active excretion of phosphates. Thus its functions are antagonistic to those of calcitonin.

PANCREATIC ISLETS OF LANGERHANS

The endocrine portion of the pancreas gland is the islets of Langerhans (Fig. 13-11). Three cell types are found in islet tissue: alpha cells, containing alcohol-insoluble granules; beta cells, the most common, containing alcohol-soluble granules; and D cells or clear cells, the least common of the three cell types.

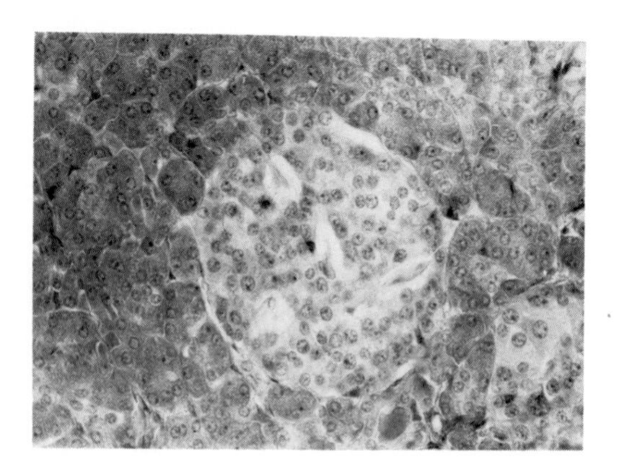

Fig. 13-11. Microscopic section of the mammalian pancreas, showing an islet of Langerhans surrounded by exocrine pancreas. Note the intense vascularization of the islet of Langerhans.

Although islet tissue occurs in all vertebrates, there is some variation in its composition and a great deal of variation in its distribution. In both the larval and adult lamprey aggregates of islet cells, primarily beta cells, are found within the intestinal epithelium. In the hagfish a compact organ of beta cells lies in the intestinal walls. Gnathostomes have all three cell types. In most jawed fishes islet tissue occurs diffusely throughout the pancreas, but in a few teleosts such as the catfish it is separate from the exocrine portion of the pancreas. In all tetrapods it occurs in separate islets within the exocrine pancreas.

HORMONES OF THE ISLETS

The alpha cells secrete glucagon, a polypeptide hormone of 29 amino acids; beta cells secrete insulin, a proteinaceous hormone of 51 amino acids. Both regulate carbohydrate metabolism, but in opposite ways.

Glucagon mobilizes glycogen. Its target organ is the liver, where it causes stored glycogen to be broken down into glucose molecules which are released into the general circulation. Insulin, on the other hand, facilitates the efficient use of the glucose in the blood and prevents the excessive breakdown of glycogen. When insulin is not produced, as in the metabolic disorder called diabetes, most of the blood's glucose passes out with the urine without being used in cell metabolism or stored as glycogen.

ADRENAL GLANDS AND TISSUES

In mammals, birds, and reptiles the adrenals are compound glands with two types of tissues. Chromaffin tissue, so named because its cells stain dark brown with chromates, composes the inner portion, or adrenal medulla. Steroidogenic tissue, so named because it produces steroid hormones, composes the outer portion, or adrenal cortex. In reptiles and birds there is often some cortical tissue interspersed with the chromaffin tissue of the medulla.

In anamniotes, both chromaffin and steroidogenic tissues are present in the region of the kidneys, but they are usually separate from one another at all developmental stages, and no discrete adrenal glands are formed. When steroidogenic tissue is isolated from chromaffin tissue in this way it can also be called interrenal tissue; it acts as does the steroidogenic tissue of discrete adrenal glands.

In amphibians the chromaffin tissue is embedded in the ventral portion of the kidney along with steroidogenic tissue. In actinopterygians, separate, small aggregates of chromaffin and steroidogenic tissue are usually found along the veins in the trunk region and in the anterior portion of the kidney. In chondrichthyeans, chromaffin tissue is in small, paired patches on the medial side of the kidneys, and steroidogenic tissue is in a midline organ lying along the mediodorsal blood vessels. In cyclostomes there are isolated, microscopic patches of separate chromaffin and steroidogenic tissue along the walls of the blood vessels and kidneys (Fig. 13-12).

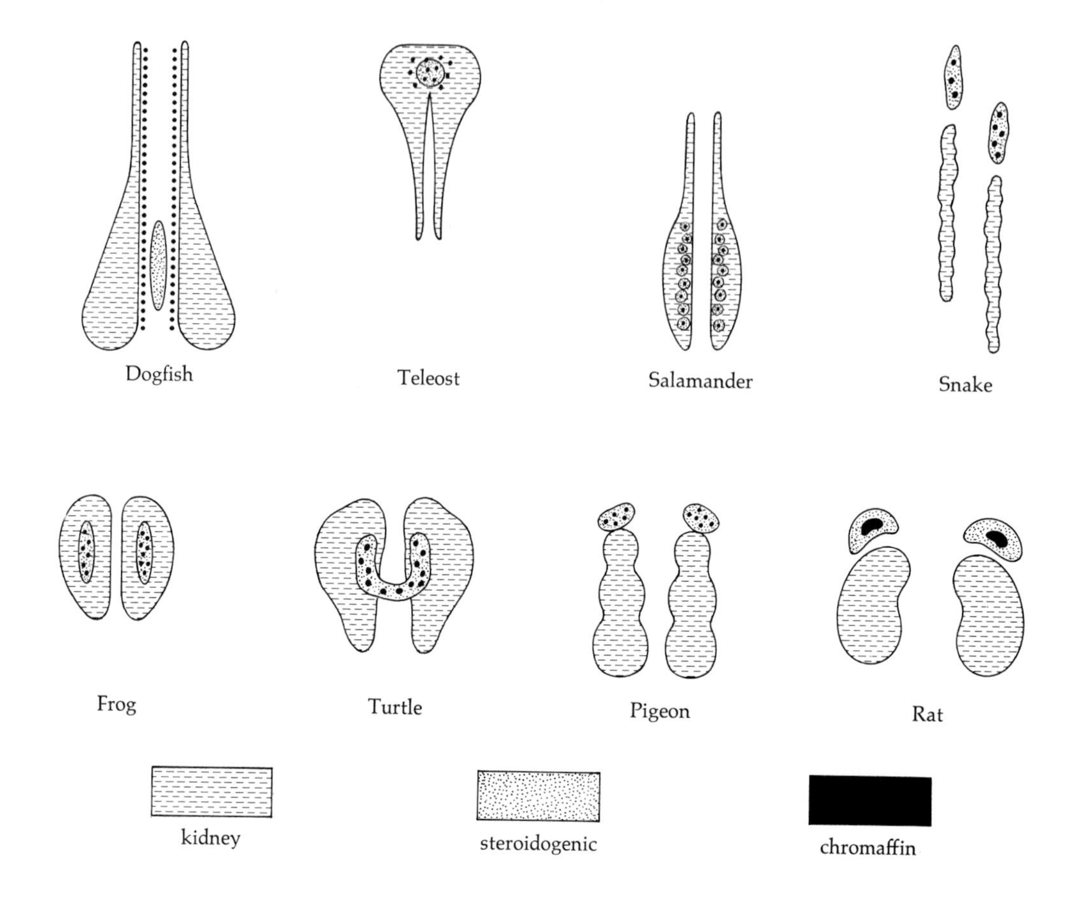

Fig. 13-12. The distribution of steroidogenic and chromaffin tissues in relation to the kidneys in diverse vertebrates.

Chromaffin tissues

These tissues are of neural crest origin; specifically, their polyhedral cells are modified postganglionic cells of the sympathetic nervous system (Fig. 13-13). Thus they are innervated by, and their activity is controlled by, preganglionic sympathetic cells. Chromaffin cells, however, secrete large amounts of epinephrine (plus a minimal amount of norepinephrine), while other postganglionic sympathetic cells secrete primarily norepinephrine.

Both the epinephrine from chromaffin tissue and the norepinephrine from the sympathetic system prepare the animal for periods of stress, with some differences. Epinephrine, for instance, dilates arterioles in muscles and liver and thus greatly increases the flow of blood through these organs; norepinephrine has no effect on this blood flow. Epinephrine has a much greater effect on increasing the heart rate than does norepinephrine.

Steroidogenic tissues

The steroidogenic (or interrenal) tissues of all vertebrates are composed of polyhedral epithelioid cells (derived from mesoderm) containing secretory granules.

These tissues have been most extensively studied in mammals. In the mammalian adrenal cortex the steroidogenic tissues can be divided into zones on the basis of the arrangement of cells and their relationships to the numerous blood sinusoids (Fig. 13-14). In the outermost, or zona glomerulosa, the cells are in irregular aggregates. In a wide zone just deep to this, the zona fasciculata, the cells form long parallel cords separated from each other by vascular sinusoids. In the innermost zona reticularis, the cells are in irregular cords interspersed with sinusoids. A fourth zone is present in fetal mammals, called zone X or the fetal cortex; its cells are irregularly placed between the medulla and the zona reticularis. It is evidently an active fetal tissue but is resorbed before birth.

Nearly 50 steroids of the adrenal cortex have been isolated and identified in mammals. Some have no known function; others have very similar functions and can be considered in groups. There are two such groups, found widely in living vertebrates, that are produced only by the adrenal cortex and regulated by ACTH.

The first group, from the zona fasciculata and zona reticularis, regulates the intracellular metabolism of carbohydrates and proteins; cortisone is the most familiar of these. The second group, from the zona glomerulosa, is important in regulating water and electrolyte balance; these steroids act primarily on kidney tubules during urine formation. The most potent of these is aldosterone, which is particularly important for osmotic balance in seabirds and in those fishes which can live in either fresh or salt water.

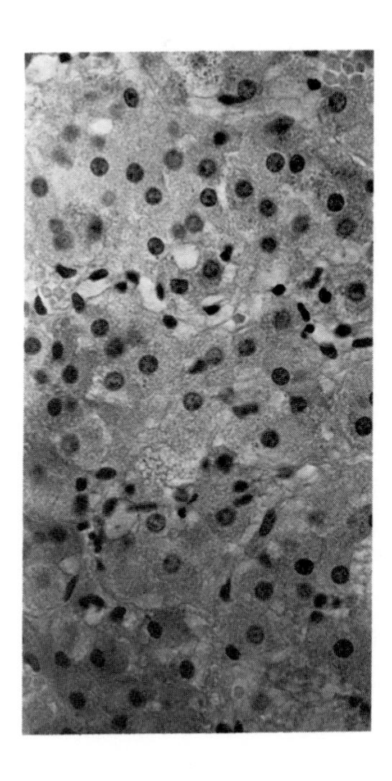

Fig. 13-13. Histology of the mammalian adrenal medulla with its granular chromaffin cells.

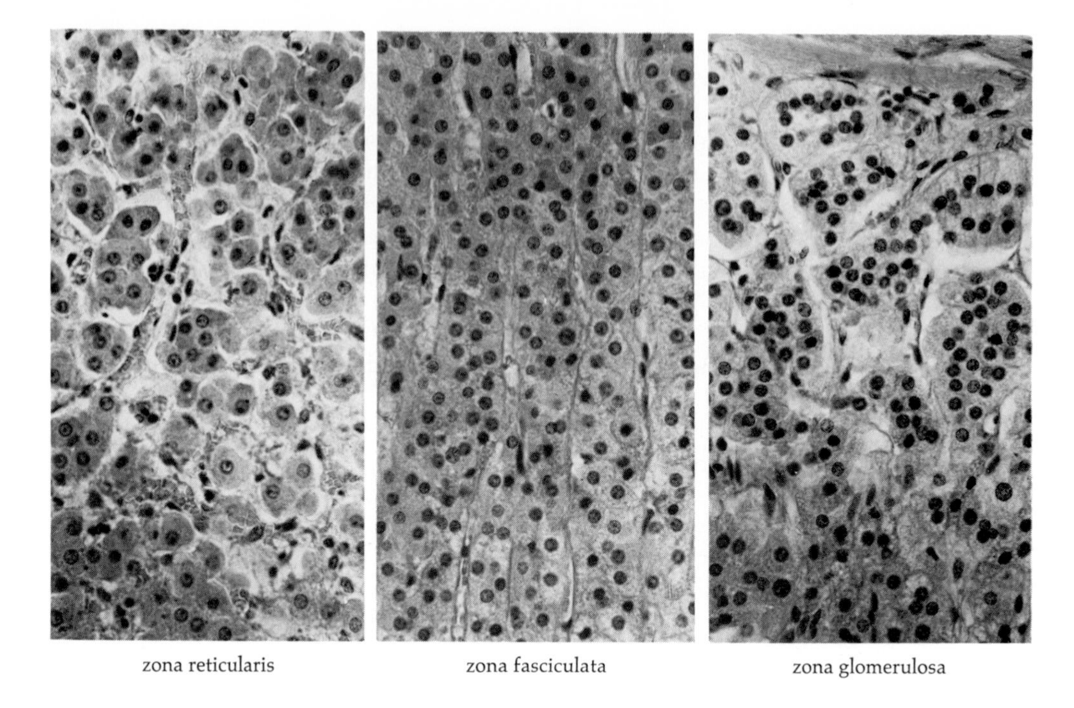

Fig. 13-14. Histological structure of each of the three zones of the mammalian adrenal cortex.

In addition to these the adrenal cortex also produces small amounts of all those steroids that are produced in much larger quantities by the reproductive organs.

GASTROINTESTINAL HORMONES

Three hormones produced by cells in the epithelium of the mammalian stomach or duodenum help to regulate digestion. Cholecystokinin, from the duodenum, causes contraction of the gallbladder and release of digestive enzymes from the pancreatic cells. Secretin, also from the duodenum, stimulates the flow of pancreatic juice. Gastrin, from the stomach, stimulates the fundic glands of the stomach to secrete hydrochloric acid. These three polypeptide hormones have similar amino acid sequences that are also similar to that of glucagon; it is interesting that all four of these hormones are products of the endoderm of a restricted part of the alimentary canal.

MINOR AND LESS WELL UNDERSTOOD ENDOCRINES

PINEAL

The tetrapod pineal, an outgrowth from the brain, was long suspected of being an endocrine gland but without experimental proof. Now it has been demonstrated that in at least a few vertebrates its secretions can affect both the gonads and the pigment cells. For instance, in larval amphibians a known pineal hormone, melatonin, causes the skin's melanophores to contract and lighten the tadpole's color. However, the significance of the pineal gland as a whole remains obscure, and despite much experimentation we have only fragmented, inconclusive evidence for its role as an endocrine organ.

THYMUS

Derived from the pharyngeal pouches, the thymus gland is unusual in that it reaches its maximum size in very young animals and then is largely resorbed and replaced by fat and connective tissue as the animal reaches maturity. A very young animal whose thymus is removed will not produce antibodies. If thymus tissue is subsequently injected, or an exogenous thymus gland transplanted, antibody formation will occur. It is thus suggested that the thymus produces some as yet unknown factor, perhaps a hormone, which stimulates the lymphoid tissue to produce specific antibodies.

UROPHYSIS

In teleosts, chondrosteans, and elasmobranchs, the very posterior part of the spinal cord contains a group of large neurons, called Dahlgren cells, whose axons extend into vascular tissue. In teleosts this tissue forms a compact vascular organ called the urophysis (Fig. 13-15); in the other fishes the urophysis is more diffuse. These Dahlgren cells are clearly neurosecretory cells; when appropriately stained their secretion can be observed both in the axons and emptying into the urophysis. It is known that the urophysis releases a protein or polypeptide hormone which elicits rhythmic contraction of smooth musculature. The

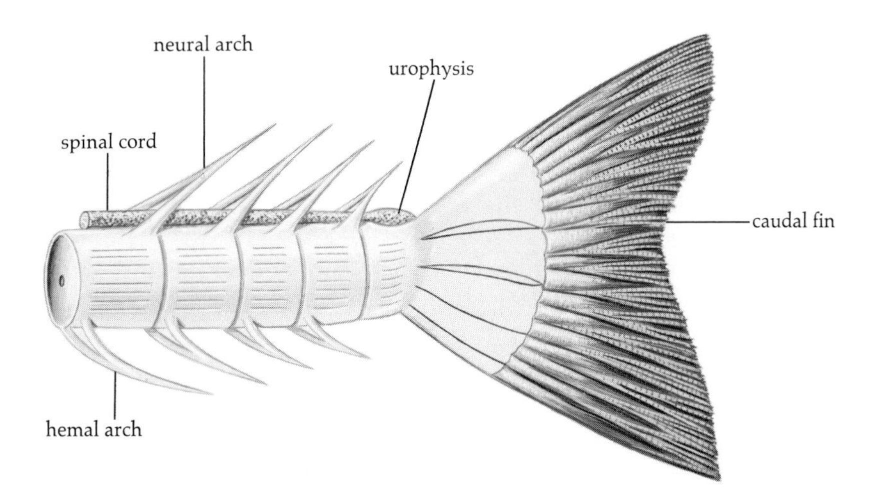

Fig. 13-15. The last caudal vertebrae and the caudal fin of a perch, showing the end of the spinal cord and the urophysis.

precise significance of this hormone and why it is found only in certain fishes are not yet understood.

ANGIOTENSIN

In mammals, and very likely in other vertebrates, the kidneys, liver, and vascular system have an interesting relationship. The kidney releases a proteolytic enzyme, renin, into the blood; the liver releases a globular protein into the blood; the renin partially digests the globular protein, thus producing a ten-amino acid polypeptide called angiotensin I. Further enzymatic action in the bloodstream converts this ten-amino acid polypeptide to an eight-amino acid polypeptide, called angiotensin II. Angiotensin II in turn causes constriction of the arterioles and thus an increase in blood pressure and stimulates the zona glomerulosa of the adrenal gland to produce aldosterone. Aldosterone causes kidney tubules to resorb water and sodium during urine formation and also inhibits the further production of renin. This type of intimate feedback, in which organs are interrelated by specific messenger chemicals (hormones), is responsible for much of the body's chemical coordination.

REPRODUCTIVE HORMONES

In addition to the reproductive hormones secreted by the pituitary gland there are powerful hormones produced by ovary, testis, and accessory reproductive organs. These will be discussed along with the reproductive system.

SUGGESTED READING

Bern, H. A. (1967). Hormones and endocrine glands of fishes. *Science* **158,** 455–462.

Legios, M. D. (1970). The median eminence of the bowfin, *Amia calva* L. *Gen. Comp. Endocrinol.* **15,** 453–463.

Lederis, K. (1970). Teleost urophysis I. Bioassay of an active urophysial principle in the isolated urinary bladder of the rainbow trout, *Salmo gairdnerii;* II. Biological characterization of the bladder-contracting activity. *Gen. Comp. Endocrinol.* **14,** 417–437.

Malvern, P. V. (1970). Interaction between endocrine and nervous systems. *BioScience* **20,** 595–601.

Turner, C. D., and Bagnara, J. T. (1971). "General Endocrinology." Saunders, Philadelphia, Pennsylvania.

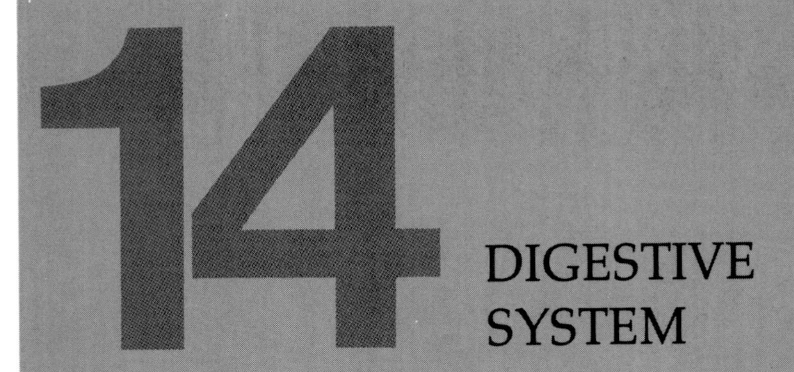

14 DIGESTIVE SYSTEM

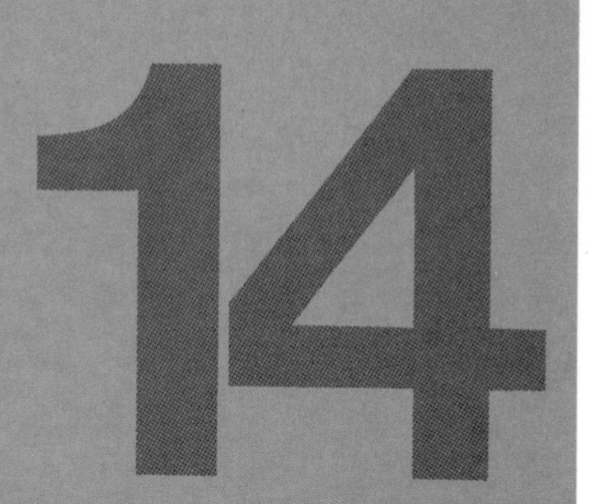

Since only plants can synthesize organic molecules from inorganic carbon dioxide and water, all animals depend, either directly or indirectly, on plants for their food. The ingestion, digestion, and absorption of food and the formation and egestion of fecal materials are the functions of the digestive system.

Ingestion brings food into the mouth and pharynx. Diverse plants and animals can serve as food, and equally diverse mechanisms have evolved for ingestion, that adapt species either to general or to very specific diets. Irregardless of structural variation, however, the mouth and pharynx are basically homologous in all vertebrates.

From the mouth and pharynx food is passed (usually by muscular action) to the stomach and intestines, where digestion and absorption occur. Digestion usually involves mechanical breakdown of the food into smaller pieces (which have a greater surface area), followed by chemical breakdown by digestive enzymes into molecules which can be absorbed through the alimentary canal into the bloodstream. Material that cannot be digested and absorbed is formed into feces in the last part of the intestines, and egested at the posterior end of the alimentary canal.

Food is ingested and swallowed by striated muscles in the jaws and pharynx. It is moved on by rhythmic contractions (peristalsis) of the smooth muscles in the remainder of the alimentary canal. To accommodate these contractions, as well as the mass of the ingested food, most of the alimentary canal is suspended by membranes in the fluid-filled body cavities; changes in diameter and some limited movement are thus possible.

SELECTIVE FACTORS IN EVOLUTION

Several factors have been important in the evolution of digestive system morphology. At any point in time, for instance, development of new adaptations has been influenced by the food then available and by the features that had already developed in any animal's digestive system. The combination of these two factors has led to the evolution of highly specialized diet–morphology relationships such as those exemplified by teeth. Herbivores, for instance, whose diet contains tough plant cellulose, have grinding teeth; carnivores have sharp stabbing and cutting teeth.

Another important factor has been the ratio of the absorptive surface area (alimentary canal wall) to the volume needing nutrients (animal's mass). As this ratio improves, an animal's metabolic rate can become higher. Therefore, various complexities evolved in the absorptive surface to keep this ratio constant or to improve it as species evolved to larger size (since volume increases by a cube factor and area by a square factor). These developments included a lengthening and folding of the alimentary tube, an increase in its diameter, and foldings of the canal lining or its cells. Some or all of these modifications are to be found in larger species.

Metabolic rate itself has been important in another way. The higher the metabolic rate, the more rapidly the absorbed food is utilized and the sooner more food is required. Vertebrates with high metabolic rates require efficient mechanical and chemical digestive processes, as well as large surface areas for absorption.

The genetic makeup of each vertebrate line has been a final, limiting influence on the evolution of alimentary canal adaptations. For instance, the gills of plankton-eating fishes have an elaborate strainer mechanism which traps plankton during respiration and concentrates them in the pharynx before they are swallowed. Plankton-eating whales, on the other hand, as described in Chapter 8, have evolved a quite different strainer mechanism of highly modified keratin hairs. It would be as unlikely for a mammal to evolve a gill filtering system as for a fish to evolve a hair filtering system.

In short, the vertebrate's alimentary morphology is usually adapted to a specific diet by a large number of modifications, based on its genetic potential and related to its size, metabolic rate, and habitat. In this way the digestive system is like every other system.

DEVELOPMENT AND GENERAL STRUCTURE

The lining of the entire alimentary canal is formed from epithelial tissue. An ectodermal invagination called the stomodeum forms the oral cavity; another, called the proctodeum, forms the anal canal epithelium; between the two is endoderm, and endodermal outpocketings form the liver, pancreas, gallbladder, and paired pharyngeal pouches. In fishes these pharyngeal pouches unite with ectodermal inpocketings to form gill slits; in tetrapods they form the middle ear cavity, tonsils, parathyroid glands, thymus, and ultimobranchial bodies. Anteriorly, a ventral outgrowth of endoderm forms the thyroid gland. In the stomodeal ectoderm a dorsal outpocketing forms Rathke's pouch, which contributes to the pituitary; in amniotes other ectodermal outpocketings form salivary glands. From the embryonic digestive system are also derived various respiratory organs: lungs, gas bladder, and gills, which will be discussed in the next chapter.

This epithelial lining of the alimentary canal is surrounded by tissues of mesodermal origin. During very early development the hypaxial mesoderm splits into a lateral, somatic mesoderm and a medial, splanchnic or visceral mesoderm. The splanchnic mesoderm gives rise to the inner connective tissue and outer muscular tissue which at any level make up the bulk of the alimentary canal (Fig. 14-1).

Between the somatic and splanchnic mesoderm develop the body cavities. Except in mammals and birds there are two body cavities: a

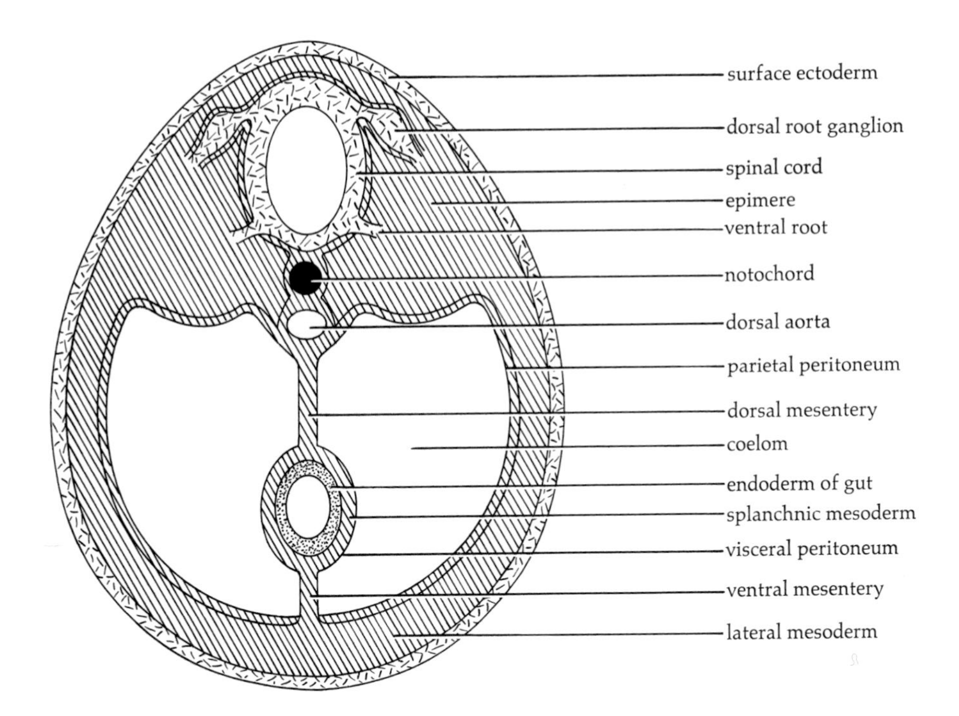

Fig. 14-1. Diagramatic cross section through the trunk of a developing vertebrate showing the relationships of coelom, mesenteries, splanchnic mesoderm, and lateral mesoderm.

large pleuroperitoneal cavity, which contains most of the viscera and the lungs if present, and a smaller pericardial cavity, which contains only the heart. In birds and mammals the lungs lie in pleural cavities, separate from both the pericardial and peritoneal cavities. The body cavities together are called coeloms; they are filled with coelomic fluid and lined with a shiny serous membrane, called mesothelium, which is formed from epithelial-type cells of mesodermal origin.

These membranes are named peritoneum when lining the peritoneal or pleuroperitoneal cavity, pericardium when lining the pericardial cavity, and pleura when lining the pleural cavities of mammals and birds. These membranes are further characterized as parietal or visceral according to whether they line the somatic wall or the viscera. The parietal peritoneum is thus the serous membrane lining the outer wall of the peritoneal or pleuroperitoneal cavity, and the visceral pleura is the membrane covering the lung. These names are merely topographical designations, however, for all serous membranes are of similar structure and are continuous within each body cavity.

Between visceral and parietal serous membranes is another continuous subdivision: double-walled membranes called mesenteries. The mesenteries hold the viscera in place while permitting some movement. They allow for communication and nourishment by virtue of the blood vessels and nerves which travel through them. They also frequently contain lymph nodes and variable fat deposits.

Although each region of the alimentary canal (e.g., esophagus, stomach, etc.) has specific morphological characteristics, there is a general layered organization which is common to all the regions lined by endoderm (Fig. 14-2). The innermost or lining layer is epithelium derived from endoderm which during development also forms the bulk of the pancreas. This layer contains glands, which vary according to species and region of the gut; they are unicellular when lining the lumen, and multicellular where they dip into connective tissue layers. There are also multicellular glands derived from this same endoderm lying outside the gut proper, for instance, the liver and pancreas.

Immediately outside the epithelial lining is a layer of vascular connective tissue, called the lamina propria, containing the capillaries where digested food enters the circulatory system. The lamina propria plus the epithelial lining are collectively called the mucosa, which is the absorptive layer of the gut.

Separating the lamina propria from deeper, denser, and less vascular connective tissue is a thin, smooth muscle layer, the muscularis mucosa. The denser connective tissue layer beyond the mucosa is the submucosa.

External to the submucosa are two layers of smooth muscles; an inner, circularly oriented layer and an outer, longitudinally oriented layer. The coordinated contractions of these muscles, controlled by the autonomic nervous system and endocrines, is peristalsis.

Finally, the outermost layer of the gut is the visceral peritoneum which has already been described.

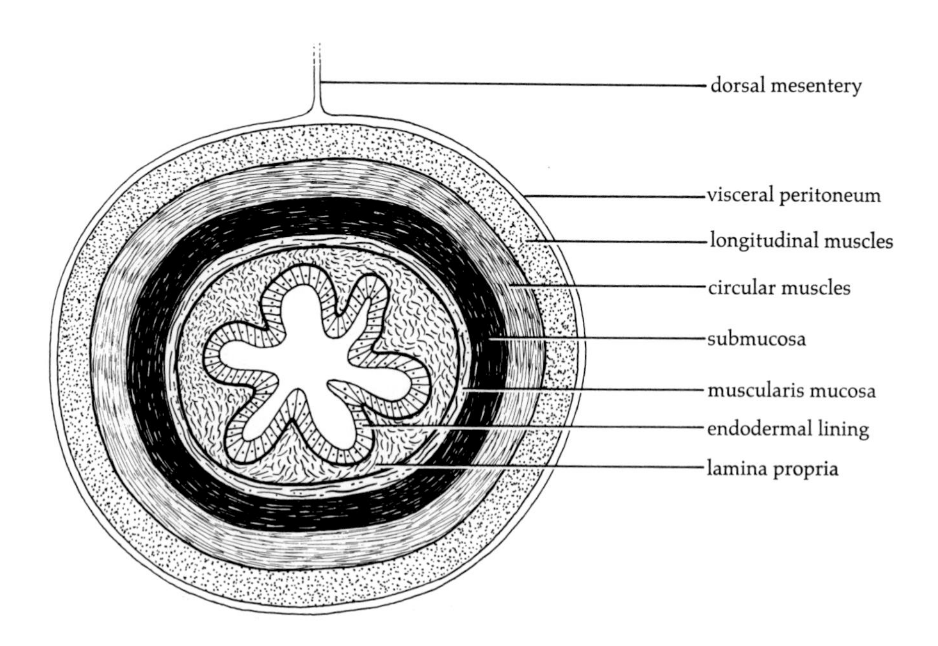

Fig. 14-2. An idealized section through the alimentary canal, showing its basic components.

Fig. 14-3. Mammalian distensible cheek pouches, such as those found in many monkeys, are controlled by complex buccinator muscles.

ORAL CAVITY

The stomodeum develops into the oral cavity, which includes the area from the lips to the pharynx. Since this part of the alimentary canal has intimate contact with the environment and receives food in its rawest state, it has evolved with many extreme modifications to diet. In addition, there are in the oral cavity some features that are constant within large taxa and are used to characterize them, for instance, the amphibians' sticky extensible tongue, the cyclostomes' rasping tongue, the amniotes' salivary glands, the mammals' four distinct tooth types, and the rodents' curved incisors.

Vestibule

In mammals the oral cavity includes the oral cavity proper and, anteriorly, the vestibule—that portion bounded externally by mouth, lips, and cheeks and internally by teeth and gums. During mastication (chewing), food squeezed between the teeth falls either into the oral cavity proper or into the vestibule, from where it can be brought back between the teeth by contractions of the platysma musculature lining the cheeks. In nonmammalian vertebrates there is no distinct vestibule. The teeth are frequently on the margins of the mouth and food outside the oral cavity proper falls to the ground or into the water.

Labial glands opening into the vestibule at the base of the teeth secrete mucus which lubricates the food. The duct of the parotid gland, called Stenson's duct, also opens into the vestibule; the parotid is a major salivary gland which helps start the digestion of starch.

Some mammals, such as squirrels and many monkeys, have internal cheek pouches—expandable dilations inside the cheeks in which they carry or temporarily store food. Not surprisingly, the buccinator muscles (that portion of the platysma filling the cheek region) of these mammals are particularly well developed (Fig. 14-3).

Oral cavity proper

In mammals the oral cavity is separated from the nasal chambers by the secondary palate, whose anterior portion, the hard palate, is formed by extensions of the premaxilla, maxilla, and palatine bones, and whose posterior portion, the soft palate, is formed by a tough but pliable connective tissue membrane. The entire secondary palate is covered by the stratified squamous epithelium of the oral cavity. An adaptive value of the secondary palate is to extend the internal nares caudally so that they open into the pharynx rather than the oral cavity; this permits simultaneous mastication and respiration. Among reptiles only the crocodilians have a complete secondary palate; it is bony its entire length. Other reptiles and birds have an incomplete, soft secondary palate formed of tissue folds that do not meet medially. Anamniotes have no secondary palate (Fig. 14-4).

Salivary glands (unicellular in fish and multicellular in tetrapods) open into the oral cavity proper. Their secretions are stimulated by au-

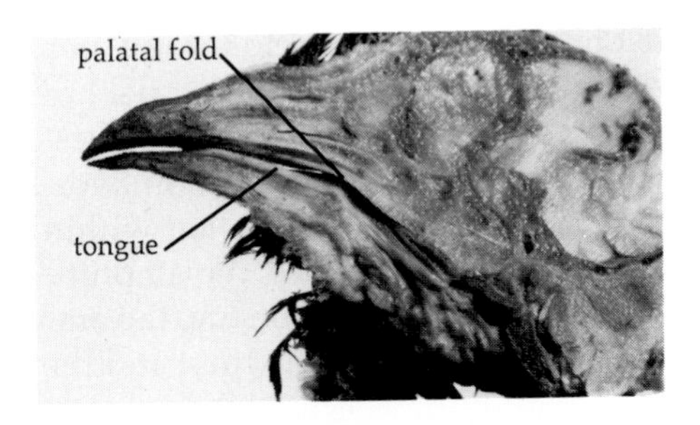

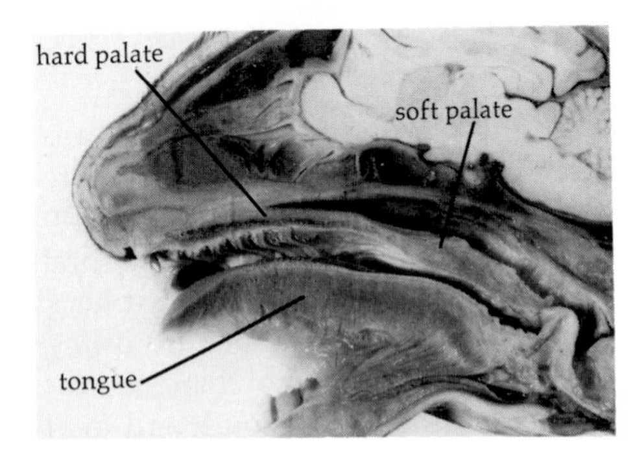

Fig. 14-4. Midline sections through the heads of a pigeon (top) and a cat (bottom), to demonstrate the complete secondary palate of a mammal as compared to the incomplete palatal folds of a bird.

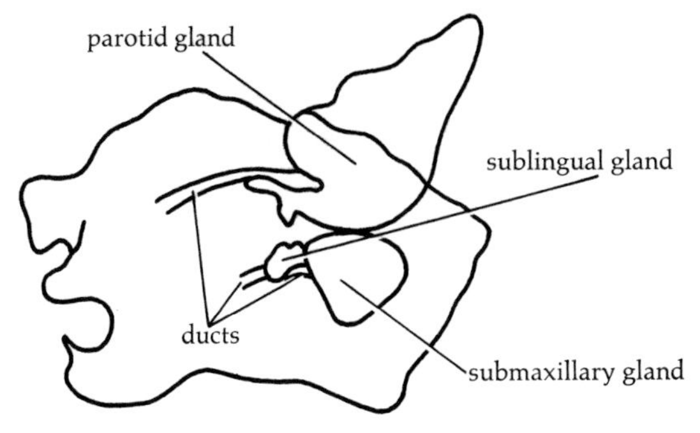

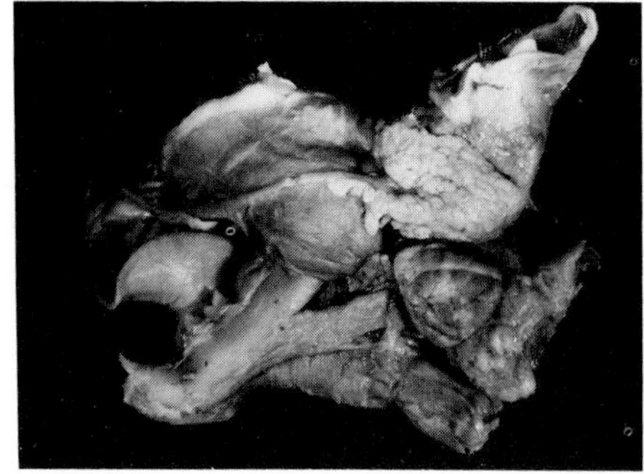

Fig. 14-5. Ventrolateral view of a cat's head, showing the major salivary glands and their ducts.

tonomic innervation and pushed along the duct of the gland into the oral cavity by muscles (usually the smooth muscles of the duct). Most of these glands secrete mucus, which acts as both a lubricant and a solvent. In tetrapods the secretion facilitates taste by dissolving food; in some cases the secretion also contains the enzyme ptyalin, which initiates carbohydrate digestion. In amphibians and many lizards special lingual glands produce the sticky mucus used for catching insects. In amniotes there are variable numbers of large salivary glands, called "mixed salivary glands," which contain clumps of both mucous and serous cells; these are frequently located some distance from the opening of their ducts (Fig. 14-5).

The viper poison gland is a modified salivary gland that is enfolded in the masticatory muscles; contraction of these muscles simultaneously forces the gland's secretion to the duct's termination at the base of the large hollow fangs and closes the jaws. The poison passes down the center of the fang and enters the prey at the deepest part of the bite. It is thus injected almost as if by a hypodermic syringe.

TONGUE

Anamniotes

Most fishes and larval amphibians have only a primary tongue. This structure does not have the fleshy element of most tetrapod tongues; it is little more than a fold of mucous membrane covering the basihyal and ceratohyal elements. It is moved by hypobranchial muscles, particularly the coracohyoid (Fig. 14-6). There are taste buds on the primary tongue, as well as in the pharynx and other parts of the oral cavity.

The cyclostome tongue bears hard, sharp, keratin teeth anteriorly. It is supported by a complex group of lingual cartilages derived from the splanchnocranium and lingual muscles derived from the branchiomeric musculature (Fig. 14-7). With this arrangement the cyclostome tongue is quite mobile and can be drawn in and out of the mouth for considerable distances. It is used to rasp away the flesh of the fishes

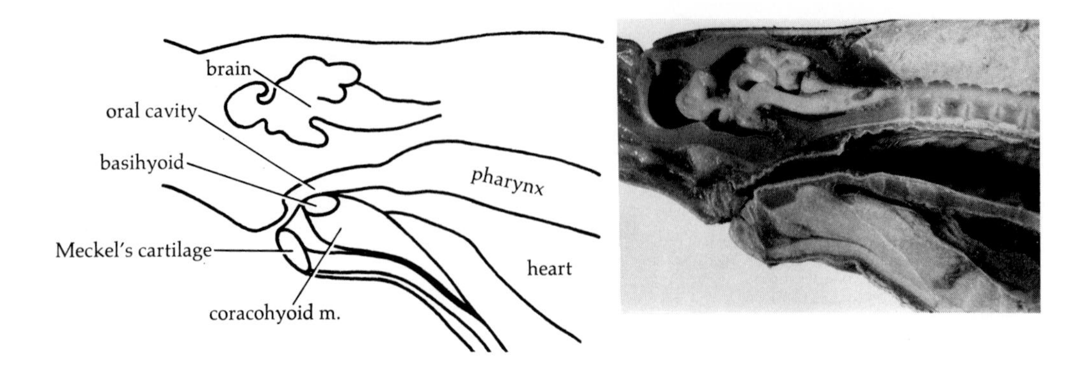

Fig. 14-6. Midline section of a dogfish head, showing the primary tongue formed by the basihyoid element in the floor of the oral cavity.

on which the cyclostomes prey, and it is a remarkable adaptation for a specific diet.

In amphibians the tongue undergoes two major changes during metamorphosis. As functional gills are lost there is a corresponding reduction in the branchial skeleton; some hypobranchial muscles are therefore freed to move the primary tongue. A secondary, fleshy tongue also develops. Just behind the mandibular arch, in the anterior oral cavity, a glandular field develops in relationship to the basihyal. This glandular area, which has many sticky mucous glands over underlying lymphoid tissue, raises the tongue above the hyoid and produces the mass of the secondary tongue (Fig. 14-8). It has little intrinsic musculature; extrinsic muscles, the hyoglossus and genioglossus, control tongue movement.

Particularly in frogs and toads the tongue can be extended a considerable distance; the base remains in approximately the same position in the anterior oral cavity, while the tip, which normally lies back in the pharynx, is flipped over and out of the mouth (Fig. 14-8). Insects are caught on its sticky surface, and it is then flipped back into the mouth.

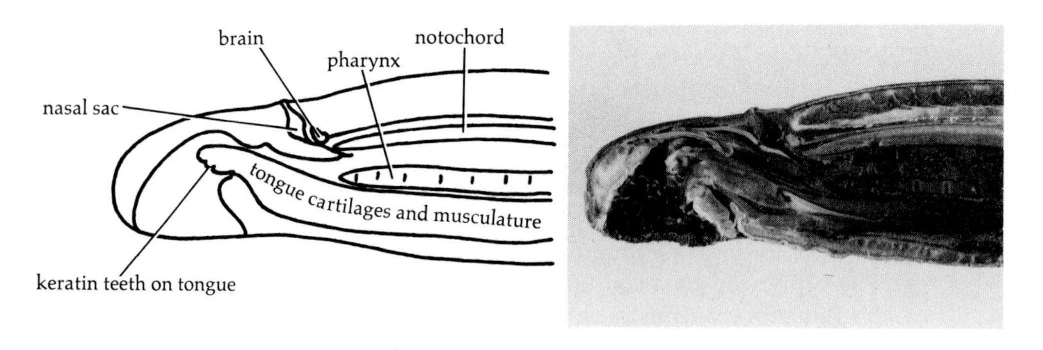

Fig. 14-7. Midline section of a lamprey head and pharynx, showing circular mouth and complex tongue.

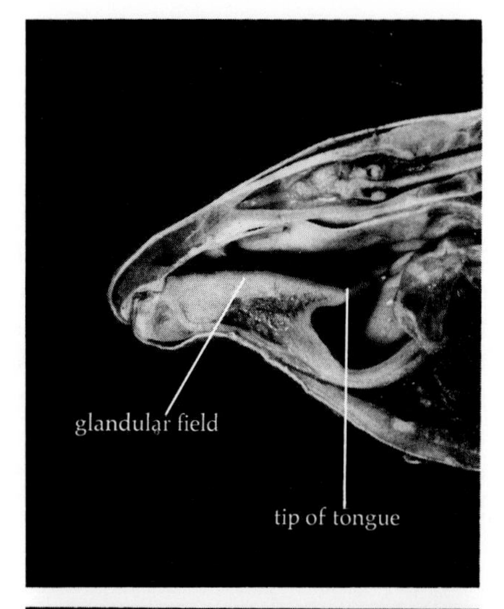

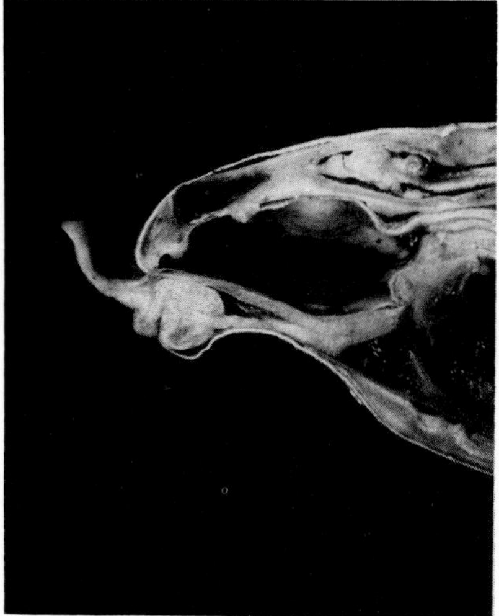

Fig. 14-8. Sagittal sections through the heads of two bullfrogs. Above is the "normal" position, with the tongue tip caudally in the pharynx; below the tongue is extended after having been "flipped over."

Amniotes

The hyoid apparatus of amniotes has migrated caudally, and a neck region has formed. As a result the base of the tongue lies far back in the pharyngeal region. Considerable diversity in the shape of the amniote tongue fits it for different functions (Fig. 14-9). Most mammals, birds, and reptiles use the tongue for lapping up liquids, for taste, for touch, and (particularly mammals) for food manipulation. Many mammals use it as a curry comb; in these, it bears numerous small keratin structures called filiform papillae, which improve its ability to groom the fur. The chameleon tongue is so extensible that it can be protruded from the mouth for a distance greater than body length. The forked tongue of snakes can be flashed in and out very rapidly and is highly sensitive to chemical and physical stimuli. Anteaters have long, very narrow tongues with which they reach deep into the holes of termites (which, contrary to their name, are their dietary staple). Among humans the tongue is extremely important for formation of the complex sounds of speech.

For these functions the tongue needs taste buds, innervation, and fine muscular control. In reptiles and mammals there is much intrinsic musculature. In birds, however, the tongue is a relatively hard structure built around a forward extension of the hyoid apparatus; it bears taste buds and a few glands. Bird songs are produced by the syrinx, and the tongue has little intrinsic musculature. Although it lacks the flexibility of the mammalian tongue, the bird tongue is nevertheless highly mobile because of its extrinsic musculature, which is primarily hypobranchial in origin.

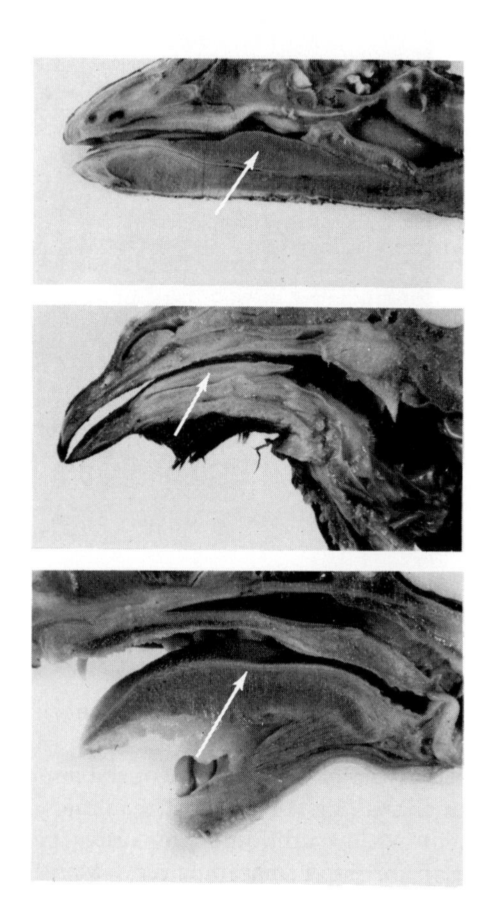

Fig. 14-9. Midline sections through the heads of a lizard (top), pigeon (center), and cat (bottom), showing the extent of their tongues (arrows). Note the tiny "currycomb" keratinous bristles on the cat's tongue.

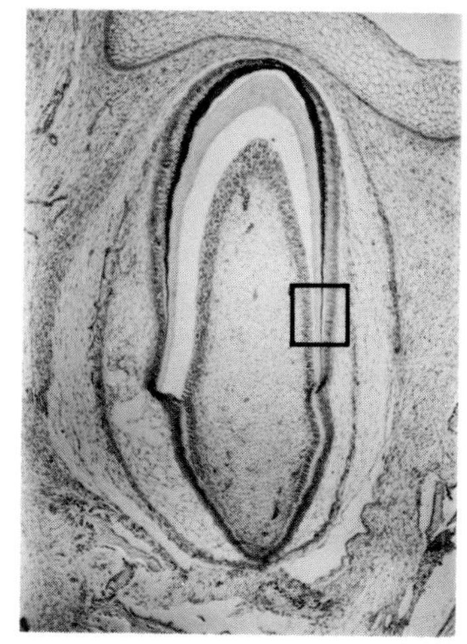

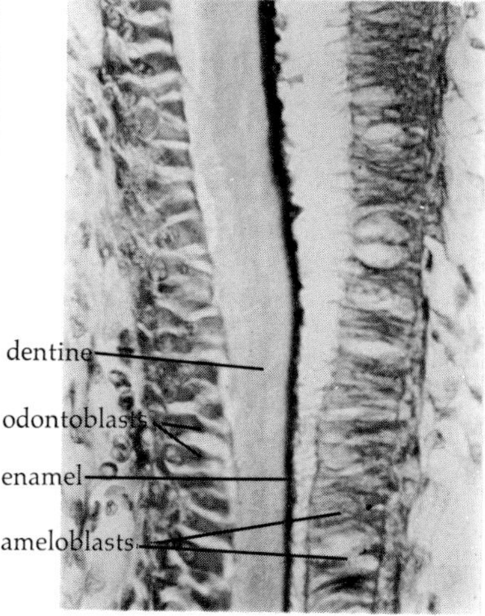

Fig. 14-10. On the left is a low-power photomicrograph of a developing mammalian tooth. On the right is a high-power photomicrograph of the area indicated by the box. The separations between odontoblasts and dentine and ameloblasts and enamel are artifacts created when these hard tissues were prepared for microscopic examination.

TEETH

True vertebrate teeth, as opposed to the cornified structures of cyclostomes, for instance, are structurally and developmentally similar to the placoid denticles of chondrichthyeans. The fossil record does not indicate that teeth evolved from placoid denticles, however, but rather that both evolved from hard denticles on the outer surface of superficial dermal bone. In chondrichthyeans the deep dermal bone has been lost, and the denticles remain as placoid denticles. In some jawed fishes the teeth, like tubercles, protrude from large dermal bones in the mouth.

DEVELOPMENT OF DENTITION

The embryological development of vertebrate teeth starts with either a thickening or an inpocketing of the stomodeal ectoderm or, in some fishes, of the pharyngeal endoderm. Deep to the thickening or inpocketing, a condensation of mesoderm forms a vascularized dermal papilla. The outer part of this papilla differentiates into epithelial-type cells called odontoblasts which secrete dentine, an acellular substance harder than bone but of similar structure which makes up the bulk of the tooth (Fig. 14-10).

The ectodermal thickening or inpocketing forms the enamel organ; its cells differentiate into secretory cells called ameloblasts, which secrete hard enamel over the tooth.

Depending on taxon, the enamel-capped portion of the tooth erupts either during or after development. It may be held into a socket by dentine roots below the surface, or it may be attached to the dermal bone by collagenous fibers or by the secretion of a hard, cementlike substance called cementum.

CHONDRICHTHYES

Most chondrichthyeans are elasmobranchs, usually predaceous fish whose large, powerful jaws are lined with many rows of very sharp teeth that can tear large pieces of flesh from prey. The teeth have neither sockets nor roots but are held onto the jaws by collagenous fibers (acrodont dentition). Throughout life teeth are continually lost and replaced (polyphyodont dentition). In most elasmobranchs all the teeth are the same shape (homodont dentition) and are distinguished only by the smaller size of the newer ones. However, several sharks, skates, and rays, whose diet is hard-shelled molluscs, have heterodont dentition; their anterior teeth are very sharp and pointed, like those of most chondrichthyeans, but the more medial and posterior teeth are flattened to form a crushing plate. In the rare and bizarre ratfish of the chondrichthyean suborder Holocephalii, there is usually one pair of large tooth plates in each jaw. The teeth, far fewer than those of sharks, are denticles upon these plates (Fig. 14-11).

ACTINOPTERYGIANS

The morphological diversity of actinopterygians noted in other systems extends also to dentition. There are toothless forms (e.g., seahorse), and forms with many teeth on the jaw margins, palate, branchial arches, and pharynx. Homodont dentition is the rule, but heterodont dentition is not infrequent. Most actinopterygians have acrodont dentition, with the teeth attached to the bone by a collagenous fiber hinge along the inner side of the tooth only. The teeth bend inward when a prey is caught, and the prey slips off into the oral cavity. However, in some actinopterygians the teeth are set in deep sockets (thecodont dentition) and well rooted into the jaw and palate skeleton (Fig. 14-12).

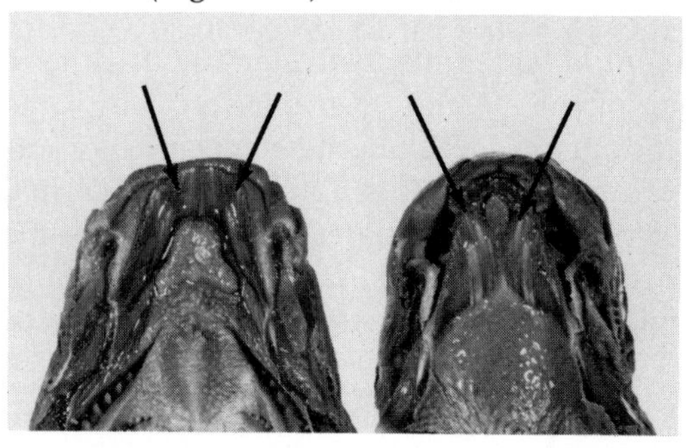

Fig. 14-11. Views of the upper jaw (right) and lower jaw (left) of a ratfish, showing the four large tooth plates (arrows) and their lighter colored denticles.

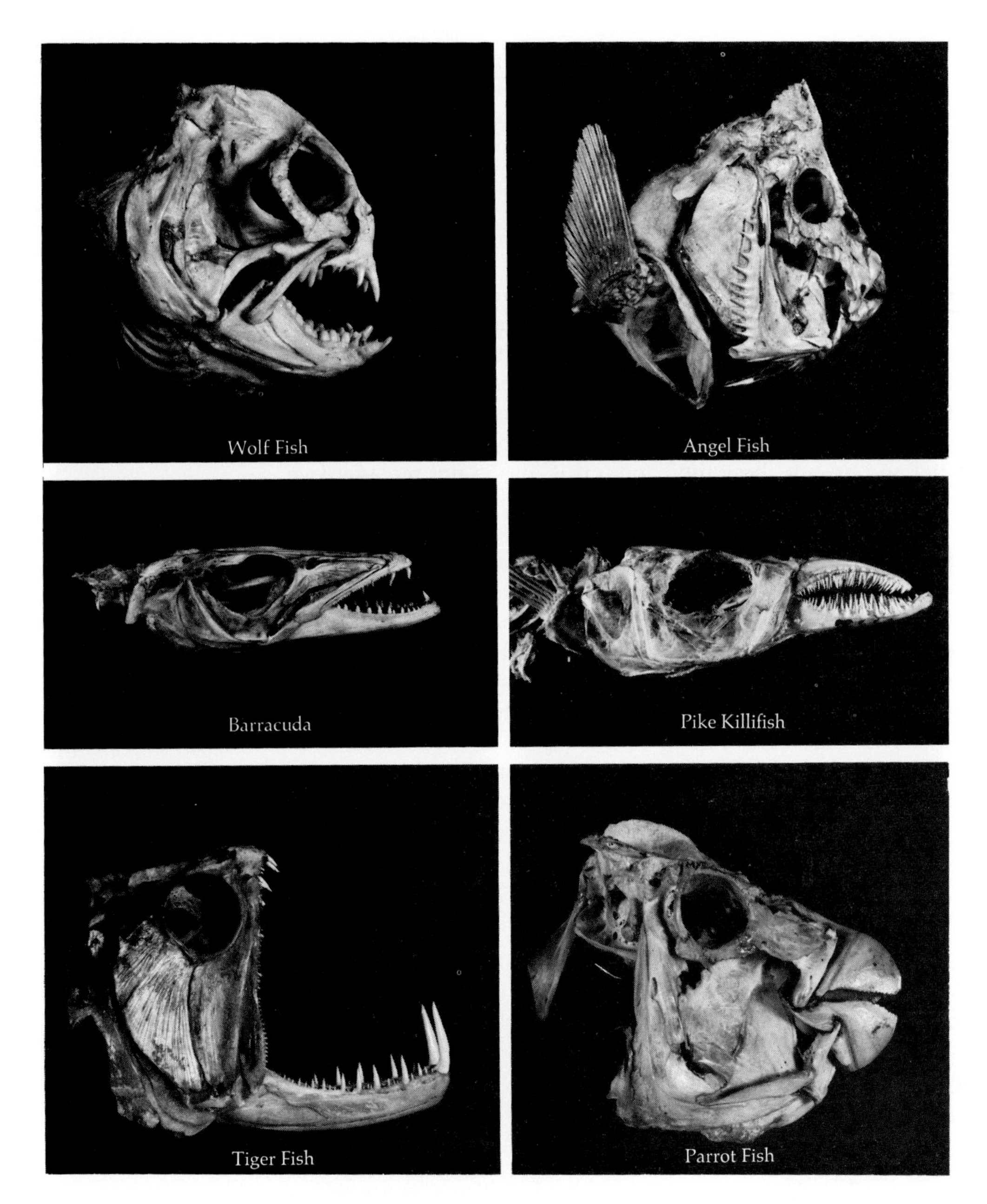

Fig. 14-12. These skulls demonstrate some of the great diversity of tooth structure found among teleosts.

SARCOPTERYGIANS

Lungfish have a pair of large tooth plates in the upper jaw and another in the lower jaw. On each tooth plate there are several cusps, which form a jagged saw edge. In contrast, the only living coelacanth, *Latimeria,* has many small conical acrodont teeth. The ancient crossopterygian fish which gave rise to the first tetrapods had yet another type of dentition. Their many small conical teeth were characterized by deep grooves infolded from the surface, which gave the tooth a labyrinthine appearance in cross section (Fig. 14-13). This labyrinthodont tooth was characteristic of the earliest amphibians, and, in fact, is the source of their name, Labyrinthodontia.

TETRAPODS EXCEPT MAMMALS

Living amphibians have small conical acrodont teeth. In urodeles and apodans they are located on the jaw margins and on the palate; in some anurans, on the upper jaw and the vomer bones of the palate; other anurans are toothless. At the base of most amphibian teeth there is a zone of weakness owing to uncalcified dentine. This results in frequent tooth loss, but since most amphibians have a relatively soft diet and do not chew their food, this loss is not maladaptive.

There is a great diversity of tooth types in reptiles (Fig. 14-14). Crocodilians have thecodont teeth; tooth replacement, which is frequent, usually involves only every other tooth at any one time, so that there are still many, evenly distributed, functional teeth. In lizards, attachment to the bone is at the base and along one side of the tooth (pleurodont dentition). Lizards and crocodilians have teeth only on the jaw margins, but *Sphenodon* and snakes have them also on the palate. Turtles compensate for their toothlessness with a hard keratin bill. The large hollow fang of the vipers is a modified incisor. When the jaws are closed the fangs are folded back into the mouth. When the mouth is widely opened the fangs are protruded.

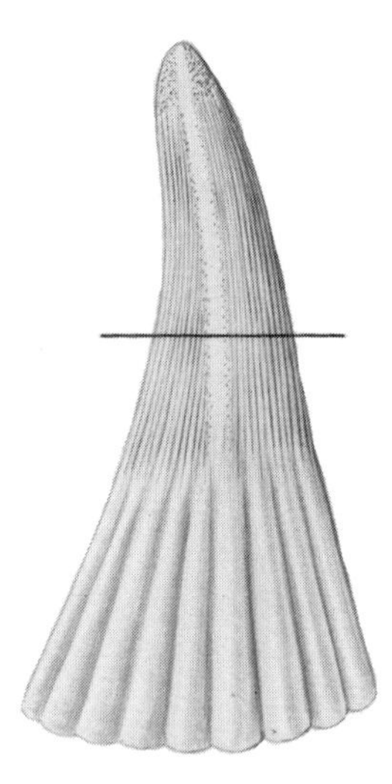

Fig. 14-13. Crossopterygian fish and labyrinthodonts possessed conical teeth with deep infoldings of dentine, as seen in the cross section. Living garpikes have similar although not as extreme infoldings of dentine. From M. Jollie (1962). "Chordate Morphology." © 1962 by Litton Educational Publishing, Inc. Reprinted by permission of Van Nostrand-Reinhold.

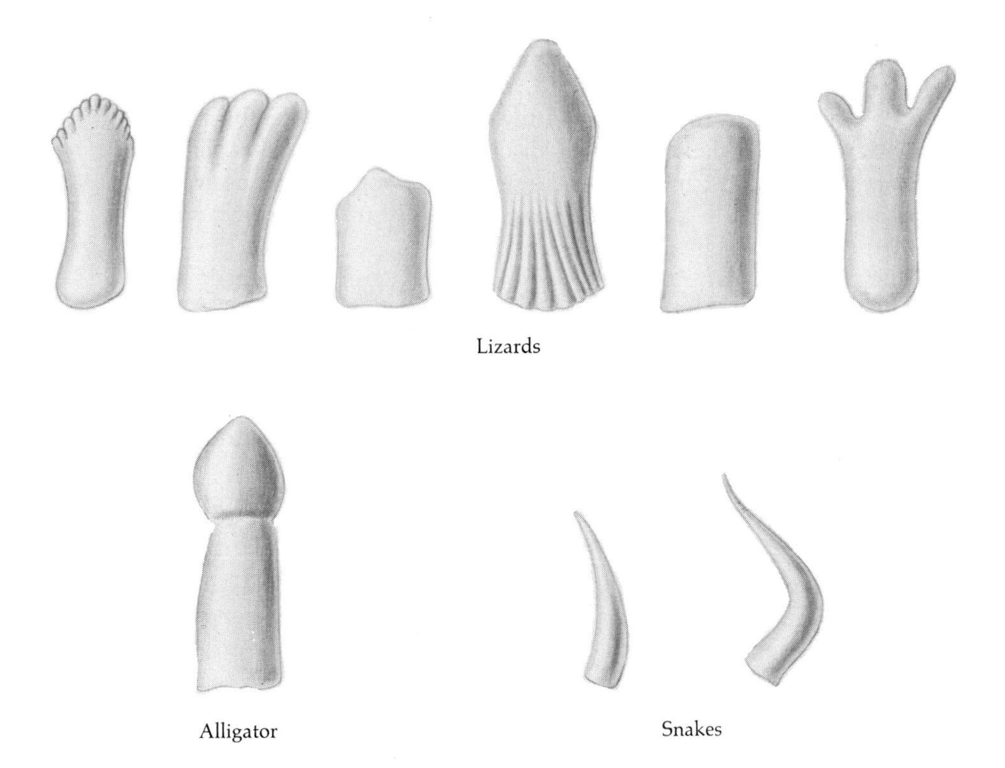

Fig. 14-14. Some of the structural diversity of reptilian teeth.

Although present in some fossil birds, teeth have been completely lost in living birds and replaced by a hard keratin beak. This is much lighter than teeth and, therefore, more adaptive to flight, and its structure is highly adapted to specific diets, for example, a short, stout beak for seeds, a long, narrow beak for nectar. The hatchling's "egg tooth" is an additional hard keratin portion on the tip of the beak with which the about-to-hatch bird forces its entrance to the world; it is soon lost.

MAMMALS

Except for some edentates, Sirenia, and cetaceans, mammals have heterodont, thecodont dentition with four tooth types in different jaw regions. The most anterior are the upper incisors on the premaxilla bones and the lower incisors on the dentary bones. These teeth have a narrow cutting surface and in some mammals are further specialized for nipping or chiseling. Behind the incisors is a single pair of conical canines, which are particularly sharp and prominent in carnivores who use them to stab and hold prey. Behind the canines are premolars and molars, which grind and crush. The molar teeth are nearest the angle of the jaw and therefore exert the greatest force; they are also broader than the premolars. They have a triangular surface in generalized mammals, but often a square or rectangular surface in specialized mammals.

Mammals have typically two sets of teeth. The first, called deciduous, baby, or milk teeth, include incisors, canines, and premolars, but no molars; the second, or permanent teeth, replace these three types and add the molars. Except in cetaceans there is no capacity to produce a third set or even a single third tooth when permanent teeth are lost. This loss of regenerative capacity may be one result of tooth specialization.

Much taxonomy of both living and extinct mammals is based upon dentition. Early (extinct) placentals and generalized living placentals have in both upper and lower jaws three pairs of incisors, one of canines, four of premolars, and three of molars. This is expressed in a shorthand called the dental formula, in which the number of each type is designated as a fraction whose parts represent upper and lower jaws; the generalized placental dental formula is

$$\frac{3\text{–}1\text{–}4\text{–}3}{3\text{–}1\text{–}4\text{–}3}$$

for one-half of the upper and one-half of the lower jaw. The total number of teeth is this times 2, or in this case, 44. The generalized marsupial dental formula is

$$\frac{5\text{–}1\text{–}3\text{–}4}{4\text{–}1\text{–}3\text{–}4}$$

which multiplied by 2 gives a total of 50 teeth (Fig. 14-15). In most mammals the adult teeth either follow the generalized placental or marsupial formula or are a reduction from it. Seldom has there been an increase. In toothed whales, however, along with the development of regenerative ability and the loss of tooth type specialization, there has been a great increase in the number of teeth. This is strong evidence that very many homodont polyphyodont teeth are highly adaptive in an aquatic situation.

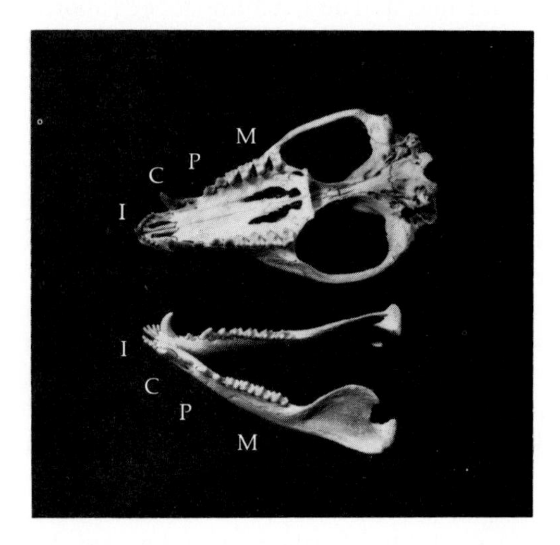

Fig. 14-15. The Virginia opossum has very generalized dentition, with the basic marsupial number of each type of tooth. I, incisor; C, canine; P, premolar; M, molar.

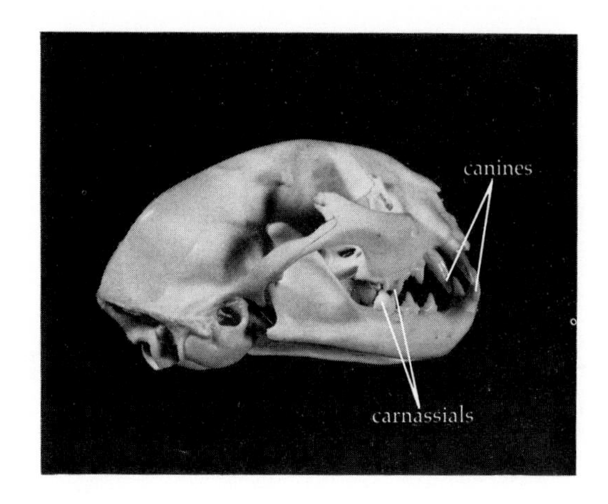

Fig. 14-16. Lateral view of a cat skull showing the specialized carnassial and canine teeth which adapt these animals to a carnivorous diet.

The four basic tooth types have often undergone further specialization, for diet or other requirements. In the order Carnivora there are specialized, sharp, shearing teeth called carnassial teeth, which are modifications of the last upper premolars and the first lower molars (Fig. 14-16). Carnivora, as mentioned, also have particularly well-developed canines; these reached their extreme in the now-extinct saber-toothed tiger with the interesting generic name of *Smilodon.*

In many herbivorous mammals the canine teeth have been lost, and the premolars have become so molarized that they are often difficult to distinguish. In those whose diet is grass or other harsh vegetation the cusps grow very high, enamel grows down between them, and cemen-

Fig. 14-17. Cheek teeth of a cow. On the right is a medial view of one of these high-crowned selenodont teeth; on the left is a similar tooth sectioned through its crown, showing the alternate layers of cement, dentine, and enamel.

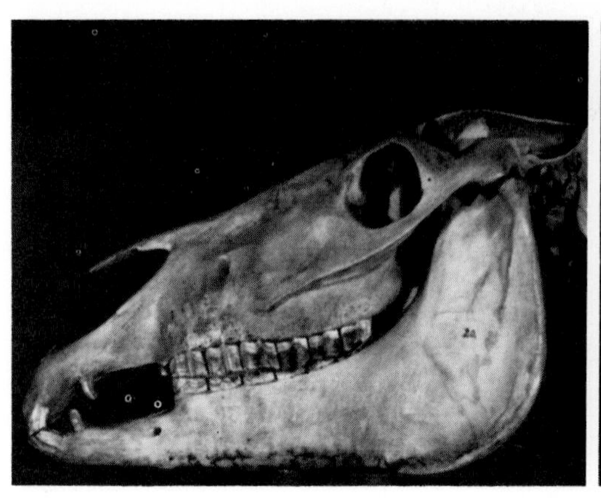

Fig. 14-18. Skulls of a horse (left) and a goat (right), each adapted to a herbivorous diet. Note that both premolars and molars are large and high-crowned. The horse has prognathous incisors and reduced canines. The goat lacks upper incisors and all canines.

tum grows up over the enamel. A very high-crowned tooth is thus formed, which when worn down has alternating layers of enamel, cement, and dentine; because of hardness differences in these three structures ridges are formed and maintained (Fig. 14-17). Such a tooth is called high-crowned, or hypsodont; if the ridges are crescent-shaped it is further characterized as selenodont, and if transverse, lophodont.

Herbivorous mammals frequently have specialized incisors. In the horse, for instance, the incisors are prognathous, that is, they stick out obliquely from the jaws and are thus adaptive for cropping grass (Fig.

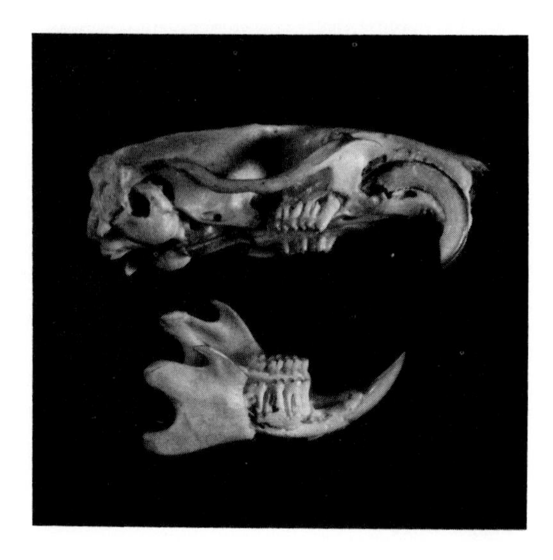

Fig. 14-19. The specialized dentition of rodents is seen in this preparation of a rat skull in which the roots of the teeth have been exposed. Note the large, ever growing, gnawing incisor teeth and the smaller, grinding molars and premolars.

14-18). In artiodactyls (even-toed ungulates such as sheep, deer, cows), the upper incisors are replaced by a horny pad on the premaxilla bones, against which the lower incisors crop the grass. Rodents have a single pair of upper and a single pair of lower incisors, greatly enlarged and curved to form powerful gnawing teeth with enamel only on the outer surface. Because the enamel is harder than the rest of the tooth, the outer surface wears away much more slowly than the inner surface, and a sharp edge is maintained. Unlike most mammalian teeth, rodents' incisor teeth continue growing throughout life (Fig. 14-19).

Pigs, people, and other omnivorous mammals have more generalized dentition. There are no high-crowned teeth. The molars are squared-off (bunodont), rather than triangular as they are in generalized Eutheria and Metatheria (Fig. 14-20).

The most bizarre tooth modification in mammals is the tusk. Elephant tusks are formed by the upper incisor teeth; they have lost their enamel, and their "ivory" is pure dentine. The lower incisors and both

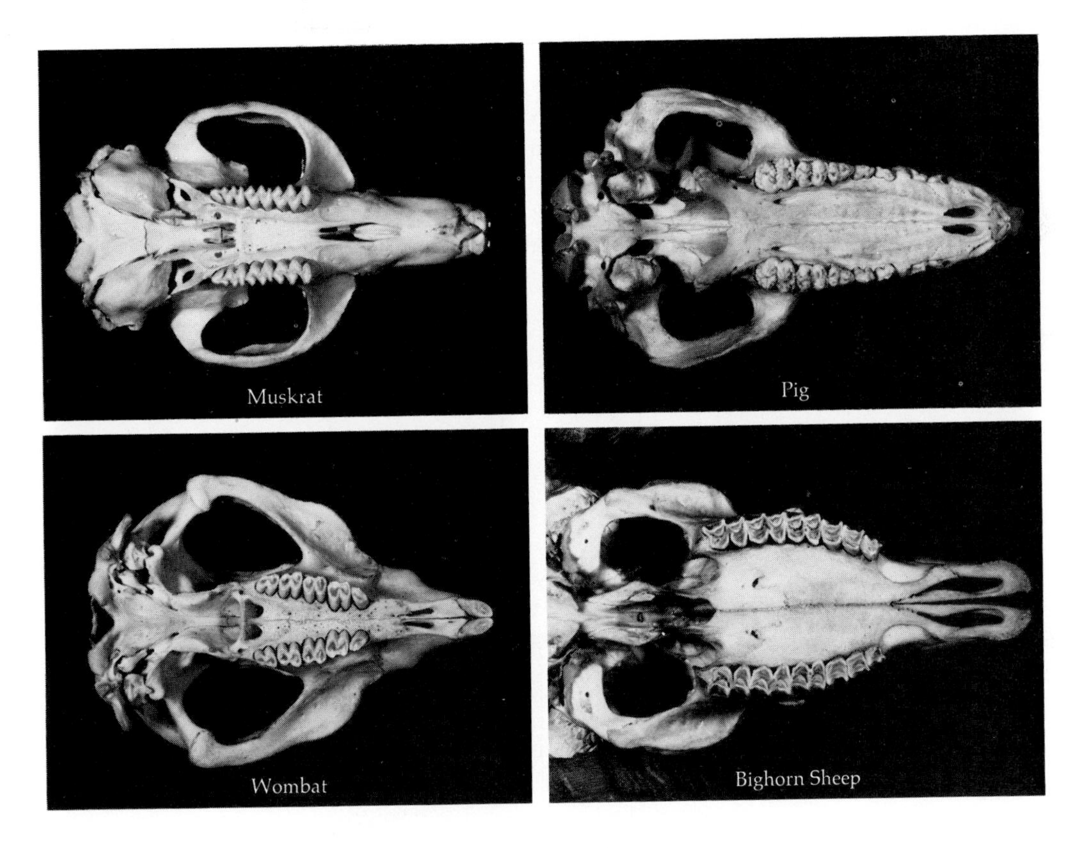

Fig. 14-20. Palatal views of the skulls of four mammals showing their dentition. Note that the marsupial wombat and the rodent muskrat have similar incisor teeth adapted for gnawing, and premolar and molar teeth adapted for grinding. The low-crowned, squared molars of the pig necessitate a relatively soft diet, while the high-crowned, selenodont dentition of the bighorn sheep make possible a diet of harsh plants.

pairs of canines are lost. Tusks are found in both male and female African elephants, but in only male Indian elephants. They continue to grow throughout life, and in a large African elephant they may be 8 feet long and weigh as much as 150 pounds. This is an amazing amount of weight for an animal to carry around at the end of its neck, and is also a considerable drain on its calcium metabolism.

The canine teeth produce tusks in walruses (upper canines), boars (upper canines recurved upward, plus lower), and an unusual Arctic cetacean, the narwhal. In this last animal the tusk develops from the male's left upper canine, and may grow up to 8 feet straight out the front of the mouth, with a characteristic groove spiraling along its entire length. It is said that in Medieval times narwhal tusks were greatly treasured as the horns of unicorns. What possible function this tusk has—other than to spark legends and pique curiosity—is not known.

PHARYNX

The pharynx extends from the back of the oral cavity caudally to the posterior limit of the last gill slit in fishes, or to the point where the trachea begins in tetrapods. In other words, it is the portion of the endodermally lined digestive tract which is common to both the digestive and respiratory systems. In crocodilians, birds, and mammals it can be divided into three regions: above the secondary palate is the nasal pharynx; below it is the oral pharynx; and caudally, opening into both esophagus and larynx, is the laryngeal pharynx (Fig. 14-21).

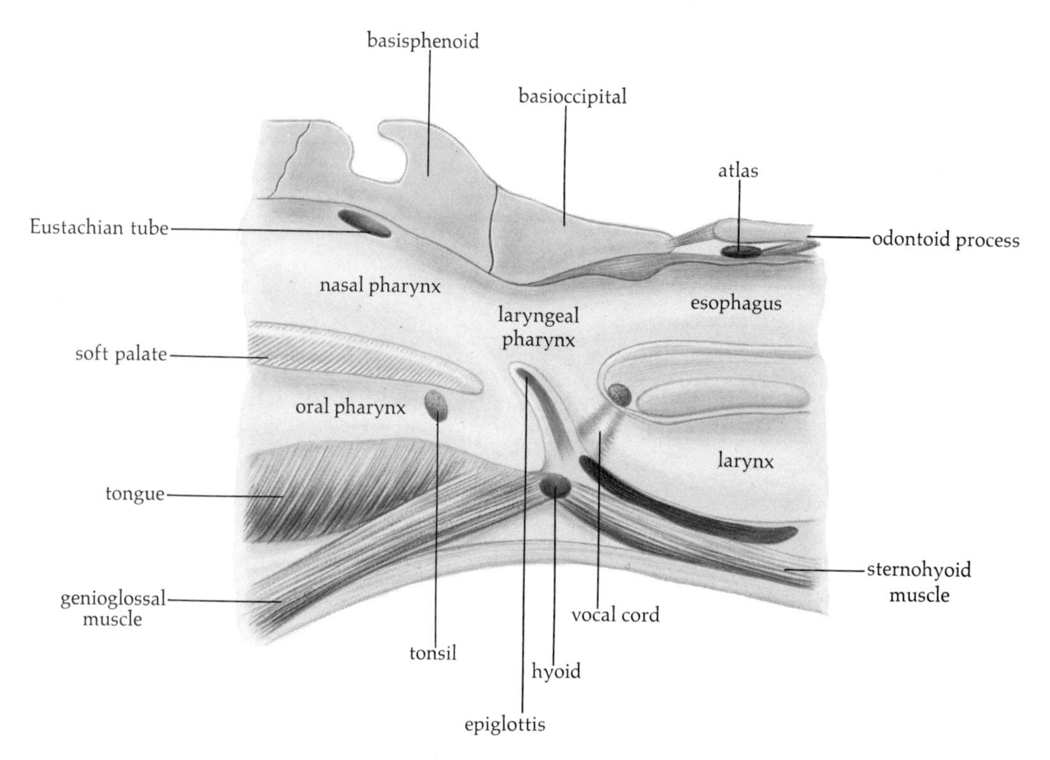

Fig. 14-21. Sagittal section through the larynx of a mammal.

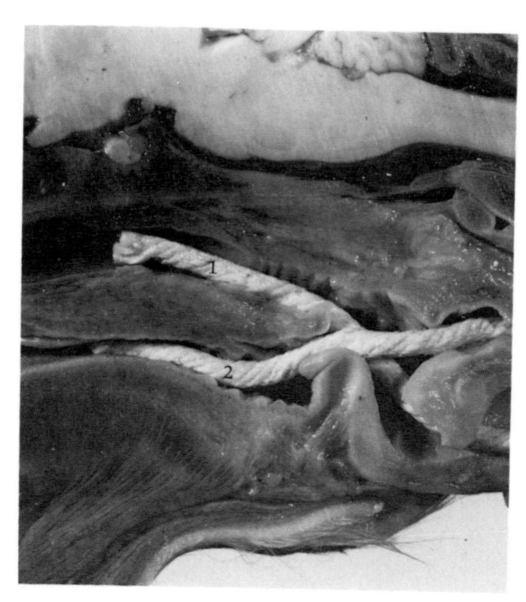

Fig. 14-22. Sagittal section through the pharynx of a cat, with crossing strings demonstrating the pharyngeal chiasma. The string labeled 1 passes from the nasopharynx ventrally into the larynx. The string labeled 2 passes out of the oropharynx (bending the epiglottis here) and passes into the dorsally located esophagus.

In fishes, food and water pass together through the pharynx; most of the water is then passed out through the gill slits while the food continues into the esophagus. In tetrapods, air may reach the pharynx from either the mouth or the nostrils. In either case the air enters the pharynx dorsally and must pass ventrally through the pharynx to reach the larynx and trachea. At the same point the food must pass dorsally to enter the esophagus. This area of the laryngeal pharynx, where the air current crosses the food current, is called the pharyngeal chiasma (Fig. 14-22). No engineer would have designed it; it is an accident of evolution which, when food lodges in the trachea or larynx, can cause great discomfort or even asphyxiation. Only in whales is there a different and presumably better system; the most anterior part of the larynx, the epiglottis, wraps around the pharynx and inserts into the internal nostrils, eliminating the pharyngeal chiasma.

In all vertebrates, of course, six pairs of pharyngeal pouches develop in the pharyngeal region (except for cyclostomes, which have many more). In fishes the pharyngeal pouches (gill pouches) unite with ectodermal invaginations to form gill slits which break through the surface. In tetrapods the endodermal lining of the pharyngeal pouches forms several different structures (Fig. 14-23).

The first pouch forms the middle ear cavity and the Eustachian tube through which it retains continuity with the pharynx. The second forms the palatine tonsils—lymphatic tissue on each side of the pharynx that retains continuity with the pharynx. The third, fourth, fifth, and sixth lose their continuity with the pharynx, and the en-

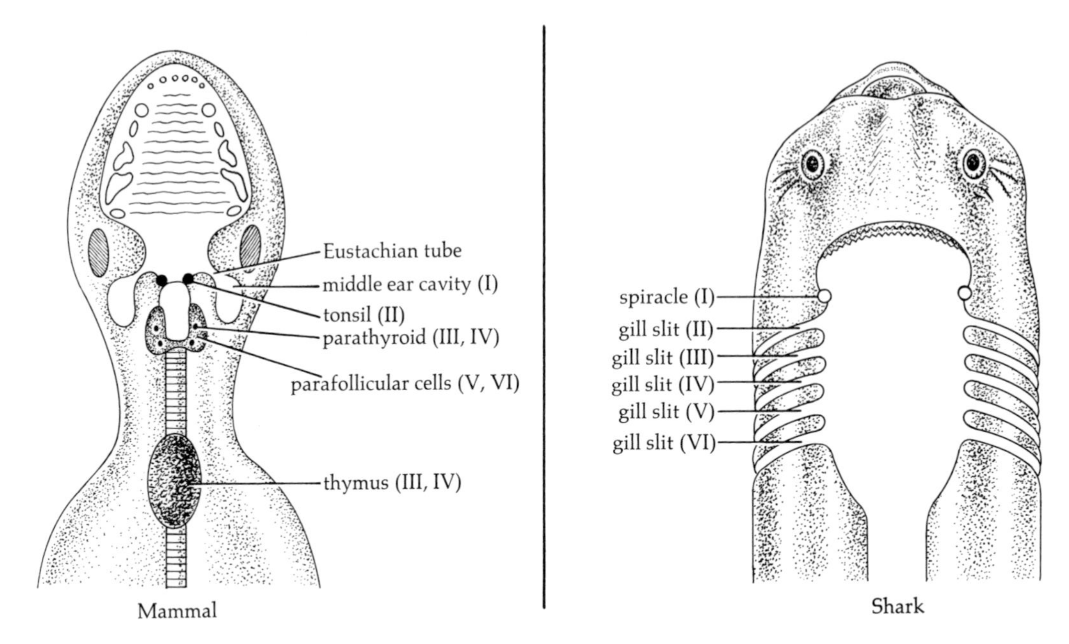

Fig. 14-23. Diagramatic representation of the derivatives of pharyngeal pouches in a mammal and a shark. Roman numerals indicate pouch numbers.

doderm migrates caudally and ventrally; the third and fourth contribute to the thymus and parathyroid glands, and the fifth and sixth to the ultimobranchial bodies, which in the mammal form the parafollicular cells of thyroid and parathyroid.

The thyroid gland forms from a midventral outpocketing of the anterior pharynx. The only prominent glands in the pharynx itself are the lymphoid tissue of the palatine tonsils and adenoids.

ESOPHAGUS

Peristalsis begins at the esophagus, which in most vertebrates is a relatively unspecialized portion of the alimentary canal between the pharynx and the stomach. The endodermal lining of the esophagus is frequently stratified epithelium which contains mucous glands whose secretions help lubricate the ingested food (Fig. 14-24). The upper area of the esophagus is wrapped by striated musculature and its lower by smooth musculature, except in ruminants, whose entire esophagus is walled by striated musculature. Ruminants, therefore, have more control over their esophagus than do most vertebrates and can regurgitate food (the "cud") to be rechewed.

The esophagus has a good deal of elastic tissue, enabling it to stretch enough to accommodate large pieces of food. This is particularly true in fishes and in some tetrapods such as snakes, which can swallow pieces of food larger than their own normal diameter.

In most birds the posterior part of the esophagus forms a sac, called the crop, which is particularly large in seed eaters. The crop acts primarily as a storage depot; food is eaten rapidly and then passed on gradually for digestion. In pigeons and a few other birds the crop sloughs off its squamous epithelium when there are nestlings. This cast off material forms a highly nutritious "pigeon's milk," or "crop milk," which is regurgitated and fed to the young. Both males and females produce crop milk, apparently at the mere sight of nestlings.

STOMACH

The vertebrate stomach is lined with simple columnar epithelium which secretes mucus. It varies from an unspecialized portion of the alimentary canal in some fishes to a huge, four-chambered affair in ruminant artiodactyls (Fig. 14-25). In most vertebrates it performs three functions: storage, mechanical digestion, and chemical digestion. It is modified for these functions as required by the animal's diet.

Storage

The portion of the stomach which is solely for storage is that closest to the esophagus, and is called the cardiac stomach (Fig. 14-26). Here the diameter of the alimentary canal is increased—a great deal in vertebrates that eat discontinuously and much less or not at all in those fish whose easily digested diet is continuously consumed.

Storage also occurs in the remainder of the stomach. In most stomachs, bands of muscular tissue in the external layer cause folds, called rugae, in the lining; these folds provide the potential for additional distension (Fig. 14-26). When the stomach is filled, internal pressure stretches this lining and obliterates the rugae.

A large stomach contributes significantly to a more mobile life style. Having to eat only occasionally leaves time for other activities. For many terrestrial animals it is advantageous to eat rapidly and then hide from potential predators while digestion occurs. For predators it allows the animal or the pack to "stock up" while a kill is fresh.

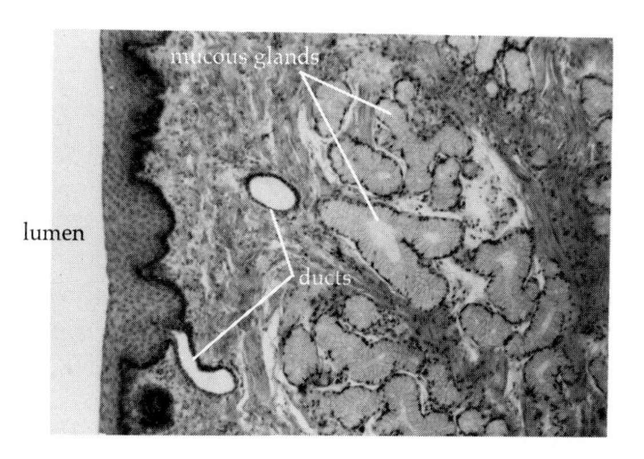

Fig. 14-24. Cross section of a small portion of the mucosa and lamina propria of a mammalian esophagus showing the stratified surface epithelium and the mucous glands in the lamina propria.

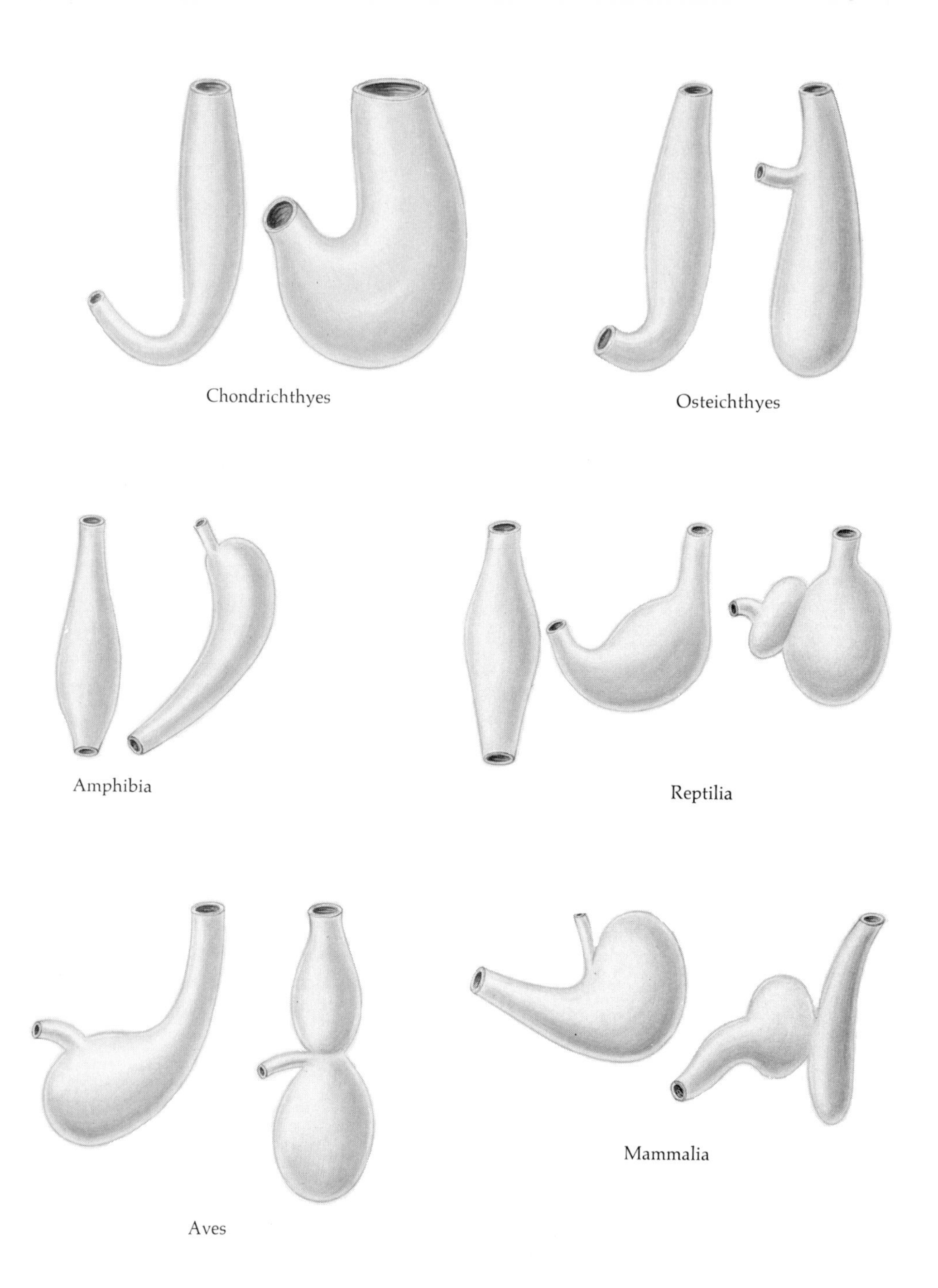

Fig. 14-25. Some of the diversity of stomach shapes in different vertebrates.

Mechanical digestion

The breaking up of food by mechanical digestion provides a greater surface area on which digestive enzymes can work. The walls of the stomach are more muscular than those of other parts of the alimentary canal. There are often encircling oblique bands of muscle, which by powerful rhythmical contractions (independent of peristalsis) churn and break up the food. This mechanical action plus the effect of the gastric juice reduces the food to a semiliquid state.

Chemical digestion

Protein digestion begins in the stomach, with secretions from the gastric glands in the central (fundic) region (Fig. 14-26). In mammals the gastric glands have two secretory cell types: chief cells, which secrete several proteolytic enzymes collectively called pepsin, and parietal cells, which secrete hydrochloric acid for the acidic medium without which pepsin will not function. In nonmammals the gastric glands have only one secretory cell type which evidently secretes both hydrochloric acid and pepsin.

Mucous glands, present throughout the alimentary canal, are particularly prominent and usually multicellular in the pyloric region of the stomach (nearest the small intestine). Their secretions lubricate the food and protect the delicate epithelium of the stomach from mechanical damage or digestion by its own enzymes.

Within this general pattern are many refinements and specializations to a diversity of diets, only some of which can be discussed here.

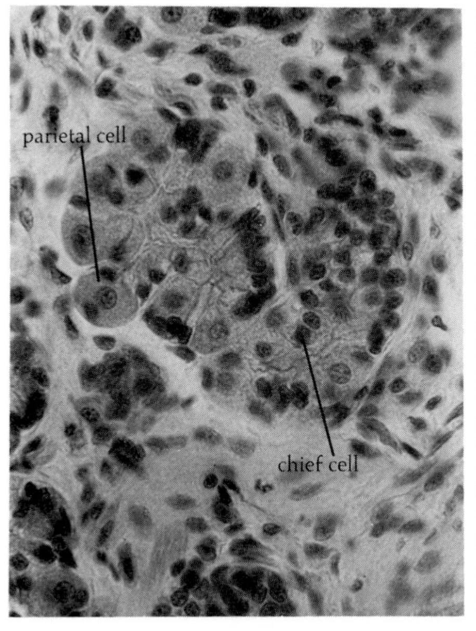

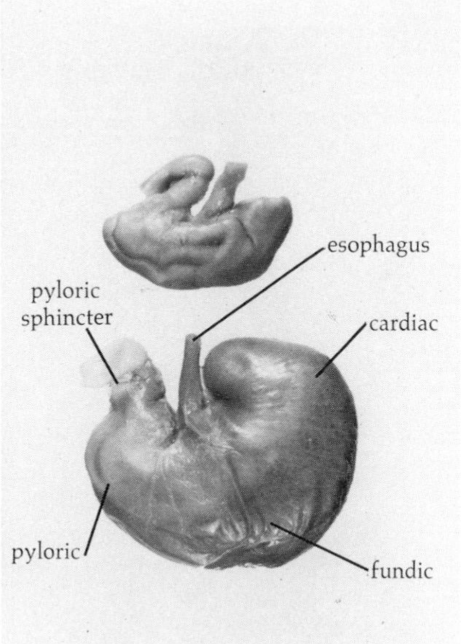

Fig. 14-26. On the left is a high-power photomicrograph of mammalian gastric glands showing chief and parietal cells. The two rat stomachs on the right, in the empty and the full condition, demonstrate the stomach's distensibility and the three major mammalian stomach regions.

FISHES

In the mucosa are many depressions, called gastric pits, lined with the columnar epithelium which also lines the stomach. In both the cardiac region and the fundic (central) region of the stomach, tubular fundic glands open into the base of each gastric pit. These glands lie within the lamina propria and contain secretory cells, called body cells, which evidently secrete both pepsin and hydrochloric acid. In the pyloric region (that region nearest the duodenum) instead of fundic glands there are mucus-secreting pyloric glands, also opening into the base of the gastric pits.

AMPHIBIANS AND REPTILES

The stomachs of amphibians and most reptiles, when examined grossly, are relatively simple dilations of the alimentary canal. With microscopic examination they can be divided into regions by gland

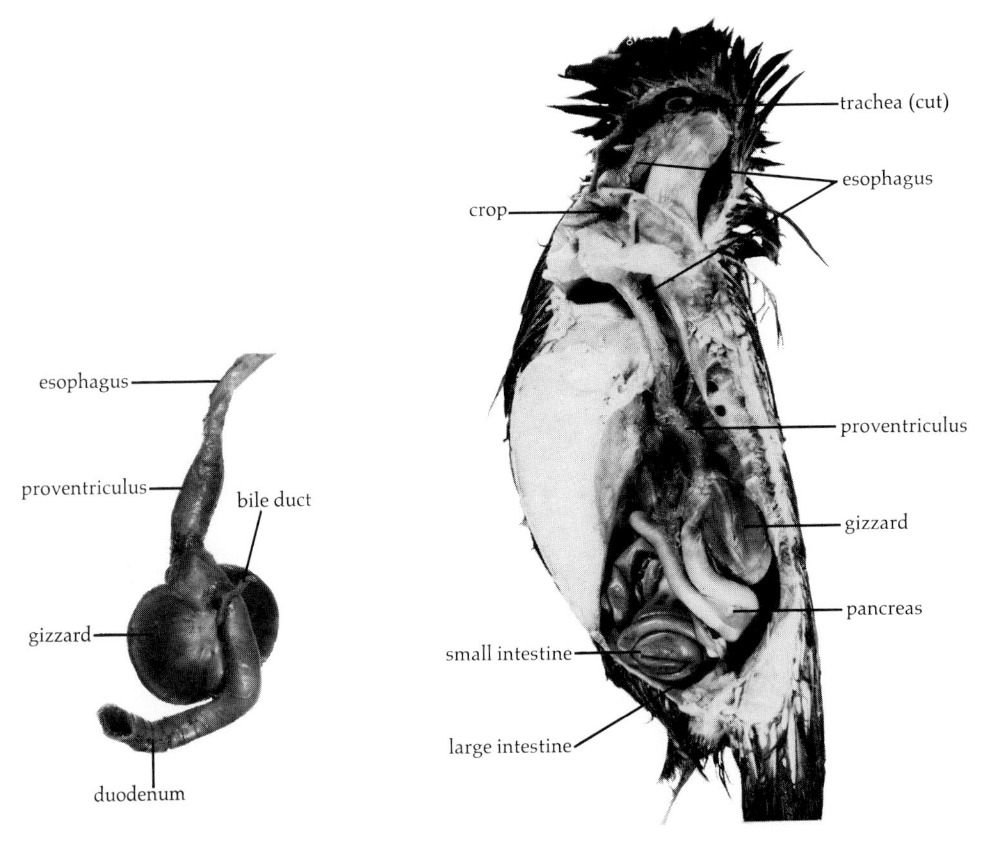

Fig. 14-27. On the right is a dissection of a pigeon's alimentary canal in its normal position but with other viscera such as heart, liver, and lungs removed. On the left is the isolated stomach of a pigeon, with a small section of esophagus and small intestine.

types, much as in fishes. In the crocodilian stomach the smooth musculature is thickened and extremely powerful; it functions analogously to the avian gizzard, violently and efficiently macerating food.

BIRDS

The avian stomach is modified into two sections, the glandular portion, or proventriculus, and the masticatory portion, or gizzard (Fig. 14-27). Food is stored in the crop at the distal end of the esophagus and periodically released into the stomach at the junction between gizzard and proventriculus.

The gizzard, a unique modification for flight, is found only in birds, although the thickened stomach walls of crocodiles, which function similarly, are interesting in this regard since crocodiles descended with birds from a common ancestor. The avian gizzard has extremely thick, powerful muscular walls and a glandular mucosa that produces a very hard, keratin-like protein called koilin. The koilin is analogous to teeth: when the gizzard muscles contract the food is crushed. The gizzard is another means for accomplishing mechanical digestion (essential for an animal with a high metabolic rate). If instead of a gizzard birds had a mammalian-type masticatory apparatus with teeth, muscles, and skeletal supports, the weight of the head would create perilous aerodynamic problems.

The extent of muscular development in the gizzard depends on both heredity and diet. For instance, if herring gulls are fed their usual soft fish diet their gizzards remain thin-walled and relatively inconspicuous. If they are forced to eat grain, however, the gizzard walls soon proliferate and strengthen.

MAMMALS

Like birds, mammals have a high metabolic rate and require an efficient digestive system. There is no need for a masticatory apparatus in the stomach since mammalian heterodont teeth provide grinding surfaces in the oral cavity. The stomach however must act as a large storage reservoir, and in most mammals the cardiac portion is distensible.

In small insectivorous mammals, such as the grasshopper mouse, the cardiac portion is lined with keratinized, stratified squamous epithelium; this protects the stomach from damage by the chitinous exoskeletons of insects. The gastric glands, instead of opening through numerous gastric pits, are all funneled into one short duct (Fig. 14-28). Since this duct is the only place where they contact the lumen, the opportunity for damage to the glands is greatly minimized.

The most dramatic vertebrate stomach, however, is surely that of the ruminant artiodactyl, a four-chambered structure with a capacity of up to 60 gallons (Fig. 14-29). The first and largest part of this stomach, into which the food enters, is the rumen. Its simple epithelial wall is filled

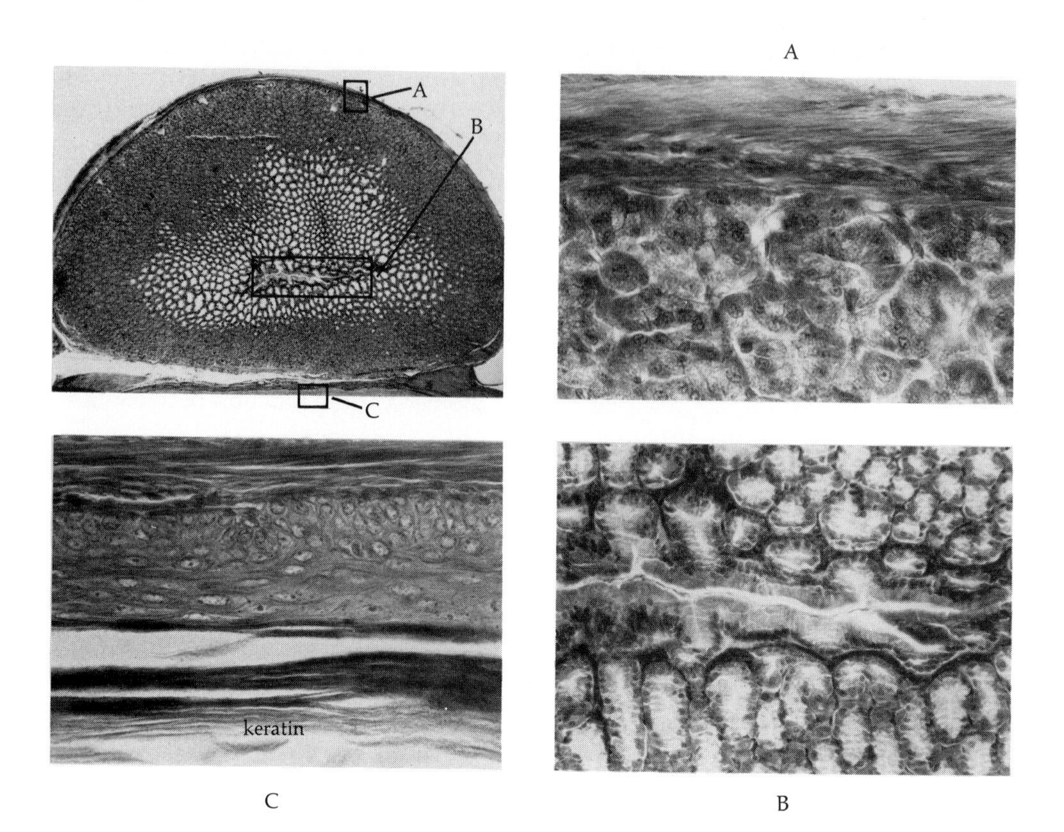

Fig. 14-28. The insect-eating grasshopper mouse has an unusual stomach. All the gastric glands (A) empty into a single duct (B), which opens into the main lumen of the stomarch. For further protection the stomach lumen is lined by a stratified, keratinized epithelium (C).

with symbiotic bacteria which can digest plant cellulose. The mucosal lining of the next chamber usually also contains symbiotic bacteria. This lining is folded in a way which produces geometric configurations similar to honeycomb. Food is frequently regurgitated from both rumen and reticulum to the oral cavity; there it is resalivated, rechewed, and reswallowed to pass once more through the rumen and reticulum for further cellulose digestion by the symbiotic bacteria. During this process much of the food's water and sugar are resorbed through the walls of the rumen and reticulum into the circulatory system. (In most vertebrates no absorption of food takes place in the stomach.)

From the reticulum the food is passed to the omasum, a heavy muscular structure with longitudinal muscle folds. The omasum walls subject the food to mechanical churning, the purpose of which is unclear. Surgical experimentation has shown, for example, that cows suffer no ill effect if the omasum is removed.

The fourth chamber is the abomasum, a typical glandular portion in which the initial digestion of proteins occurs.

PYLORIC SPHINCTER

At the junction of stomach and intestine the circular smooth muscles are greatly thickened, forming the pyloric sphincter. When this sphincter contracts it completely closes the alimentary canal; thus it controls the rate at which food passes from stomach to intestine. The pyloric sphincter itself is controlled by the autonomic nervous system and by endocrines.

INTESTINES

The rest of digestion as well as absorption into the bloodstream and formation of fecal material all occur in the intestines. Food enters the intestines in a fluid or semifluid state called chyme, after having undergone mechanical digestion and some chemical digestion. In the intestines are the necessary enzymes for lipid digestion and for the final digestion of carbohydrates (sometimes started in the mouth) and proteins (started in the stomach). Some of these enzymes are produced and secreted by the intestinal walls; others are produced in the pancreas and carried to the intestines by the pancreatic ducts. There are also mucus-secreting cells in the intestines, particularly goblet cells which concentrate the mucus into a large globule before releasing it (Fig. 14-30). As in the stomach, the mucus helps protect the intestinal epithelium from autodigestion.

The intestines' large surface area is richly supplied with capillaries so that absorption of nutrients and water can occur. The columnar absorptive cells of the intestine are similar in all vertebrates. Their most striking characteristic is what light microscopists have described as a brush border on the free surface, at the lumen of the intestine. With electron microscopy the brush border of a single cell is seen to be comprised of as many as 2000 microvilli, or long, fingerlike projections

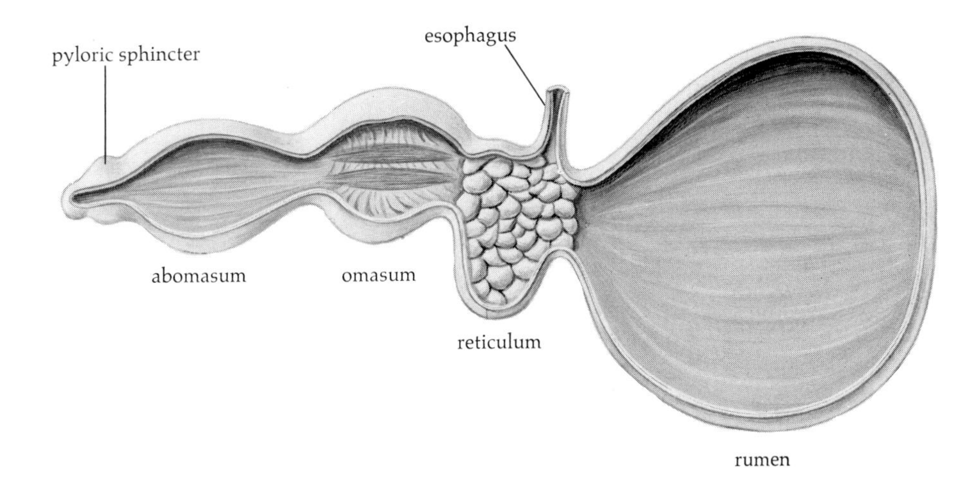

Fig. 14-29. Diagramatic section through the four-chambered stomach of a cow.

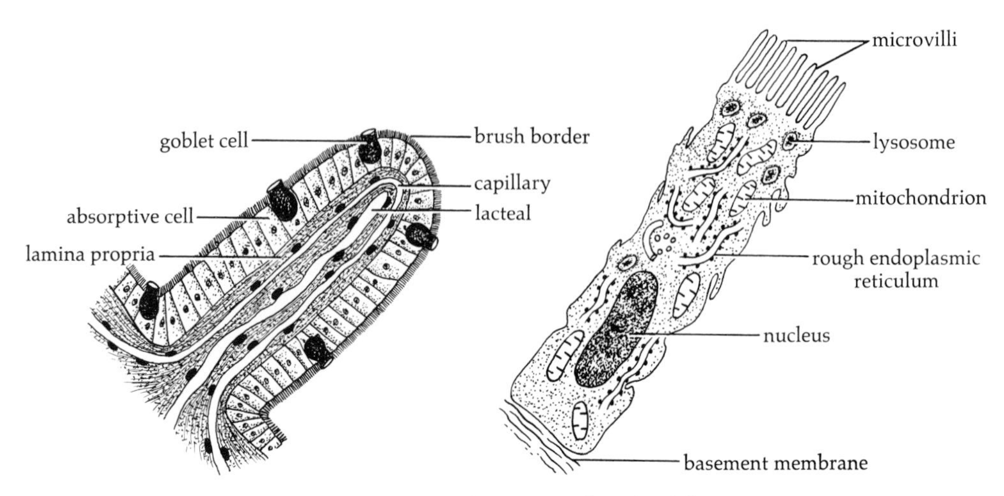

Fig. 14-30. On the left is a typical, short intestinal villus, showing the distribution of goblet and absorptive cells and their relation to capillaries and lacteals in the lamina propria. On the right is a single absorptive cell with its microvilli which greatly increase the surface area.

from the apical end of each cell; microvilli are a major and universal way of extending surface area (Fig. 14-30).

Absorption takes place through the cells, not between them. The digested material (primarily amino acids and simple fats and sugars) passes directly into the capillaries in the underlying lamina propria.

FISHES

In contrast to mammals there is no real distinction between large and small intestines in fishes. Frequently, however, the proximal portion, where the pancreatic ducts open, is called the duodenum; the most posterior portion, where fecal material is formed, is called the large intestine; and the intermediate portion is called the small intestine.

In fishes there have evolved four major mechanisms for further increasing the absorptive surface area (Fig. 14-31). In cyclostomes a simple fold, called the typhosole, runs the length of the intestinal wall. In elasmobranchs, holosteans, chondrosteans, and Dipnoi the mucosa and submucosa form a prominent spiral fold, called the spiral valve, throughout most of the intestine. This fold forces the food to spiral on its course through the intestine, rather than passing directly down its length, thus prolonging its contact with the intestinal walls. In most teleosts blind diverticuli, called pyloric caeca (singular, caecum), come off the intestinal lumen just past the pyloric sphincter; in some, such as the mackerel, there may be as many as 200 of these pyloric caeca. In other teleosts, however, the intestine is greatly lengthened, and coiled or curved back upon itself.

AMPHIBIANS AND REPTILES

In amphibians and reptiles intestinal length is also greatly increased, particularly in herbivorous forms. In the herbivorous anuran tadpole, for instance, the intestine is more than five times the length of the intestine in the insectivorous adult (Fig. 14-32). In reptiles the epithelial lining is deeply folded as well; the folds involve not only the mucosa but the entire depth of the lamina propria and some of the submucosa.

Large and small intestines are differentiated in amphibians and reptiles. Both have mucus-secreting goblet cells and absorptive columnar cells with microvilli very much like those of fishes, but in the large intestine there are many more goblet cells secreting mucus which helps to form the fecal material.

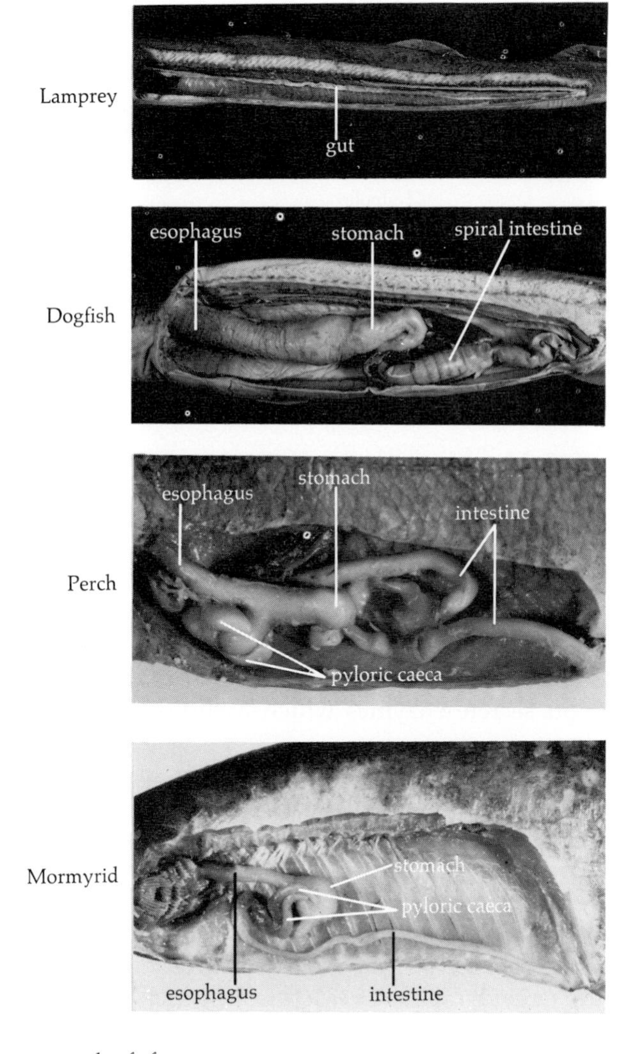

Fig. 14-31. Lateral views of the alimentary canal of four fishes, with all major viscera except the gut proper removed.

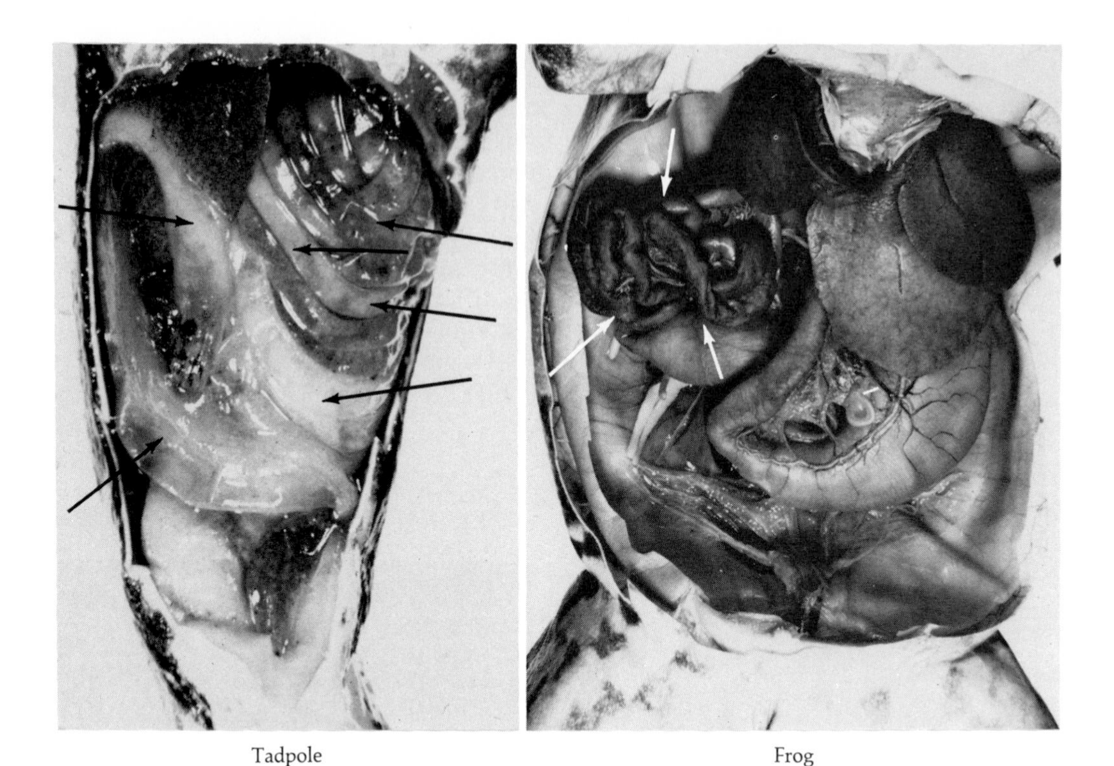

Fig. 14-32. Ventral views of the organs of the body cavity of an adult bullfrog (right) and of a bullfrog tadpole (left). Note that the intestine (arrows) is far greater in the herbivorous tadpole than in the insectivorous adult.

BIRDS

The intestinal epithelium of birds contains columnar and goblet cells similar to those of reptiles, but also has two sharply distinguishing features (Fig. 14-33). In the mucosa are extremely deep grooves, the crypts of Lieberkühn, which carry the epithelium throughout the very thick lamina propria and give the inner surface of the intestine a folded appearance. Large, branched tubular glands at the base of these crypts secrete enzymes which hydrolyze carbohydrates, proteins, and fats.

All over the intestine there are also villi, barely macroscopic fingerlike processes, which extend out into the lumen and give the folds a velvety appearance. A villus has a core of lamina propria completely covered by columnar epithelial and goblet cells. The lamina propria core contains a capillary network where absorbed amino acids and carbohydrates enter the blood; it also has a single, blind lymph vessel called a lacteal which absorbs digested fats and passes them into larger lymphatic vessels from which they eventually drain into the main circulatory system.

At the junction of the large and small intestine there are usually paired caeca, which in herbivorous forms may contain symbiotic bacteria and masses of lymphoid tissue. In the large intestine, which is quite short in birds, the number of goblet cells increases and the number of villi decreases.

MAMMALS

The large and small intestines are clearly distinguished in mammals. The small intestine itself can be divided (proximally to distally) into three portions: duodenum, jejunum, and ileum (Fig. 14-34). Throughout all three portions the surface area is increased much as it is in birds by both villi and crypts of Lieberkühn and also by large folds of the submucosa, called the valves of Kerkring, which extend into the lumen. The three sections are distinguished morphologically. The duodenum has glands of Brunner, alveolar glands secreting both mucus and a proteolytic enzyme (not pepsin), which pierce through the muscularis mucosa and lie in the submucosa. The jejunum and ileum, which lack glands of Brunner, are quite similar to one another; the ileum has more goblet cells and smaller villi than the jejunum.

At the junction between the small and large intestines there is usually a prominent colic caecum—a blind diverticulum structurally similar to the small intestine but with more lymphatic tissue and larger concentrations of goblet cells. Occasionally there is also a small blind pouch, the vermiform appendix, extending from the caecum; the vermiform appendix lacks villi but contains many tonsil-like lymphatic nodules.

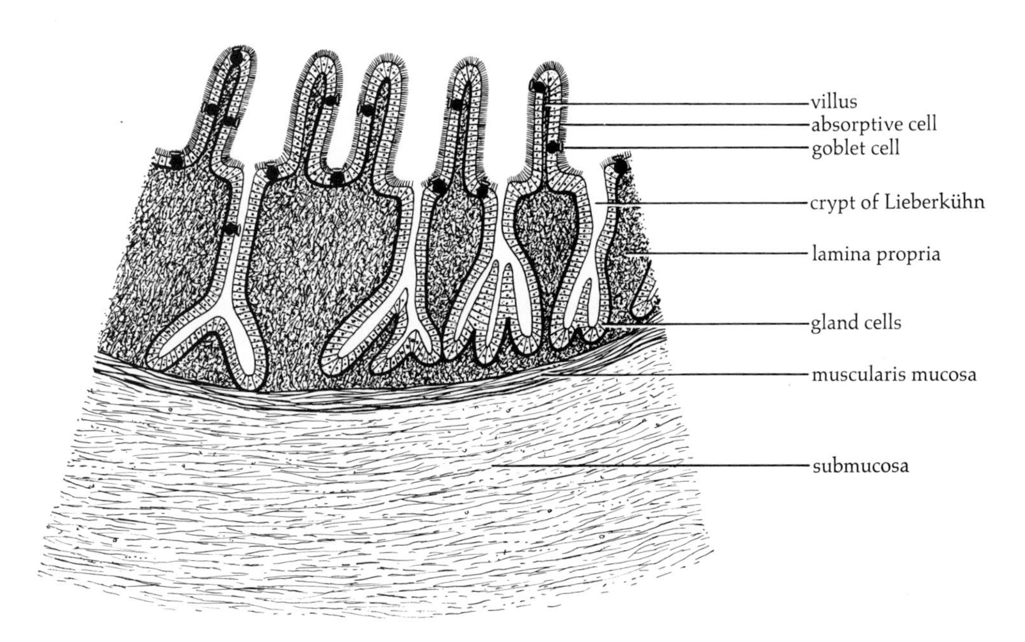

Fig. 14-33. A transverse segment of the mucosa, lamina propria, and submucosa of the avian small intestine.

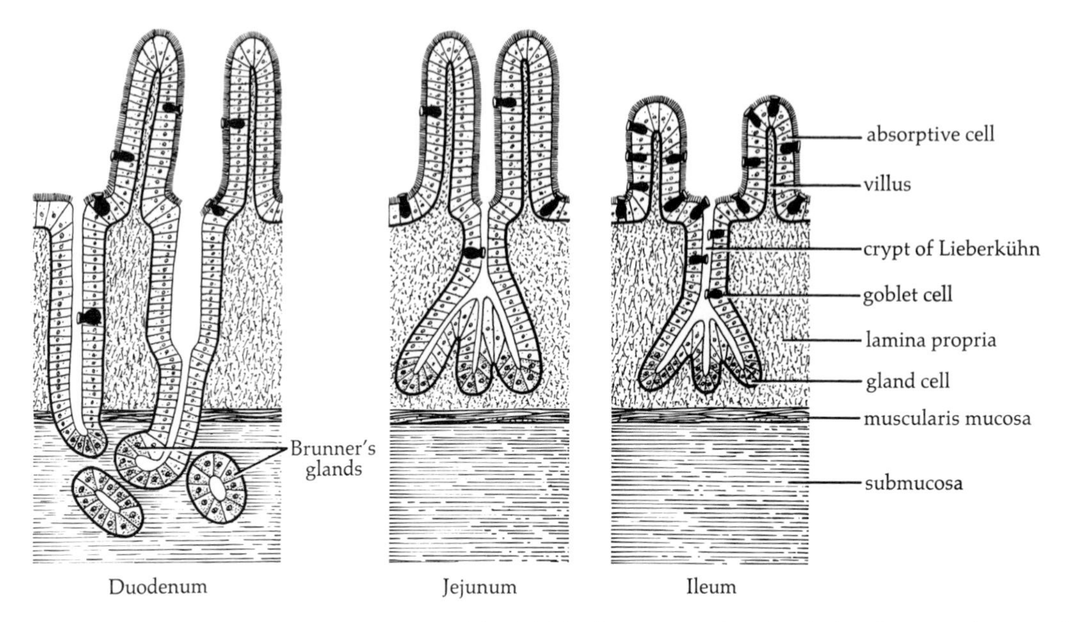

Fig. 14-34. A segment of each of the three portions of the mammalian small intestine showing similarities and differences in the villi, glands, and goblet cells.

The large intestine is much shorter, but has a larger diameter. It has more lymphoid tissue than the small intestine, and usually no villi. Long tubular glands of Lieberkühn secrete copious amounts of mucus. The large intestine resorbs a great deal of the water still in the undigested material and forms the remainder into feces.

CAUDAL TERMINATION OF THE ALIMENTARY CANAL

The alimentary canal ends as it began, in a derivative of an ectodermal invagination. The posterior chamber, developed from the embryonic proctodeum, is the cloaca (in most vertebrates) or the anus (in mammals and most teleosts). The cloaca receives not only undigested food (feces) but also urine and reproductive discharges. The anus, on the other hand, carries only feces and opens separately from the urinary and reproductive tracts.

PANCREAS

The alimentary canal by itself is not enough; digestion will not occur without digestive enzymes. Although several are produced in the alimentary canal, most enter the intestine from the pancreas. The pancreas is formed by two or frequently three outgrowths of endodermal epithelium just caudal to the presumptive stomach, which retain their embryological connections as the pancreatic ducts. In most vertebrates the pancreas is both an exocrine and an endocrine gland; the endocrine portion has already been described.

The exocrine portion is a complex, tubular alveolar gland whose secretions contain enzymes for the digestion of proteins, carbohydrates, and fats. In fact, the pancreatic enzymes alone are capable of breaking down all these major food components into units that are small enough to be absorbed by the intestinal epithelium.

In cyclostomes and several bony fishes there is no compact pancreas; instead, microscopic bits of pancreatic tissue line the walls of the gut and adjoining organs and are connected by ducts which empty into the intestine. In other vertebrates the pancreas is a relatively compact, two-part organ; sometimes there is secondary fusion of the dorsal and the ventral pancreas to form a single adult structure (Fig. 14-35). From the dorsal pancreas there is the duct of Santorini; from the ventral pancreas there is the duct of Wirsung, usually a paired branched duct draining into the intestine next to or in conjunction with the bile duct.

LIVER AND GALLBLADDER

During development paired endodermal cords grow out of the alimentary canal, close to the point of origin of the pancreas. The terminal ends of these endodermal cords proliferate and form the bulk of the liver and the gallbladder. The cords become hollowed, and in the adult are the cystic and hepatic ducts, draining the gallbladder and the liver directly into the intestine. In many vertebrates the distal portions, nearest the duodenum, fuse into a common bile duct (Fig. 14-36).

The gallbladder receives the bile, as the liver's secretions and excretions together are called, concentrates it, and stores it until hormonally

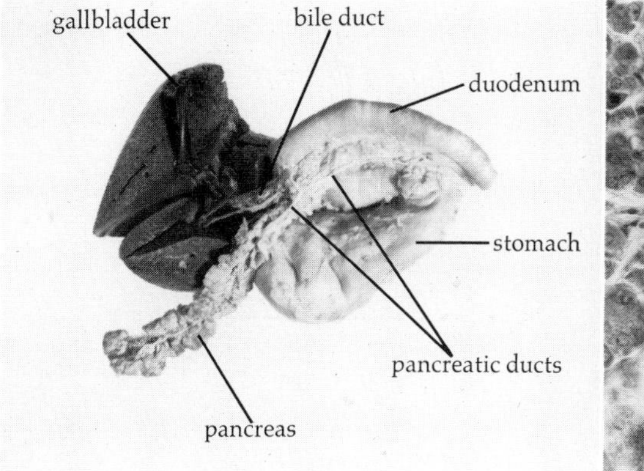

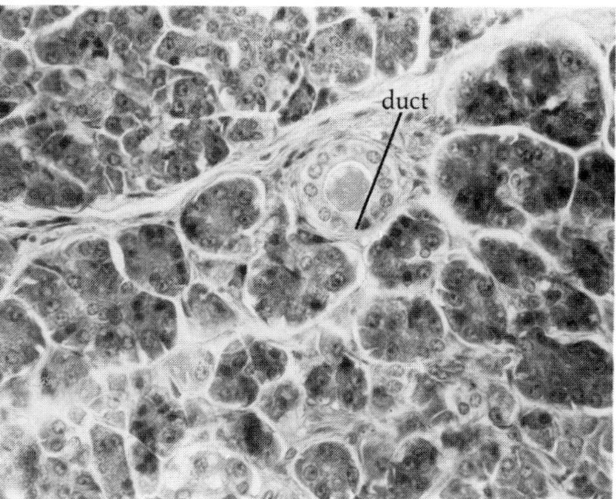

Fig. 14-35. On the left is a preparation of a cat's pancreas, duodenum, stomach, gallbladder, and part of the liver. The pancreatic ducts are exposed, demonstrating their courses through the pancreas to enter the duodenum with the common bile duct. On the right is a microscopic preparation of the exocrine pancreas.

stimulated to release it into the intestine. However, the gallbladder is not essential; in many species of various classes there is none, and in others it can be surgically removed without serious damage to the animal.

The liver, on the other hand, which is the largest gland and the filtering system for the blood, is absolutely essential for life. Its many functions are made possible because of its unique vascular supply. Hepatic arteries bring it oxygenated blood. Most of its blood, however, comes not from arteries but from portal veins which drain the entire alimentary canal; these break up into sinusoids and distribute this alimentary canal venous blood throughout the liver. Both the arterial capillary beds and the hepatic sinusoids are drained by hepatic veins (Fig. 14-37).

Through this filtering mechanism passes the entire blood supply—in humans this occurs several times an hour. Toxic materials, including bacteria, are removed from the blood and detoxified, and then either excreted with bile or passed back to the bloodstream in a harmless form and excreted by way of the kidneys. Potentially dangerous nitrogenous molecules are transformed into urea, which is only slightly toxic and can be readily excreted by the kidneys.

The blood from the alimentary canal is rich in glucose; much of this glucose is removed in the liver and converted into glycogen for storage. When the supply of blood sugar is low, glycogen is mobilized and glucose put back into the bloodstream for general body use. Both these processes are hormonally controlled.

The liver also produces fibrinogen, the protein necessary for blood clotting. It destroys worn-out red blood cells, and it synthesizes new ones during early development and sometimes also in the adult.

The liver produces bile which is emptied into the duodenum from the cystic duct of the gallbladder. Bile is partially an excretion, eliminating wastes such as detoxified bacteria and excess iron from red

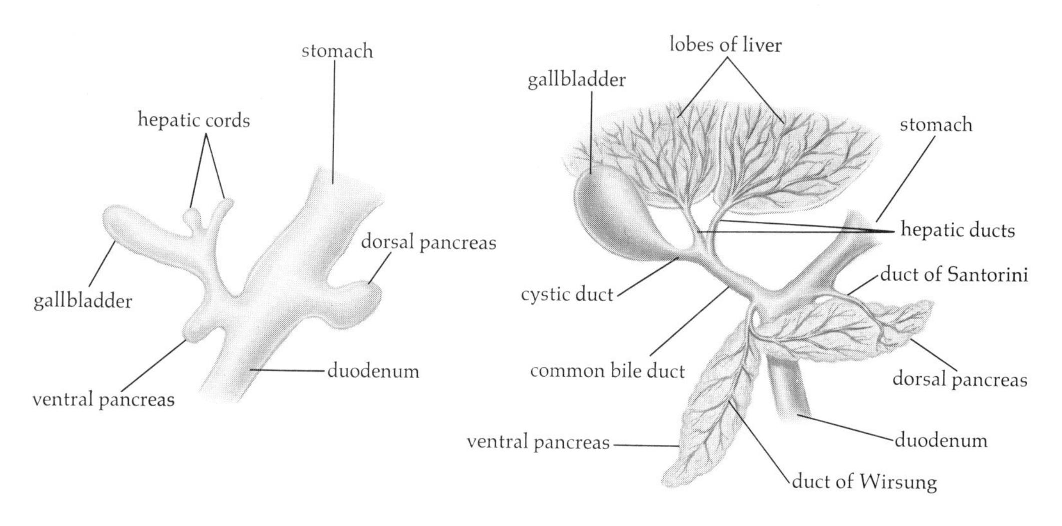

Fig. 14-36. Two stages in the development of the liver, gallbladder, and pancreas.

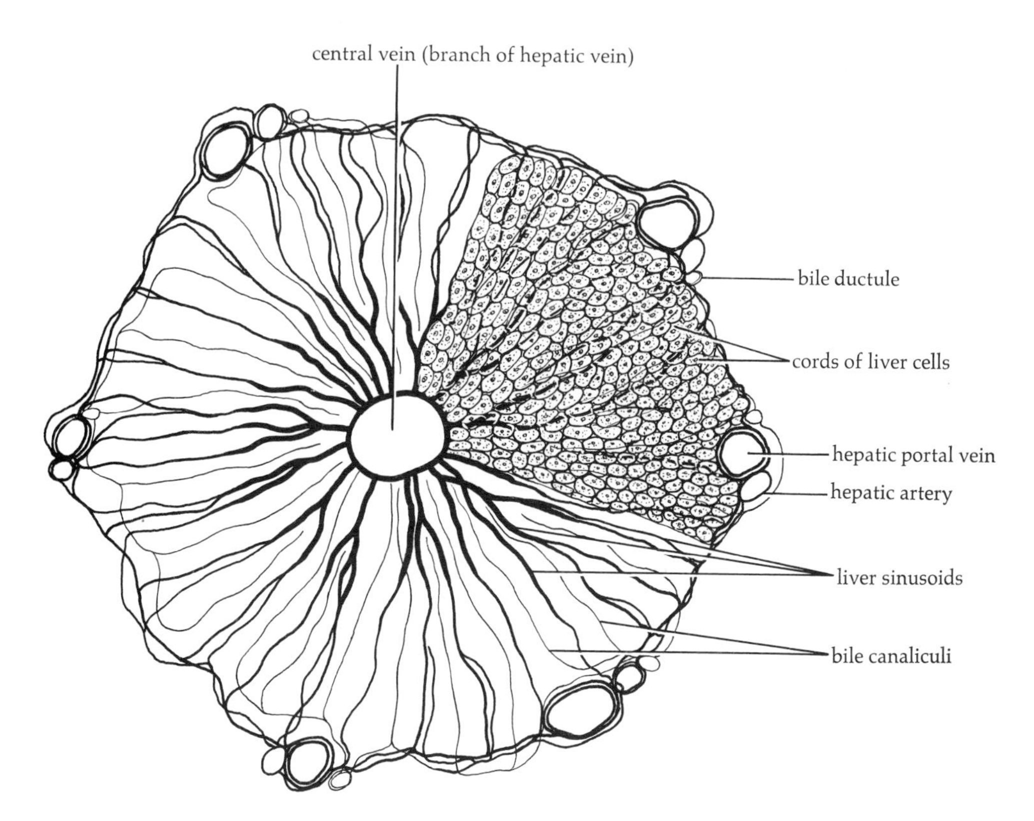

Fig. 14-37. A single lobule of the liver cut in cross section to show its vascular supply and drainage. Each lobe of the liver has thousands of such lobules. Blood enters each lobule through branches of the hepatic portal vein and the hepatic artery. Within the lobule blood flows centrally in sinusoids and is drained by a central vein, a branch of the hepatic vein which drains the entire liver. Bile is formed by liver cells and passes into bile canaliculi which carry it to peripheral bile ductules; these join together to form the hepatic ducts which drain the bile from the liver. The liver cords are shown in only one quadrant of the lobule to facilitate recognition of the vascular and bile channels.

blood cells. Bile is also partially a secretion which emulsifies fats, making them more susceptible to digestive enzymes.

Thus this large and important gland, embryologically derived from the alimentary canal, assists in digestion and also controls a great deal of the body's metabolism by its selective filtration and secretion and its unique blood supply.

SUGGESTED READING

Barrington, E. J. W. (1957). The alimentary canal and digestion. *In* "The Physiology of Fishes" (M. E. Brown, ed.), Vol. 1, pp. 109–162. Academic Press, New York.

Brauer, R. W. (1963). Liver circulation and function. *Physiol. Rev.* **43,** 115–213.

Diaconescu, N. (1971). On the liver "visceralization" in the vertebrate series. *Acta Anat.* **78,** 74–83.

Farner, D. S. (1960). Digestion and the digestive system. *In* "Biology and Comparative Physiology of Birds" (A. J. Marshall, ed.), Vol. 1, pp. 411–468. Academic Press, New York.

Horner, E. B., Taylor, J. M., and Padykula, H. A. (1965). Food habits and gastric morphology of the grasshopper mouse. *J. Mammal.* **45,** 513–535.

Peyer, B. (1968). "Comparative Odontology." Univ. of Chicago Press, Chicago, Illinois.

15 RESPIRATORY SYSTEM

15

In oxidative metabolism, oxygen is supplied to body tissues and carbon dioxide is removed from them. Vertebrates have too much mass for this exchange of gases to occur by diffusion through the external surface. Instead, it occurs by means of the respiratory and circulatory systems. When gases are exchanged between the external milieu (air or water) and the circulatory system it is termed external respiration; when between blood and tissues, internal respiration. Only the former is involved in what we study as the respiratory system, that is, the structures which get oxygen into and carbon dioxide out of the bloodstream.

In order for gases to pass between the environment and the blood the first requirement is a thin membrane. The membrane must be moist, since the substances must be dissolved on it before they can diffuse. The membrane's surface area must be great enough to supply the animal's requirements; the same considerations of surface area and volume relationships are involved here as in the digestive system, and surface area is also increased in the same ways—primarily by internal folding. The membrane must also be richly vascularized, so that there is enough blood to accomplish the exchange.

The capacity of the blood to carry oxygen is increased because of the carrier molecule, hemoglobin, which loosely binds much of the oxygen in the blood. The Antarctic ice fish, however, have no hemoglobin. They are thin fish with a large surface area compared to their mass, and they live in cold water, which has a high oxygen content; enough gases can diffuse through their surface to supply their needs.

In order for the process of diffusion to occur a higher concentration of oxygen and a lower concentration of carbon dioxide must be maintained outside the membrane than in the blood of the respiratory membrane. Therefore the second requirement for a respiratory system is a ventilation mechanism to move the respiratory medium (air or water) past the respiratory membrane. In most vertebrates this is accomplished by a muscular pump, operating by either suction or pressure.

Air contains 200 cm^3 of oxygen per liter; fresh water contains no more than 9 cm^3, and extremely stagnant water as little as 0.02 cm^3 of dissolved oxygen per liter. Thus there is relatively little oxygen available for aquatic animals, and air-breathing vertebrates would apparently have a great advantage. However, carbon dioxide elimination is equally important as oxygen intake. Carbon dioxide is carried dissolved in the plasma, usually as a bicarbonate ion, to the respiratory membrane and then eliminated. Both water and air normally have a low carbon dioxide tension, but because carbon dioxide is much more soluble in water than in air its elimination occurs more rapidly in

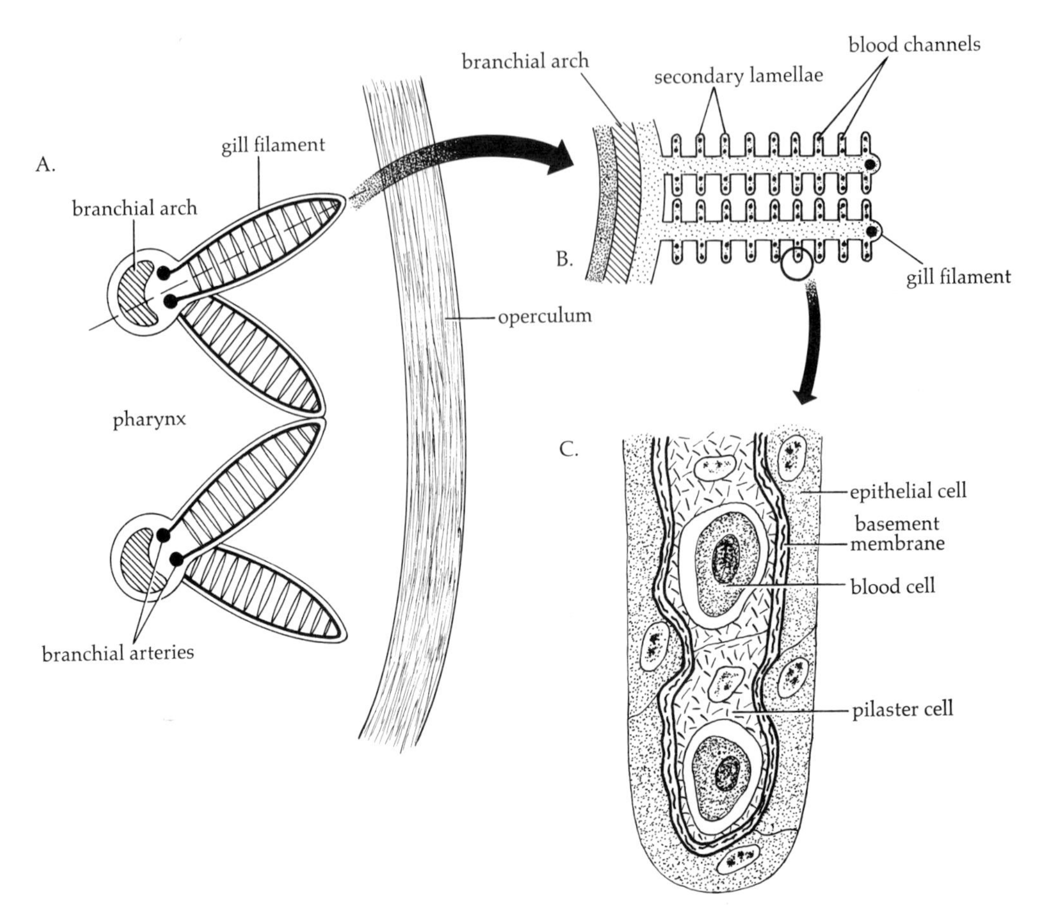

Fig. 15-1. The structure of teleost gills. In part A is a horizontal section through the pharynx, including two holobranchs. Part B shows a section along the dashed line of A, and includes two gill filaments. Part C is a high-power magnification of a single secondary lamella.

fishes than in tetrapods. Respiratory diffusion is limited in most fishes by the rate at which the respiratory membrane can absorb the limited available oxygen; in tetrapods it is limited by the rate at which the membrane can expel carbon dioxide. In addition the tetrapods' respiratory membrane must be kept moist.

GILL RESPIRATION IN FISHES

The primary and usually the only respiratory organs of most fishes and larval amphibians are gills. Internal gills, characteristic of adult fish, differentiate from the endodermal lining of the pharyngeal pouches; external gills, characteristic of embryonic and larval fish and larval amphibians, differentiate from the ectodermal portion of the gill slit. The significance of this developmental difference, if any, is not known.

TELEOSTS

The most abundant fish, teleosts also have the most efficient gill system. There are usually four pairs of branchial arches lying deep in the walls of the pharynx, each supporting a double row of gill filaments which are not parallel but diverge as they extend toward the body surface. Each double row of filaments supported by a single arch is called a holobranch; each single row, a demibranch. The rows are at oblique angles to the branchial arch; the anterior demibranch of one gill arch interdigitates with the posterior demibranch of the arch in front of it. Water must pass through a narrow space between filaments; this space is made even narrower by secondary lamellae, or folds, on the dorsal and ventral surfaces of each filament (Fig. 15-1). It is these secondary lamellae that constitute the vascularized respiratory membrane.

Each secondary lamella is covered by thin epithelial cells on a fine basement membrane. Deep to the basement membrane are unusual cells called pilaster cells, which form both the surface epithelium of the gills and the walls of the respiratory membrane's fine blood channels. The blood channels so formed are just large enough for the passage of one blood cell at a time and thus provide maximum surface area for

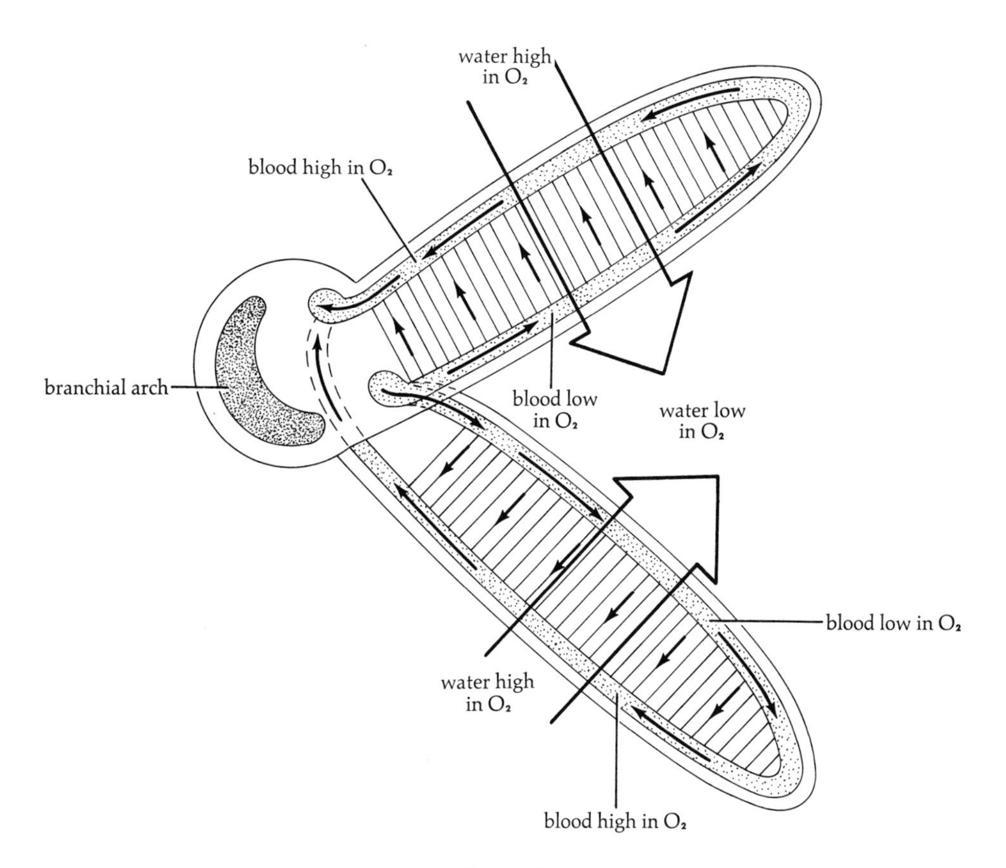

Fig. 15-2. Diagram demonstrating countercurrent flow of water and blood through a teleost's holobranch.

gas diffusion. Furthermore, diffusion occurs between the blood cells and the pilaster cells on both sides of the secondary lamellae (Fig. 15-1).

The gill filaments also contain occasional acidophilic cells, called chloride cells. These function not in respiration but in the excretion of salt.

The efficiency of the gill apparatus is increased by a countercurrent distribution system. The water entering the gills, bearing the most oxygen and the least carbon dioxide, passes the blood leaving the gills, also bearing the most oxygen and the least carbon dioxide—while the water leaving and the blood entering the gills contain the most carbon dioxide and the least oxygen (Fig. 15-2). This maintains a constant diffusion gradient between blood and water by which as much as 80% of the dissolved oxygen is diffused into the blood. If the flow of water is experimentally reversed, the efficiency is reduced from 80% to 10%!

Ventilation

A double pump involving both suction and pressure mechanisms is used to move water past the gills (Fig. 15-3), while the esophagus is kept closed by a sphincter muscle.

The suction pump functions with the operculum closed in the following manner. The oropharyngeal cavity is expanded by opening the mouth and by contracting the sternohyoideus muscle which lowers the floor of the cavity. This draws water into the oral cavity. At the same time the opercular cavity (between gills and operculum) is ex-

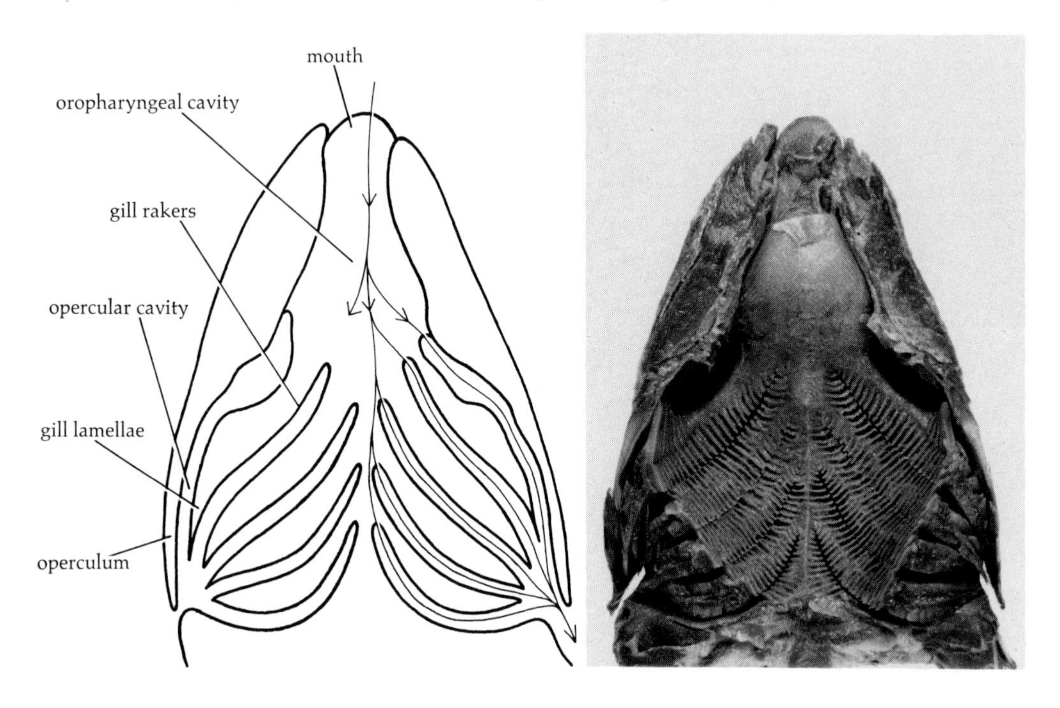

Fig. 15-3. Horizontal section through the mouth and pharynx of a teleost, the buffalo fish. Arrows indicate the water flow through one set of gills.

panded, primarily by the levator hyoidei and dilator operculi muscles; this draws water past the gills and into the opercular cavity. Thus water is simultaneously drawn into the mouth and past the gill lamellae by the suction pump.

The pressure pump functions with the mouth closed (by the adductor mandibulae muscles) and the opercular valve opened (by the adductor operculi muscles). Closing the mouth reduces the volume of the oropharyngeal cavity, which forces more water past the gill lamellae. The volume of the oropharyngeal cavity is further reduced by the geniohyoideus muscle, which raises its floor; this forces still more water past the lamellae.

Throughout this cycle water is passed almost continuously from the pharynx to the opercular chamber, always passing the respiratory membrane in the same direction. Of course the double pump requires some oxygen to run it, but it is so efficient that much more oxygen is gained than is utilized. During rapid swimming, however, the teleost opens both mouth and opercular valve. The fish's own forward motion then forces water into the mouth, through the oropharyngeal cavity, past the gills, and out through the opercular cavity without the use of the pump mechanism.

CHONDRICHTHYEANS

Holocephali, the aberrant subclass of Chondrichthyes, have an operculum. Although there have been few functional studies, their morphology indicates mechanisms similar to but simpler than those described for teleosts.

The subclass Elasmobranchii has no operculum, but has individually flapped valves which cover each gill slit externally. The space between the gill lamellae and the flap valve of each gill slit is called the parabranchial chamber; it functions analogously to the teleost's opercular chamber (Fig. 15-4). The parabranchial chambers are separated from one another by gill septae, which extend from each branchial arch to the surface skin where the flap valves lie.

Elasmobranchs also have a spiracle, the modified anterior gill slit of the hyoid arch which is completely closed in teleosts. It is not a respiratory structure but is used to bring water into the oropharyngeal cavity. Only oxygenated blood is brought to the spiracle, and its lamellae, which do not function in gas exchange, form a pseudobranch.

There is a double pump mechanism similar to that in teleosts (Fig. 15-4). The coracomandibularis muscles are the primary muscles that open the mouth. The expansion of the oropharyngeal cavity draws water in through both the mouth and the spiracle. With the gill flaps closed the coracohyoid and coracobranchial muscles are contracted, further expanding the oropharyngeal cavity. This forces water past the gills and also expands the parabranchial chambers, which are highly elastic due to the cartilaginous nature of the chondrichthyean branchial basket.

The gill flaps are then opened, and the mouth and spiracle are closed (by the adductor mandibulae and spiracularis muscles, respectively);

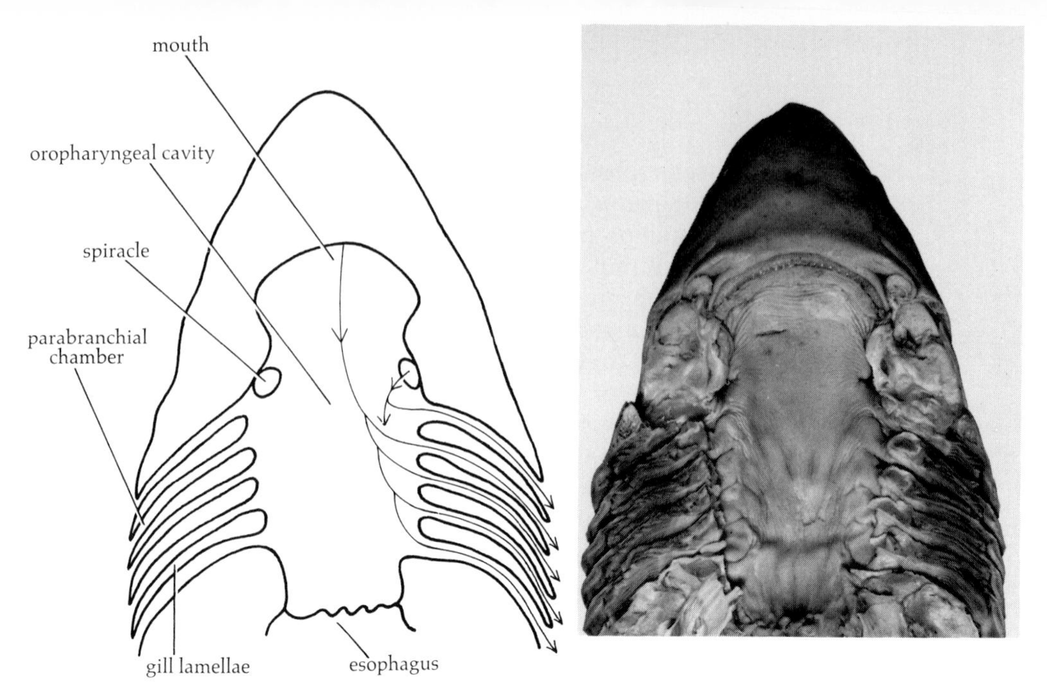

Fig. 15-4. Horizontal section through the mouth and pharynx of a dogfish. Arrows indicate the flow of water, in through the mouth and spiracle and out past the gill lamellae.

this forces more water past the gills. The epihyoideus and cuccularis muscles contract; this raises the branchial arches and the floor of the mouth and further reduces the size of the oral cavity. Finally the superficial constrictors compress the parabranchial chambers, forcing water out of them, and the cycle begins again. The structure is not as tightly knit in Chondrichthyes as it is in Teleostei, and there is only about 50% oxygen utilization.

CYCLOSTOMES

There is an independently evolved gill system in cyclostomes, with the gills and their muscles and blood vessels all contained in an elastic branchial basket. There are more gill slits than in any other vertebrate. Internally they open to a branchial chamber just ventral to and separate from the esophagus. Externally the gill slits open separately in the lamprey, but are covered in most hagfish and lead to a single common opening on each side. On the hagfish's left side, just behind the last gill slit, there is a cutaneous esophageal canal; it has no respiratory membrane but it is a direct connection between the branchial cavity and the outside.

In both lampreys and hagfish the gill lamellae for each gill slit lie within a muscular pouch, which also contains the respiratory membrane and its vascularization (Fig. 15-5). In the hagfish water is brought in through the nostril by special muscles attached to the velum of the oral cavity, and water is pumped out through the gill pouches by contraction of their intrinsic muscles. Thus the hagfish has a continual one-way flow.

In lampreys, on the other hand, water moves in and out of the pouches in a tidal ebb and flow; it is drawn in by their expansion (caused by the elasticity of the branchial basket) and pumped out by contractions of their intrinsic muscles. Valves in the pouch ensure that water goes through the lamellae and past the respiratory membrane only while water is being pumped out. This tidal flow is in contrast to the continuous flow of hagfish, teleosts, and chondrichthyeans.

In both lampreys and hagfish the skin is vascularized and scaleless. A certain amount of gas exchange occurs through the skin; this is called cutaneous respiration.

THE GAS BLADDER

In most teleosts a gas-filled sac called the gas bladder (or swim bladder) lies just dorsal to the serous membrane lining the body cavity, in what is called a retroperitoneal position. The gas bladder develops as a diverticulum off the embryonic gut, and in some teleosts it remains connected to the esophagus by the pneumatic duct while in others it has no adult connection; these are called the physostomous and physoclistous conditions, respectively.

Its internal structure is similar to that of the postpharyngeal alimentary canal, with layers of epithelium, lamina propria, muscularis mucosa, submucosa, and muscles (Fig. 15-6). Its outer muscle layer is made up of tough, collagenous connective tissue, embedded with

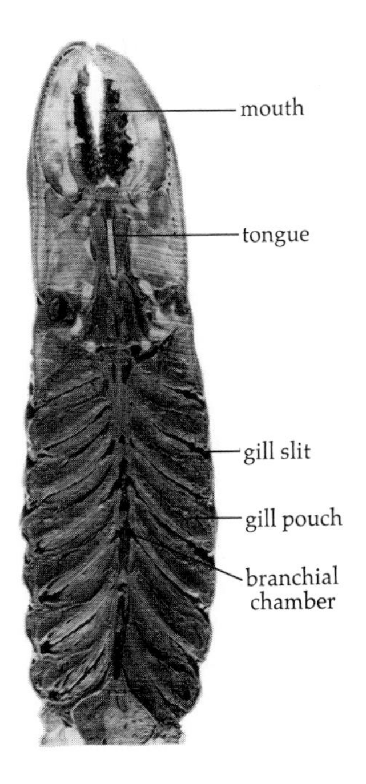

Fig. 15-5. Horizontal section through the mouth and pharynx of a lamprey. Note the seven pairs of gill slits, each with a muscular branchial pouch.

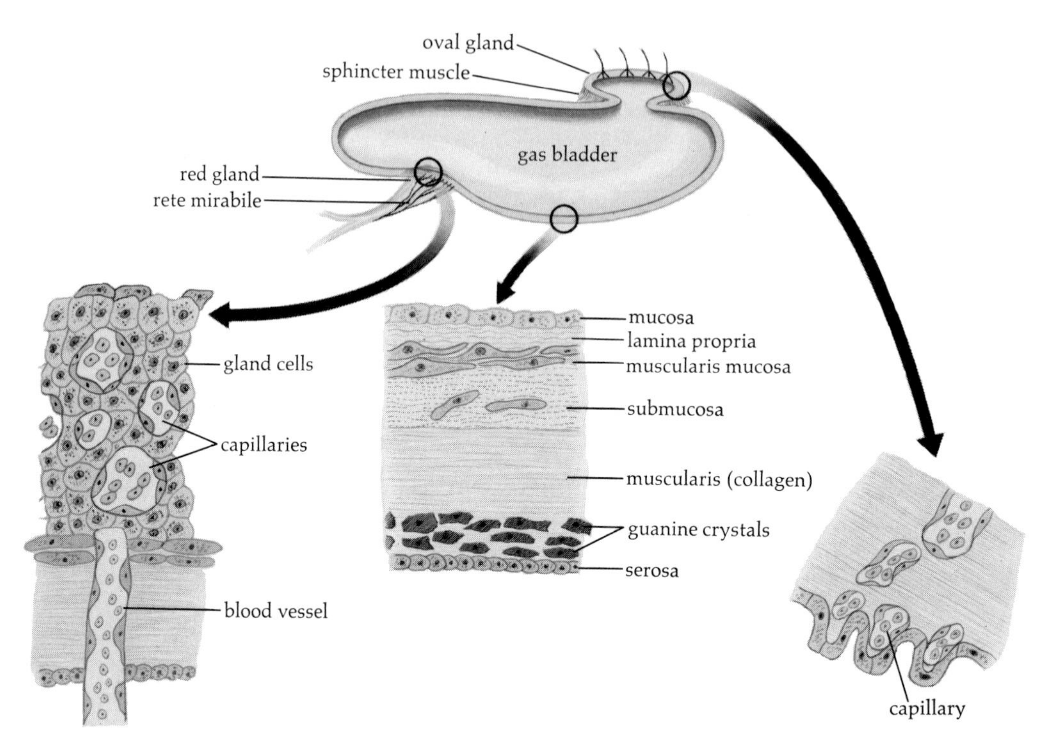

Fig. 15-6. Sagittal section through the gas bladder of a physoclistous teleost, with details of its red gland, oval gland, and unspecialized wall.

many guanine crystals which give it a silvery appearance in gross dissections. This layer makes the gas bladder impermeable to diffusion, although it is not clear whether it does so by its guanine or because its collagenous fibers are very tightly packed.

Anteriorly the epithelium is modified into the red gland, which is made up of folded columnar epithelium and heavily vascularized by long loops of densely packed capillaries (Fig. 15-6). This type of structure, wherever it occurs, is called a rete mirabile (Latin for "miraculous network"). The rete mirabile here produces gas, which the red gland secretes into the gas bladder. In many teleost species this gas closely resembles air, but in some it is almost pure oxygen and in others almost pure nitrogen.

In physoclistous fishes a posterior and usually dorsal part of the gas bladder is modified as the oval gland, which is surrounded by a sphincter muscle. When this sphincter is contracted the oval gland is shut off from the rest of the gas bladder; when the sphincter is open, the oval gland absorbs gas from the bladder.

In physostomous fishes the amount of gas in the bladder can be more rapidly decreased by contractions of the bladder musculature which force gas out the pneumatic duct, esophagus, oropharyngeal cavity, and mouth. Conversely, by gulping in air physostomous fishes can get additional gas into the gas bladder more rapidly than the red gland can secrete it. In general, therefore, physostomous fishes have no oval gland and a smaller red gland than do physoclistous fishes.

This unusual structure, the gas bladder, functions primarily as a hydrostatic organ. By changing the amount of gas in its gas bladder a fish changes its specific gravity and thus its buoyancy. When it has the same specific gravity as the water around it, it can maintain its level without expending metabolic energy. It is not surprising that most bottom-dwelling teleosts, whose protective and food-gathering mechanisms depend on their staying at the bottom, do not have a gas bladder. It is also usually lost or greatly reduced in certain other teleosts whose functioning would be impeded by a large bubble of gas, for instance, in freshwater fishes that live in turbulent streams and in the most rapidly swimming marine fishes such as the mackerel.

The gas bladder also has evolved other functions. As discussed in Chapter 10, it facilitates auditory reception in several teleosts. In others it is involved in sound production. In still others its silvery outer layer reflects bioluminescence through the translucent body tissues.

AIR BREATHING IN FISHES

Most fishes must breathe oxygen dissolved in water, for without the water's buoyancy their gills collapse. Fishes living in environments that are unusually deficient in oxygen have a special problem. Warm water, for instance, holds only a fraction of the oxygen in cold water, and stagnant water contains additional carbon dioxide because of the decay occurring in it. Moreover, such areas often have a heavy vegetational cover which cuts down the sunlight and thus the photosynthesis and oxygen-production of plants living in the water. As a result, swamp water may have as little as 0.2 to 2.0 cm^3 of dissolved oxygen per liter.

Some teleosts in these environments dwell at the surface, where they may take advantage of the higher percentage of dissolved oxygen in the water or of the opportunity for cutaneous respiration with the air. Cutaneous respiration is particularly useful in very long or narrow fish with a favorable mass-to-surface area ratio. The common eel, *Anguilla,* for instance, needs only cutaneous respiration and commonly slithers overland from pool to pool.

Other fishes—usually with poorly developed gills—have evolved adaptations for breathing the air's oxygen directly (Fig. 15-7). All these gulp in air through the mouth; even when internal nares are present, as they are in lungfish, air is not breathed through them. In these fish, most carbon dioxide elimination occurs through the gills and most oxygen uptake in other, diversely adapted structures.

For instance, the gills of the electric eel, *Electrophorus,* are quite degenerate. The mucous membrane in its mouth is raised into large oral papillae, each one of which is richly vascularized. Oxygen uptake occurs between the air gulped into the oral cavity and the vascularization of the oral papillae. These oral papillae are very fragile and might be damaged by an active prey; however the electric eel kills or stuns its prey with its electric discharge and then swallows it when it is immobile.

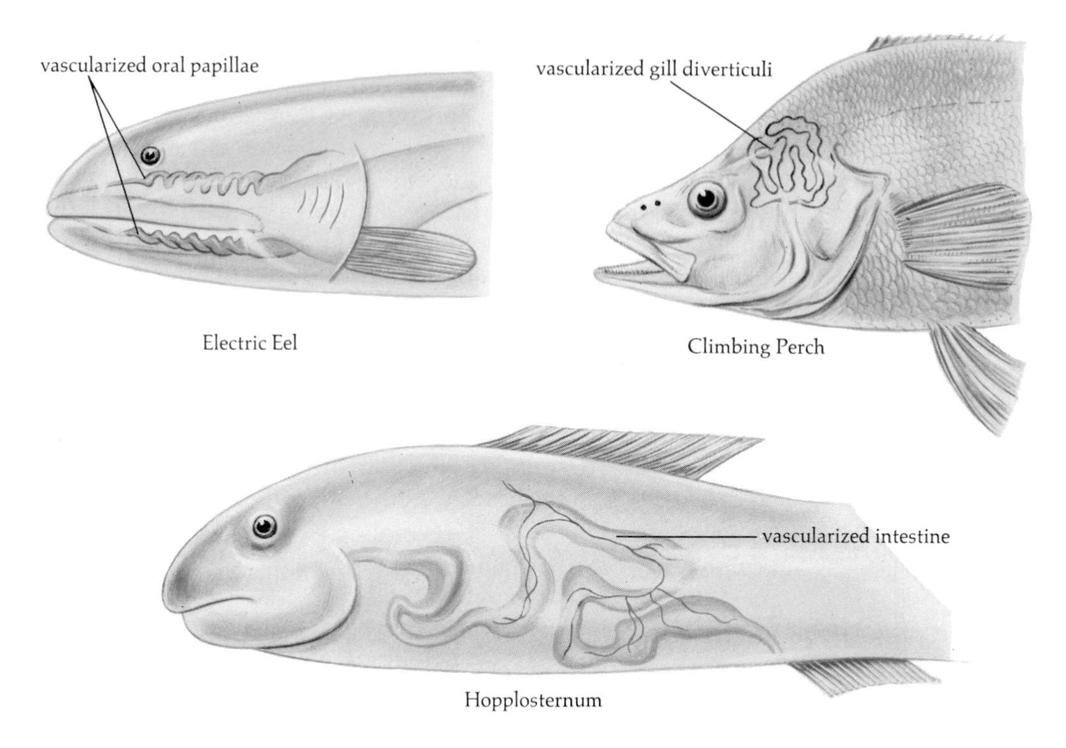

Fig. 15-7. Three of the many adaptations facilitating aerial respiration in specialized teleosts.

In several teleosts of tropical swamps other portions of the alimentary canal have been modified and oxygen uptake occurs in structures which are not normally respiratory in function. For instance, a South American teleost, *Hopplosternum,* swallows air. Peristaltic contractions move the air along to the unusually long intestine where oxygen uptake occurs, facilitated by the fact that most of the intestine is thin and densely vascularized and that only a small portion is used for digestion.

In other teleosts gulped-in air goes from the oropharynx to dorsal diverticuli of the pharynx, located above the gills, where there are extensions of the highly folded vascularized gill membranes. These diverticuli are found in the climbing perch (*Anabas*) of stagnant southeast Asian swamps, and, even more complexly, in the mud-skipper (*Periophthalmus*) which lives in similar environments.

The gas bladder has a respiratory function in some fish (Fig. 15-8). Among teleosts, for example, the genera *Gymnarchus, Erythrinus,* and *Umbra* have a richly vascularized gas bladder used as a lung, usually in addition to gills. In two holosteans, bowfin (*Amia*) and garplike (*Lepisosteus*), the gills alone are sufficient for respiration in cold water, which in addition to having a higher oxygen content also keeps the fish's metabolic rate relatively low. In warm or otherwise oxygen-deficient water, however, these holosteans must use their gas bladders as supplementary respiratory organs. The same is true for at least one teleost, *Erythrinus.*

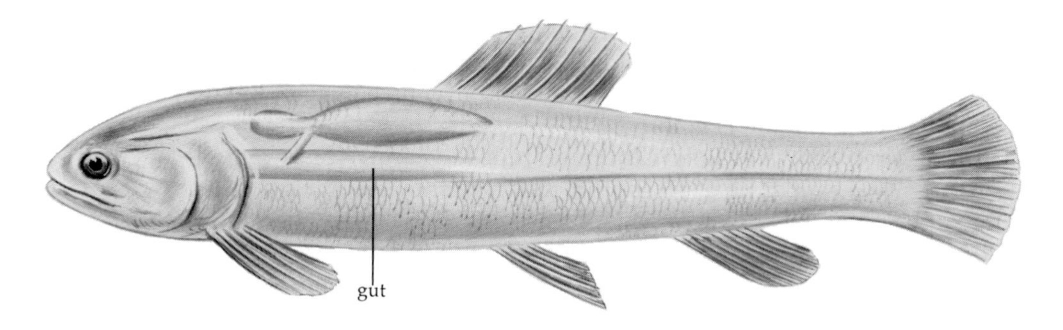

Erythrinus (teleost)

Lepisosteus (holostean)

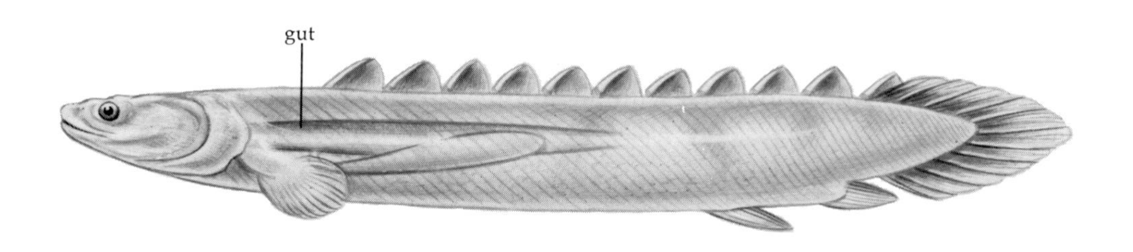

Polypterus (chondrostean)

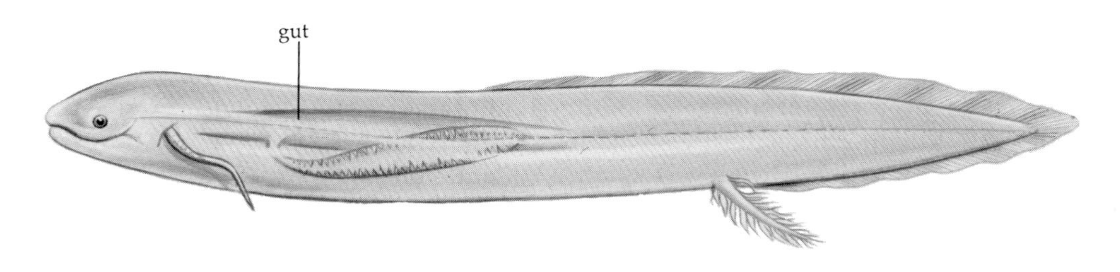

Lepidosiren (lungfish)

Fig. 15-8. Four widely divergent Osteichthyes with gas bladder or lungs that are used for aerial respiration.

The most generalized of all living actinopterygians, *Polypterus,* has two densely vascularized evaginations from the caudal pharynx, which are variously called gas bladders or lungs and in which oxygen uptake occurs. Carbon dioxide is eliminated by the gills.

In the three genera of lungfish there are also vascularized evaginations from the caudal pharynx, but in this case they are always called lungs (Fig. 15-8). Of the three genera, the gills of the Australian lungfish, *Neoceratodus,* are the most efficient and will satisfy normal respiratory needs; lungs are used in addition to gills in oxygen-deficient waters only. The gills are poor, however, in both *Lepidosiren* (the South American lungfish) and *Protopterus* (the African lungfish), and these fish always use lungs for oxygen uptake. The waters in which they live dry up periodically and during these times the fish aestivate: that is, their metabolic rate, and hence their respiration, are vastly lowered. *Lepidosiren* burrows into the mud; *Protopterus* not only burrows but also secretes a mucous cocoon which hardens and protects the animal from desiccation even when the surrounding mud is dried. Until the next rainy season *Protopterus* stays in this cocoon and, under experimental conditions, has remained there for up to five years. The same sort of aestivation behavior in response to drying has been noted in the North American holostean, the bowfin *Amia.* For the oxygen uptake and the carbon dioxide elimination required to maintain life during aestivation, these fish are completely dependent upon their "lungs"; if they did not possess "lungs" this adaptive habit would not be possible.

The only other living sarcopterygian is the coelacanth, *Latimeria,* which lives in extremely deep marine waters off the island of Madagascar. It has lungs which are filled with fat and have ossified walls. Apparently they are present only as a functionless ancestral heritage.

ARE LUNGS AND GAS BLADDERS HOMOLOGOUS?

Although only relatively few living fishes—all in oxygen-deficient waters—use either lungs or gas bladder as respiratory organs, air-breathing is apparently not a recent specialization to a stressful environment but an ancient feature. Some air-breathing Osteichthyes (e.g., *Polypterus, Amia, Lepisosteus*) are generalized, not specialized, in other structures, and their skeletal characteristics are close to those of fossil forms. Moreover, careful studies of fossil fish impressions suggest that lungs evolved with the earliest Osteichthyes, or possibly even in some Placodermi. These lungs were very likely an adaptation to the oxygen-depleted environments in which the earliest Osteichthyes probably lived. It is quite possible that for fishes moving into more adequately oxygenated waters there would be a strong adaptive value not for an (unneeded) aerial respiratory organ, but for a hydrostatic organ, and that the teleost gas bladder represents not a precursor of the lung but a more adaptive modification of it. This is the view most generally accepted by biologists today.

Why is it not universally accepted? Because it depends upon the gas bladder and the lung being homologous, and yet between the lungs of lungfish and the gas bladders of teleosts there are apparent inconsistencies in development, adult connections, and interrelationships with other parts. Because of these inconsistencies, some biologists have felt that the gas bladder and the lung cannot be homologous.

The fact that there is but a single gas bladder while lungs are usually paired is not a major problem. Living lungfish have but a single lung that is bilobed in two genera. The "lung" of *Polypterus* is paired, although one is considerably larger than the other. The "lungs" of the bowfin and the garpike are both single organs.

Problems arise in other areas. If lungs and gas bladders are homologous one would expect them to have a similar arterial supply and venous drainage (Fig. 15-9). The respiratory alimentary organs of the three genera of lungfish, of *Polypterus,* of the bowfin, and of one teleost (*Gymnarchus*), are all supplied by the artery to the last gill arch; this vessel, called the pulmonary artery, is homologous to the pulmonary artery of tetrapods. However, the arterial supply to the gas bladder in other teleosts and in the garpike, *Lepisosteus,* is by branches of the dorsal aorta. The venous drainage of the lung is the same in

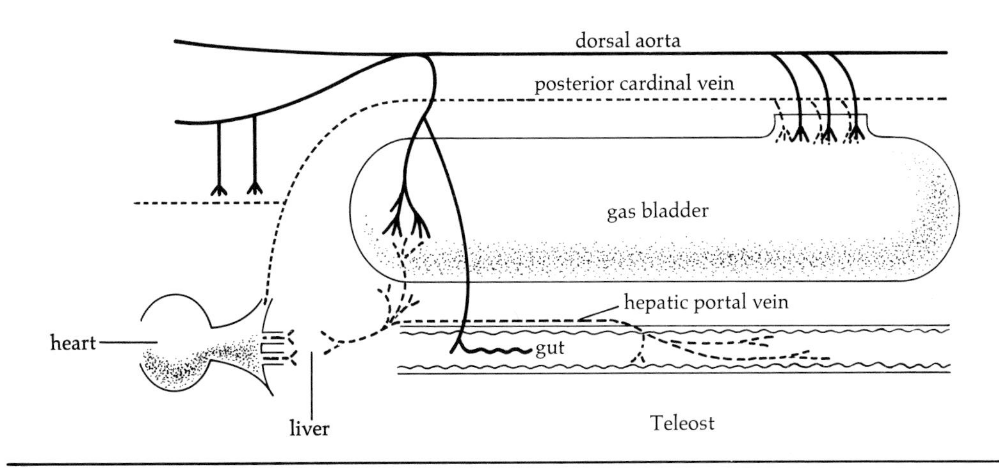

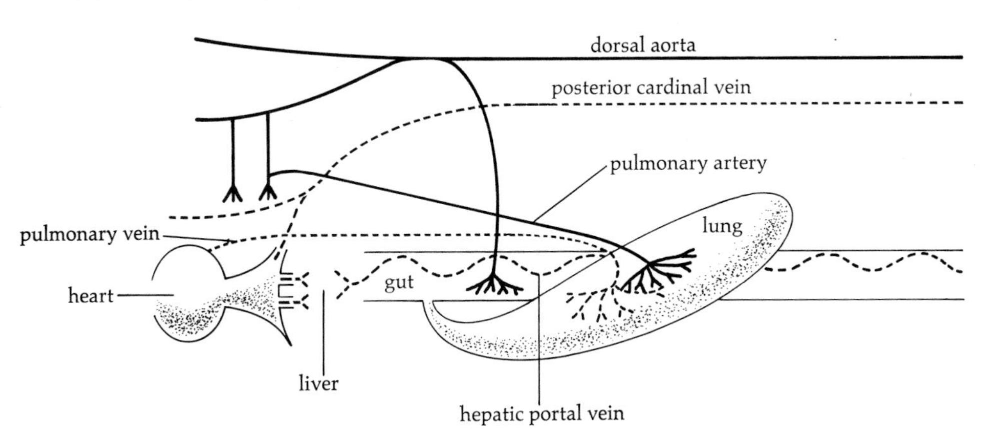

Fig. 15-9. The arterial supply and veinous drainage of the teleost gas bladder differ fundamentally from those of the lungfish lung.

lungfish as in tetrapods—a pulmonary vein going directly from the lung to the heart. There is a chaotic distribution of veins in other air breathers, none of which has a true pulmonary vein.

Lungs and gas bladders also differ in microscopic structure. The adult teleost gas bladder has the typical epithelium, connective tissue, and muscle layers described for the alimentary canal. In the lungs of tetrapods and lungfish, however, connective tissue and smooth muscles are very scant and are never arranged in the characteristic layers found in most of the alimentary canal.

As usual the best clue to possible homologies is embryology (Fig. 15-10). Both gas bladders and lungs develop as foregut diverticuli, but they involve different areas of the foregut. In sarcopterygians and tetrapods, the lungs develop from ventral diverticuli of the foregut of the pharyngeo-esophageal border, or possibly even from a posterior pair of pharyngeal pouches. In their early development, lungs are histologically more similar to the pharynx than to any other part of the alimentary canal. In teleosts, however, the gas bladder very clearly arises as a more posterior, dorsal diverticulum from the esophagus or even from the presumptive stomach region; in the larval codfish the gas bladder even has peptic glands! The gas bladder's adult structure also resembles posterior rather than pharyngeal parts of the alimentary canal.

Considering all these factors, it seems likely that the earliest Osteichthyes had "lungs," adaptive to an oxygen-deficient aquatic environment, and that these have been retained in the very generalized living actinopterygians (e.g., *Polypterus*) as well as in the sarcopterygians and tetrapods. During the evolution of those actinopterygians that successfully inhabited the increasing number of adequately oxygenated freshwater and salt-water regions, it is also likely that

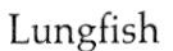

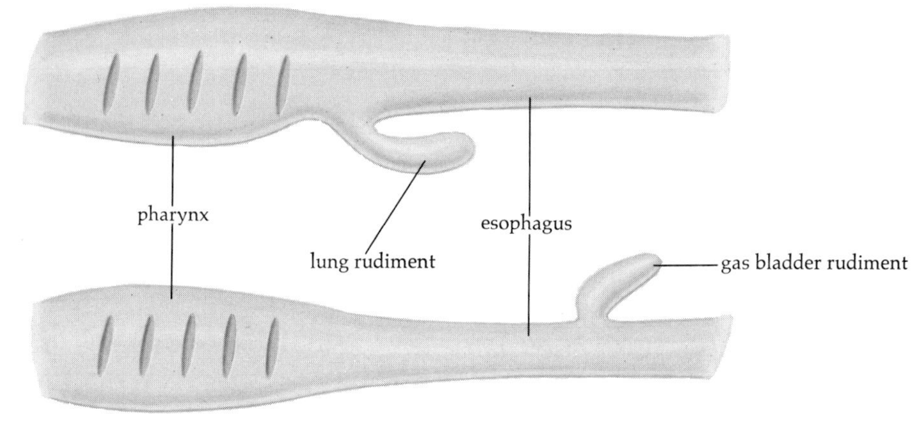

Fig. 15-10. Lateral views of the developing pharynx and esophagus in a lungfish and a teleost. Note that the lung develops ventrally at the border of the pharynx and esophagus, whereas the gas bladder develops dorsally and far more caudally.

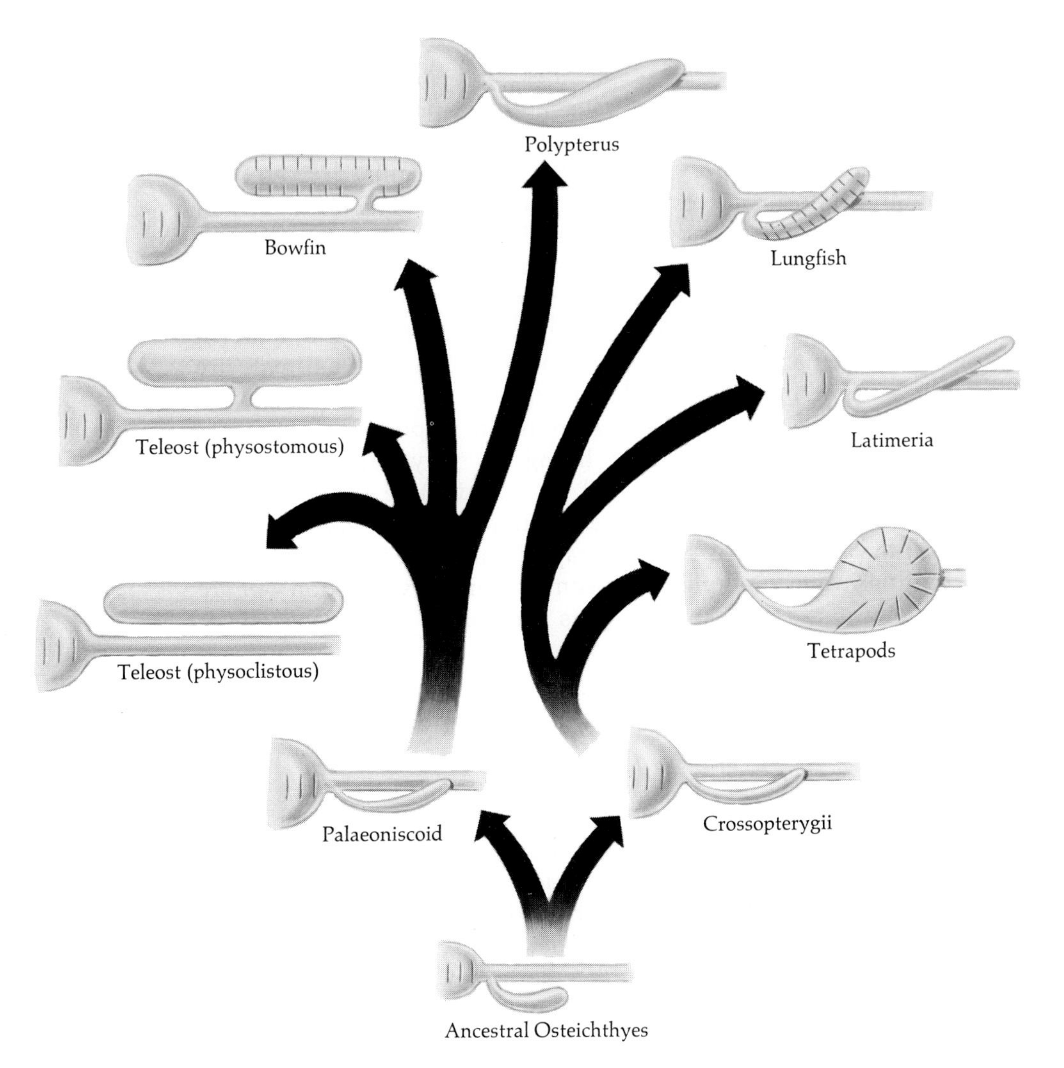

Fig. 15-11. A possible interpretation of the evolution of lungs and gas bladders in diverse fishes.

an "air bubble" lung—especially one in the ventral body cavity—would be maladaptive. Natural selection would then favor individuals with a reduced lung, or even no lung. A hydrostatic organ, on the other hand, would no doubt have been a great advantage during the remarkable adaptive radiation of teleosts. The gas bladder, then, may be neither a precursor nor a modification of the lung, but may have evolved quite independently for its own adaptive value. During subsequent radiations, as some actinopterygians came to inhabit areas with oxygen-depleted waters, the hydrostatic organ may have secondarily evolved a respiratory function, providing another uninhabited niche to colonize (Fig. 15-11).

If this be so, the histological and developmental differences between the teleost gas bladder and the lungfish and tetrapod lungs are explained. At present, however, it is still an open question whether or not gas bladders and lungs are homologous.

AMPHIBIAN RESPIRATION

Although the respiratory organ of most adult amphibians is the lung, at different times in their life cycles amphibians also utilize gills, skin, and the walls of the oropharyngeal cavity. External gills, derived from ectoderm, are the primary respiratory organs of the larval tadpoles of all amphibians. In anuran tadpoles a larval operculum covers the gills, while in salamanders and caecilians there are external gills which project directly into the water and are rhythmically moved by the branchial muscles for ventilation. External gills persist throughout life in the neotenic salamanders (e.g., *Necturus*), which gain sexual maturity while otherwise retaining larval characteristics.

In many larvae and adults the skin is an important respiratory organ. For instance, salamanders living in very rapid mountain streams, where an air bubble would be maladaptive to locomotion, have lost the lungs in the adult; all respiration is cutaneous, facilitated by the fact that the water is cold and rapidly moving. For salamanders living in quiet water the lungs have some respiratory function but are more important as hydrostatic organs similar to the teleost swim bladder.

THE AMPHIBIAN LUNG

As in all tetrapods, the amphibian lungs develop from the floor of the alimentary canal just behind the gill slit region. They are supplied with blood by derivatives of the arteries to the last gill arch and drained by the pulmonary veins which go directly to the heart.

The lungs lie within the pleuroperitoneal cavity and are covered by visceral peritoneum except anteriorly where they are attached to the anterior walls of the body cavity. The lumen of each lung is continuous, either directly or through very short bronchi, with the laryngeal chamber, a midline structure deep to the floor of the pharynx lined with muscles and small cricoid and arytenoid cartilages. The cartilages surround a slitlike opening into the pharynx, called the glottis, which is controlled by laryngeal muscles attached to the laryngeal cartilages. In most tetrapods the glottis opens in the floor of the pharynx, just anterior to the opening of the esophagus. In amphibians, however, there is a deep concavity in the floor of the oropharynx (particularly prominent in anurans), and the position of the glottis is shifted to a posterior vertical surface of this concavity, just ventral to the esophagus. During respiration the esophagus is closed by a sphincter muscle.

The internal complexity of the lung and hence the extent of its respiratory membrane vary according to order. Septae support the vascularized respiratory membrane and divide the lung into intercommunicating air spaces called infundibuli. Each infundibulum is lined by a thin epithelium with capillary beds lying deep to it in connective tissue. There are also smooth muscles just deep to the vascularized layer. In frogs and toads the septae are more complex than in salamanders and caecilians; the septae divide the lung into more and smaller air spaces (here called alveoli), giving a larger total surface area for the respiratory membrane. Toads, whose keratinized skin permits little cutaneous respiration, have the most complex alveoli and the largest respiratory membrane.

Ventilation

Among amphibians ventilation of the lung has been studied in the greatest detail in the bullfrog (Fig. 15-12). Its pressure pump mechanism is unique to amphibians, although recent studies indicate that similar respiratory movements ventilate the lungs of lungfish. The mouth remains closed during all parts of respiration, and the glottis

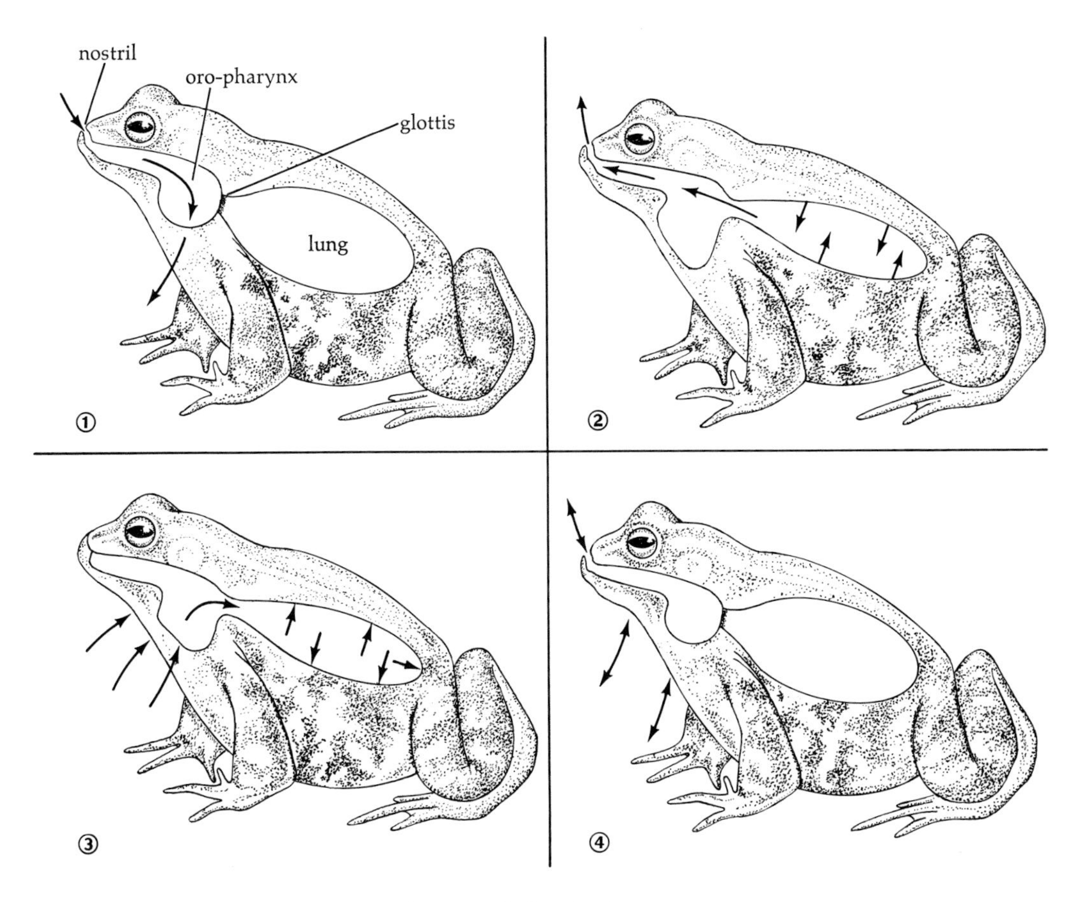

Fig. 15-12. Ventilation of the lungs in a frog. In (1) the glottis is closed; lowering the floor of the oropharynx draws air into the cavity. In (2) the glottis is opened and lung muscles contracted, forcing air out through the glottis, oropharynx, and nostrils. In (3) the nostrils are closed and the floor of the oropharynx raised, forcing air into the lungs. In (4) the glottis is closed and the nostrils opened; oscillatory movements of the oropharynx bring fresh air into this cavity. After Gans, C., DeJongh, H. J., and Farber, J. (1969). Science **163,** *1223–1225. © 1969 by the American Association for the Advancement of Science.*

remains closed except when air is actually being pumped into or out of the lung (Fig. 15-12).

The most rapid movements of respiration do not cause ventilation. They are rhythmic, oscillatory pumping movements of the floor of the oropharynx, which force air into and out of the nostrils in a tidal flow. Because the glottis is closed air moves only between the atmosphere and the oropharyngeal cavity and not past the respiratory membrane. These oscillatory movements remove carbon dioxide from the pharynx and replace it with oxygen-rich air.

Ventilation occurs in the following manner. The floor of the oropharynx is depressed deeper than it is during oscillatory movements; this draws a large amount of air down into the ventral concavity of the

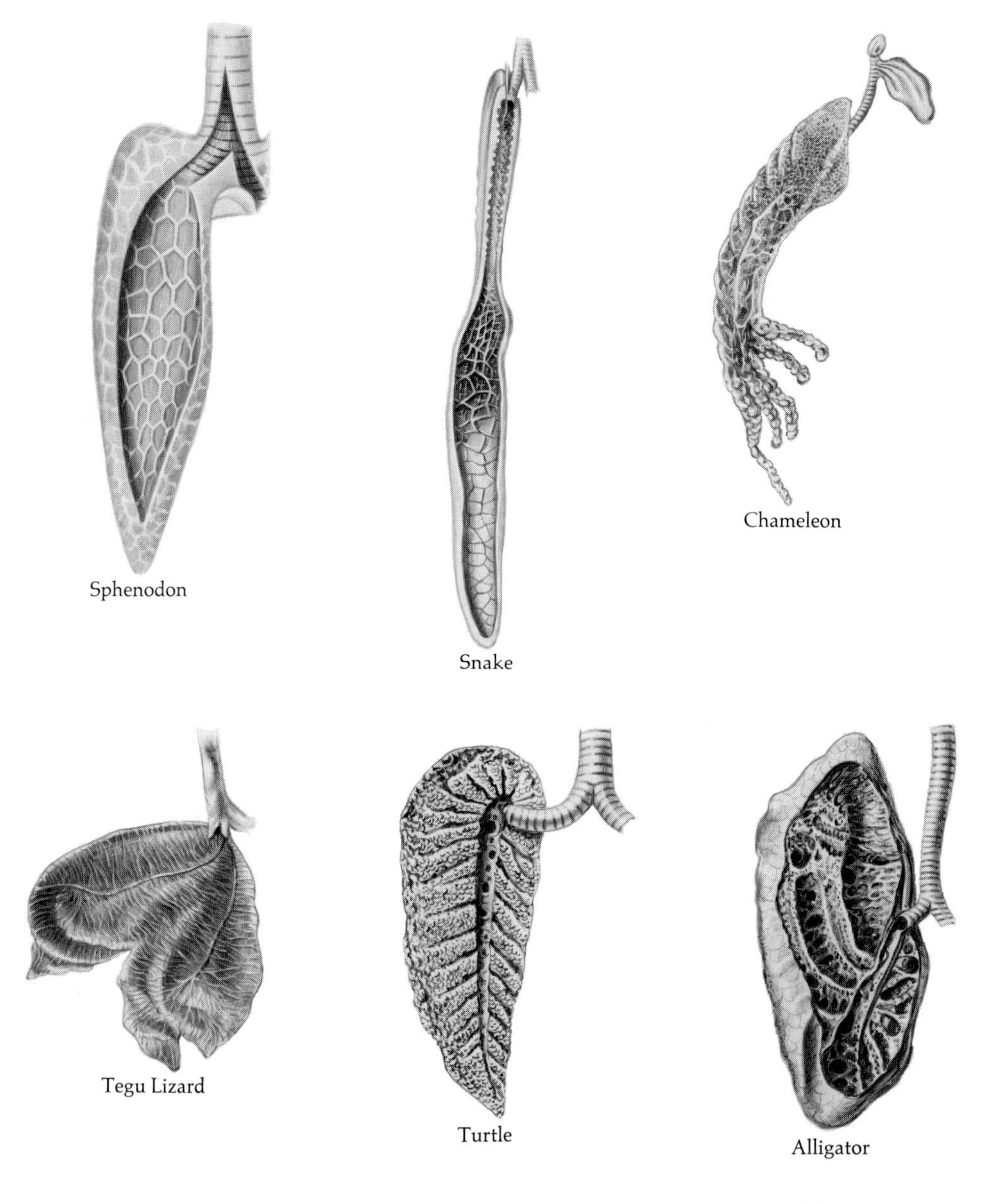

Fig. 15-13. Diversity of lung structure in reptiles.

oropharynx. The glottis is then opened and the lung musculature contracted, forcing the air already in the lungs to pass directly over the depression of the oropharynx and out the nostrils. The nostrils are then closed and the floor of the oropharynx is rapidly raised, which forces air into the lungs. Immediately then the glottis is closed and the nostrils opened, and the oscillatory movements of the oropharyngeal floor are resumed. In this way the lungs always maintain a positive internal pressure, even immediately after expiration. As will be seen, this is not the case in any amniotes, all of which use suction pumps to draw air into the lungs.

REPTILIAN RESPIRATION

Like other reptilian systems, the respiratory system is characterized by diverse morphology and hence diverse functions (Fig. 15-13). The simplest reptilian lungs are in the one living rhynchocephalian, *Sphenodon,* with but few, small septae and a simple, general form. In ophidians (snakes) the left lung is rudimentary and the right lung is long, narrow, and functional, with moderately developed septae and alveoli. The lacertilians (lizards) have a more complex lung, with definitive bronchioles coming off the trachea and leading to alveoli. In a few forms (e.g., chameleon) air sacs or thin membranes lead from the lungs into the body cavity. These structures, like similar structures in birds, supply additional air spaces but are completely unvascularized. In Chelonia (turtles), although the lungs occupy much of the body cavity, the internal foldings consist of large, separated infundibuli (or alveoli), so that the total respiratory surface area is not great. The large lungs of crocodilians have a more complex internal structure; their bronchioles and very small alveoli give a large respiratory surface area. Only in crocodilians are the pleural cavities separated from the posterior peritoneal cavity by a transverse septum.

In all reptiles air passes through either the nostrils or the mouth, then through the glottis into a distinct larynx and to the trachea, which is supported by cartilaginous rings. The trachea then bifurcates, sending a bronchus to each lung.

Ventilation

In all reptiles studied, ventilation of the lungs is caused by suction rather than by pressure, that is, air is sucked in rather than being pushed in to the lungs. However, the mechanisms of the suction pump vary considerably among reptilian groups (Fig. 15-14).

In lepidosaurians—both Rhynchocephalia and Lacertilia—the costal musculature moves the ribs and expands the thoracic region of the body cavity. Since the cavity is filled with incompressible fluid, this expansion puts negative pressure on the contained organs; this in turn expands the size of the lungs and thus draws air into them. Contraction of the abdominal muscles compresses the body cavity and helps in exhalation, which is also assisted by contractions of the lungs' intrinsic smooth musculature.

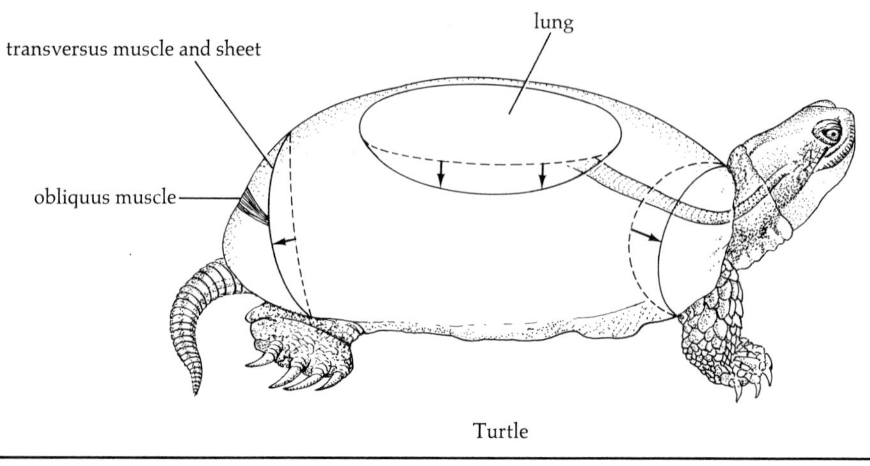

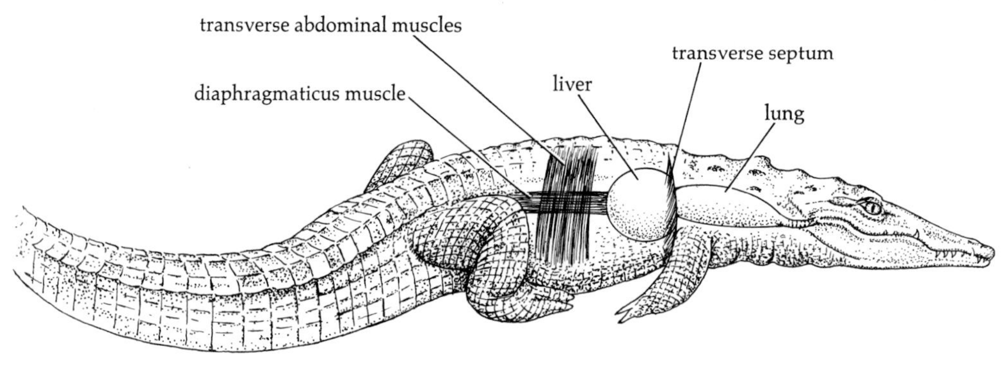

Fig. 15-14. Diagramatic representation of ventilation mechanisms in a turtle and an alligator. In the turtle the entire body cavity is expanded by moving the pectoral girdle outward and by contracting the obliquus muscle; the resulting negative pressure draws air into the lungs. Moving the pectoral girdle inward and contracting the transversus muscle compresses the body cavity, forcing air out of the lungs. In crocodilians, contraction of the diaphragmaticus pulls on the liver, which is attached to the transverse septum; this expands the pleural cavities and lungs and draws air in. Contraction of the transverse abdominal muscles pushes the transverse septum forward, reducing the size of the pleural cavities and forcing air out of the lungs. After Gans, C., (1970). Respiration in early tetrapods—the frog is a red herring. Evolution **24,** *723–734.*

Ophidians, with their long narrow bodies, have trunk musculature specializations that facilitate ventilation. A dorsolateral muscle sheet runs from the ventral surface of the vertebrae to the medial surface of the ribs. A second ventrolateral muscle sheet runs from the medial surface of the ribs to the midventral skin. Contraction of these two muscles compresses the pleuroperitoneal cavity, causing exhalation. Inhalation is caused by levator costalis muscles (one per rib). These muscles arise from the surface of each prezygapophysis and extend obliquely and caudally to insert on the anterior surfaces of the ribs. Their contractions pull the ribs anteriorly and thus into a more vertical position. This expands the pleuroperitoneal cavity, creating negative pressure in the lungs so that air is inhaled.

In chelonians most of the trunk is covered by the bone of the carapace and plastron, which cannot expand or contract to change the pressure in the body cavity (Fig. 15-14). However there is a transversus muscle attached to an externally convex connective tissue sheet; this sheet lies just deep to the skin over the openings for the hindlimbs and tail. When the transversus muscle is contracted it flattens the sheet and thus reduces the volume of the body cavity. This forces air out through the glottis. Its antagonist, the obliquus, pulls in the other direction and increases the curvature of the skin; it therefore acts as a suction pump drawing air back into the lungs. An analogous situation exists at the forelimbs. The pectoral girdle is completely contained within the shell and is loosely articulated with it. By withdrawing the limb and pulling in on the pectoral girdle the volume of the body cavity is reduced and air is forced out. The opposite movements expand the body cavity and fill the lungs.

In crocodilians there is an even more unusual condition (Fig. 15-14). Running from the pelvic girdle and inserting on the liver is a muscle called the diaphragmaticus (although it has no relationship to the mammalian diaphragm). The liver is attached to the transverse septum, which separates the pleural from the peritoneal cavity. Contraction of the diaphragmaticus pulls both the liver and the transverse septum caudally, compressing the peritoneal cavity and expanding the pleural cavities and thus drawing air into the lungs. The transverse abdominal muscles are its antagonists. These muscles compress the peritoneal cavity, which pushes the transverse septum forward and promotes exhalation. These movements are augmented by coordinated rib movements caused by contractions of the intercostal muscles.

AVIAN RESPIRATION

Because of their high metabolic rate (highest of any vertebrate group) birds require large amounts of oxygen and a rapid and efficient exchange of oxygen and carbon dioxide. They have evolved the most extraordinary respiratory apparatus (Fig. 15-15).

The lungs lie dorsally against the ribs in small pleural cavities. They contain all of the respiratory membrane, and it is here that all gas exchange takes place. They are nevertheless small and compact and but a fraction of the total respiratory system.

From the oropharynx air passes through a glottis, into a relatively simple larynx with laryngeal cartilages, and down through a cartilage-supported trachea. The voice box is not the larynx, as it is in other tetrapods, but the syrinx, which is located at the base of the trachea where the trachea bifurcates into two primary bronchi which lead to the lungs.

When a primary bronchus enters the lung it is called a mesobronchus, and as such it continues directly through the lungs, giving

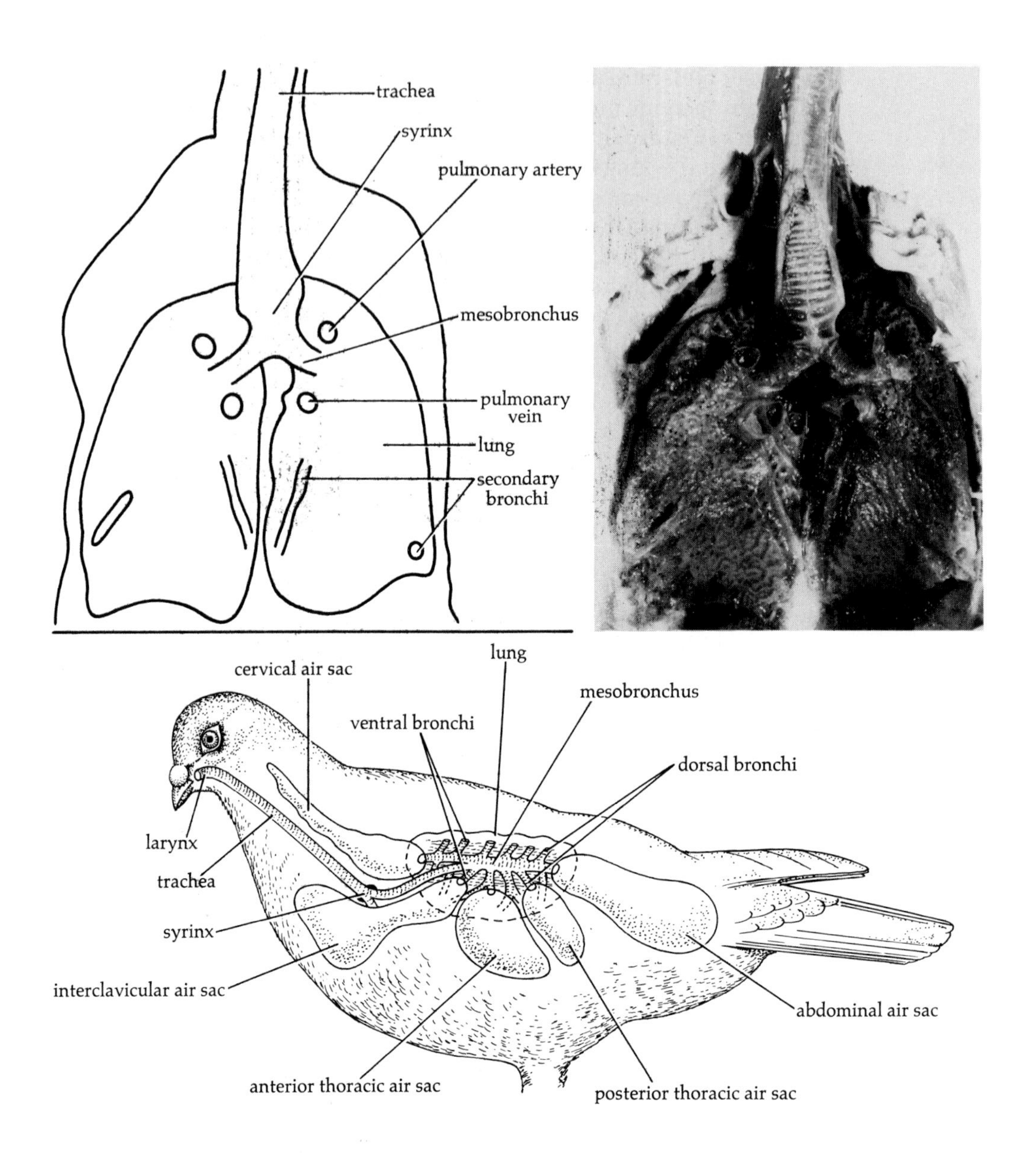

Fig. 15-15. Top: a dissection of a pigeon's respiratory system after most viscera and the air sacs have been removed. The tiny holes and tubes in the lungs are parabronchi. Bottom: Diagram showing the gross relationships of the parts of a bird's respiratory system. Note the extensive air sac system.

off secondary bronchi on its way. The secondary bronchi are of two types: posteriorly they are called dorsal bronchi, although some of them come off ventrally, and anteriorly they are called ventral bronchi, which may come off dorsally—thus adding unnecessary confusion to the terminology. Some secondary bronchi terminate in thin-walled air sacs, but all send branches into the lungs.

The dorsal bronchi and the mesobronchus communicate with the more posterior air sacs: these are the paired abdominal air sacs and the paired posterior thoracic air sacs. The ventral bronchi communicate with anterior air sacs: these are the paired cervical and anterior thoracic air sacs and the midline interclavicular air sac. These air sacs, whose general positions are indicated by their names, have little vascularization and are not the site of gas exchange. They are extra air spaces necessary for the complicated process of ventilation, to be described. Diverticuli of these air sacs enter all portions of the body except the head and legs, sometimes including the hollow bones.

In the lungs are tertiary bronchi, called parabronchi, which interconnect dorsal and ventral bronchi and facilitate gas exchange. Along their entire length are tiny diverticuli, called air capillaries, and surrounding them a network of fine blood capillaries. This arrangement puts the air capillaries into extremely close apposition to the bloodstream (Fig. 15-16), and it is here that gas exchange occurs.

Ventilation

Inhalation involves an aspiration pump analogous to that of reptiles. Contractions of the external intercostal, subcostal, scalene, and costal levator muscles straighten the ribs, which in birds are jointed midway between their vertebral and sternal connections. This straightening moves the sternum and procoracoid anteriorly and ventrally and ex-

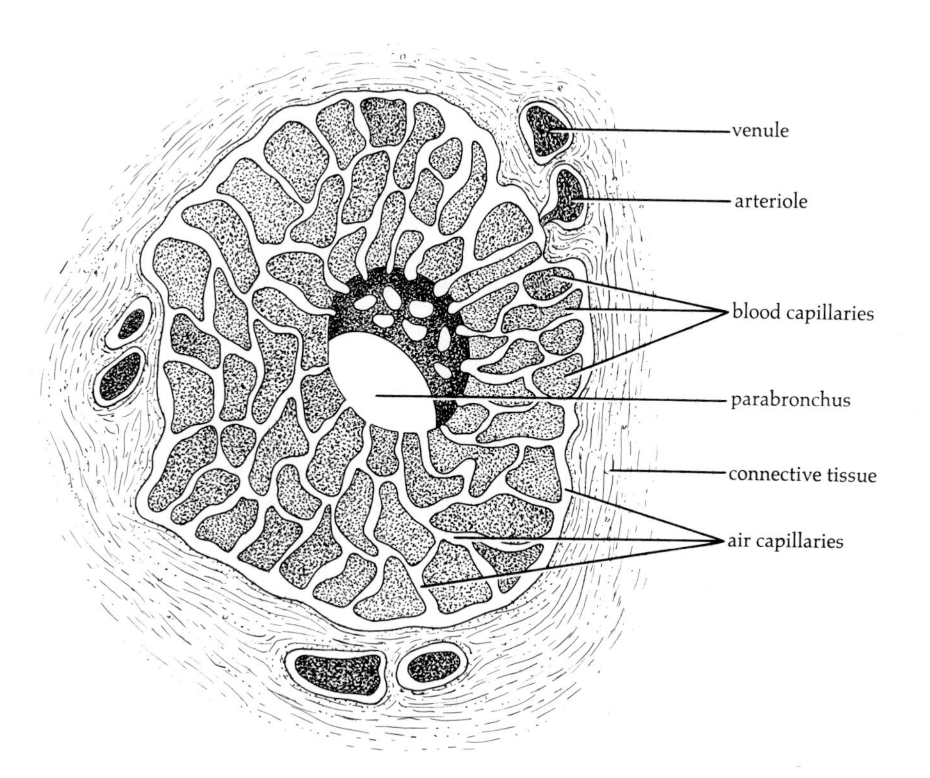

Fig. 15-16. Section through a small segment of a parabronchus in the avian lung.

pands the thoracic cavity; this creates suction and draws in air, most of which passes into the posterior air sacs. At the same time that inhalation draws air into the body, other air is, surprisingly, forced *out* of the lungs and into the anterior air sacs. Lung volume is changed by muscles attached to the oblique septum. Contraction of these muscles pulls on the oblique septum, producing a negative pressure around the lungs and, therefore, causing the lungs to expand. When these muscles are relaxed the lungs' elasticity causes the opposite movement (Fig. 15-17). Thus inhalation causes some air to move into the posterior air sacs and dorsal bronchi while other air moves out of the lungs (into the anterior air sac.

During exhalation those processes are reversed. The internal intercostals contract, as do the abdominal muscles and the muscles tensing (flattening) the oblique septum. This puts positive pressure in the abdominal and thoracic cavities and they contract. The lungs, however, are under negative pressure and expand. The air in the posterior air sacs moves into the lungs and that in the anterior air sacs passes out the ventral bronchi to the mesobronchus and through the trachea to the outside. During flight, when more oxygen is required for flight muscles, the respiratory rate is increased, but the rate is usually a whole number of respiratory cycles per wing beat, indicating coordination between the two.

The lung itself is such a small part of this system that at any instant it contains only about 4% of the air in the system. The posterior air sacs provide a reserve supply of air that is high in oxygen and low in carbon dioxide, while the anterior air sacs act as collection and disposal depots for air that is low in oxygen and high in carbon dioxide. This system is the most efficient that has evolved among tetrapods. It allows a constant movement of air through the parabronchi and air capillaries. Furthermore, this movement is in one direction only, which makes possible a countercurrent flow (of air versus blood) for efficient gas exchange analogous to that in fish gills. The fact that this system contains no dead air space along the respiratory membranes also makes it more efficient than the tidal-flow mechanisms of other tetrapods.

MAMMALIAN RESPIRATION

The high metabolic rates of mammals also require large respiratory membranes and rapid, efficient ventilation. The mammalian lung is structurally more similar to that of reptiles than of birds. It is basically

Fig. 15-17. Ventilation mechanisms in birds. Above are shown the muscles involved in inspiration and the movement of air through lungs and air sacs; below are shown the movements and air flow during expiration. Note that oxygen-rich air passes the respiratory membranes during both inspiration and expiration.

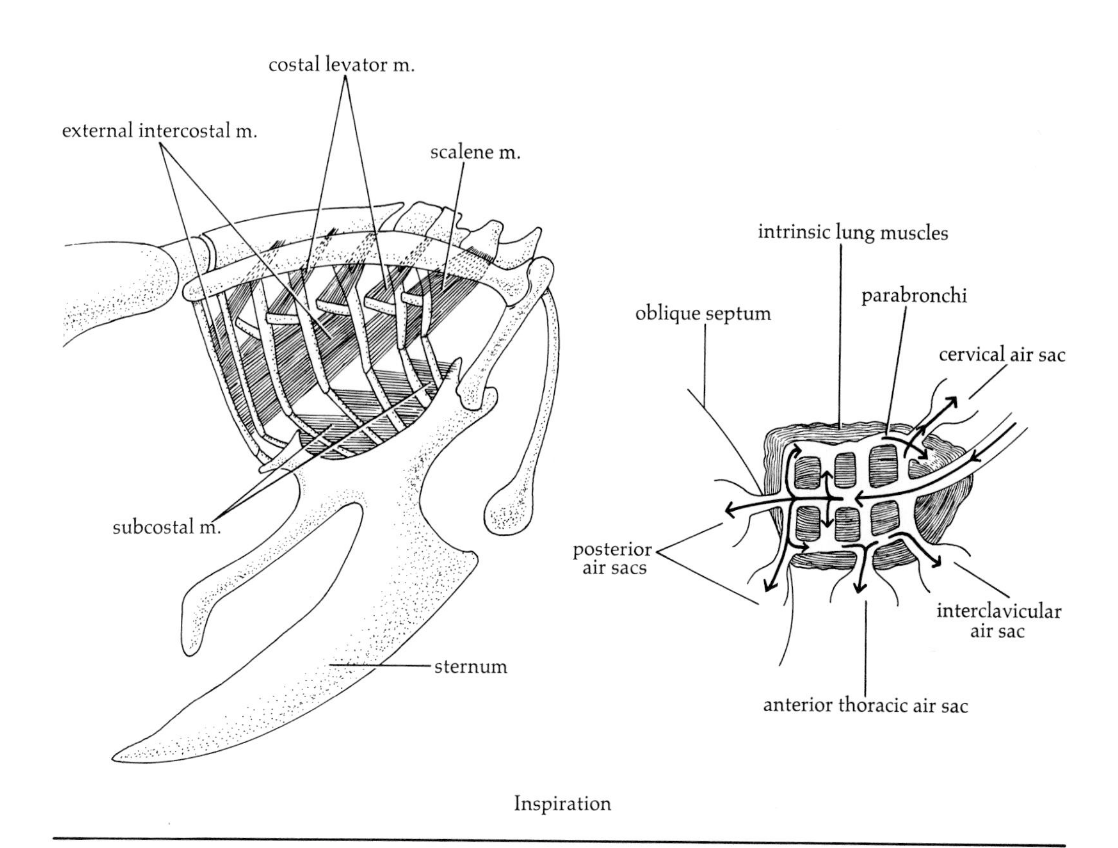
costal levator m.
external intercostal m.
scalene m.
subcostal m.
sternum
intrinsic lung muscles
parabronchi
oblique septum
cervical air sac
posterior air sacs
interclavicular air sac
anterior thoracic air sac

Inspiration

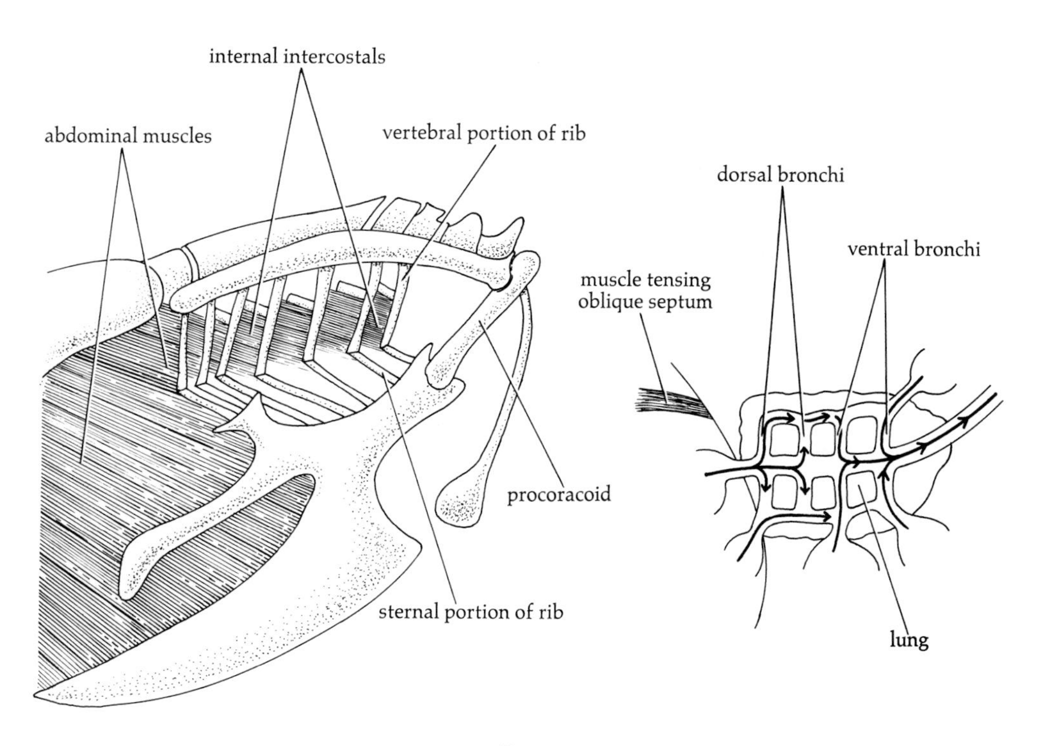
internal intercostals
abdominal muscles
vertebral portion of rib
procoracoid
sternal portion of rib
dorsal bronchi
ventral bronchi
muscle tensing oblique septum
lung

Expiration

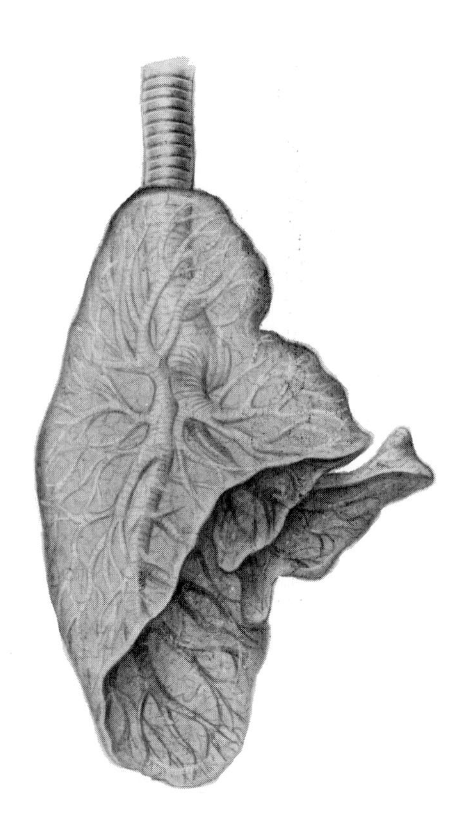

Fig. 15-18. Mammalian lungs have an extremely complex bronchial tree extending small branches to the alveoli.

an alveolar lung in which each entering bronchus divides into smaller and smaller bronchioles which finally terminate in microscopic alveoli (Fig. 15-18). These alveoli are so small that they would be collapsed by the fluid moistening the respiratory membranes were it not that this fluid, unlike water, has an extremely low surface tension.

Ventilation

As is inevitable with this type of structure, there is a tidal rather than a continual flow of air past the membrane and a good deal of dead air in the lung. Ventilating movements must therefore be powerful and rapid; these movements are produced by the unique mammalian structure, the diaphragm.

The diaphragm, a large sheet of muscle derived from cervical myotomes, fills the septum separating the pleural and peritoneal cavities. It is peripherally attached to the lower ribs and the vertebral column, with its fibers converging to insert on a broad central tendon of the diaphragm. When relaxed, the diaphragm arches anteriorly; when contracted, it flattens and thus increases the space within the pleural cavities, causing a partial vacuum which draws air into the lungs. The intercostal muscles also help enlarge the pleural cavities, but in most mammals it is the diaphragm that plays the major role. Expiration is caused by the relaxation of the diaphragm, the elasticity of the lung tissues, and the action of the internal intercostal muscles and abdominal muscles.

The mammalian respiratory system, therefore, while large, bulky, and inefficient, is nevertheless powerful. Many mammals have great speed and endurance, supported in part by this system.

SOUND PRODUCTION

In vertebrates sound is usually produced by modifications of the respiratory system: air is forced past a constriction and certain structures are set into vibration. Frequently there are also resonators, which are more peripheral structures that amplify the vibratory energy.

Many fishes produce various sounds. Some teleosts, called grunters, produce sound by forcing air out through the pneumatic duct from the gas bladder. Others, called drumming fishes, rapidly contract certain muscles which strike the gas bladder causing it to vibrate and, because of its volume, to resonate.

Reptiles, and among amphibians both caecilians and salamanders, are largely mute; anurans however have powerful voices and vocalization plays a particular role in their mating behavior. On each side of the glottis, arytenoid and cricoid cartilages support the vocal cords, which are ligaments in the laryngeal chamber covered by tissue folds. When these ligaments are held taut by the muscles of the pharynx, air forced out of the lung sets them into vibration; the sound is amplified by the large resonating chamber provided by the oropharyngeal cavity.

Both mammals and birds produce a variety of sounds of great significance in their social behavior. Their voice boxes are quite different: in

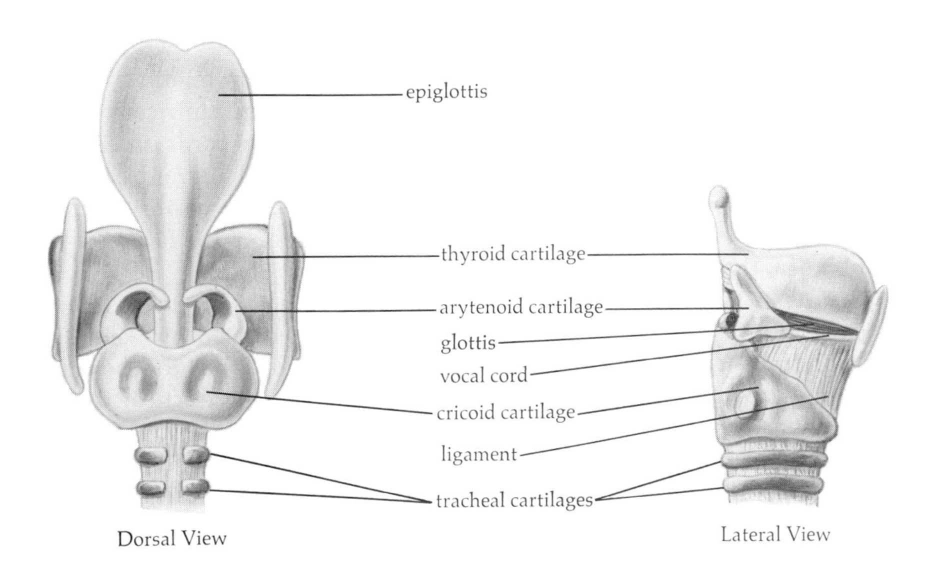

Fig. 15-19. The mammalian larynx with muscles and most ligaments removed. In the lateral view, the epiglottis and right half of the thyroid cartilage are removed.

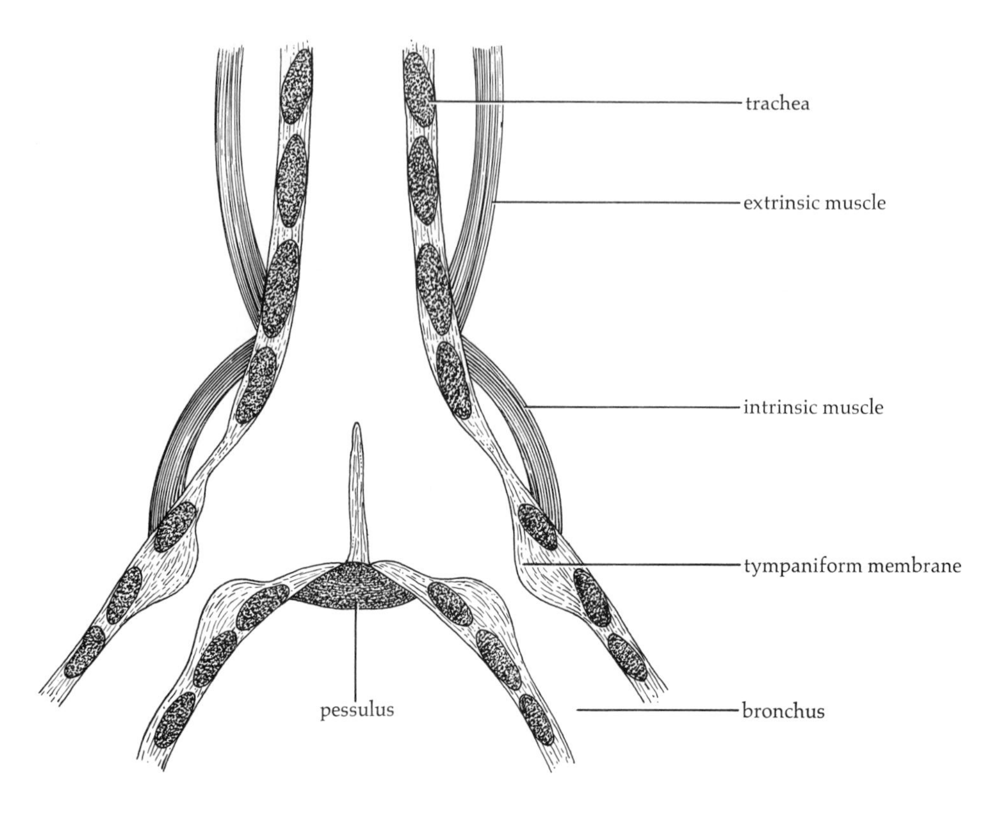

Fig. 15-20. A section through a generalized avian syrinx.

mammals it is the larynx, and in birds the syrinx (at the point where trachea and bronchi join).

The mammalian larynx is composed of the paired arytenoid cartilages, the cricoid cartilage, the (larger) thyroid cartilage, intricate intrinsic laryngeal musculature, and the ligamentous vocal cords on each side of the glottis running between the arytenoid and thyroid cartilages (Fig. 15-19). When these vocal cords are tensed, air passing them produces sounds, which are modified by the structure of the larynx itself and by the mouth and tongue.

The avian syrinx is even more complex. The opening between the trachea and the bronchi on each side is partially constricted by a small tympaniform membrane. A third membrane extends from the point between the two bronchi directly into the lumen of the trachea; this membrane is usually supported by a small bone called the pessulus (Fig. 15-20). Tension of parts of the syrinx can be minutely controlled by a group of variable, always complex, extrinsic and intrinsic syringeal muscles. When contracted these muscles place the membranes under sufficient tension to produce sounds, whose frequencies are varied by the amount of tension. The large dilations which are often found in the syrinx walls are resonating chambers. With this structure, birds produce their remarkable variety and range of songs.

SUGGESTED READING

Berger, M., Roy, O. Z., and Hart, J. S. (1970). The co-ordination between respiration and wing beats in birds. *Z. Vergl. Physiol.* **66,** 190–200.

Gans, C. (1970). Respiration in early tetrapods—the frog is a red herring. *Evolution* **24,** 723–734.

Hughes, G. M. (1963). "Comparative Physiology of Vertebrate Respiration." Harvard Univ. Press, Cambridge, Massachusetts.

Johansen, K. (1971). Comparative physiology: gas exchange and circulation in fishes. *Ann. Rev. Physiol.* **33,** 569–612.

Johansen, K., Hanson, D., and Lenfant, C. (1970). Respiration in a primitive air breather, *Amia calva. Resp. Physiol.* **9,** 162–174.

Johansen, K., Lenfant, C., and Hanson, D. (1970). Phylogenetic development of pulmonary circulation. *Fed. Proc.* **29,** 1135–1140.

King, A. S. (1966). Structural and functional aspects of the avian lung and air sacs. *Int. Rev. Gen. Exp. Zool.* **2,** 171–267.

Rahn, H., Rahn, K. B., Howell, B. J., Gans, C., and Tenney, S. M. (1971). Air breathing of the garfish (*Lepisosteus osseus*). *Resp. Physiol.* **11,** 285–307.

Scheid, P., and Piiper, J. (1971). Direct measurement of the pathway of respired gas in duck lungs. *Resp. Physiol.* **11,** 308–314.

Schmidt-Nielsen, K. (1971). How birds breathe. *Sci. Amer.* **225**(No. 6), 72–79.

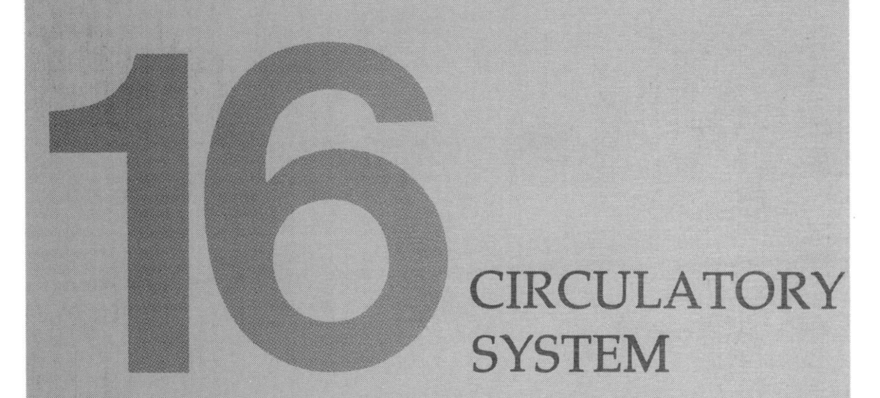

16 CIRCULATORY SYSTEM

16

The major morphological units of the circulatory system are the blood vessels and the heart; its chief functional unit is the blood itself. With these components, the circulatory system integrates all of the body's metabolic activities.

This system coordinates the body's chemical actions and reactions. It helps to maintain a constant water and ion content in interstitial fluids and thus helps to keep the animal osmotically independent of its environment. It carries and transmits hormones from the endocrine glands to their various target organs.

The blood carries oxygen, which is necessary for the production of energy. Most of the oxygen is attached to hemoglobin molecules in red blood cells, forming oxyhemoglobin. In an area relatively high in carbon dioxide the oxyhemoglobin gives up its oxygen, which diffuses into the tissues. Oxidative metabolism occurs in the tissues and carbon dioxide is carried away in the plasma, mostly as a bicarbonate ion.

The circulatory system is of major importance in maintaining a constant body temperature in homeotherms. Similar mechanisms function in poikilotherms. Body heat—most of it produced by muscular contraction—is distributed by way of the circulatory system. When there is an excess of heat, for instance during muscular exertion, some of the superficial blood vessels dilate and thus bring blood nearer the surface where heat may be dissipated to the environment. When the external environment is very cold the superficial blood vessels constrict; little blood then approaches the surface but is circulated instead in the deeper parts of the system, preserving essential heat (Fig. 16-1). In aquatic mammals and birds of the Arctic and Antarctic the temperature of the extremities is kept just above freezing by the heat of arterial blood. Returning venous blood is rewarmed, often by being wrapped about by the outgoing arteries, before entering the deeper parts of the body.

The circulatory system carries food from the digestive system to all parts of the body and waste materials back to the excretory organs. The system can spread disease by carrying infective microorganisms throughout the body, and it can combat disease by both producing and transmitting antibodies.

All these functions are performed by the blood itself—a fluid connective tissue composed of red and white blood cells suspended in a liquid matrix called plasma. In comparative morphology, however, our major attention is directed to the structure and function of the vessels and heart.

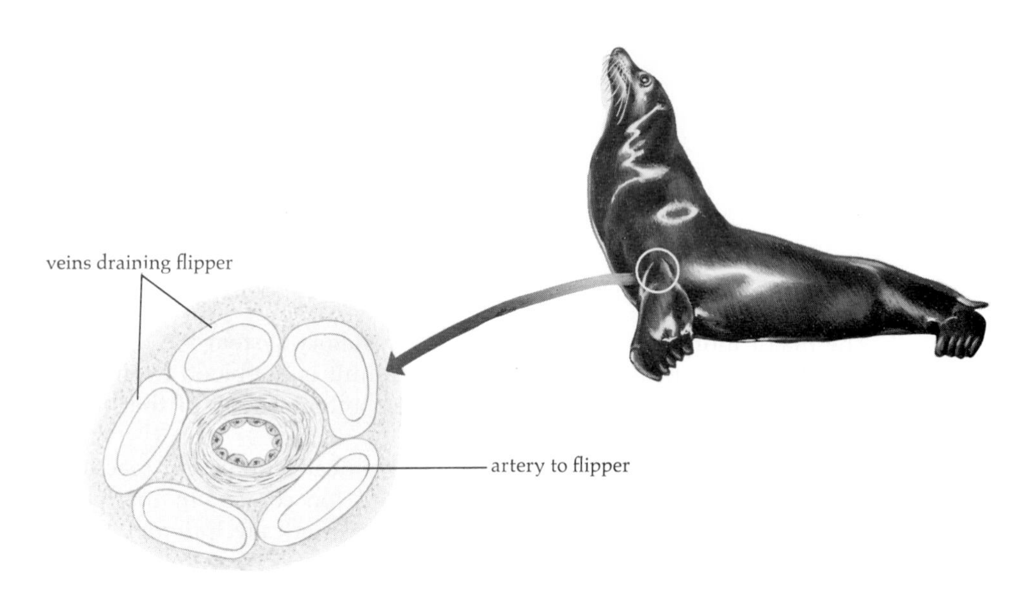

Fig. 16-1. The broad, flat flipper of a seal is an adaptive specialization for swimming but also presents a large surface area for heat loss to the cold water. This potential metabolic waste is minimized by a countercurrent heat distribution which keeps the flipper significantly cooler than the rest of the body. Veins draining the flipper wrap around the artery that supplies the flipper. This arrangement causes the venous blood returning to the body to be warmed and the arterial blood going to the flipper to be cooled.

FORMATION OF BLOOD (HEMOPOIESIS)

The first blood cells and blood vessels in the developing embryo form from mesodermal blood islands which lie along the mesoderm bordering the yolk sac. These blood islands give rise to an anastomosing (interconnecting) network of blood vessels and to the earliest hemopoietic tissue whose mesenchymal cells differentiate into hemocytoblasts, which form the blood cells (Fig. 16-2). In the older embryo and the adult, hemopoiesis is taken over by a variety of organs including liver, kidney, spleen, lymph nodes, intestinal submucosa, and bone marrow; hemopoiesis continues throughout life. "Worn out" blood cells are also continually destroyed, primarily in the spleen, and secondarily in the kidney and liver.

In adult vertebrates, red blood cells (erythrocytes) and granular white blood cells (granular leukocytes) are produced in myeloid tissue, and agranular white cells (agranular leukocytes) in lymphoid tissue. In different vertebrate groups these tissues may be either physically separated from one another or part of the same organ.

In adult teleosts, erythrocytes form mainly in the kidneys and spleen, granular leukocytes form mainly in the intestinal submucosa and the spleen, and agranular leukocytes form in several mesodermally derived tissues. In the garpike and bowfin, which are more generalized actinopterygians, hemopoiesis also occurs within skeletal tissues. In most amphibians, leukocytes (both granular and agranular) are formed in bone marrow, and most erythrocytes are formed in the spleen. The situation is variable in reptiles; for instance, the spleen is the primary site of hemopoiesis in the horned toad, and the bone marrow is the primary site in other lizards.

In birds and mammals nearly all agranular leukocytes are produced in lymph nodes and spleen, and nearly all erythrocytes and granular leukocytes are produced in red bone marrow. (Yellow marrow is mainly for deposition of fat.) However, under duress, other tissues will produce blood cells. In response to anemia or excessive blood loss, yellow marrow is converted to red marrow and the spleen, liver, kidneys, and intestines can all start producing red and agranular white blood cells. It is apparent that the potential for hemopoiesis is present in a great variety of mesodermally derived tissues, but under normal circumstances this potential is expressed by only a few.

STRUCTURE OF BLOOD VESSELS

The blood vessels of all vertebrates are structurally similar (Fig. 16-3). The innermost layer, called the tunica intima, is made up of squamous endothelial cells lying on a thin basement membrane which may contain elastic fibers as well as reticular fibers. Superficial to the tunica intima is a muscular layer, the tunica media; in addition to its smooth muscle cells the tunica media contains connective tissue cells and elastic fibers, whose numbers vary in different vessels. Superficial to

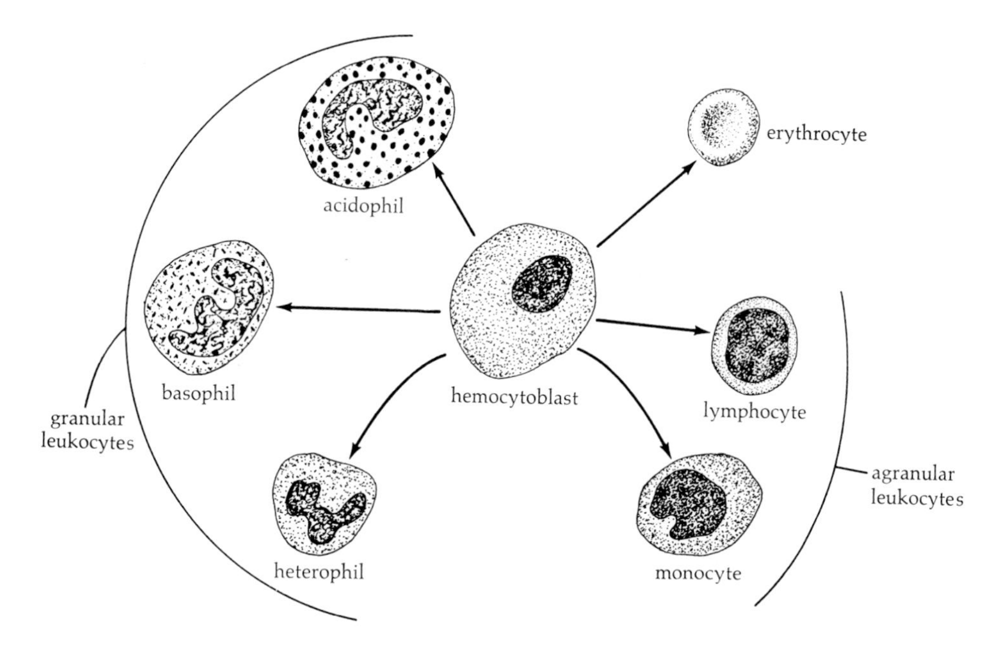

Fig. 16-2. The varied types of blood cells all differentiate from hemocytoblasts.

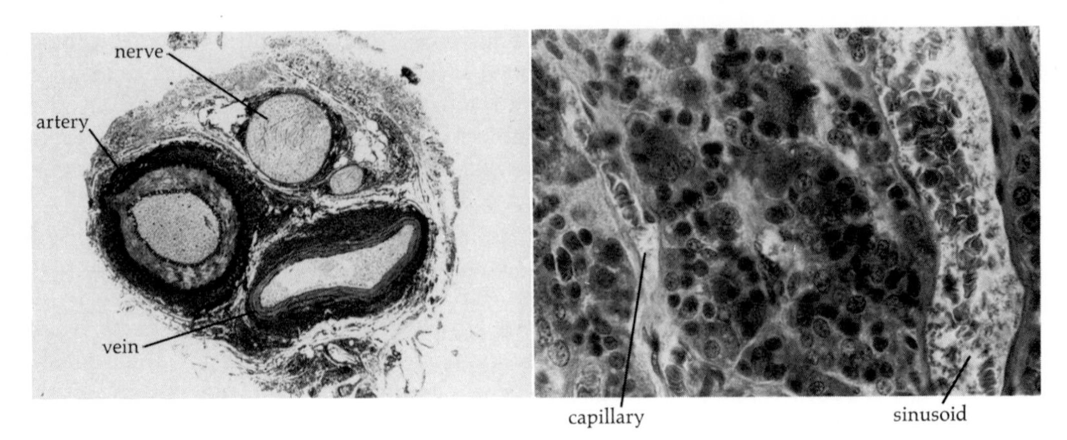

Fig. 16-3. On the left is a photomicrograph of a large artery, a vein, and a nerve, magnified 5 times. On the right is a photomicrograph of the pituitary, showing a capillary and a sinusoid, magnified 500 times.

the tunica media is the outermost coat, the tunica adventitia; it is composed of loose connective tissue and some elastic fibers. The outer portion of the tunica adventitia is continuous with the surrounding connective tissue.

Blood vessels are arteries or veins, or diminutives of these (arterioles and venules), or capillaries. Arteries and veins differ structurally, first in the tunica media which is very thin in veins and very thick in arteries. In both arteries and veins, the tunica media is innervated by autonomic nerves which influence the degree of contraction and thus (within limits) the diameter of the lumen. When arteries and veins run side by side, however, the vein is larger both in its external diameter and in its lumen diameter.

A second major structural difference is that veins contain valves at irregular intervals, which allow the blood to flow only toward the heart. Arteries have no valves; the blood is propelled by muscular action and cannot backflow.

As arteries run toward capillary beds they branch profusely and become smaller in diameter, being then called arterioles. Closer to the capillary bed their smooth musculature is thinned and eventually both the tunica adventitia and the tunica media are lost; only endothelial cells and an incomplete basement membrane form the structure of the capillary bed. The smallest veins into which the blood flows upon leaving the capillaries are called venules; several venules then join together to form a vein.

The lumen of most capillaries is just large enough to pass a single blood cell at a time, similar to the situation in fish gills. Although this may seem awkward, it is, as in the gills, adaptive: it gives a maximum surface area, and it greatly slows the blood (by friction) and reduces its pressure. Both these factors facilitate the diffusion of substances between the blood and the interstitial fluids.

Because the blood passes much more slowly in capillaries than in arteries (by a factor of 300 compared to the mammalian dorsal aorta), the capillary beds must be capable of handling more volume than the arteries. It has been estimated that a man's capillaries, if untangled and laid end to end, would reach around the world two and a half times. The complexity and extent of the capillary network is difficult to comprehend.

In the liver and certain other organs the capillaries have a much larger diameter and are called sinusoids, rather than capillaries. Their walls are still formed only of endothelial cells and an incomplete basement membrane.

In general the capillary network lies between arterioles and venules. However, in gills it usually lies between two sets of arteries, and in portal systems, such as those of kidney, liver, and hypothalamus, it lies between two sets of veins.

There are frequent short-cut shunts, or anastomoses, between arterioles and venules (Fig. 16-4). If these anastomoses are open the blood bypasses the capillary network almost entirely; if they are closed, the entire blood flow is forced through the capillary bed. Anastomotic vessels also frequently occur between adjacent arteries or adjacent veins. Such anastomoses allow for collateral circulation, when, for instance, a vessel is blocked or injured.

VERTEBRATE HEART

The heart is the primary pump for this complex system. As might be expected the vertebrate heart is developmentally and structurally an extremely modified blood vessel. Endothelium (here called endocardium), comprising a compact tunica intima, lines its lumen. Its tunica adventitia, which frequently contains adipose tissue, is covered by the visceral pericardium of the pericardial cavity. Its tunica media

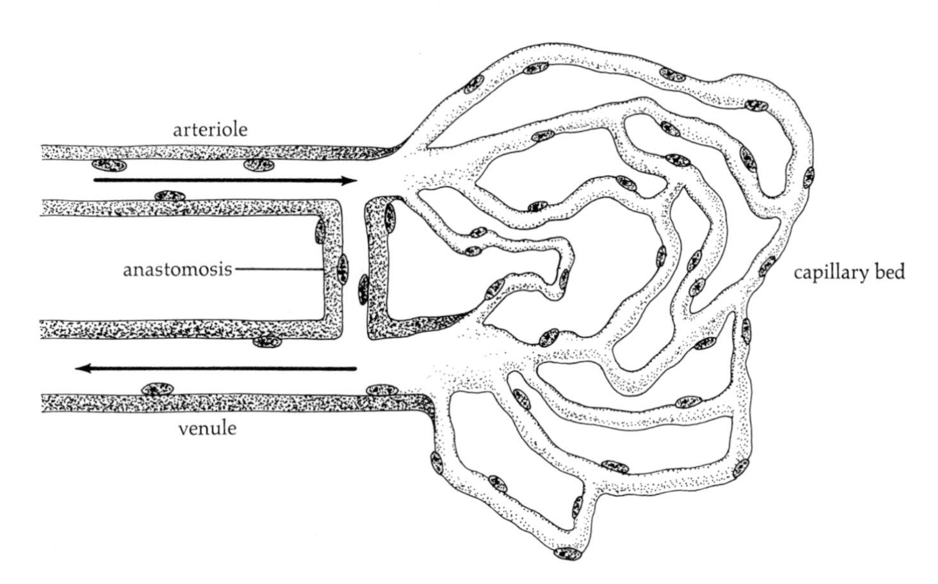

Fig. 16-4. Diagram of a capillary bed with an arteriole–venule anastomosis.

is greatly thickened and is the most modified of its parts. Instead of smooth muscles the tunica media contains striated cardiac muscles with intercalated discs as described in Chapter 6. These cardiac muscles, collectively called the myocardium, comprise the great bulk of the vertebrate heart.

The rate of contraction of the myocardium is regulated by the vagus nerve; parasympathetic innervation causes the rate to decrease and sympathetic innervation causes it to increase. Contraction always begins in the sinus venosus or its homologue, where the systemic veins (i.e., veins draining all of the body except the lungs) come together before entering the heart. This is followed by contraction of the atrium (or atria, if there are two) and then contraction of the ventricle(s). The ventricular contraction begins at the point furthest from where the vessels leave the heart, and continues smoothly toward the openings into the arteries or, in fish, to an anterior chamber of the heart called the conus arteriosus.

These progressive cardiac muscle contractions are controlled and coordinated by specialized, elongated neuromuscular cells called Purkinje fibers. There is a group of these cells in the sinus venosus or its homologue, called the sinoatrial node or pacemaker; this initiates each heart contraction by activating the atrial musculature and conducting the excitation to a second group of Purkinje fibers, the atrioventricular node, which lies between the atrium and the ventricle.

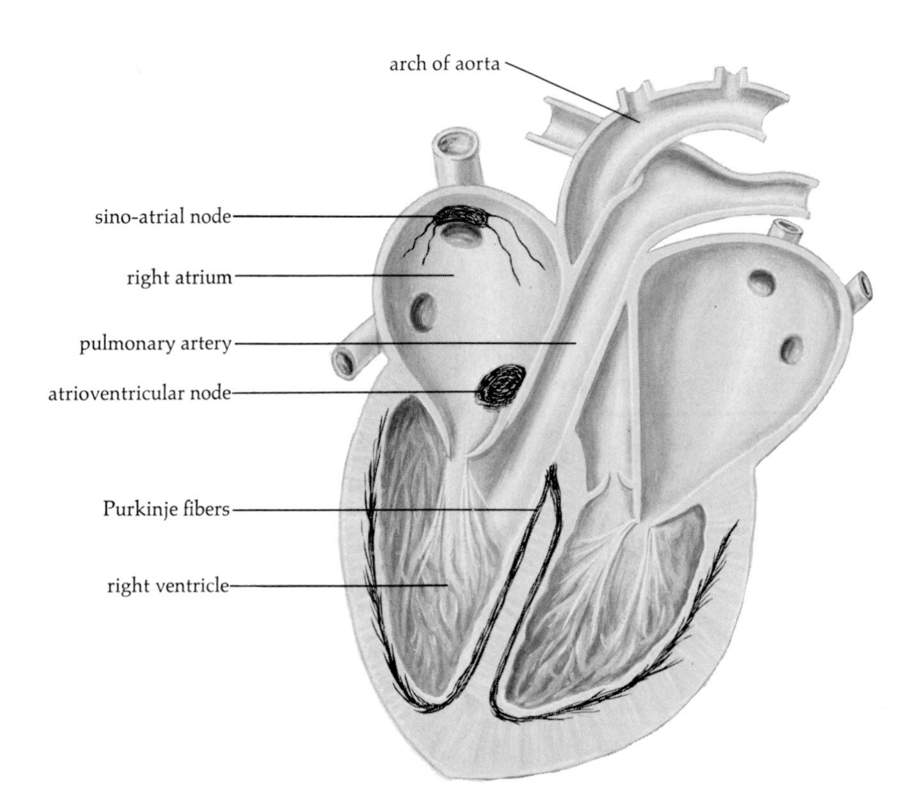

Fig. 16-5. The specialized neuromuscular coordinating and conducting system, as seen in a sectioned mammalian heart.

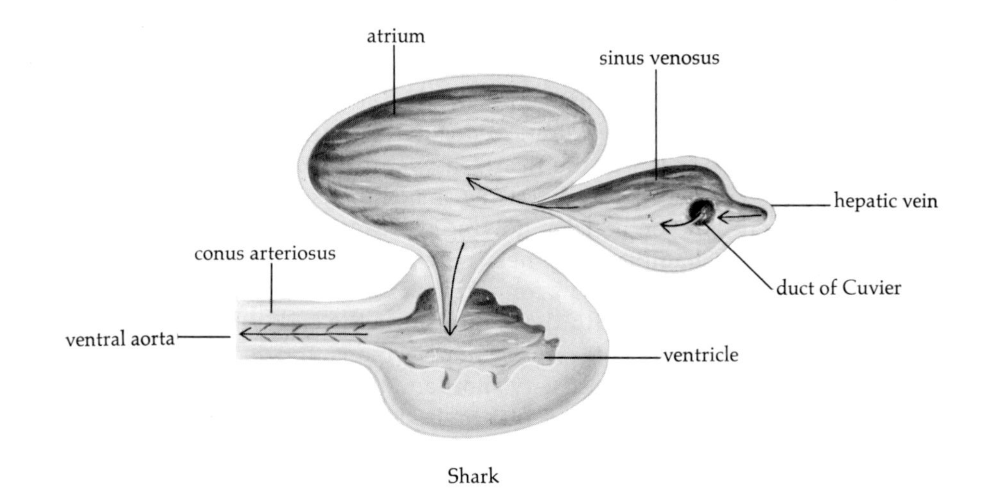

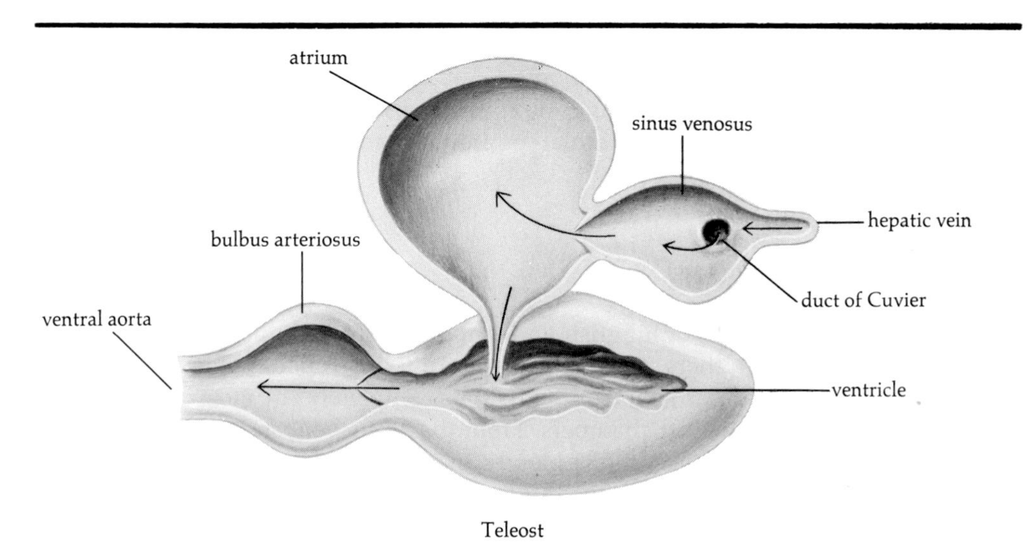

Fig. 16-6. Sagittal sections through the hearts of a shark and a teleost. Arrows indicate direction of blood flow.

From the atrioventricular node other Purkinje fibers spread into the ventricles and cause them to contract (Fig. 16-5).

FISHES (EXCEPT LUNGFISH)

In all fishes except sarcopterygians the heart is composed of a folded tube having thick muscular walls and divided into four chambers (Fig. 16-6). At the posterior part of the pericardial cavity is its thinnest-walled portion, the sinus venosus. Venous blood from all parts of the body enters the sinus venosus through paired common cardinal veins (ducts of Cuvier) and paired hepatic veins. Here it is under relatively low pressure; in fact, it is drawn into the sinus venosus partly by the negative pressure of the pericardial cavity.

When the sinus venosus contracts, blood is forced into the next chamber, the atrium, which is larger and more muscular although still relatively thin walled. When the atrium contracts, the blood, prevented by a valve from passing back to the sinus venosus, is forced into the next and thickest walled portion of the heart, the ventricle. When the ventricle's thick walls contract it produces the major pressure in the system. The blood, again prevented from flowing backward by a valve, is forced out to the fourth chamber of the heart, the conus arteriosus. In elasmobranchs the conus arteriosus is a long, tubular, muscular structure that is continuous anteriorly with the ventral aorta through which the blood passes on its way to the gills. In actinopterygians, the ventral aorta and conus arteriosus together are very short and bulb shaped and are usually called the bulbus arteriosus. In both elasmobranchs and actinopterygians, semilunar valves on the walls of the conus arteriosus prevent the return of blood to the ventricle.

These chambers are not laid out linearly but are folded one upon another in an S-like configuration (Fig. 16-6). This makes the heart more compact, so that it fits into a relatively small pericardial cavity. In chondrichthyeans, the pericardial cavity is connected with the pleuroperitoneal cavity by a pair of pericardial–peritoneal canals which are not present in teleosts.

LUNGFISH AND AMPHIBIANS

In many respects the hearts of lungfish and amphibians are similar to those of most fish, but there are some significant differences. The atrium is divided, either partially (in lungfish) or completely (in most amphibians) into right and left chambers. From the sinus venosus the right atrium receives systemic venous blood—that is, blood from the veins draining all of the body except the lungs, which is, therefore, high in carbon dioxide and low in oxygen. The left atrium receives blood from the lungs via the pulmonary vein, with a high oxygen and low carbon dioxide content (Fig. 16-7).

The walls of both atria and especially of the ventricle are more heavily muscular than those of most fishes, with heavy folds (trabeculae) which are more pronounced in amphibians. In both lungfish and amphibians a spiral fold in the conus arteriosus separates it into two portions along much of its length. In lungfish a partial septum incompletely divides the ventricle into two portions.

From the gross study of a dead heart one would think that the oxygenated blood from the left atrium and the unoxygenated blood from the right atrium would mix in the ventricle, giving the blood in the conus arteriosus relatively large amounts of both carbon dioxide and oxygen. However, in the living heart the muscular trabeculated walls of the ventricle could contract differentially, thus *functionally* dividing the ventricle to receive either primarily oxygenated or primarily unoxygenated blood.

That is apparently what happens. Functional studies using X-ray opaque materials in the living heart and studies taking blood samples from various points show that, in fact, unoxygenated blood from the

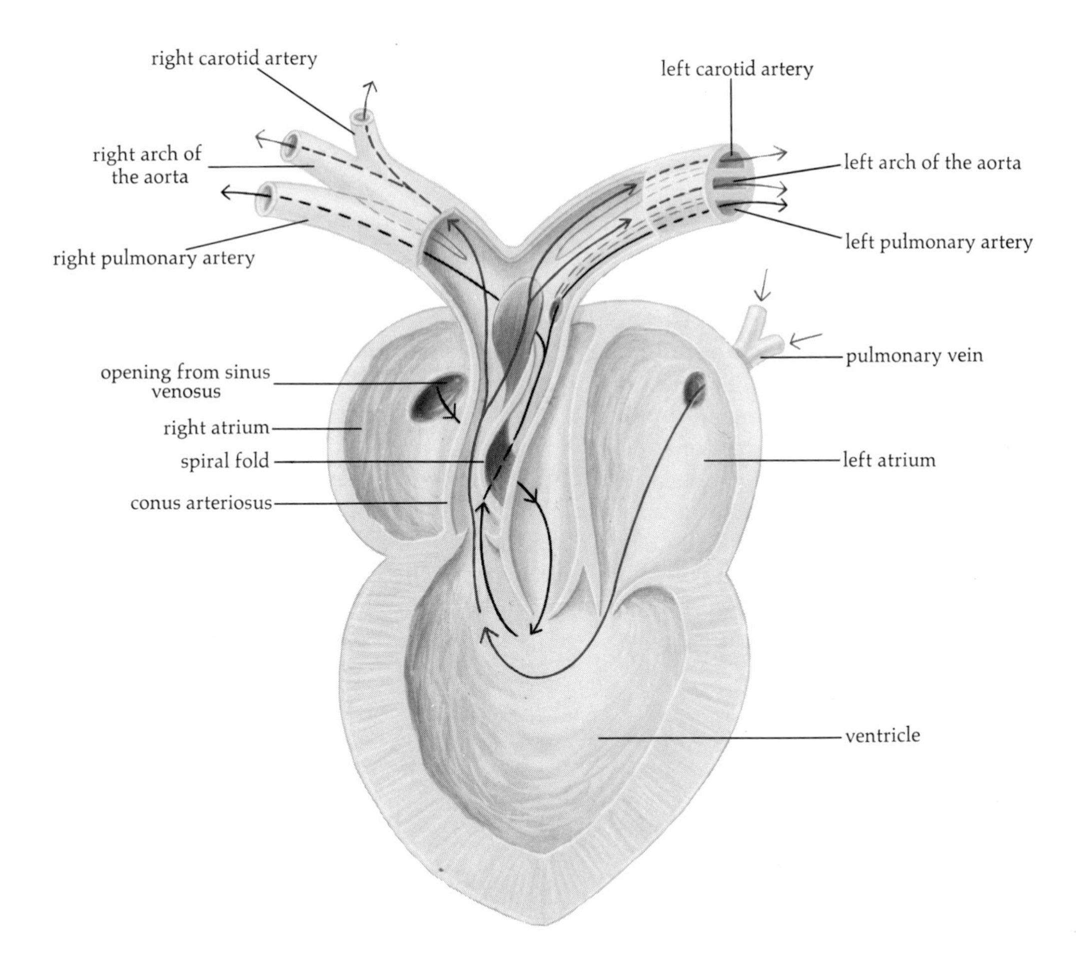

Fig. 16-7. A dissection of the ventral aspect of a frog heart, showing chambers, major blood vessels, and flow of blood through the heart.

sinus venosus, the right atrium, and the right side of the ventricle is directed preferentially into the pulmonary circulation by way of the pulmonary arteries; there it is carried to the lungs, and oxygenated blood is returned to the left atrium (or left side of the single atrium). This oxygenated blood passes into the ventricle and preferentially out to the arteries leading to the head and other somatic portions of the body. Thus, although there is not a solid septum dividing systemic from pulmonary circulation in lungfish and amphibia, the functional attributes of the heart cause a preferential distribution of oxygenated and unoxygenated bloods.

REPTILES

Among reptiles there are two distinct heart types, which differ in the morphology of their ventricles and of the large arteries leading from them. In both types, however, the two atria are completely divided by an interatrial septum. The right atrium receives unoxygenated systemic venous blood from the sinus venosus; the left atrium receives oxygenated blood from the pulmonary veins.

In Lepidosauria and Chelonia there is a single ventricle. A large, ventrally placed muscular septum partially divides it into communicating right and left chambers. The left side, which receives blood from the left atrium, is called the cavum arteriosum; the right side which receives blood from the right atrium is called the cavum venosum; the narrow portion between them is called the interventricular canal (Fig. 16-8). The cavum venosum has a ventral diverticulum, called the cavum pulmonale, from which the pulmonary artery goes to the lungs. Exiting directly from the cavum venosum are two vessels, the right and left aortae, which carry blood through the systemic arteries to the rest of the body.

When the atria contract, the large, flaplike valves between atria and ventricle open and occlude the interventricular canal. This causes all of the oxygenated blood from the left atrium to pass into the cavum arteriosum, and all of the unoxygenated blood from the right atrium to pass into the cavum venosum and cavum pulmonale; while the atria are contracted the blood in the two atria cannot mix. When the atria

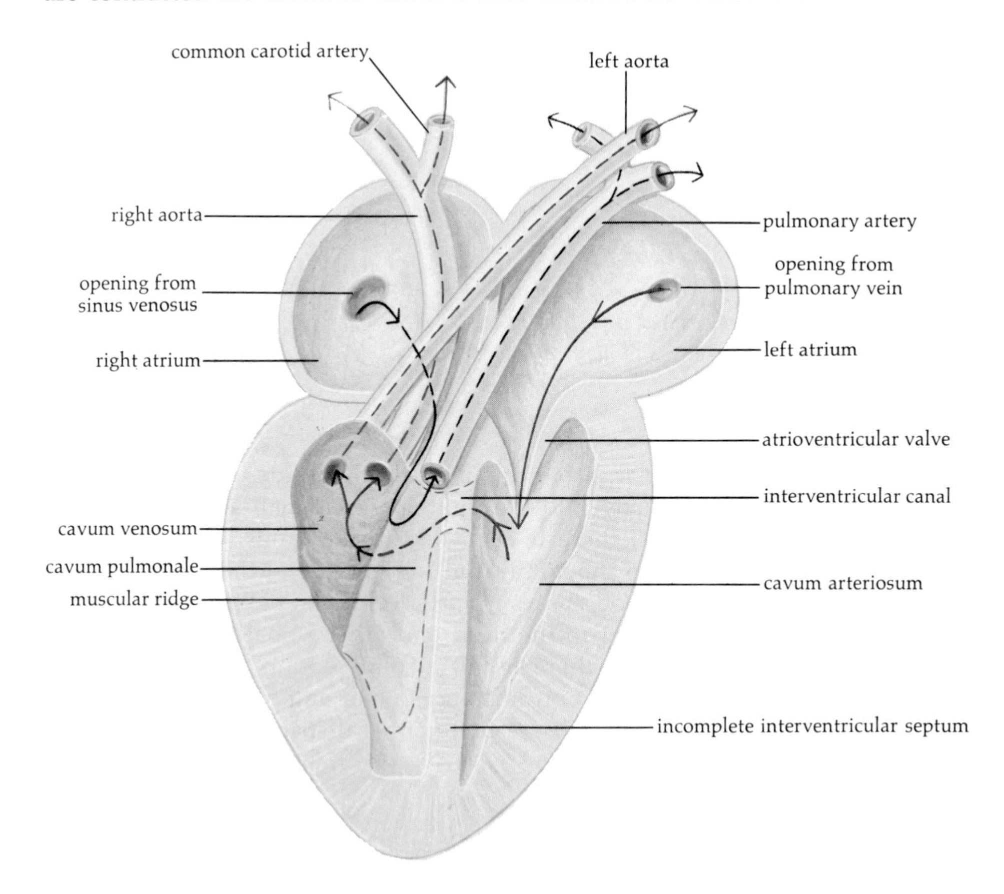

Fig. 16-8. A dissection of the ventral view of the heart of a tegu lizard, showing chambers and major blood vessels. Arrows indicate the flow of blood through the heart during normal respiration.

relax and the ventricle contracts, however, the flap valves are forced closed and the interventricular canal is opened. Contraction of the ventricle also reduces the volume of its cavities and forces out the blood, which could pass through either of the aortae or the pulmonary artery. However, because there is less resistance to flow in the pulmonary artery than in the aortae, the first blood to leave the ventricle passes into the pulmonary artery. This is unoxygenated blood which has entered from the right atrium and passed to the cavum venosum and cavum pulmonale. As the ventricle continues to contract, a muscular ridge is brought against the opposite ventricular wall and the cavum pulmonale is thus closed off. As the cavum venosum is emptied, more blood enters it through the interventricular canal from the cavum arteriosum; this is oxygenated blood from the left atrium. Because the pulmonary artery is closed off, this blood is forced into the right and left aortae and travels to all of the body except the lungs. There is, thus, a highly efficient separation of oxygenated and unoxygenated bloods in the systemic and pulmonary circulation of Lepidosauria and Chelonia.

Unlike other reptiles, crocodilians have a complete interventricular septum (Fig. 16-9), so that bloods from the right and left atria remain separate in the ventricles. Like other reptiles they have no conus arteriosus, but its homologue forms the base of three large blood vessels: the pulmonary artery, coming from the right ventricle and dividing to send an artery to each lung; and the left and right aortae, from the right and left ventricles, respectively. After leaving the ventricles, the two aortae cross, and at that point there is a hole, the foramen of Panizzae, through which their lumens communicate.

The crocodilian heart functions in the following manner. Atrial contraction forces blood from the right atrium into the right ventricle, and from the left atrium into the left ventricle. When the atria relax and the ventricles contract, the atrioventricular valves are closed and blood is forced out of the ventricles through the pulmonary artery and right and left aortae. The right ventricle, containing unoxygenated blood, could conceivably force blood out through both the pulmonary artery and the left aorta. However under normal circumstances nearly all the blood passes out through the pulmonary artery. The reason for this, again, is that there is less resistance in the pulmonary artery than in the systemic circulation. From the left ventricle blood passes out through the right aorta, is distributed by way of the foramen of Panizzae to the left aorta as well, and thus feeds the entire systemic arterial circulation. Thus crocodilians have the same functional attributes as other reptiles but with a different morphological basis.

The mechanisms by which oxygen-poor blood is shunted to the lungs and oxygen-rich blood is shunted to other parts of the body would seem unduly complicated in reptiles. It would certainly seem much simpler to have a complete instead of a partial interventricular septum, to omit the foramen of Panizzae entirely, and to have only the pulmonary artery coming from the right ventricle and only the systemic arteries from the left ventricle. Such is the situation in birds and mammals, and the reptilian heart, therefore, has often been called transitional. However, recent studies indicate that for most reptiles

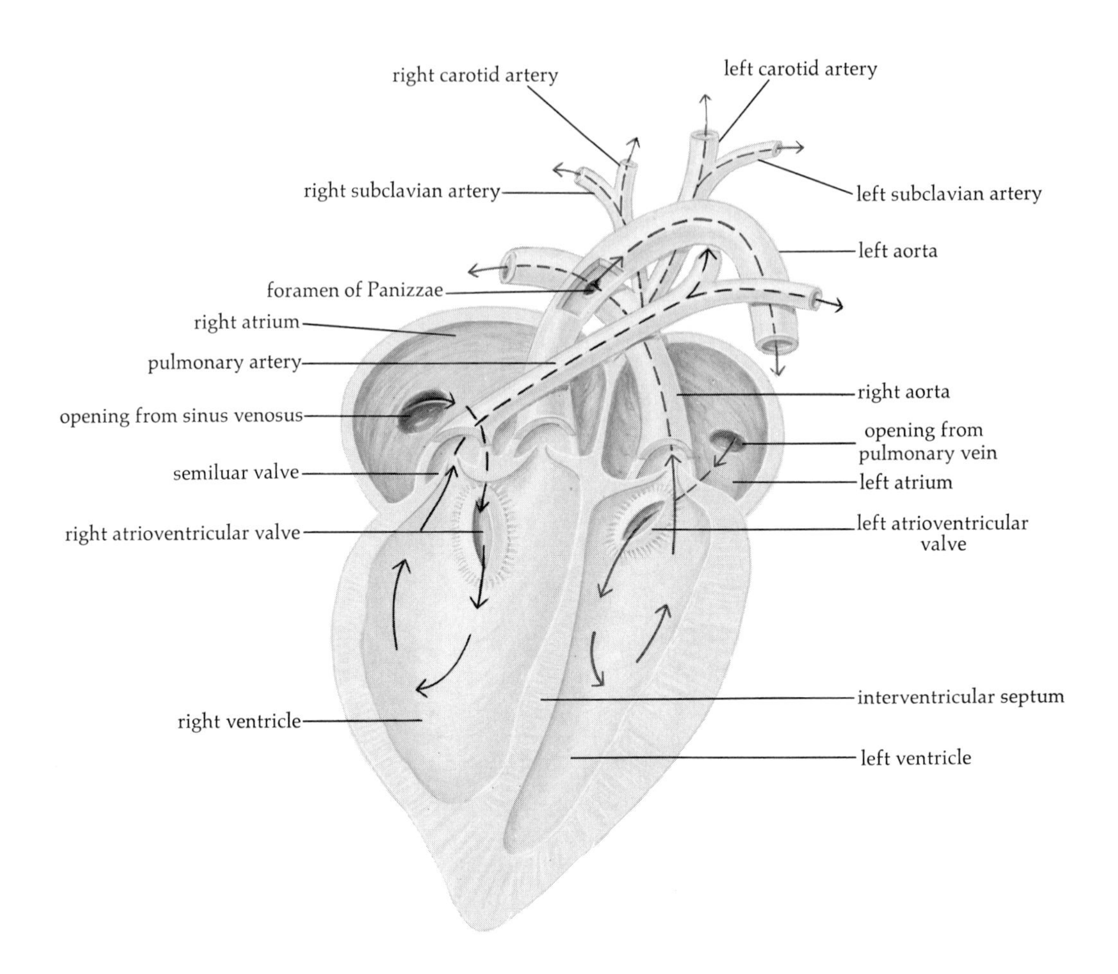

Fig. 16-9. A dissection of the ventral aspect of the heart of an alligator, showing chambers and major blood vessels. Arrows indicate the flow of blood through the heart during normal respiration.

this complex system, with its potential mixing of oxygen-rich and -poor bloods, has a unique adaptive value.

Many reptiles, including nearly all crocodilians and turtles, are semiaquatic. They spend long periods under water without breathing, thus building up an oxygen debt and a high carbon dioxide concentration in their body tissues. Because they continue to use oxygen during this time, however, they must conserve as much energy as possible.

In all diving tetrapods the vagus nerve causes the heart rate to slow considerably when the animal dives. The pulmonary arteries constrict, which greatly increases their blood pressure and their resistance to blood entering from the right ventricle. In Lepidosauria and Chelonia this increased resistance and pressure of the pulmonary arteries cause the preferential flow of blood from the cavum venosum to go to the right and left aortae; similarly in Crocodilia this change in pressure and resistance causes most of the blood from the right ventricle to pass into the left aorta, and some into the right aorta. Energy is conserved, therefore, by bypassing the pulmonary circulation—at no cost to the animal because sending blood to its lungs while under water would fulfill no purpose. When the diving tetrapod reemerges into air the heart rate increases, the smooth musculature of the pulmonary vessels relaxes, and normal circulation resumes; the carbon dioxide accumulated during the dive is then dissipated and the blood in the lungs reoxygenated.

The reptilian circulation, therefore, is not a transitional situation, but rather an adaptation to the environmental situation of most reptiles. And indeed this is not surprising, since from what we know of evolutionary processes we should not expect "intermediate" forms among living groups which have separated from one another and been evolving on their own for over 200 million years.

MAMMALS AND BIRDS

In mammals and birds, as already indicated, there is complete separation of both right and left atria and ventricles (Fig. 16-10). The sinus venosus is not a separate entity but has become incorporated into the walls of the right atrium. Systemic blood thus enters directly into the right atrium, passes through the right atrioventricular valve into the

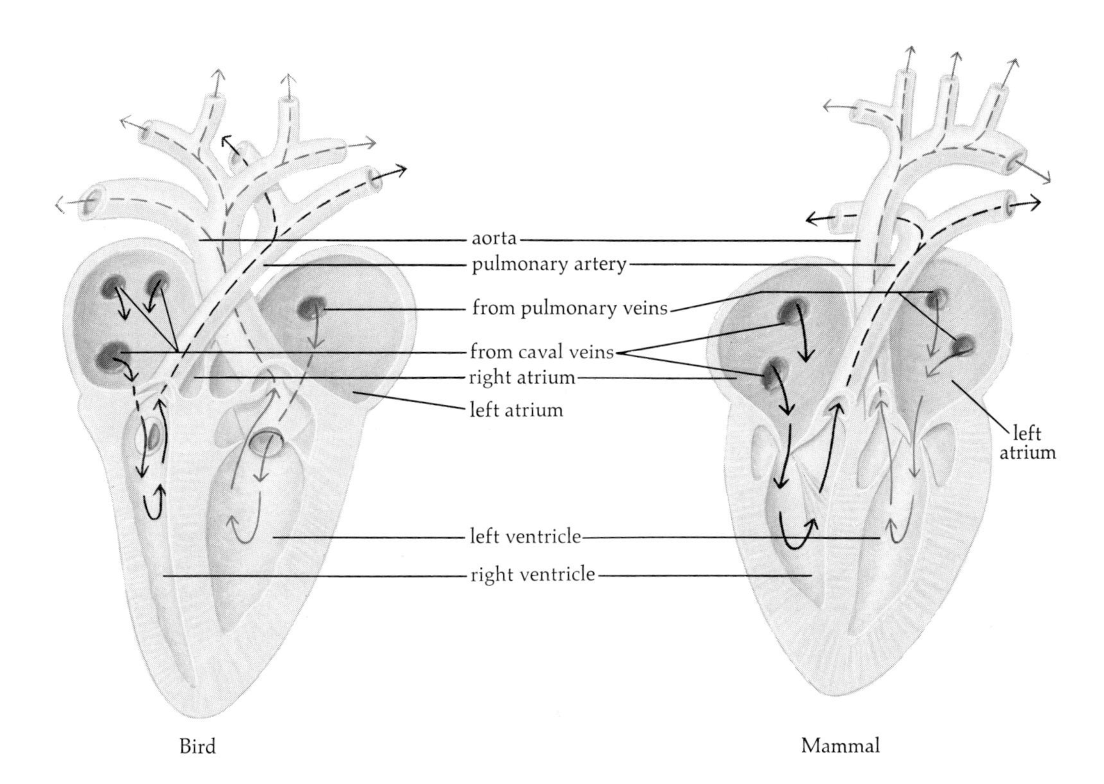

Fig. 16-10. Dissections of the ventral aspects of the hearts of a bird and a mammal, showing similar chambers and major blood vessels. Arrows indicate the flow of blood through the heart.

right ventricle, and then is pumped through the pulmonary artery to the lungs, oxygenated, returned through the pulmonary veins to the left atrium, passed through the left atrioventricular valve, and pumped out the left ventricle into the single aorta (left in mammals, right in birds) to all parts of the body except the lungs. This efficient, relatively simple, double circulatory pattern features complete separation of the two circuits. For diving mammals and birds, however, it is less efficient than the reptilian circulatory system.

During fetal life the lungs of amniotes do not function as respiratory organs, and the heart structure and circulatory pattern are different than they are in the adult. The details will be described only for mammals, although comparable situations exist in reptilian and avian fetal circulation.

The fetal mammal receives both food and oxygen at the placenta, by diffusion from the maternal circulation. From the placenta this oxygenated blood travels in the umbilical veins and then enters the systemic veins and thus the right atrium. From the right atrium it passes to the left atrium through the foramen ovale; this oval hole, present only in the fetus, is protected against backflow by a flap valve (Fig. 16-11). When the atria contract, blood passes through the atrioventricular valves to the ventricles. The ventricles are completely separated in the fetus, as they are in the adult, and their contraction forces the blood in the right ventricle up the pulmonary artery, and that in the left ventricle up the aorta. However, there is another structure found only in the fetus, the ductus arteriosus or duct of Botallus, which is a wide short vessel connecting the pulmonary artery with the aorta. The portion of the pulmonary artery beyond the ductus arteriosus is extremely small, with a high resistance to flow. Most of the blood

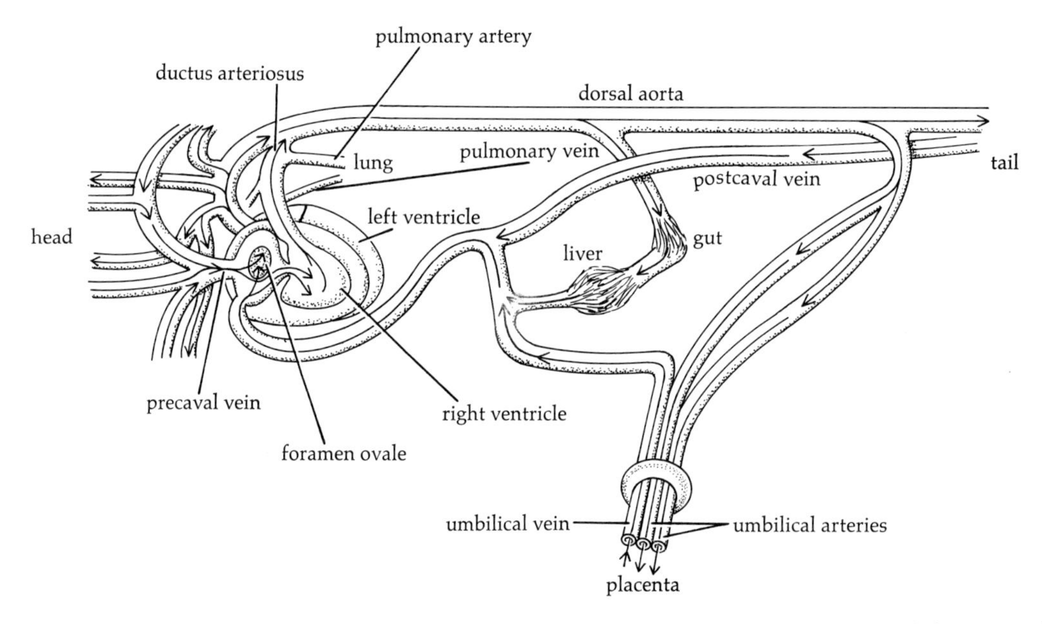

Fig. 16-11. Diagram of the major features of the mammalian fetal circulation.

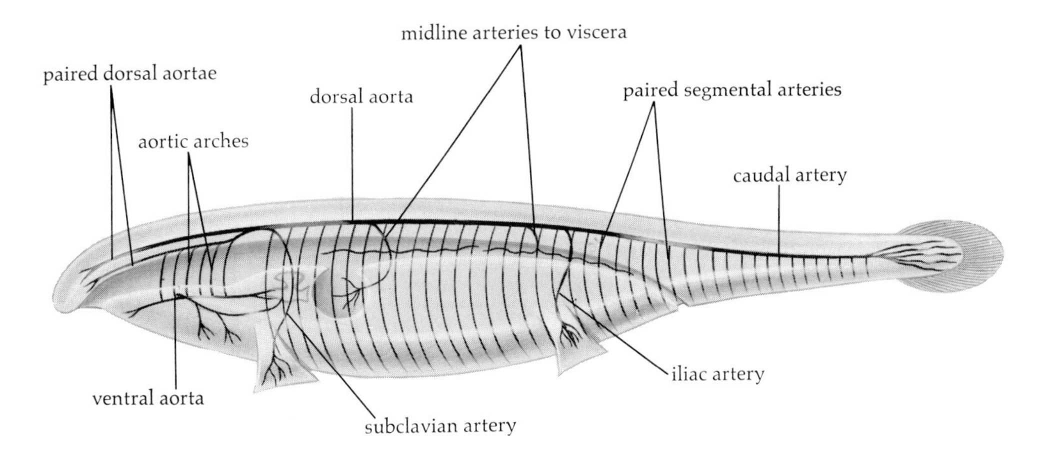

Fig. 16-12. The generalized arterial system of a vertebrate.

leaving the right ventricle, therefore, rather than passing to the lung, passes through the ductus arteriosus into the systemic circulation.

This essentially single circuit is very adaptive for the fetus, whose respiratory organ is the placenta. During birth, however, the umbilical vessels are shut off, and unless the newborn mammal starts using its lungs almost immediately its tissues will not have enough oxygen. This requires drastic alterations of the heart to direct large quantities of blood into the lungs for oxygenation.

With the closing off of the umbilical circulation the level of oxygen in the blood decreases and that of carbon dioxide increases. This acts upon respiratory centers in the brainstem, which reflexly cause the first respiratory movements of the diaphragm and intercostal muscles. These movements reduce pressure within the thoracic cavity and draw in air; these same movements also draw blood into the pulmonary artery, reducing the resistance to flow there so that much of the blood from the right ventricle, rather than flowing into the ductus arteriosus, flows directly to the lungs. The consequent flow of oxygenated blood from the lung through the pulmonary vein into the left atrium increases the pressure in the left atrium, and forces shut the valve of the foramen ovale; within twelve hours after birth there is no more mixing of blood between the atria. Although during the first few days of life this flap is held shut by pressure only, and can reopen, it soon knits to the rest of the interatrial septum and forms a permanent barrier. The ductus arteriosus remains for a much longer time, and is not completely closed in the human until several months after birth. Its remnant persists throughout life as the ligamentum arteriosum.

Arterial System

The early development of the arterial system is similar in all vertebrates (Fig. 16-12). Extending anteriorly from the conus arteriosus, just ventral to the pharynx, is the ventral aorta. This vessel gives off paired aortic arches which course dorsally around the pharynx between the developing pharyngeal pouches and then join the paired dorsal aortae. Caudal to the pharynx the paired dorsal aortae meet, fuse, and form a midline dorsal aorta running caudally just ventral to the notochord. Paired segmental arteries come off the dorsal aorta and distribute arterial blood to the somatic musculature, paired viscera, and skin; unpaired midline arteries from the dorsal aorta travel through developing mesenteries and supply the unpaired viscera. The head is supplied by paired anterior extensions of, the ventral and dorsal aortae. This basic arterial system is modified in later development to provide the diversity of arterial patterns seen in adult vertebrates.

AORTIC ARCHES AND ANTERIOR ARTERIES

In all gnathostomes there are six pairs of aortic arches, which develop in relationship to the pharyngeal pouches and skeletal branchial arches; like the latter, these aortic arches are numbered one through six. Not surprisingly the adult fate of each aortic arch is closely related to the fate of the corresponding pharyngeal pouch.

In amniotes the anterior aortic arches (one and two) are frequently lost before the posterior arches (five and six) even develop. In gill-breathing gnathostomes the first aortic arch is lost. The more posterior aortic arches are interrupted in the gills by a capillary network. Each aortic arch which supplies gills is divided into dorsal and ventral portions: the ventral portion, which carries unoxygenated blood into the gill, is called the afferent branchial artery; the dorsal portion, which drains oxygenated blood from the gill, is called the efferent branchial artery.

CHONDRICHTHYEANS

In chondrichthyeans the ventral portions of aortic arches two through six form five pairs of afferent branchial arteries traveling up their respective pharyngeal arches; each one, except arch two, vascularizes its own holobranch. The second aortic arch, giving rise to the most anterior afferent branchial artery, passes between the spiracle and the first gill slit. Since the spiracle has no respiratory membrane, this afferent branchial artery from the second arch supplies only the anterior demibranch of the first typical gill slit, which is derived from the third embryonic pharyngeal pouch.

Each gill slit has gill lamellae on either side; they are derived from the anterior and posterior demibranchs of two different gill arches. Around the gill slit, collecting from the capillaries of the gill lamellae, is a closed loop of efferent arteries (Fig. 16-13). The anterior portion of each loop is called the pretrematic efferent artery, and the posterior, the posttrematic efferent artery; together they are called the collector loop. Dorsally the entire loop is drained into the efferent branchial artery. The situation is complicated by so-called cross trunks or anastomotic connections between the pretrematic efferent branch of one loop and the posttrematic efferent branch of the next.

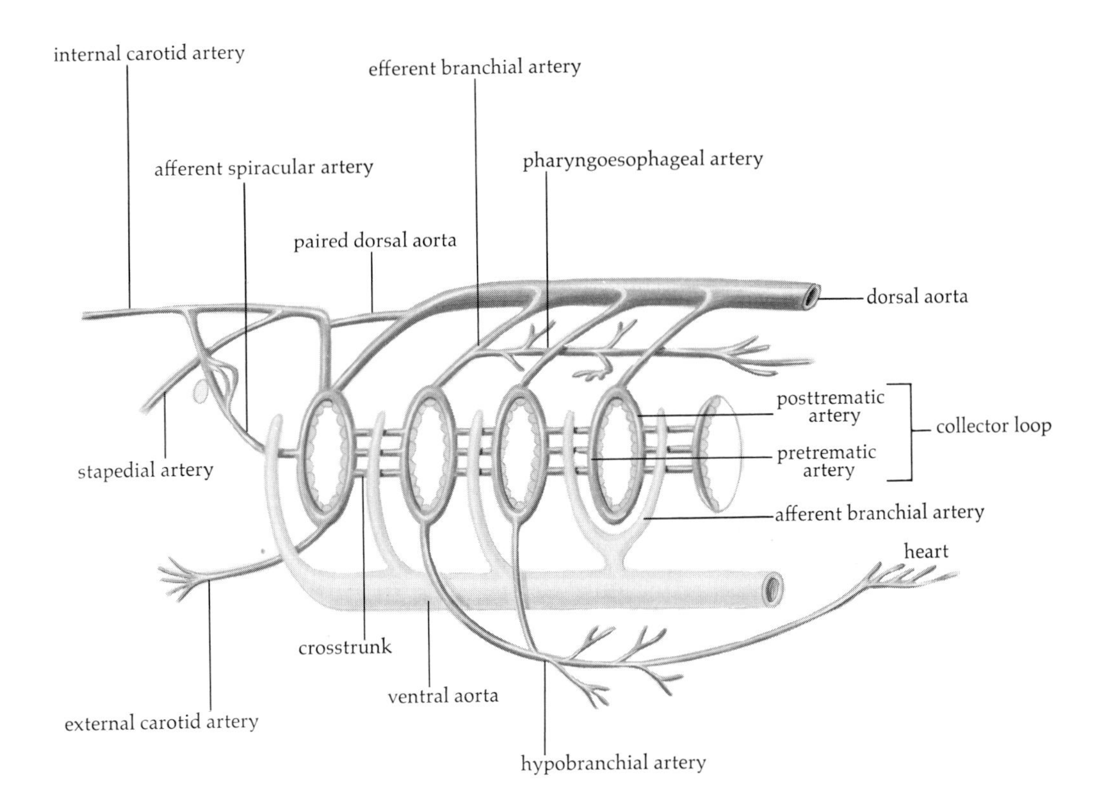

Fig. 16-13. Lateral view of the branchial circulation of a dogfish.

These collector loops give rise to four efferent branchial arteries from the dorsal portions of aortic arches three, four, five, and six. These four arteries form the descending midline dorsal aorta. From the middle of the first collector loop arises the afferent spiracular artery, which carries oxygenated blood to the spiracle. The external carotid artery arises from the ventral portion of the first collector loop and carries oxygenated blood to the lower jaw; the internal carotid artery arises from the dorsal portion of the first collector loop and carries oxygenated blood to the dorsal portion of the head and to the brain. The hypobranchial artery comes off the ventral portions of the second and third collector loops and carries oxygenated blood to the hypobranchial musculature and to the muscular walls of the heart.

TELEOSTS

In adult teleosts there are only four functional afferent branchial arteries, which are formed from the ventral portions of aortic arches three through six; these four arteries bring unoxygenated blood to the gill lamellae. The efferent branchial arteries do not form collector loops; each efferent branchial artery collects from an entire, single holobranch and travels up to one of a pair of dorsal aortae as the dorsal portion of aortic arch three, four, five, or six (Fig. 16-14).

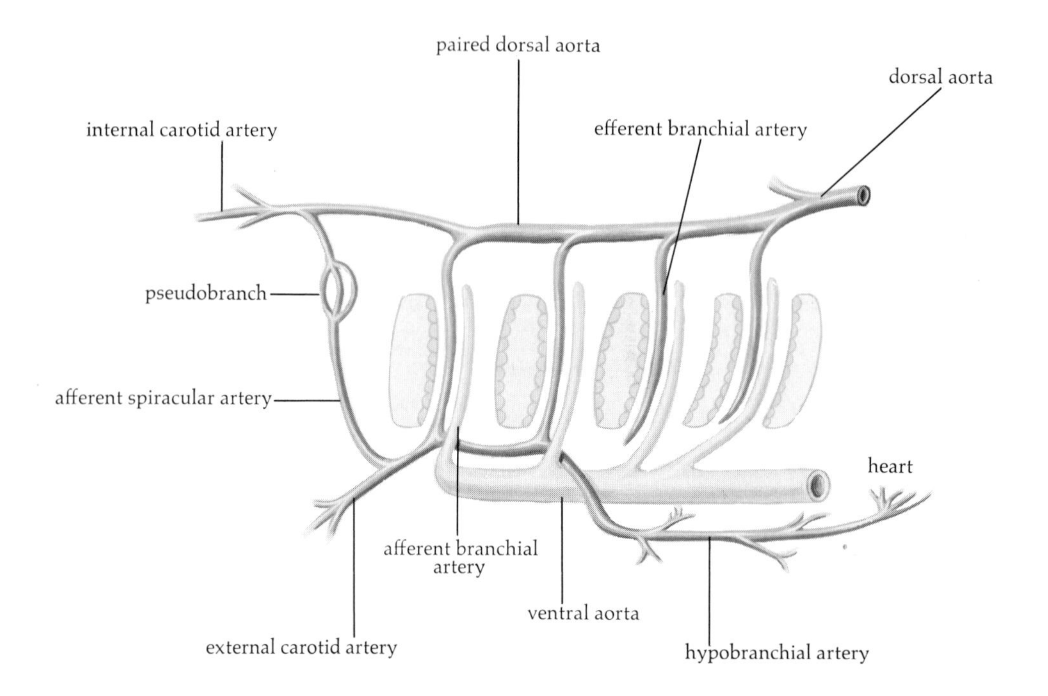

Fig. 16-14. Lateral view of the branchial circulation of a teleost.

The anterior extensions of the ventral aortae, which would carry unoxygenated blood, become separated from their more posterior parts and gain anastomotic connections with the ventral tips of the third and fourth efferent branchial arteries, from which they receive oxygenated blood. They then extend anteriorly as the external carotid arteries, carrying oxygenated blood to the lower jaw. A separate branch of the external carotid, from the second aortic arch, also carrying oxygenated blood, goes to the pseudobranch (where the spiracle would be if teleosts had a spiracle) and then continues dorsally and joins the anterior portion of the paired dorsal aortae. From this point, as the internal carotid, it extends further anteriorly into the head and brain. The efferent branchial arteries from the third and fourth aortic arches, in addition to contributing to the external carotid, also send a ventral branch to the hypobranchial and heart muscles.

LUNGFISH

As discussed in the preceding chapter, the gills of the three genera of lungfish are best developed in the Australian, *Neoceratodus,* less developed in the African, *Protopterus,* and least developed in the South American, *Lepidosiren.* In *Protopterus,* aortic arches two, five, and six form capillary networks in the gills. There are no gills for branchial arches three and four, and their aortic arches shunt the blood directly from the ventral aorta to the paired dorsal aortae (Fig. 16-15). Anterior extensions of the ventral aorta form the external carotid, and extensions of the paired dorsal aortae form the internal carotid arteries. A branch from the sixth efferent branchial artery goes to the lungs as the pulmonary artery.

AMPHIBIANS

In larval amphibians, and in neotenic adults, aortic arches three, four, and six are interrupted by capillary networks in the external gills, and aortic arches one, two, and five are lost. During metamorphosis the external gills are lost, and their aortic arches are reconstituted as complete vessels, going from the ventral aorta to the dorsal aorta.

In adult urodeles and apodans, anterior extensions of the ventral aorta form the external carotid artery; the internal carotid is formed by arch three and the anterior extension of the paired dorsal aortae. On each side, the portion of each dorsal aorta between arches three and four is called the carotid duct; it remains as an anastomosis between the dorsal aorta and the internal carotid artery (Fig. 16-16). Arches four and six persist as the systemic arch and the pulmonary arch, respectively, but both reach dorsally to join paired dorsal aortae which then unite posteriorly to form the descending aorta. The pulmonary artery to the lung comes off the pulmonary arch, as it does in lungfish.

In anurans there is a slightly different condition (Fig. 16-16). The connection between the internal carotid artery and the paired dorsal aortae, formed by the carotid duct in other amphibians, is lost in anurans. The portion of the ventral aorta between the third and fourth

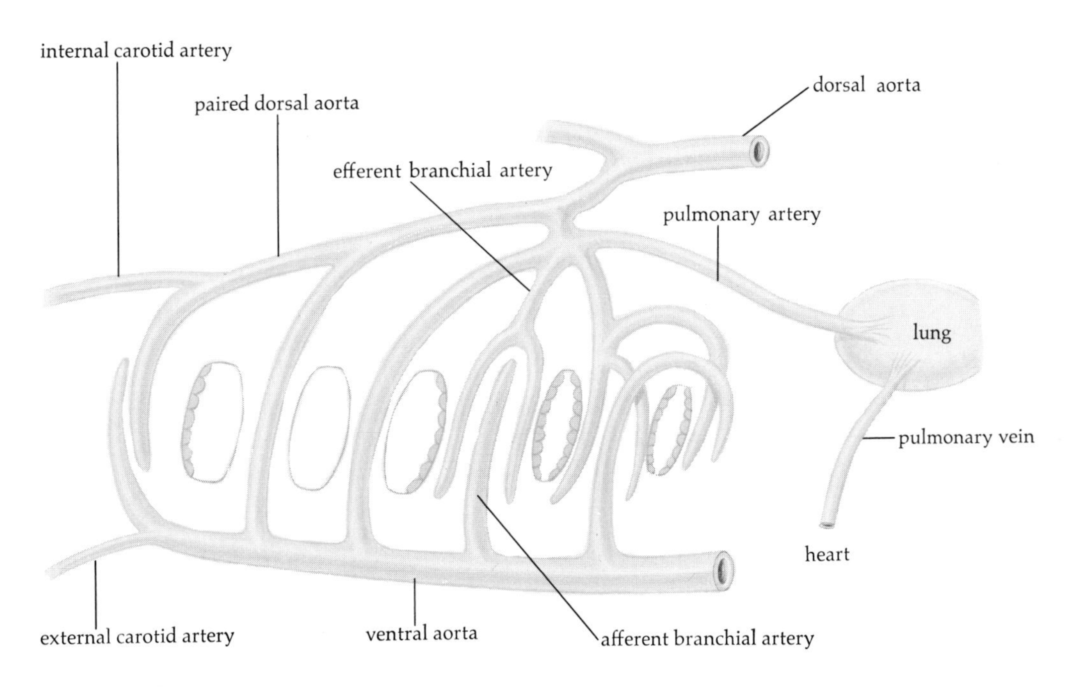

Fig. 16-15. Lateral view of the branchial and pulmonary circulation of the lungfish, Protopterus.

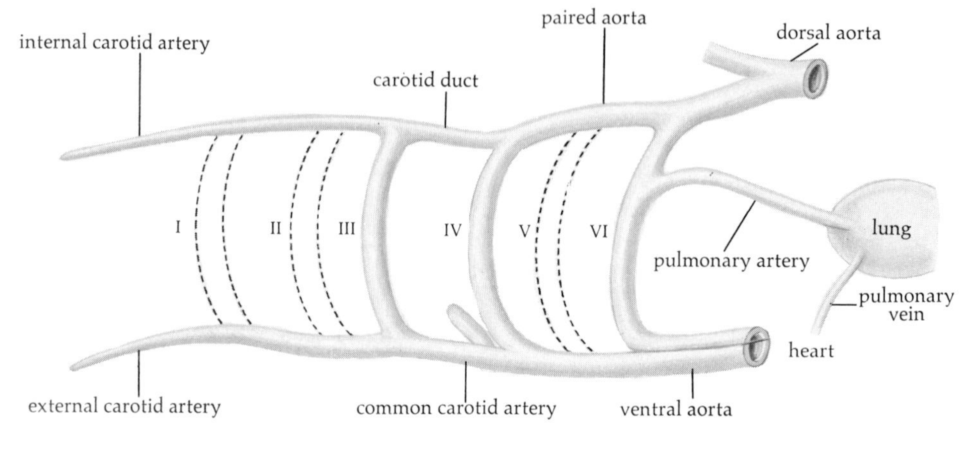

Apodans and Most Urodeles

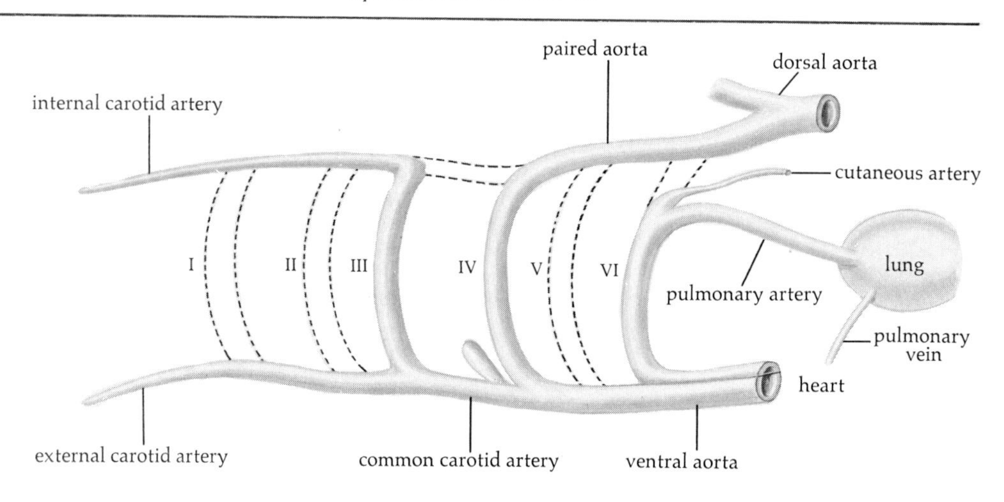

Anurans

Fig. 16-16. Lateral views of the derivatives of the aortic arches, seen in adult amphibians. Roman numerals indicate aortic arch numbers and dotted lines indicate vessels present during development but lost in the adult.

aortic arches forms the common carotid artery; the third aortic arch forms the proximal part of the internal carotid artery. The connection between the dorsal portion of the sixth aortic arch and the paired aortae is also lost; only the ventral portion of the sixth aortic arch and its posterior extension to the lung remain, forming the base of the pulmonary artery.

REPTILES

The aortic arch system of reptiles is very similar to that of anuran amphibians, except in the conus arteriosus and ventral aorta which are divided into three separate vessels coming out of the ventricles (Fig.

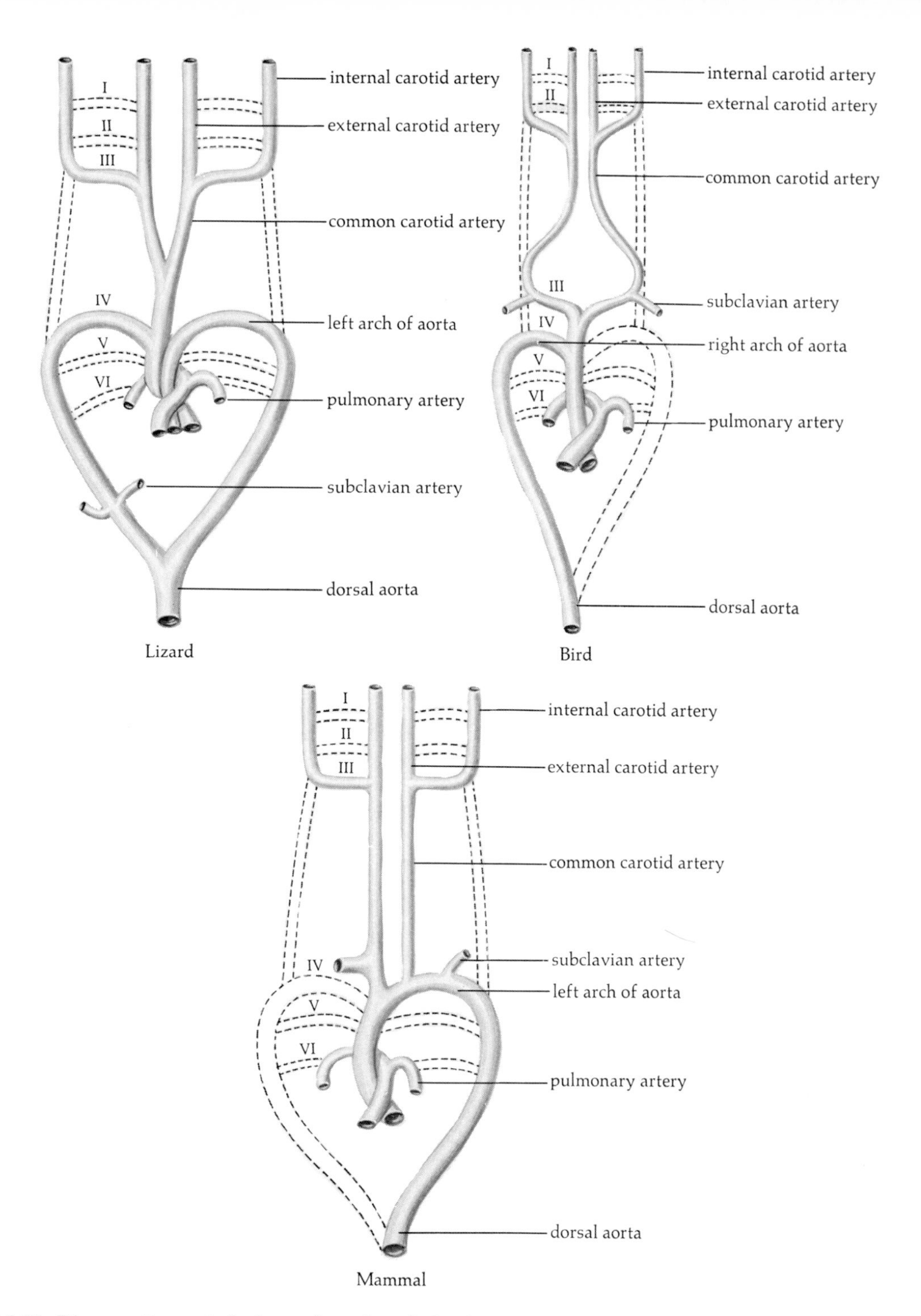

Fig. 16-17. Diagramatic ventral views of amniote derivatives of aortic arches. Dotted lines indicate vessels present during development but lost in the adult. Roman numerals indicate aortic arch numbers. Note the variations in the connections of the subclavian arteries, which carry blood to the forelimbs. Development of the neck causes a large separation between arches three and four in lizards and mammals, and between arches two and three in birds and crocodiles.

16-17). One of these vessels, derived from the sixth aortic arch, forms the two pulmonary arteries. The fifth aortic arch is lost. The fourth forms the systemic arch; except for the crocodilians' foramen of Panizzae there is no connection between the right and left systemic arches. At the base of the right fourth aortic arch arises the common carotid artery, developing from the portion of the ventral aorta which runs between aortic arches three and four. The more anterior portions of the ventral aorta form the paired external carotid arteries. The third aortic arch forms the proximal portion of the internal carotid arteries, and extensions of the paired dorsal aortae form its distal portions. Aortic arches one and two, as well as five, are lost in adult reptiles.

MAMMALS AND BIRDS

In mammals and birds the ventral aorta and the aortic arches develop very similarly to those of reptiles, with exceptions only in the systemic arches (the fourth). In birds the left systemic arch is lost; all blood to the descending aorta passes through the right systemic arch and back to the descending aorta (Fig. 16-17). In mammals just the opposite is true: the right systemic arch is lost, and the left arch persists.

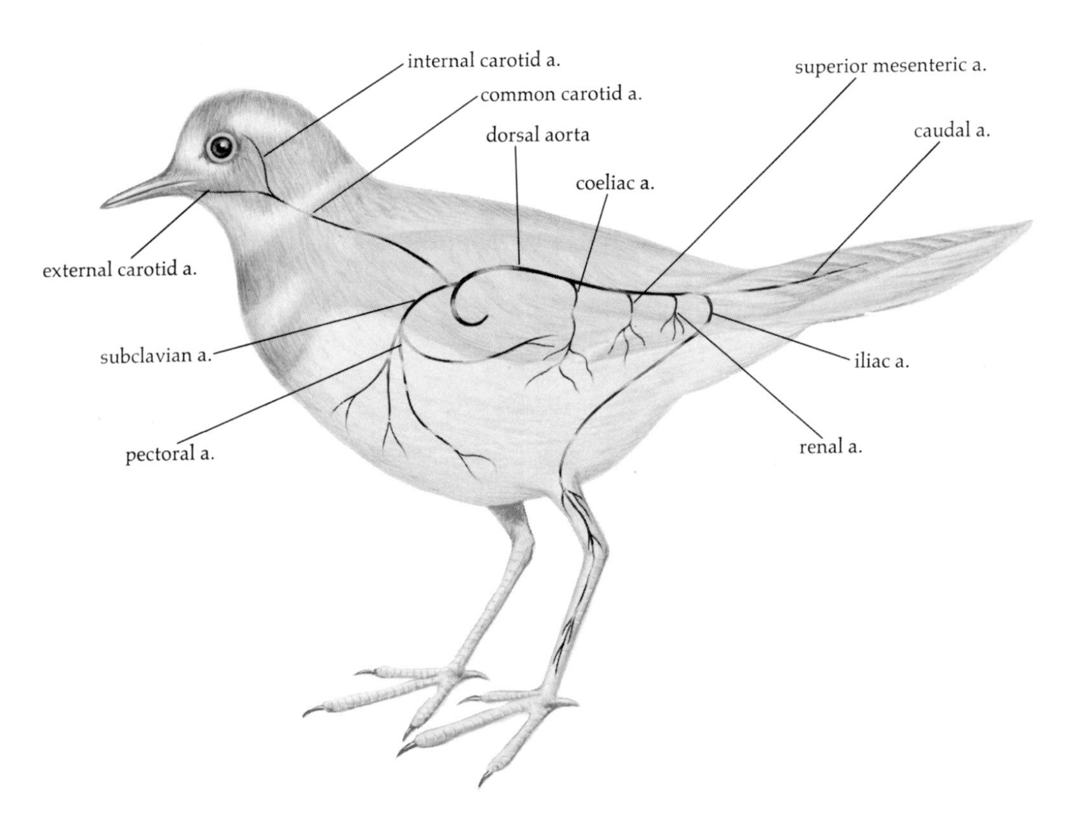

Fig. 16-19. The major arteries of a bird.

Arch three, and two when present, forms part of the internal carotid circulation to most of the head. The anterior extensions of the ventral aorta form the external carotid, going to the lower jaw. Arch four (the systemic arch) is always retained, at least on one side; in tetrapods it carries most or all of the blood to the descending aorta (Fig. 16-18, see pages following 177). In tetrapods and sarcopterygians the ventral part of arch six forms the base of the pulmonary artery; the dorsal part of arch six, which would connect the pulmonary artery and the dorsal aorta, is lost as a blood vessel but remains as a ligament (the ligamentum arteriosum) in the adult anuran and amniotes.

POSTERIOR ARTERIES

The remainder of the arterial system is made up of paired branches to the paired organs such as kidneys and gonads, to the somatic musculature, and to the limbs. The major artery to the forelimb is the subclavian artery, and to the hindlimb, the iliac artery. Midline vessels, primarily the coeliac and anterior mesenteric arteries, pass through the dorsal mesentery to supply most of the alimentary canal and its derivatives (Fig. 16-19).

Venous System

The early embryo's first veins develop from blood islands in the yolk sac. These vitelline veins, as they are called, enter the body and pass through the ventral mesentery of the gut to the sinus venosus. The liver, developing within the ventral mesentery and transverse septum, surrounds these vitelline veins. At about the same time the hepatic portal vein courses over from the gut and joins the left vitelline vein; its blood then enters the liver. Within the liver the vitelline veins anastomose with one another and form a complex of sinusoids, which are drained by the large hepatic veins. Blood coming in from either the vitelline veins or the newly developed hepatic portal vein passes through these sinusoids to the hepatic veins and then to the sinus venosus.

As these developments occur, paired anterior cardinal veins form in the dorsolateral head and drain caudally toward the heart, and paired posterior cardinal veins form in the dorsal body wall and course anteriorly toward the head. At their posterior ends, the posterior cardinal veins join with one another and receive the caudal vein from the tail. At the level of the heart, the anterior and posterior cardinal veins come together to form the paired common cardinal veins, or ducts of Cuvier, which enter the sinus venosus laterally. Several subcardinal veins form around the developing kidneys and interconnect with the nearby posterior cardinal veins (Fig. 16-20).

Long veins, the lateral abdominal veins, form in the lateral body wall, run the length of the trunk, and empty into the common cardinal

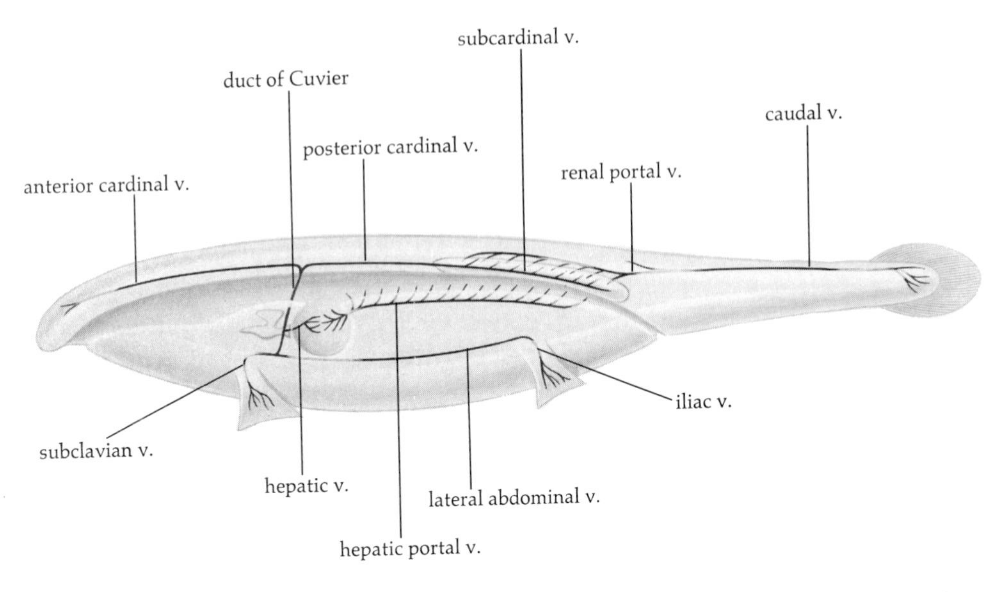

Fig. 16-20. The generalized venous system of vertebrates.

veins; in addition to draining the lateral body wall these veins also receive blood from the pelvic appendages via the iliac veins and from the pectoral appendages via the subclavian veins.

On each side, as the lateral abdominal vein is forming, the portion of the posterior cardinal vein just anterior to the level of the kidney is lost. Blood coming from the caudal region (the primary muscular region in fish) must henceforth pass into the posterior part of the posterior cardinal, then into the subcardinal, and then through the capillary beds of the kidney. The subcardinal vein then drains anteriorly into the anterior part of the posterior cardinal vein, and from there blood is carried to the duct of Cuvier and to the heart. This is a renal portal system, filtering blood from the tail through a kidney capillary system before sending it to the heart.

Such is the general pattern of the early development of venous drainage in all vertebrates.

FISHES

In adult fishes the venous drainage resembles the general developmental pattern, although with some specializations in various groups (Fig. 16-21). As the yolk sac is resorbed, the vitelline veins to the liver are reduced, and their intramesenteric portions remain only as a small ligament in the ventral mesentery. Most teleosts lack lateral abdominal veins. The iliac veins drain into either the posterior cardinals or the renal portals, and the subclavian veins drain directly into the duct of Cuvier. In chondrichthyeans there are large sinuses in the blood vascular system, particularly along the anterior and posterior cardinal veins.

AMPHIBIANS

The largest vein returning blood to the amphibian heart is the postcaval. It arises between the kidneys and passes anteriorly through the liver to the sinus venosus (Fig. 16-22); it does not itself go through the capillary network, but from the hepatic veins it receives blood which has come through the liver sinusoids. The postcaval vein has two separate embryonic origins. Anteriorly this vein develops from a caudal evagination of the right hepatic vein; posteriorly it is a continuation of the right subcardinal vein.

The posterior and anterior parts of the lateral abdominal veins become separated. The posterior parts receive blood from the iliac veins; the right and left posterior parts then move together ventrally and form the midline ventral abdominal vein, which passes through the ventral mesentery of the alimentary canal and empties into the hepatic portal vein. The anterior parts of the lateral abdominal veins remain as cutaneous veins draining the skin and superficial muscles

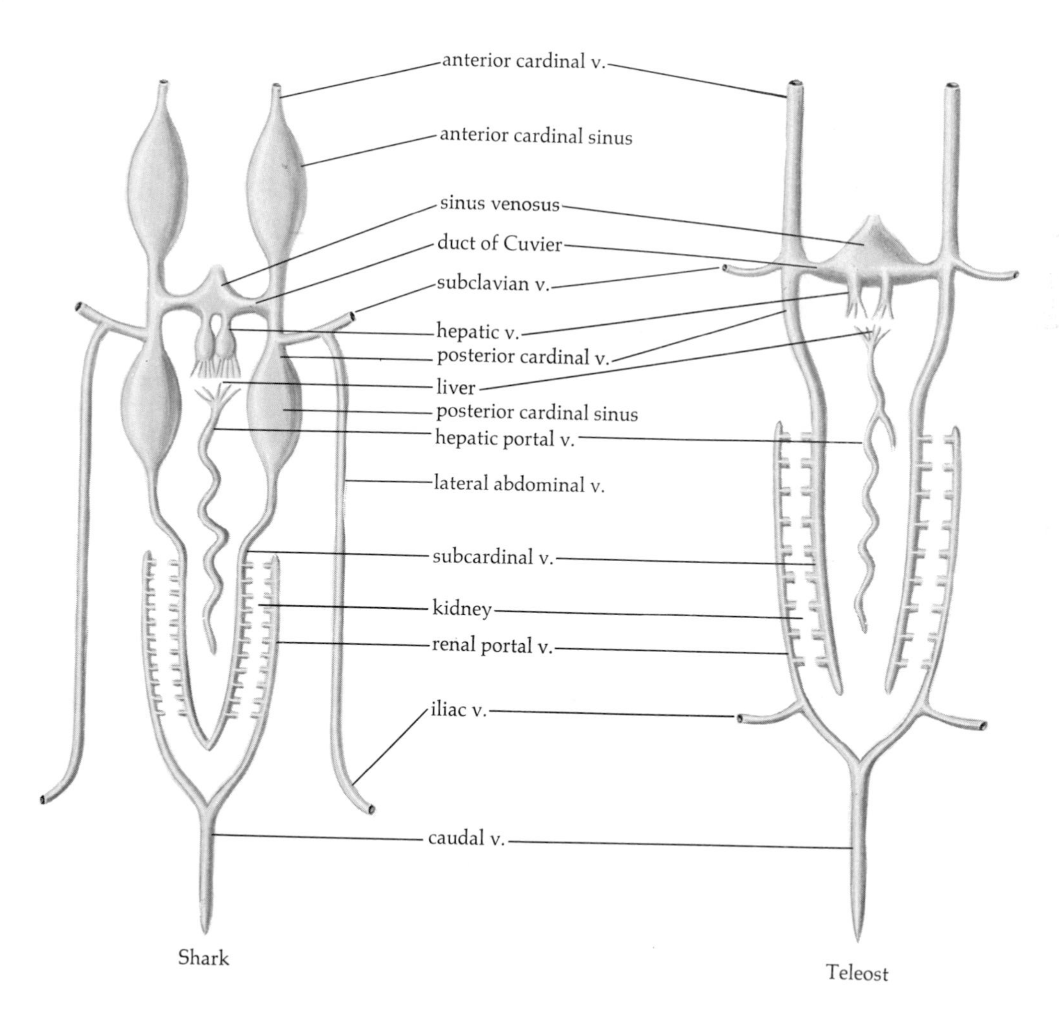

Fig. 16-21. The venous systems of a shark and a teleost.

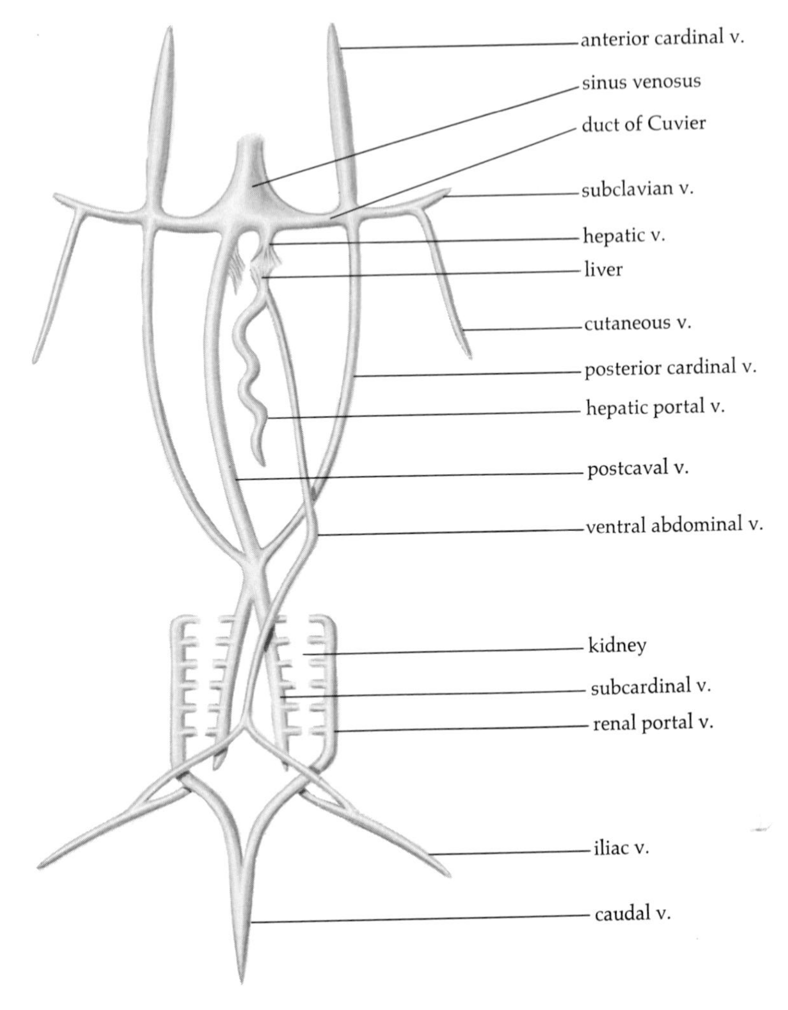

Fig. 16-22. Ventral view of the urodele venous system.

from the trunk region, and then, joined by the subclavian veins, drain into the ducts of Cuvier. In anurans the posterior cardinal veins are lost, and the postcaval vein alone drains the kidneys and most of the posterior body; in urodeles, although the posterior cardinal veins are retained, the postcaval vein nevertheless carries most of the blood from the posterior body.

REPTILES

The reptilian venous system is much like the anuran's. The postcaval vein is the primary vein of the posterior body, since the anterior portions of the posterior cardinal veins have completely disappeared. Since no posterior cardinal veins drain into the common and anterior cardinals, these last two vessels cannot be distinguished; the term precaval vein is used to denote them. The anterior cardinal vein distal to the precaval vein is called the jugular vein. Blood from the hindlimb may pass to either the renal portal vein or the hepatic portal vein by way of the ventral abdominal veins. The blood in the renal portal vein does not have to pass through the capillaries in the kidney, since there are many anastomoses between the renal portal and the postcaval veins (Fig. 16-23).

BIRDS AND MAMMALS

The venous system of birds is nearly identical to that of reptiles except that in birds the renal portal circulation is almost completely lost, and the iliac veins drain directly into the postcavals (although some branches pass through the kidney, perhaps with slight renal portal circulation; Fig. 16-23). Because of this large connection between the iliac and postcaval veins, the ventral abdominal vein, here called the inferior mesenteric, is quite small and carries little blood.

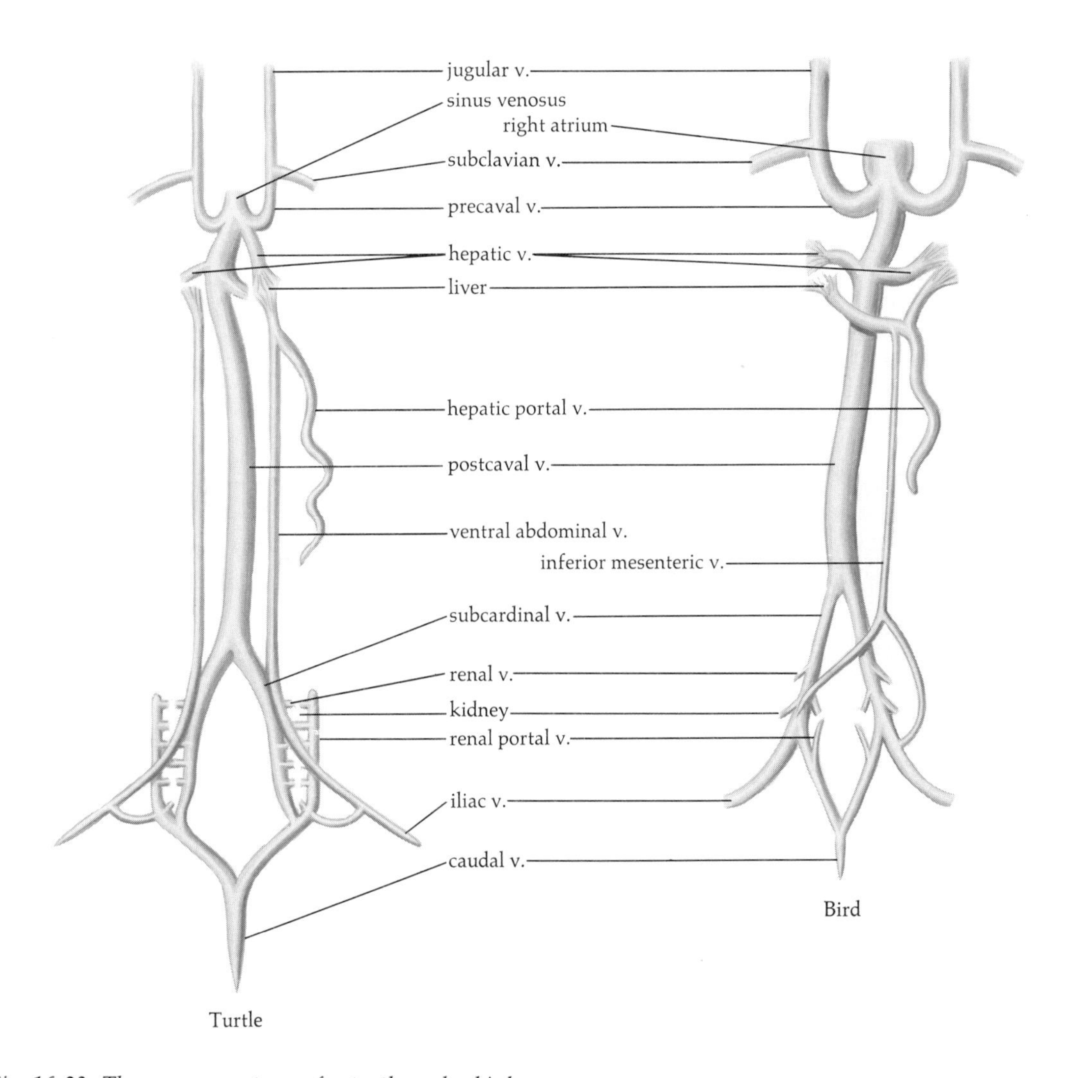

Fig. 16-23. The venous systems of a turtle and a bird.

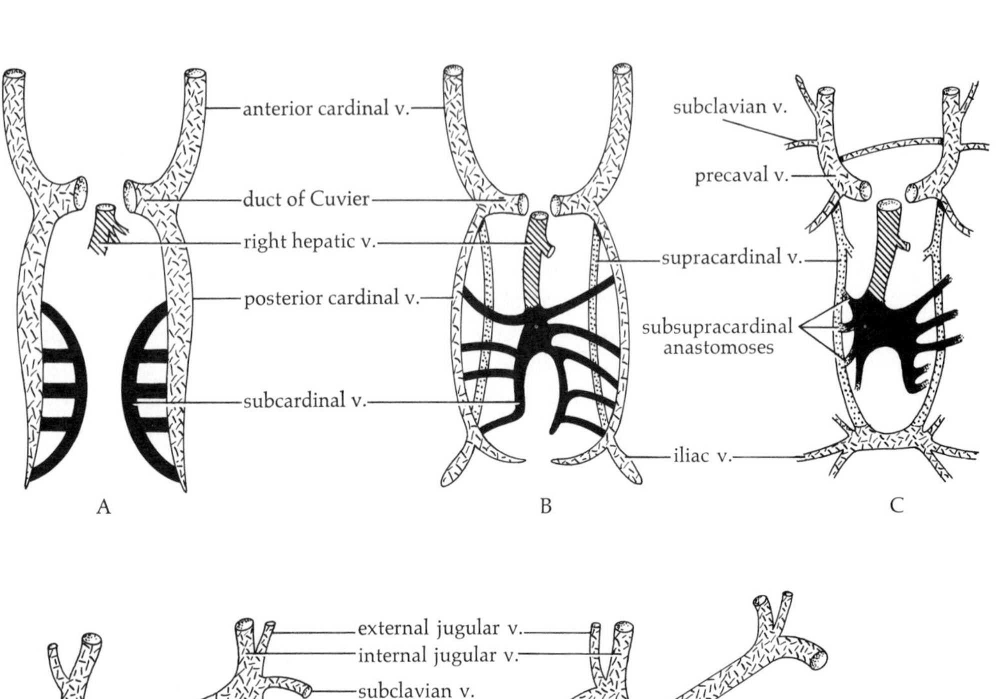
anterior cardinal v.
duct of Cuvier
right hepatic v.
posterior cardinal v.
subcardinal v.
subclavian v.
precaval v.
supracardinal v.
subsupracardinal
anastomoses
iliac v.
A
B
C

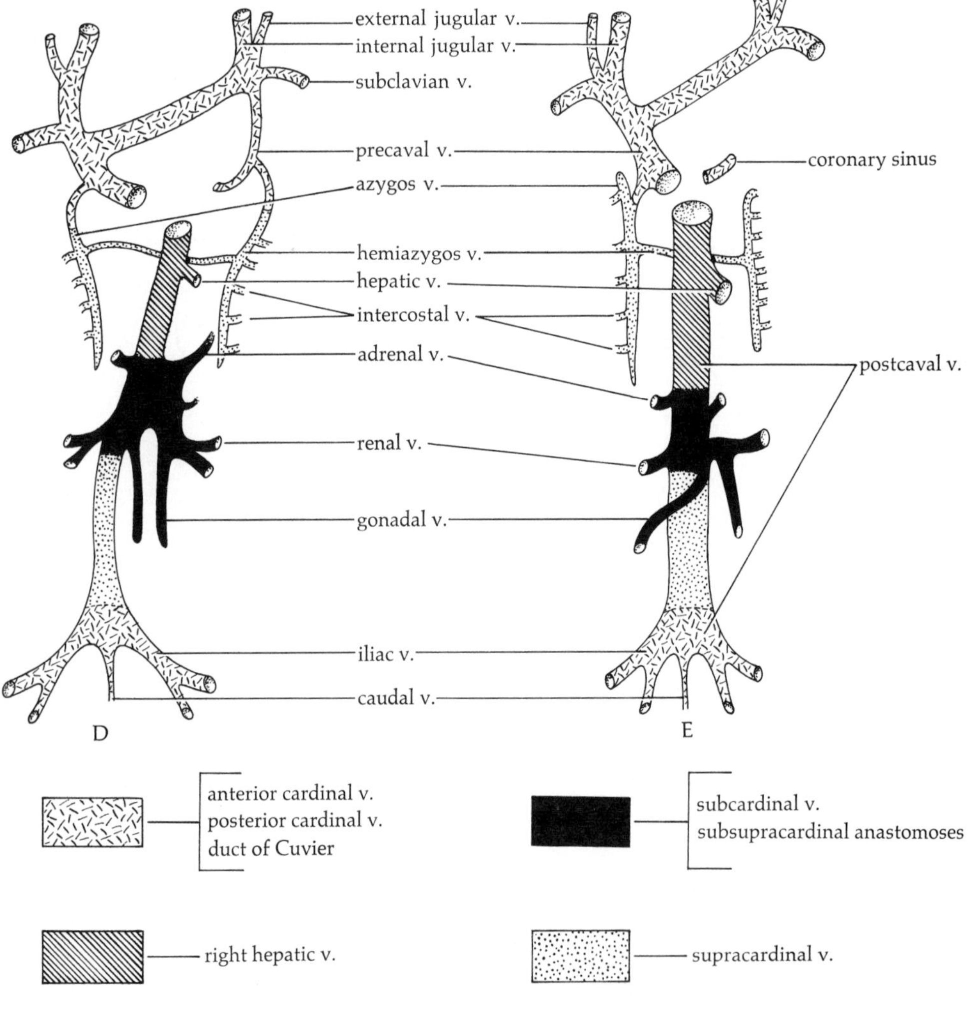
external jugular v.
internal jugular v.
subclavian v.
precaval v.
coronary sinus
azygos v.
hemiazygos v.
hepatic v.
intercostal v.
adrenal v.
postcaval v.
renal v.
gonadal v.
iliac v.
caudal v.
D
E
anterior cardinal v.
posterior cardinal v.
duct of Cuvier
subcardinal v.
subsupracardinal anastomoses
right hepatic v.
supracardinal v.

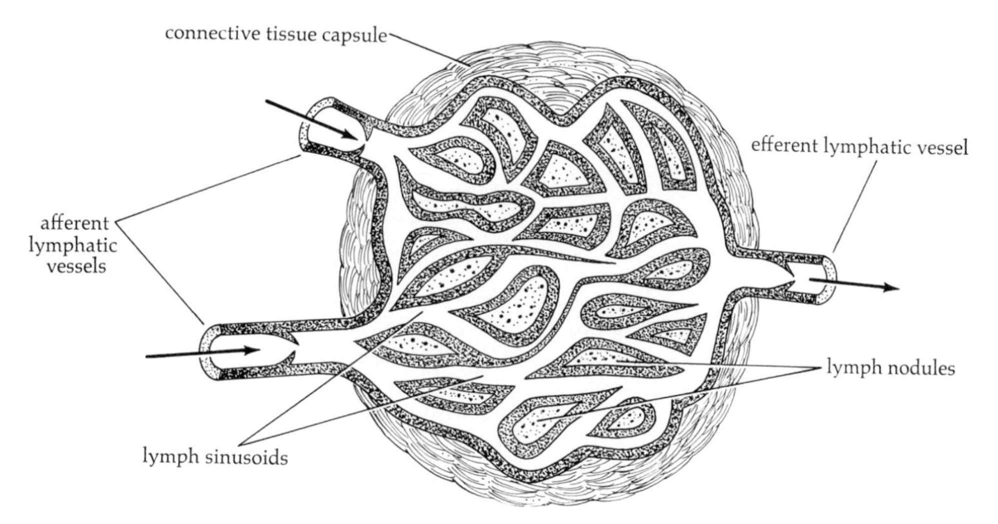

Fig. 16-25. Diagramatic representation of a small mammalian lymph node.

The mammalian system is also similar to that of reptiles; the hepatic portal is basically the same, but with increased drainage by the postcaval (Fig. 16-24). The abdominal vein is not present in adult mammals. The postcaval vein develops ontogenetically from several different vessels, including the right hepatic vein, the right subcardinal vein, the subsupracardinal anastomosis, the posterior part of the right supracardinal vein, and the posterior part of the posterior cardinal vein. The details can best be understood from the accompanying diagrams (Fig. 16-24).

Lymphatic System

The lymphatic vessels form another distinct part of the circulatory system. They contain neither red blood cells nor hemoglobin in any form. The fluid of the lymph is quite similar, or perhaps identical, to intercellular fluids. The "lymph cells" are agranular leukocytes; they are formed in the white substance of the spleen and in lymph nodes which may be scattered in various portions of the body (frequently concentrated along the mesenteries in the peritoneal or pleuroperitoneal cavity).

Lymphatic capillaries end—or rather, begin—blindly, course over relatively large distances, and anastomose extensively. Eventually these capillaries join in larger lymph vessels which may contain small amounts of smooth musculature and which have valves that prevent

Fig. 16-24. Diagrams of the rather complex sequence of events in the development of the mammalian precaval and postcaval veins.

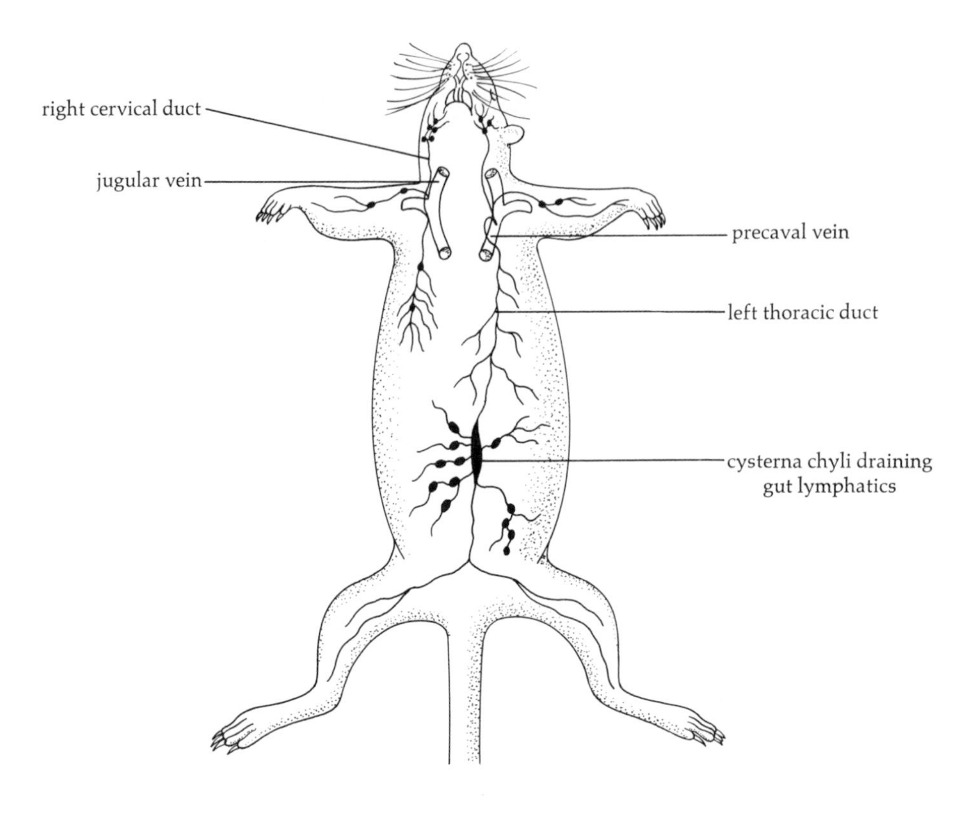

Fig. 16-26. Diagram of the major lymphatic vessels and lymph nodes of a rat.

the backflow of the lymph. Interposed along the route are lymph nodes in which the lymph vessels break up into sinusoid networks. Here old lymph cells are destroyed and new ones formed (Fig. 16-25). Eventually the lymphatic vessels empty into large veins of the blood vascular system; here the lymph mixes with the blood and is no longer distinguishable from it.

The greatest concentration of lymphatic vessels is found within the absorbing portion of the intestine. In mammals and birds each villus contains a blind lacteal, or lymph capillary. All the fat absorbed from the intestine enters these lacteals rather than the blood capillaries; the fat is then carried through the lymph channels and eventually into the venous system. Fats absorbed from the intestine are, therefore, the one food not passed directly to the liver.

The larger lymph vessels may be seen by careful dissection. In most fishes the largest, the subvertebral lymph vessels, are just ventral to the vertebral column. Similar vessels, called subvertebral trunks, are found in amphibians, and in reptiles a large medial subvertebral trunk runs along the midline and then divides anteriorly and enters the right and left precaval veins. Paired thoracic ducts run along the dorsal part of the thorax in birds; mammals typically have a left thoracic duct and a right cervical duct (Fig. 16-26). Most frequently the lymphatic vessels drain into either the precaval or the subclavian veins; however in reptiles large lymph vessels drain also into the iliac veins.

Small pulsating structures of smooth muscles, called lymph hearts, are found at the points where the lymph vessels enter veins. These lymph hearts, which help pump the lymph into the veins, have been described in each class of vertebrates except Mammalia and Chondrichthyes, but are most prominent in Amphibia.

The lymphatic system plays an integral part in the circulatory system. It absorbs fats and carries them to the general circulation so that they bypass the liver. Other particles also enter the blood from the interstitial fluids by way of the lymph. Even more important, lymph cells form antibodies which are carried by the blood throughout and combat infection.

SUGGESTED READING

Bugge, J. (1960). The heart of the African lungfish, *Protopterus. Vidensk. Medd. fra Dansk Naturh. Foren.* **123,** 193–213.

Huntington, G. S., and McClure, C. F. W. (1920). The development of the veins in the domestic cat. *Anat. Rec.* **20,** 1–31.

Johanson, K., and Hanson, D. (1968). Functional anatomy of the hearts of lungfishes and amphibians. *Am. Zool.* **8,** 191–210.

Johanson, K., Lenfant, C., and Hanson, D. (1970). Phylogenetic development of pulmonary circulation. *Fed. Proc.* **29,** 1135–1140.

Randall, D. J. (1968). Functional morphology of the heart in fishes. *Am. Zool.* **8,** 179–190.

Rosenquist, G. C. (1971). The common cardinal veins in the chick embryo: their origin and development as studied by radioautographic mapping. *Anat. Rec.* **169,** 501–508.

Rowlatt, V. (1968). Functional morphology of the heart in mammals. *Am. Zool.* **8,** 221–229.

Satchell, G. H. (1970). A functional appraisal of the fish heart. *Fed. Proc.* **29,** 1120–1123.

Sturkie, P. D. (1970). Circulation in Aves. *Fed. Proc.* **29,** 1674–1679.

White, F. N. (1968). Functional anatomy of the heart of reptiles. *Am. Zool.* **8,** 211–220.

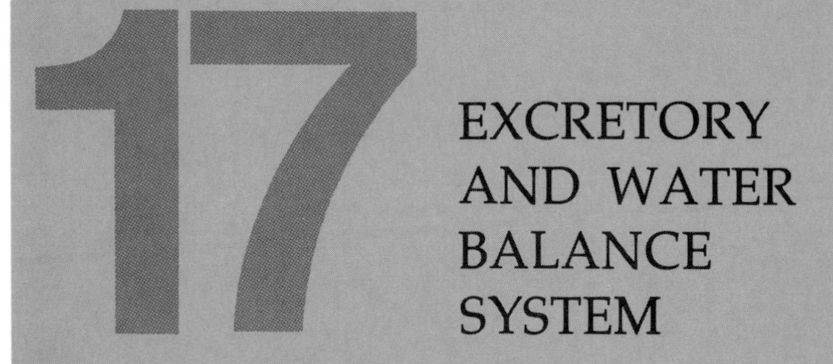

17 EXCRETORY AND WATER BALANCE SYSTEM

17

The composition of a vertebrate's internal fluid milieu is kept relatively constant and independent of the external environment by osmoregulation. Osmoregulation is imperative for a vertebrate's survival. If extracellular fluid is too concentrated, osmotic pressure causes water to move out of the cells into the extracellular fluid and permeable ions to pass into the cells. This changes the cells' internal milieu so drastically that they can no longer function, and they die. On the other hand, if the extracellular fluid is too dilute, osmotic pressure causes water to enter the cells until they literally "blow up" (cytolysis). The complex process of maintaining an optimum extracellular fluid composition involves the continual activity of lungs, gills, skin, sometimes salt-secreting glands, and, most notably, the kidneys to balance the input and output required by the animal's metabolic functions. For example, blood and interstitial fluids bring oxygen and food materials to the cells and carry away their metabolic wastes. These wastes are primarily carbon dioxide from respiration and nitrogen compounds from protein and nucleic acid metabolism; most of the former is eliminated as a gas by respiratory organs and the latter are excreted in the urine by the kidneys.

The primary nitrogenous waste produced by cell metabolism is ammonia, formed by the deamination of amino acids. Because ammonia is highly toxic it must either be eliminated very rapidly, which requires large amounts of water, or be chemically transformed into something less toxic. Tetrapods, therefore, whose osmotic balance depends upon water conservation, chemically transform ammonia into less toxic substances—most frequently, urea and uric acid.

The mechanisms of osmoregulation vary in different environments. The body fluids of freshwater vertebrates, for instance, are more concentrated than their environment, and those of marine vertebrates usually less concentrated. Osmotic pressure alone would tend to dilute the former and dessicate the latter, and active mechanisms are required to maintain proper balance. Terrestrial vertebrates, particularly those in very dry climates, must conserve water and replace the water that is lost through urine and feces, during respiration, and through the surface of the skin ("insensible water loss"); this is done by intake through the alimentary canal and by metabolic water.

Metabolic water is a bonus of cell metabolism. For every 100 gm of protein that are fully metabolized, 41.3 gm of water are produced as a by-product; for every 100 gm of carbohydrate, 55.5 gm of water; for every 100 gm of fat, 107.1 gm of water. Particularly in animals with a low protein, high lipid diet, enough metabolic water is synthesized for it to play an important role in water economy.

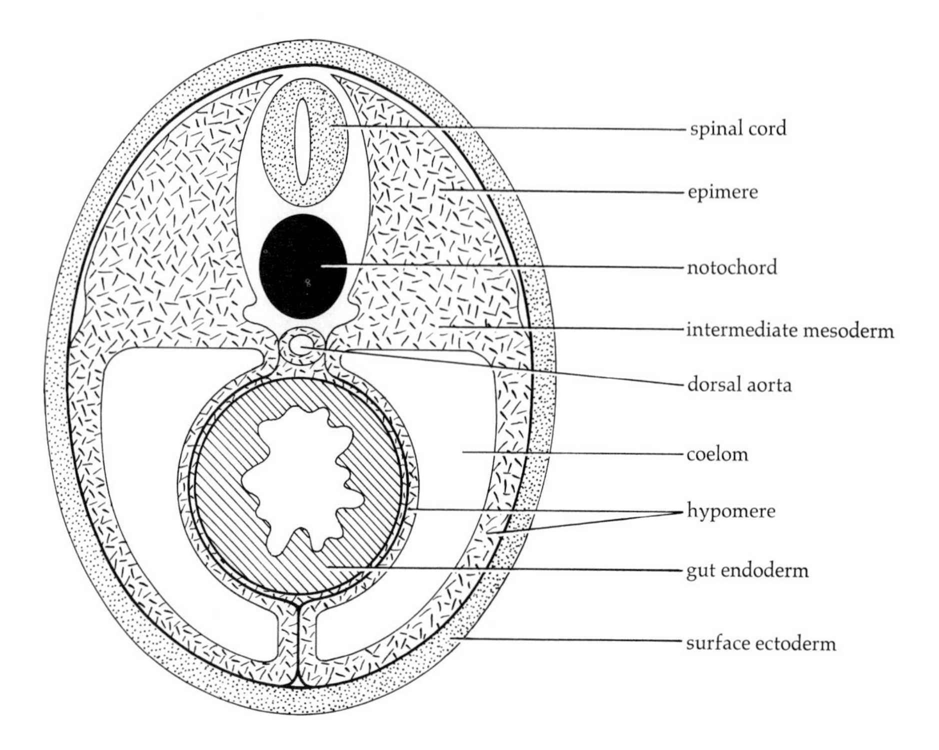

Fig. 17-1. Diagramatic representation of a cross section through a developing vertebrate to show the relationship of the nephrogenic intermediate mesoderm to other tissues.

ONTOGENY OF KIDNEYS AND DUCTS

In all vertebrates the kidneys develop embryologically from mesomeric tissue (of the intermediate mesoderm), which lies in a retroperitoneal position along the dorsal edge of the developing coelom (Fig. 17-1). Like the vertebral column and somatic muscles, the developing kidneys are segmented; like many other structures, they develop in a cephalocaudal sequence (Fig. 17-2). The most anterior nephrogenic tissue is the first to differentiate into kidney tubules; this portion is called the pronephros. The intermediate portion develops next as the mesonephros. The most posterior portion differentiates last, and in animals where only this portion persists in the adult it is called the metanephros.

It is in the kidney tubules that the urine is formed. In the pronephros of most vertebrates one kidney tubule develops for each body segment. These tubules join distally and form a large duct which grows caudally, adjacent to the undifferentiated posterior nephrogenic tissue, and then all the way to the cloaca where it terminates. Unfortunately this primary urinary duct has been given many names: pronephric, holonephric, archinephric, opisthonephric, and, as we will call it, Wolffian duct (Fig. 17-2).

In the mesonephros and metanephros more tubules usually develop per segment—even as many as several hundred. In most vertebrate groups these tubules drain into the Wolffian duct or, in amniotes, the separate, independently evolved urinary duct called the ureter.

But this is getting ahead of our story. The structure and function of the amniote kidney and ducts will be better understood after an analysis of their counterparts in other vertebrates.

THE HOLONEPHROS

The kidney of the larval hagfish is structurally and functionally so generalized that it is thought to be very similar to a phylogenetically early kidney. It is called the holonephros (Fig. 17-3). It lies in a retroperitoneal position, but extends the entire length of the body. Not surprisingly it develops in a cephalocaudal sequence and has one kidney tubule per segment throughout its length. The tubules from the pronephros portion join and form the Wolffian duct, which runs the entire length of the nephrogenic tissue and terminates at the cloaca. As

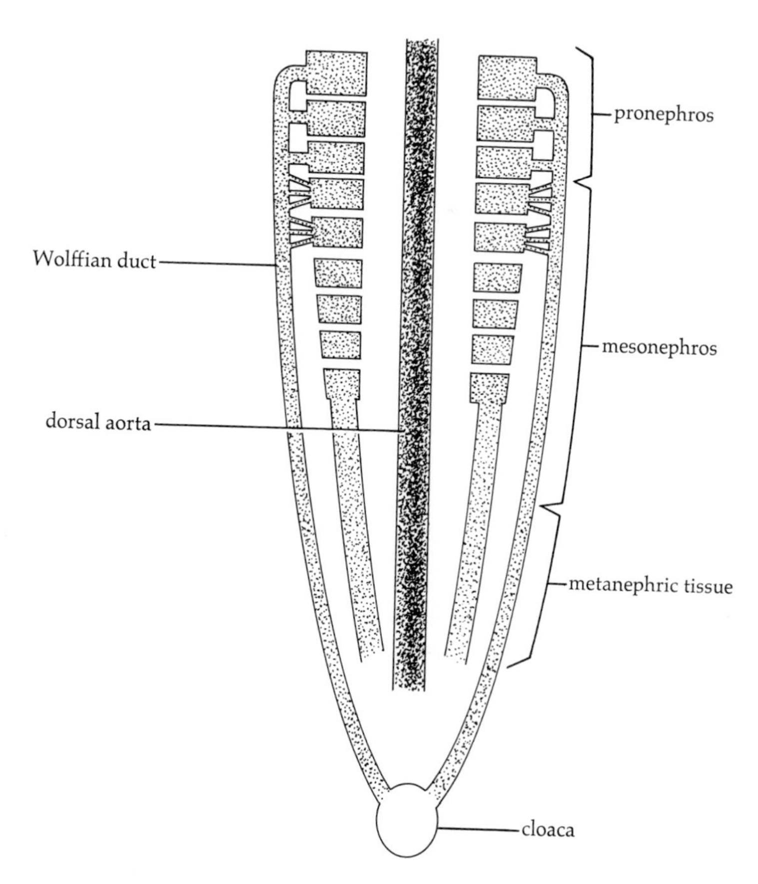

Fig. 17-2. A diagramatic horizontal section through the intermediate mesoderm of a developing vertebrate to show the segmental cephalocaudal development of kidneys. The pronephros develops first and lays down the Wolffian duct. The mesonephric and then the metanephric portions develop later.

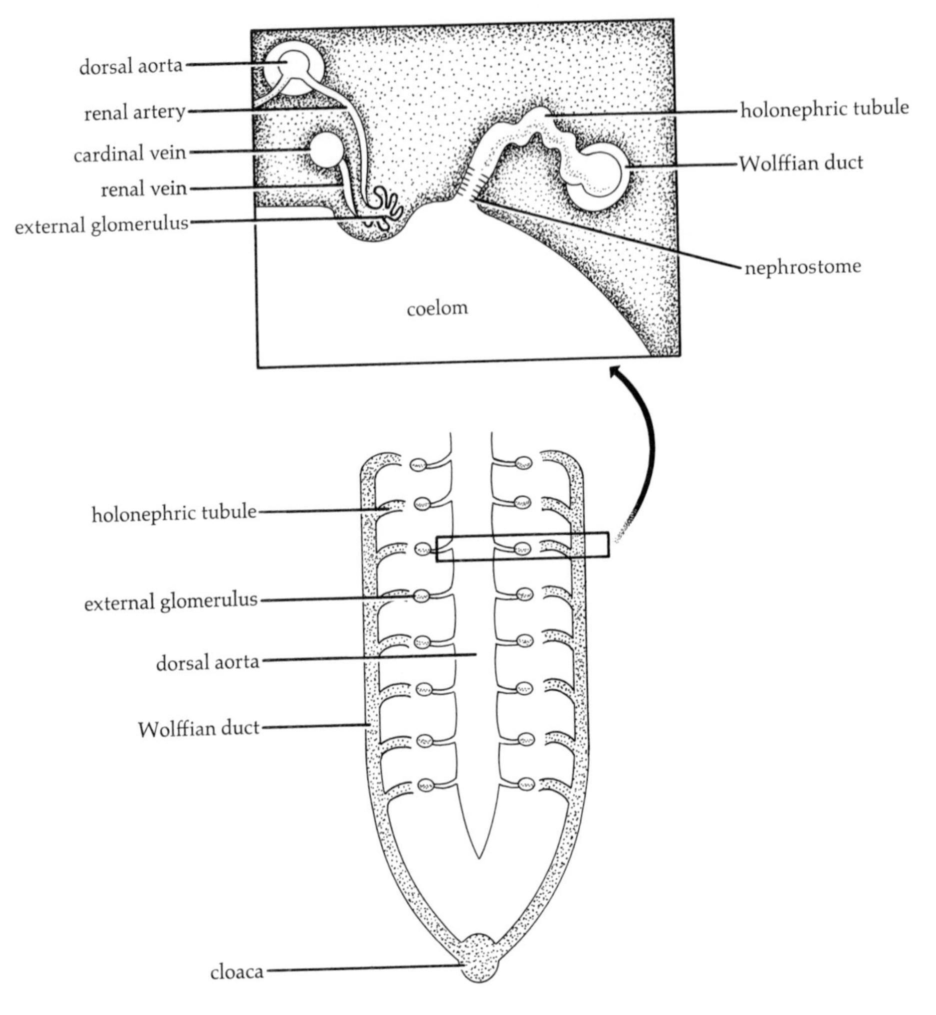

Fig. 17-3. An idealized holonephric kidney, possibly the ancestral vertebrate type.

the mesonephros and metanephros develop, their tubules also hook on to the Wolffian duct.

Each holonephric tubule originates at the body cavity; its lumen is continuous with the coelomic fluid through an opening at its neck, called the nephrostome. The tubule cells in the nephrostome are ciliated and pump coelomic fluid into the tubule (Fig. 17-3).

Blood comes to each holonephric kidney tubule in one of a pair of renal arteries from the dorsal aorta. This renal artery terminates in a capillary knot, called an external glomerulus, against the serosa of the coelom. From the glomerulus, small vessels carry blood to the walls of the tubule where another capillary network forms. A renal vein drains the tubule capillary blood into the posterior cardinal veins; there is no renal portal system.

Blood pressure is very high in the external glomerulus, and its walls are thin. Much of the blood, with its burden of nitrogenous wastes, is filtered out of the capillaries into the coelom; remaining in the blood capillaries are some plasma and the blood cells and larger proteins which cannot pass through this filter. Coelomic fluid is drawn through the nephrostome by the cilia in the neck of the tubule; as the coelomic fluid passes through the tubule some essential materials, such as glucose, are resorbed into the blood from the forming urine. The urine then passes out of the holonephric tubule into the Wolffian duct and is carried down to the cloaca for excretion.

Because the tubules are short, little water can be resorbed through them and the urine is copious. There is some active transport of salts; the urine is, therefore, slightly hypotonic to the body tissues, which is necessary in order to keep the animal's body fluids slightly hypertonic to sea water. Hagfish retain higher salt concentrations in their tissues than do other vertebrates; the adult does it by an opisthonephric kidney.

THE OPISTHONEPHROS

In anamniote embryos (other than hagfish) the most anterior mesomeric tissue forms a pronephros, which lays down the Wolffian duct as described and which may also function as the embryonic kidney. In the anamniote adult the pronephric tubules may or may not be resorbed, but in any case these tubules have no excretory function.

The adult anamniote kidney is the opisthonephros, which develops from the intermediate and posterior portions of the mesomeric tissue.

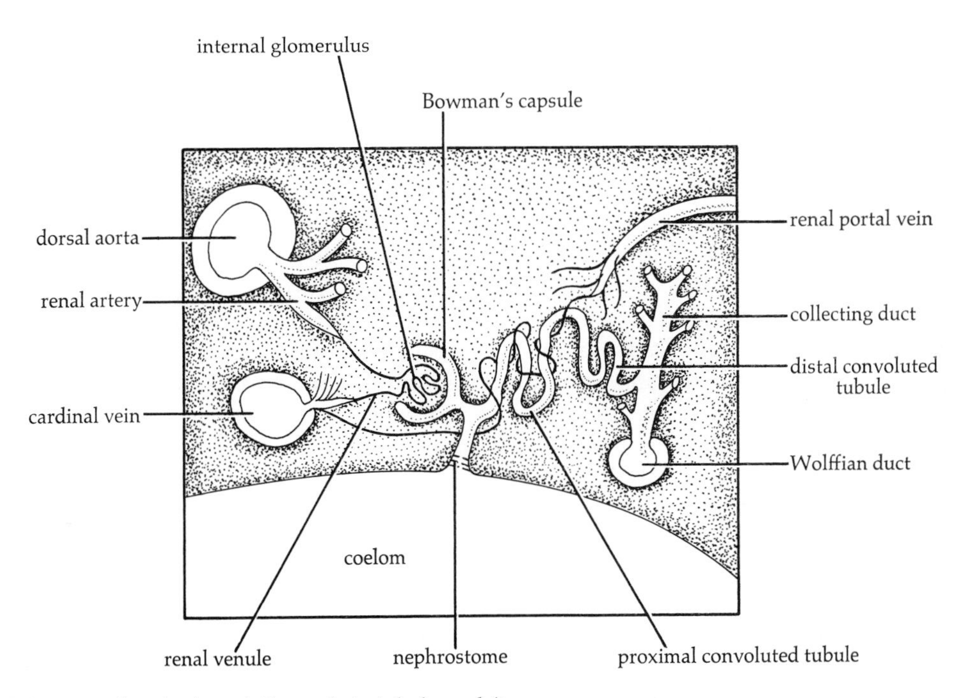

Fig. 17-4. Diagram of a single opisthonephric tubule and its vascularization. Many opisthonephric kidneys lack nephrostomes.

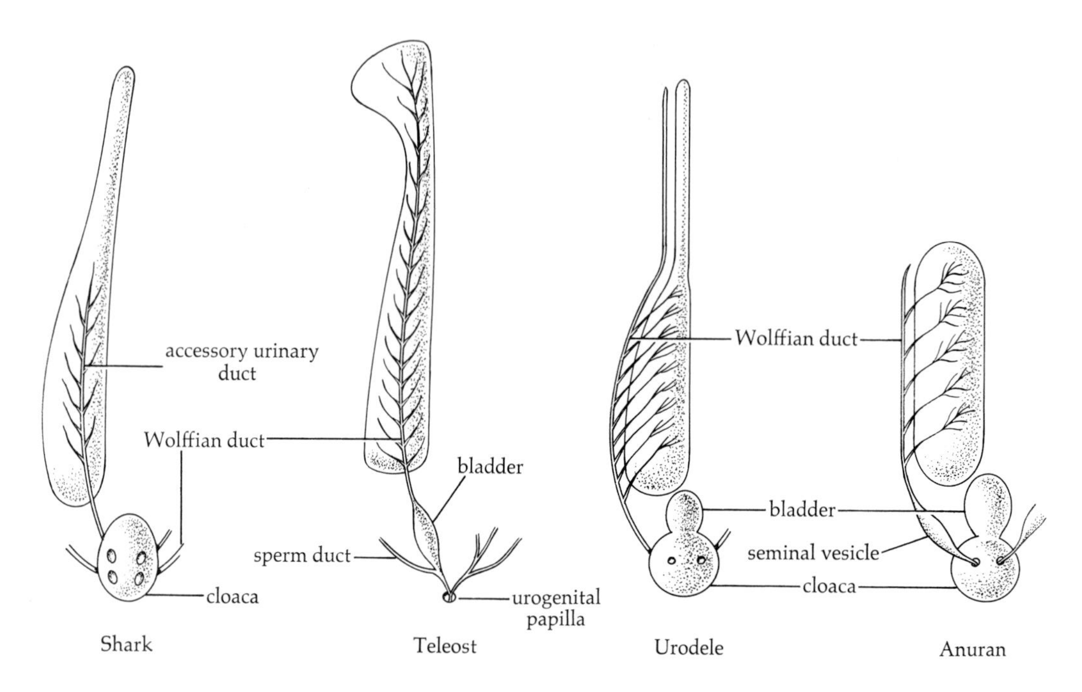

Fig. 17-5. The opisthonephric kidneys and excretory ducts of four different male anamniotes. In urodeles and anurans the Wolffian duct transports both urine and sperm. In teleosts it transports only urine and in male sharks only sperm. In female sharks the Wolffian duct transports urine and no accessory urinary duct is present.

Unlike the pronephros it has several tubules per segment, the number increasing caudally, and each tubule is longer and more convoluted than the relatively short, straight pronephric tubules (Fig. 17-4).

In the opisthonephric kidney each renal artery (or rather, each small branch of the renal artery) terminates in a tuft of capillaries which is called an internal glomerulus because it lies within a double-walled, cup-shaped termination of the kidney tubule called Bowman's capsule (Fig. 17-4). The internal glomerulus plus Bowman's capsule comprise the renal corpuscle. Electron microscopy reveals that the cells of the inner wall of Bowman's capsule, called podocytes, contain small spaces through which the filtrate from the glomerulus can pass into the kidney tubule. Sometimes but not always there is a nephrostome connecting the kidney tubule with the coelomic cavity.

Each opisthonephric tubule is composed of a Bowman's capsule, a short and relatively straight neck region, a convoluted region called the proximal convoluted tubule, another short, straight segment, and a distal convoluted tubule. Several distal convoluted tubules join in a collecting duct, which in most anamniotes empties into the Wolffian duct but in some empties into a different duct, formed by the coalescence of collecting ducts; by one or the other of these ducts the urine is emptied into the cloaca (Fig. 17-5).

The internal glomerulus receives blood from a small branch of the renal artery called the afferent arteriole and is drained by the efferent arteriole, which then forms a capillary network wrapping around the proximal and distal convoluted tubules and the intervening small, straight segment. This capillary network around the tubule also receives blood from the renal portal vein, which drains the hindlimbs and tail. However, the renal portal vein supplies no blood to the glomerulus (Fig. 17-6). The capillaries from both the efferent arteriole and the renal portal system are drained by renal venules into renal veins which then drain into either the postcardinal or postcaval veins, depending on the anamniote.

At the glomerulus the blood is filtered into the kidney tubule, except for some plasma, the blood cells, and the very largest protein molecules. The filtrate passes down the kidney tubule, and throughout this passage there is selective resorption of salts and sugars and selective secretion of additional waste products, particularly of urea and ammonia; there is very little resorption of water. Thus the final product, urine, has a distinctly different composition than the filtrate at Bowman's capsule.

THE METANEPHROS

In amniotes a pronephros develops very early and lays down a Wolffian duct; it is never a functional kidney, however. The mesonephros develops next from the intermediate mesomeric tissue; the mesonephros has a tubule structure similar to that of the anamniote opisthonephros and is drained by the Wolffian duct. This is the functional kidney of the embryo and early fetus (Fig. 17-7).

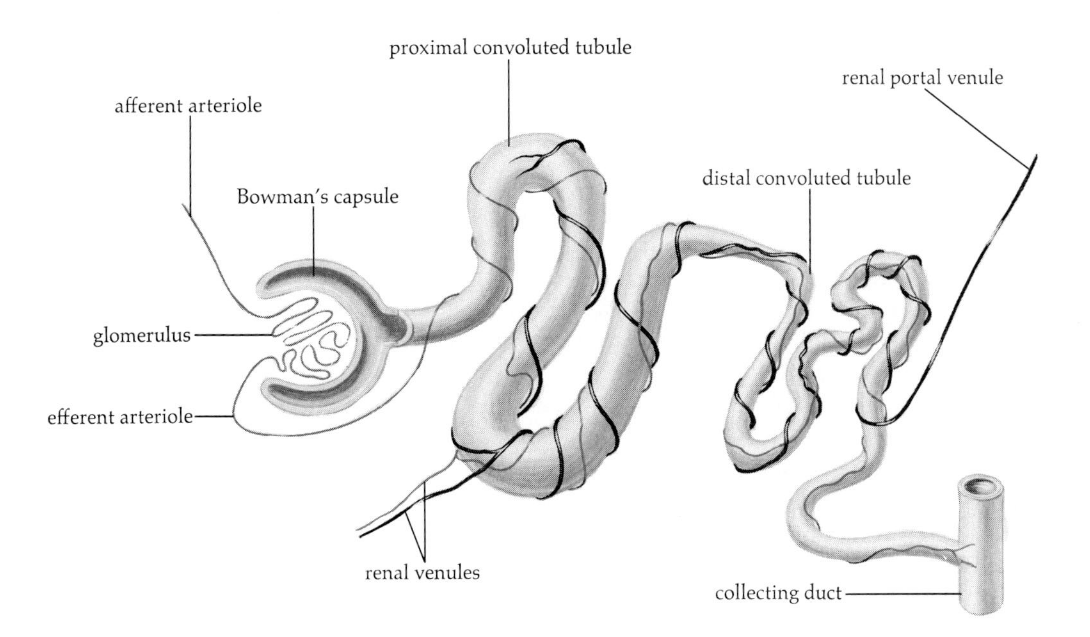

Fig. 17-6. Detail of the vascularization of a single opisthonephric tubule.

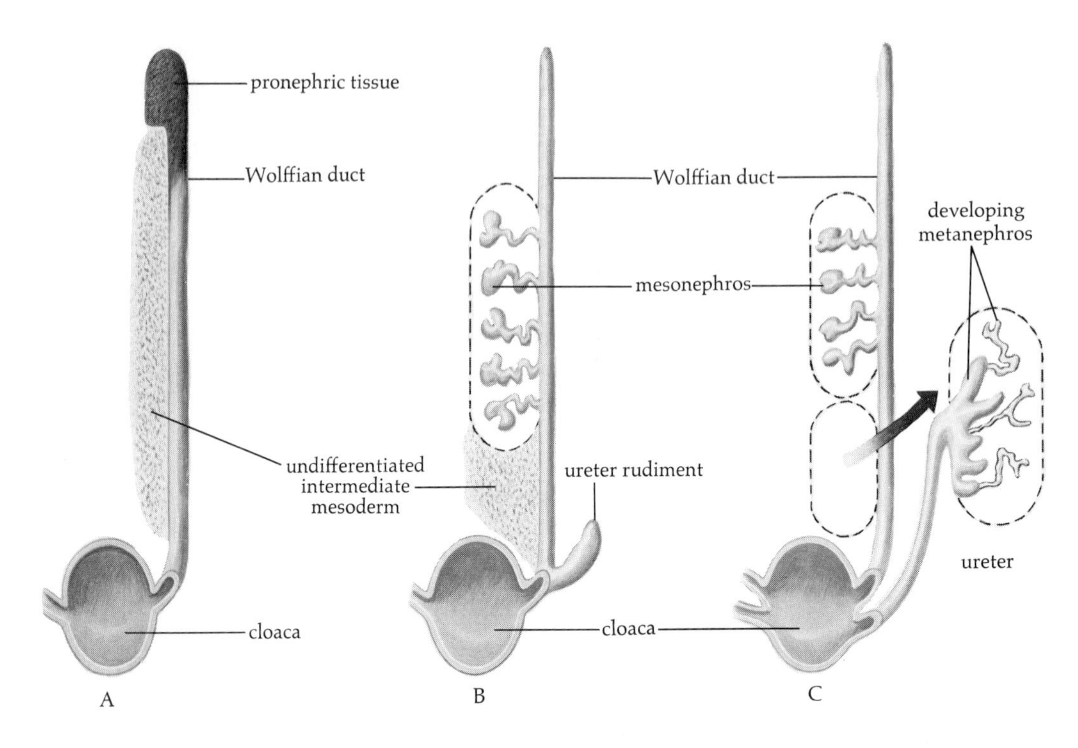

Fig. 17-7. Three stages during the development of the amniote kidney.

The kidney of the adult amniote, however, is the metanephros. It develops from the most posterior portion of the mesomere, with many more kidney tubules than in any other kidney type. Each metanephric kidney tubule, like each opisthonephric tubule, is made up of a Bowman's capsule, a very short neck region, a proximal convoluted tubule, an intermediate region, and a distal convoluted tubule; many distal convoluted tubules empty into a single collecting duct. However the collecting ducts of the metanephros do not enter the Wolffian duct. Instead, at the junction of the Wolffian duct and the cloaca an evagination grows anteriorly toward the metanephros, forming the ureter; it enters the metanephric tissues and divides into large funnel-like structures called calyces. It is into these calyces that the collecting ducts of the kidney empty (Fig. 17-8). Therefore the metanephric kidney has a dual origin—partially from the posterior nephrogenic tissue and partially from the budding of the cloaca.

There are also functional similarities between the metanephros and the opisthonephros. In the renal corpuscle, only a thin connective tissue membrane separates the glomerular capillaries from Bowman's capsule; here blood plasma is filtered through fenestrations in the capillary endothelium and spaces between the podocytes of Bowman's capsule (Fig. 17-9). This efficient mechanism filters about 90% of the blood plasma into the cavity of Bowman's capsule, from which it passes through the neck region into the proximal convoluted tubule.

The proximal convoluted tubule is made up of truncated, pyramid-shaped cells, with a strong brush border of microvilli adjacent to the lumen and large interdigitations in the plasma membrane adjacent to the thin basement membrane. These cells are metabolically very active and are responsible for selective resorption of glucose, amino acids, and various salts from the filtrate. At the same time additional nitrogenous wastes are secreted from the bloodstream into the lumen of the proximal convoluted tubule.

The intermediate segment, of variable length, will be discussed below. It has low cuboidal to squamous epithelium and is a primary site of water resorption (Fig. 17-10).

In the distal convoluted tubules, where the cells are somewhat flatter and their plasma membranes have fewer convolutions, the final formation of urine takes place; some more water is resorbed, the pH is adjusted, and there is selective sodium resorption.

The kidney's activity is adjustable over a wide range. If, for instance, the body is flooded with sodium, large amounts of sodium are excreted; if the body is low in sodium, very little is excreted. Furthermore, within large tolerances, the amounts of wastes and water that are excreted are independent: when there has been a large water intake the urine simply becomes more dilute, and when there has been a small water intake, it becomes much more concentrated.

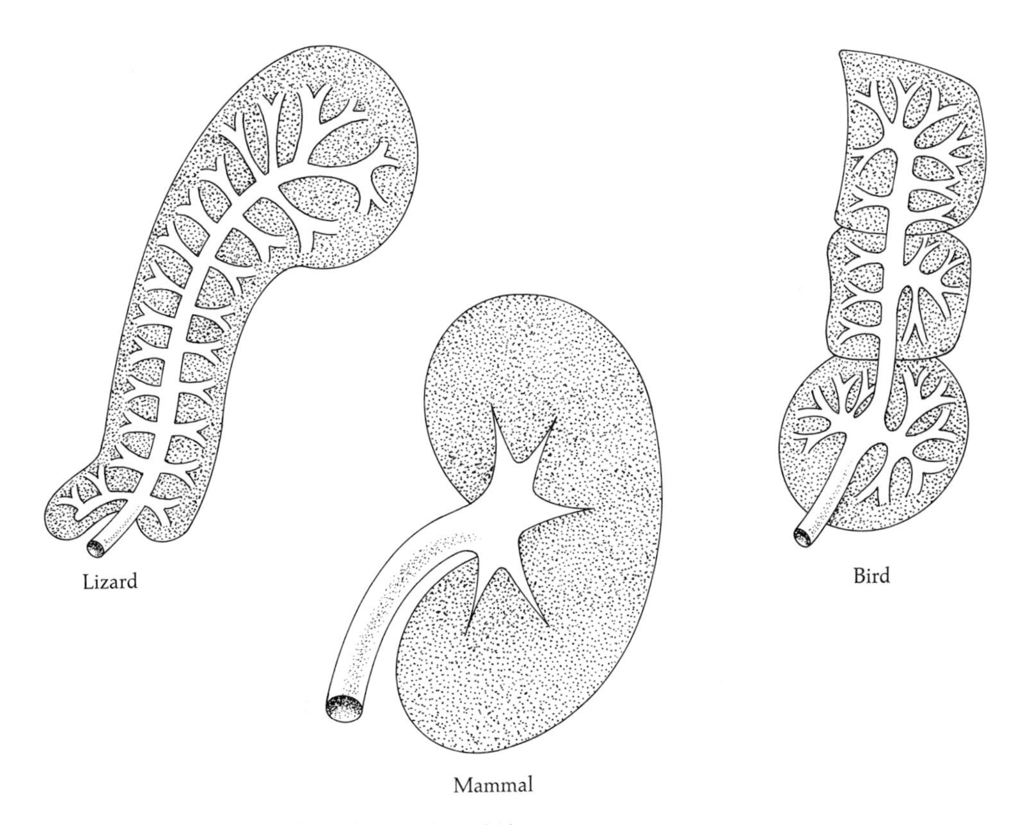

Fig. 17-8. Horizontal sections through amniote kidneys showing the distribution of the ureter and calyces within the kidney substance.

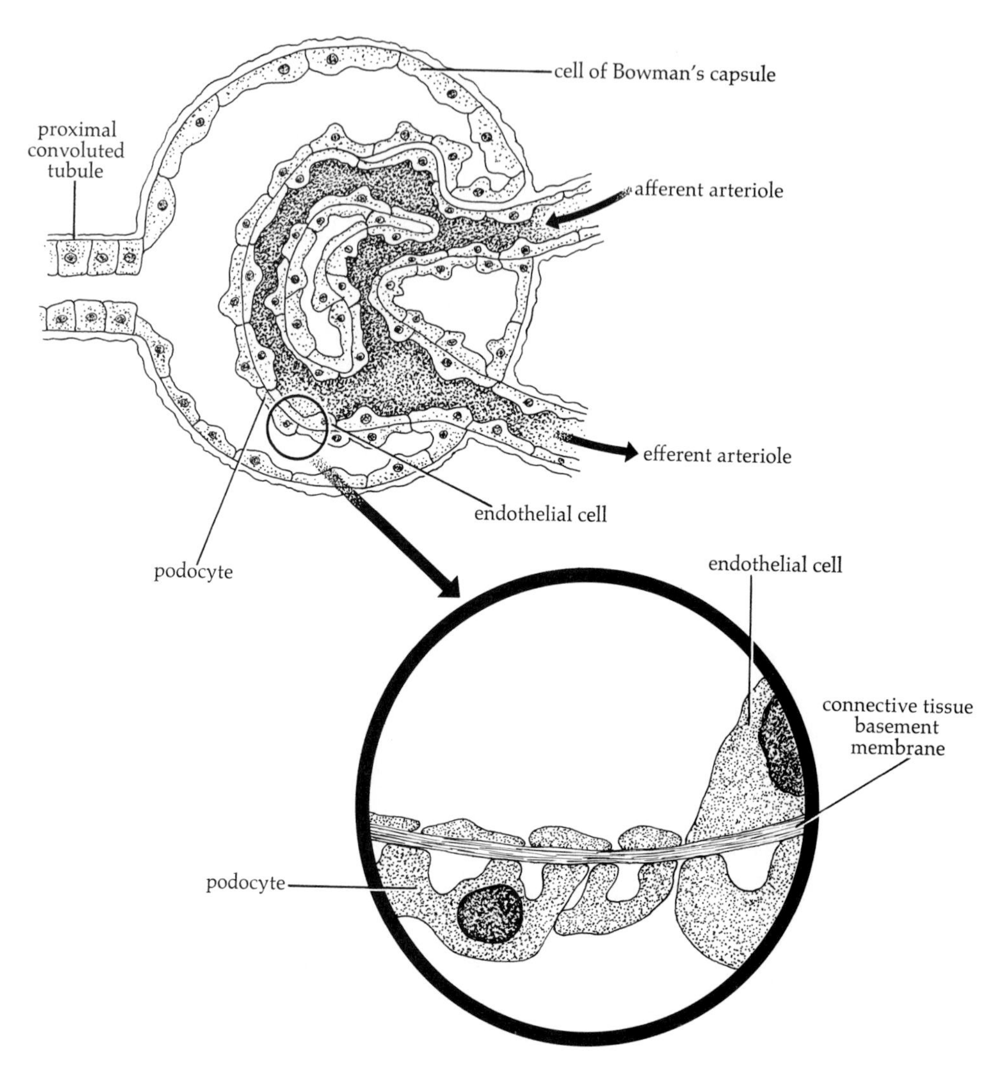

Fig. 17-9. Detailed structure of Bowman's capsule and glomerulus.

The mechanisms of this control are not fully understood, but some elements are clear. Myoepithelial cells—not unlike those around sudoriferous glands—form a juxtaglomerular apparatus which enfolds the afferent arteriole just before it enters the glomerulus. Contraction of this apparatus greatly diminishes the amount of blood flowing through the glomerulus and thus the amount of filtration that can occur. It is also thought that these myoepithelial cells secrete the enzyme renin which, it will be recalled, can transform the angiotensin I in the blood to angiotensin II, which in turn stimulates further vasoconstriction. One portion of the distal convoluted tubule, the macula densa, has specialized slender, columnar cells which apparently respond to the osmolarity of the urine and pass this information to the juxtaglomerular cells, thus influencing whether or not renin is secreted. The mechanisms of such a feedback system are unclear, and in fact the whole idea, although consistent with current experimental evidence, is still speculative.

The intermediate segment of the amniote kidney tubule is relatively short in most reptiles, long and somewhat looped in birds, and extremely long and looped in mammals where it is called the loop of Henle (Fig. 17-11). Its function is best understood in mammals.

All of the mammalian renal corpuscles, proximal convoluted tubules, and distal convoluted tubules are in the outer portion, or cortex, of the kidney; the collecting ducts, calyces, and loops of Henle are in the deeper part, or medulla. The loops of Henle pass from the proximal convoluted tubule down parallel with the collecting ducts and finally loop back, travel up, and join the distal convoluted tubules. Because the blood in the collecting ducts and the forming urine in the loops of Henle flow in opposite directions, there is a countercurrent distribu-

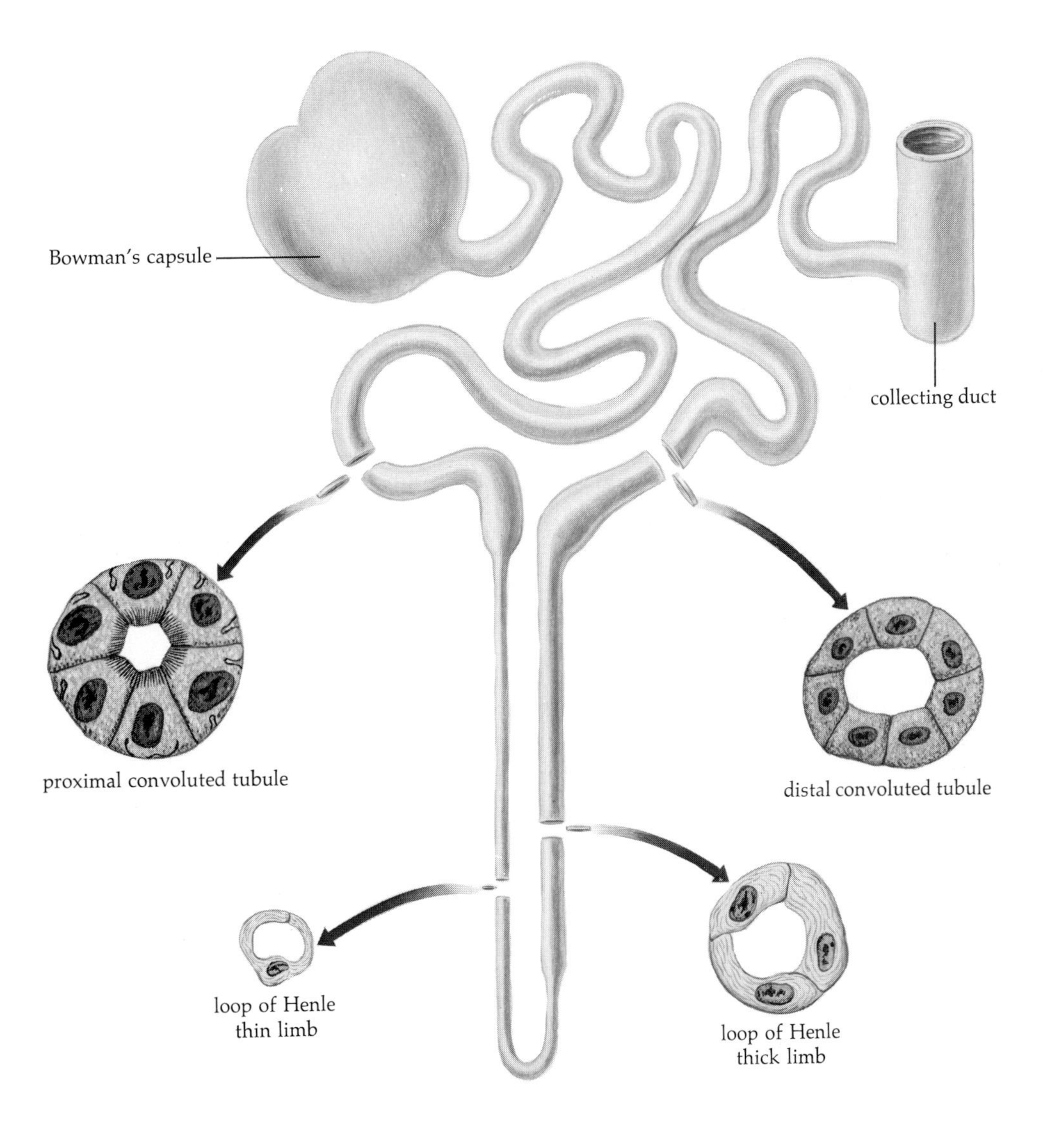

Fig. 17-10. Details of a mammalian metanephric tubule.

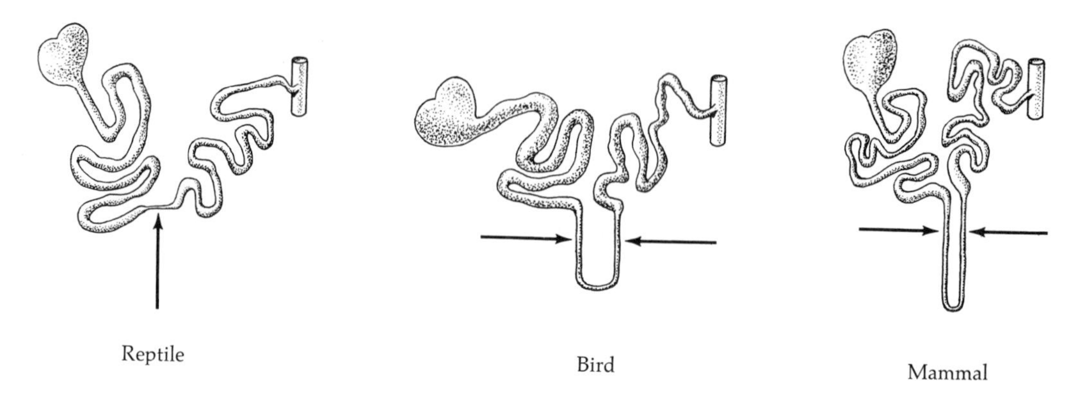

Fig. 17-11. The most variable portion of the metanephric tubule structure of amniotes is the extent of the intermediate segment, indicated here by arrows.

tion which facilitates and accelerates the absorption of water from the urine—further influenced by the fact that the interstitial fluid in the medulla is hypertonic to that of other parts of the body. The longer the loops of Henle, the more opportunity there is for water resorption to occur. They reach their extreme size in some desert mammals

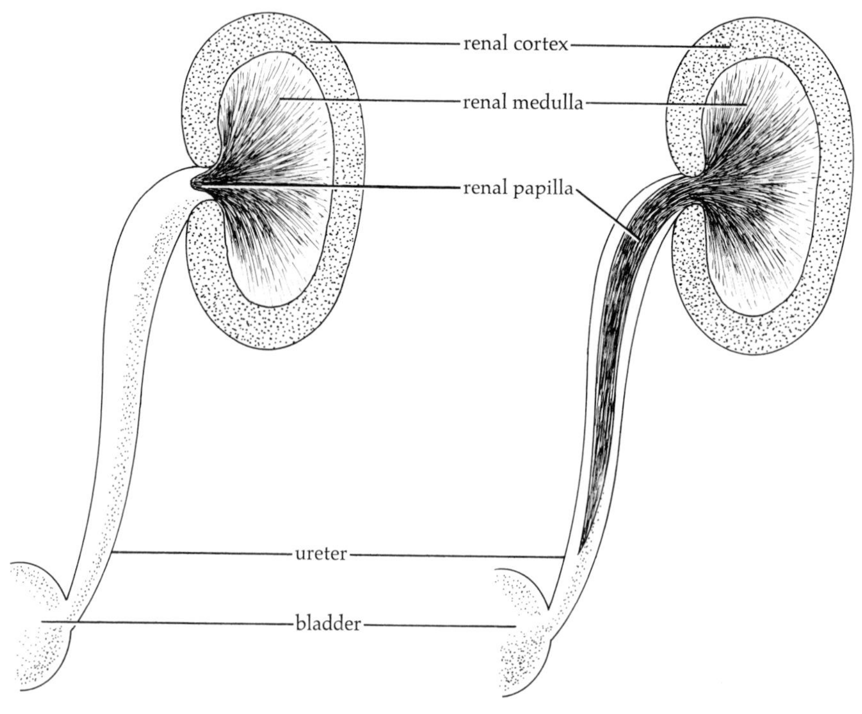

Fig. 17-12. Horizontal sections through the kidney and ureter. The loops of Henle pass from the cortex to the tip of the renal papilla, make a hairpin turn, and pass back to the cortex. Most water resorption occurs in the loops of Henle, which reach their maximum length in desert rodents.

where they pass far into the ureter before looping back up (Fig. 17-12). Thus the loop of Henle plays an extremely important role in the selective resorption of water, in mammals at least. This occurs by osmosis rather than by active transport; so far as is known there is no means for the active transport of water in vertebrates.

The intermediate segments also tend to be longer in reptiles and birds inhabiting dryer climates than in those for whom water is abundant, thus suggesting that the water-conserving role of this part of the kidney is not limited to mammals. However, since water is resorbed in the avian and reptilian cloaca, they have had less selective pressure for the development of long loops of Henle than have mammals.

MAINTENANCE OF WATER BALANCE IN MAJOR GROUPS OF VERTEBRATES

Organs other than the kidney are involved in osmoregulation. It is therefore appropriate to look at the functional anatomy of the entire water balance system, particularly as it relates to environmental stresses.

FRESHWATER FISHES

Freshwater fishes have an internal osmotic pressure greater than that of their environment, that is, they are hypertonic. This being so, there

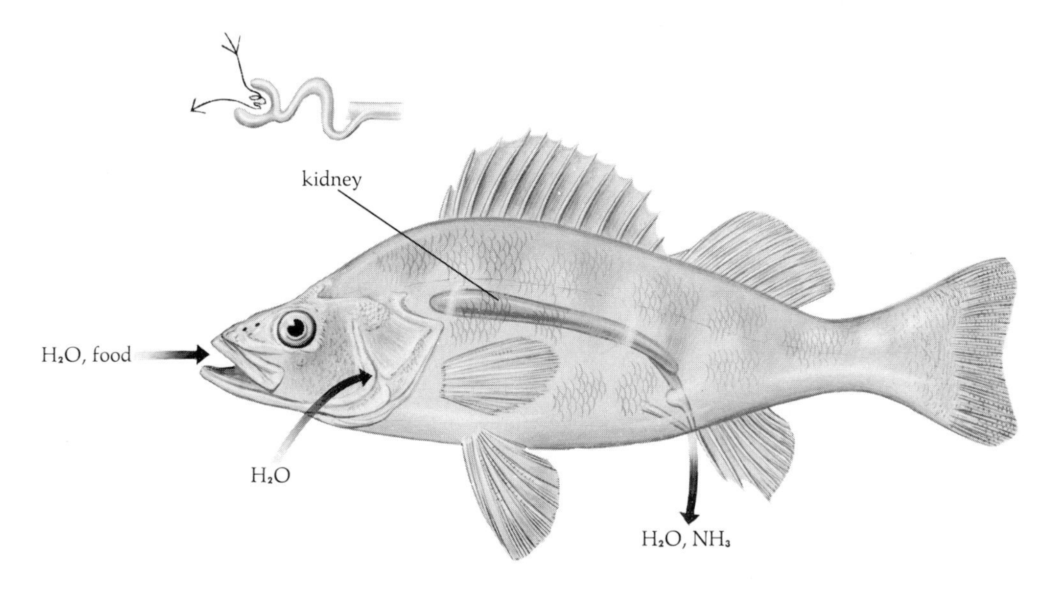

Fig. 17-13. Freshwater vertebrates must maintain a greater internal osmotic pressure than that of the water in which they swim. Teleosts do this by a kidney with large glomeruli and small tubules, which facilitates production of a copious, dilute urine that flushes water out as fast as it enters.

would be a tendency for water to diffuse into them, and if there were not some control, their water balance could not be maintained. The mucous layer over the skin renders it nearly impermeable to water. However, the alimentary canal and the gills must remain permeable to water, and through these structures water constantly diffuses into the animal. With so much water coming in it is necessary to produce a copious, dilute urine. The opisthonephric kidneys of freshwater fishes have large glomeruli for this purpose, and short tubules which resorb little if any water (Fig. 17-13). With so much water being excreted so rapidly, ammonia does not need to be transformed but can be excreted before it can build up to a toxic level.

In most freshwater fishes the chloride cells of the gills actively secrete sodium, and thus play a minor role in osmoregulation.

MARINE VERTEBRATES

Vertebrates in a salt-water environment (except the hagfish) have the opposite metabolic problem: their body fluids are hypotonic to sea water, and there is a tendency for water to diffuse out, thus drying up the animal. Moreover, sodium diffuses in through the gills and ali-

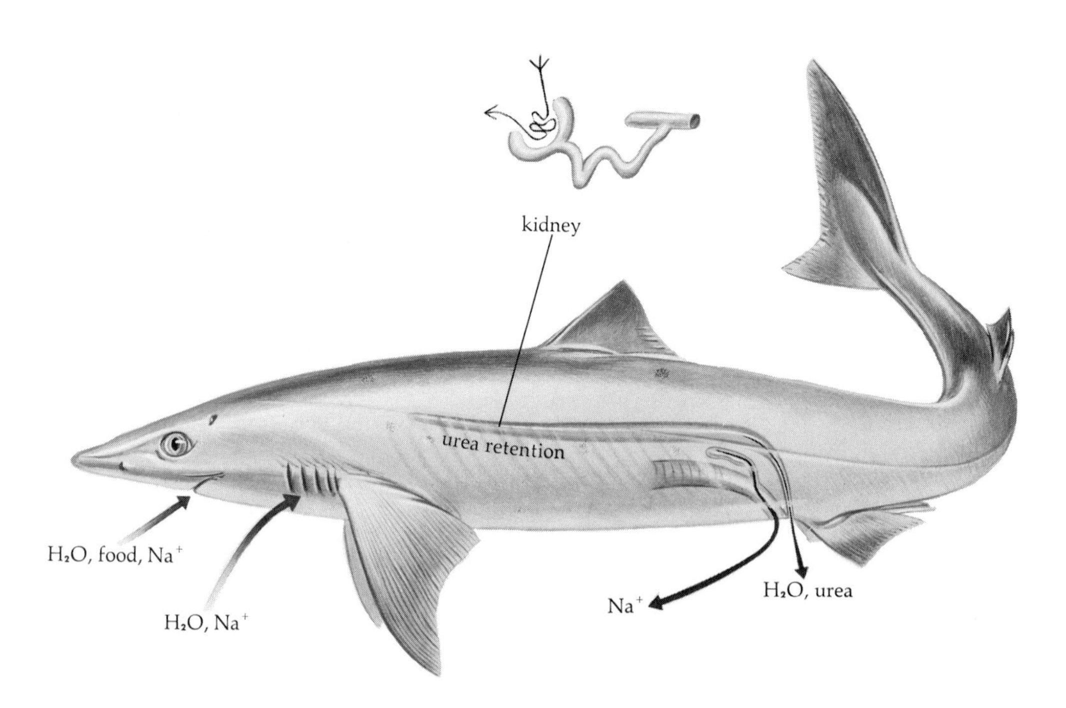

Fig. 17-14. Sharks, although in a marine environment, maintain an osmotic pressure which is greater than that of sea water by maintaining a high urea concentration in their blood and tissues. Excess water and sodium enter through the alimentary canal and gills. A large-glomerular, small-tubular kidney eliminates the excess water and the rectal gland excretes a concentrated sodium chloride solution.

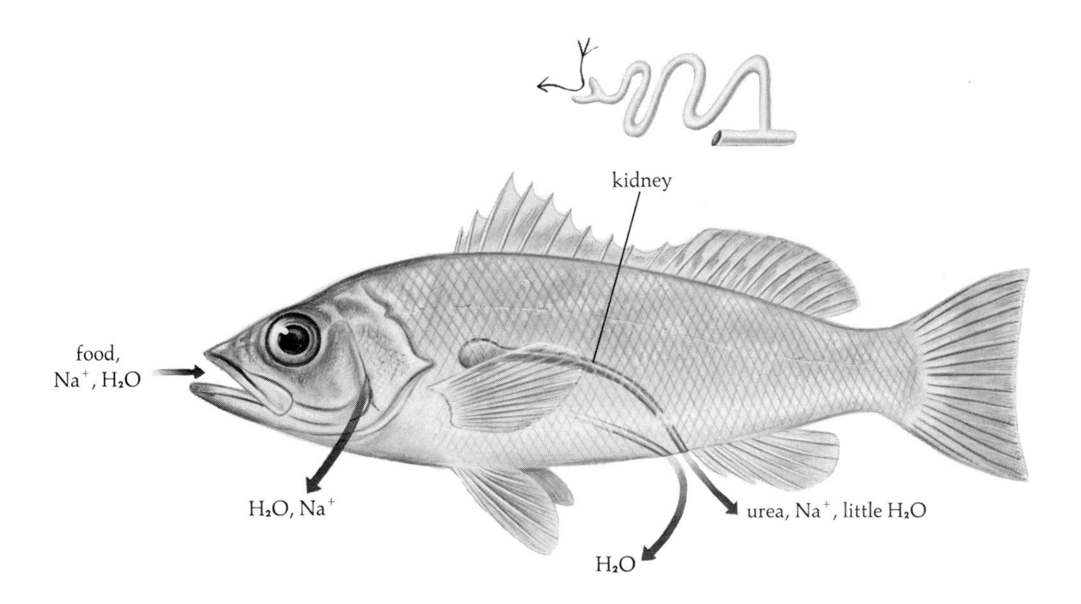

Fig. 17-15. Salt-water teleosts maintain an internal osmotic pressure which is less than that of their environment. Their small-glomerular kidneys lose little water by filtration. The longer kidney tubules secrete urea and sodium, forming a concentrated urine. Excess sodium is eliminated by the chloride cells of the gills.

mentary canal, thus potentially increasing the tonicity of the body fluids and putting an undue osmotic strain on the cells.

It is interesting that every vertebrate class has at least a few marine members, which have evolved somewhat different mechanisms to maintain metabolic balance. All these mechanisms have two characteristics in common. First, all involve relatively impermeable body surfaces which reduce the amount of water diffusing out of the body; in salt-water fishes it is most frequently the same sort of mucous layer that protects freshwater fishes, and in marine tetrapods it is a very dense dermis. Second, no marine form excretes ammonia; the necessary large volume of water would further deplete the animal's internal water reserves.

The hagfish, alone among marine vertebrates, has salt concentrations greater than sea water and an osmotic gradient similar to that of freshwater fishes. Thus it also has similar water balance problems, and it is not surprising that it has large glomeruli and a copious urine excretion. Precisely how the hagfish maintains other metabolic activities is not understood; in most vertebrates, for instance, nerves could not conduct action potentials in the presence of so much sodium.

All but a few species of chondrichthyeans live in salt water. They maintain osmotic equilibrium in a complex way. Their high protein diet yields a large amount of ammonia, which is rapidly converted to urea in the liver. The urea is then maintained in the body tissues and blood at concentrations which in most vertebrates would cause uremia and death, making the body quite hypertonic (Fig. 17-14) and causing water to diffuse into the body tissues through the gills and alimentary canal. Not surprisingly, chondrichthyeans have a very large renal corpuscle and produce a copious and dilute urine; water is thus excreted as rapidly as it enters. Excess salt is actively excreted by the rectal, or digitiform gland, which produces a concentrated sodium chloride solution and excretes it into the rectum from which it passes out through the cloaca (Fig. 17-14). It is interesting that the few freshwater chondrichthyeans also retain high concentrations of urea compared to other freshwater forms. This increases their osmotic pressure, causing more water to come into the animal and, therefore, puts a greater burden on their kidneys.

Analyses of the one living crossopterygian, the marine coelacanth *Latimeria*, indicate that it also retains a high urea content; evidently its osmoregulatory mechanisms are similar to those of chondrichthyeans.

Marine teleosts also convert most of their ammonia into urea, but the urea is then rapidly excreted by the kidneys and perhaps even by the gills. Their kidneys are smaller than those of chondrichthyeans, with very small glomeruli, or, in some forms, none at all. This mini-

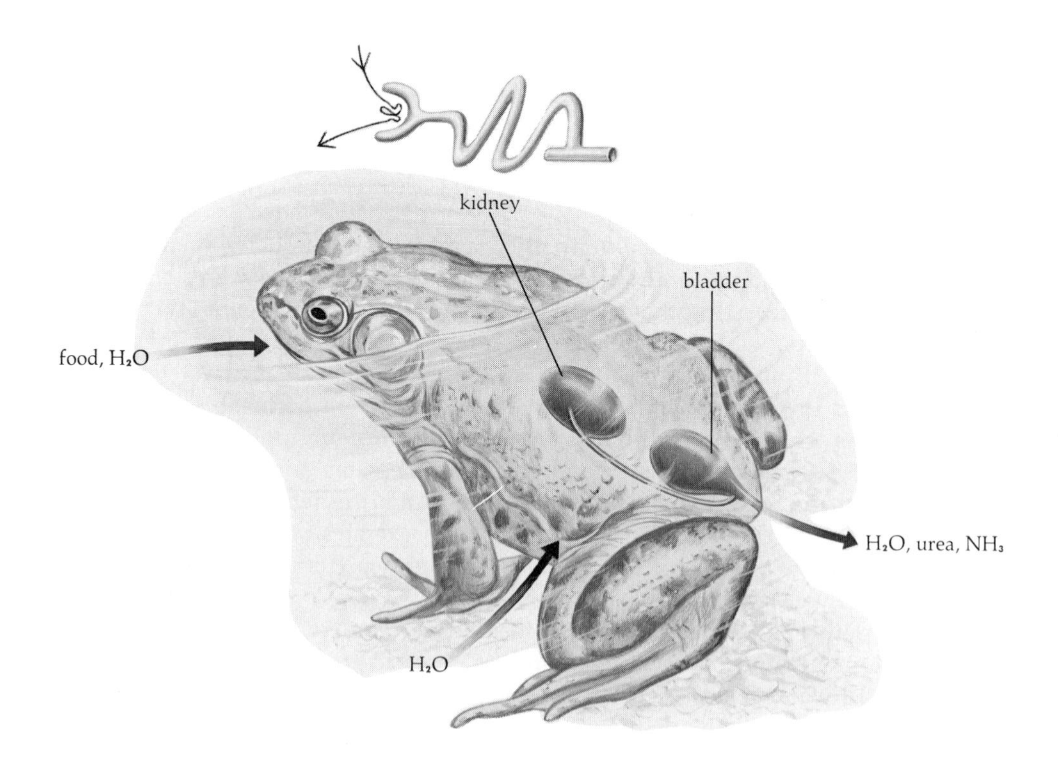

Fig. 17-16. While in water, frogs have a greater internal than external osmotic pressure; water enters through the alimentary canal and the skin, and the large-glomerular kidney produces a copious, dilute urine. On land, desiccation is a problem and water can be resorbed from the large bladder. Even with this water economy, however, the animal would perish except that it remains in moist areas where evaporation is minimized.

mizes filtration. The urine is formed by the direct secretions from the capillaries of the renal arteries and renal portal veins which surround the kidney tubules, and is very concentrated, being primarily urea with a small amount of water (Fig. 17-15).

Like all marine forms, marine teleosts must eliminate excess salt. Their chloride cells are more profuse than those of freshwater teleosts and actively excrete a hypertonic sodium chloride solution.

AMPHIBIANS

When not on land, most amphibians inhabit fresh water, and unlike fishes their skin is highly permeable to water. Their tissues are hypertonic, and thus water tends to flood into the body; however on land they tend to lose a great deal of water by evaporation from the surface.

Their kidneys have large glomeruli and produce a copious, dilute urine. This is not passed directly to the outside; after entering the cloaca it backs up into a large sac, the urinary bladder, where it may be stored for a considerable period before being vented (Fig. 17-16). Thus when the animals are on land they can offset the effects of evaporation by resorbing water from their urinary bladder. Toads, which are more terrestrial than most amphibians, have a skin which is denser, less moist, and much less permeable to water; their kidney tubules are larger and their glomeruli smaller, producing a more concentrated urine with less water loss.

One of the very few marine amphibians is a crab-eating frog of southeastern Asia whose internal milieu is slightly hypertonic to the sea. This hypertonicity is maintained by retaining slightly more sodium chloride than do most vertebrates and by retaining urea much as do chondrichthyeans and the coelacanth, *Latimeria.*

REPTILES AND BIRDS

The more completely terrestrial vertebrates—reptiles, birds, and mammals—have metanephric kidneys, which have a greater capacity for concentrating urine and thus for reducing water loss.

Reptiles (except freshwater turtles) and all birds excrete their nitrogenous wastes primarily as uric acid. Because this is almost insoluble in water it does not add to the osmotic pressure, and thus it helps in water conservation. Reptiles have reduced numbers and sizes of glomeruli (not unlike marine teleosts), and most of the urine is formed by direct secretion of wastes but very little water from the blood into the kidney tubules. The reptilian intermediate segment is very short; of the small amount of water in the wastes very little is resorbed, for there must be some water to carry the uric acid down the ureter to the cloaca or else it would "stop up the excretory plumbing." Thus the cloaca itself is the major water-resorption organ in the reptiles. There urine and fecal material are mixed, most of the water is resorbed, and a nearly dry, pasty mixture is extruded.

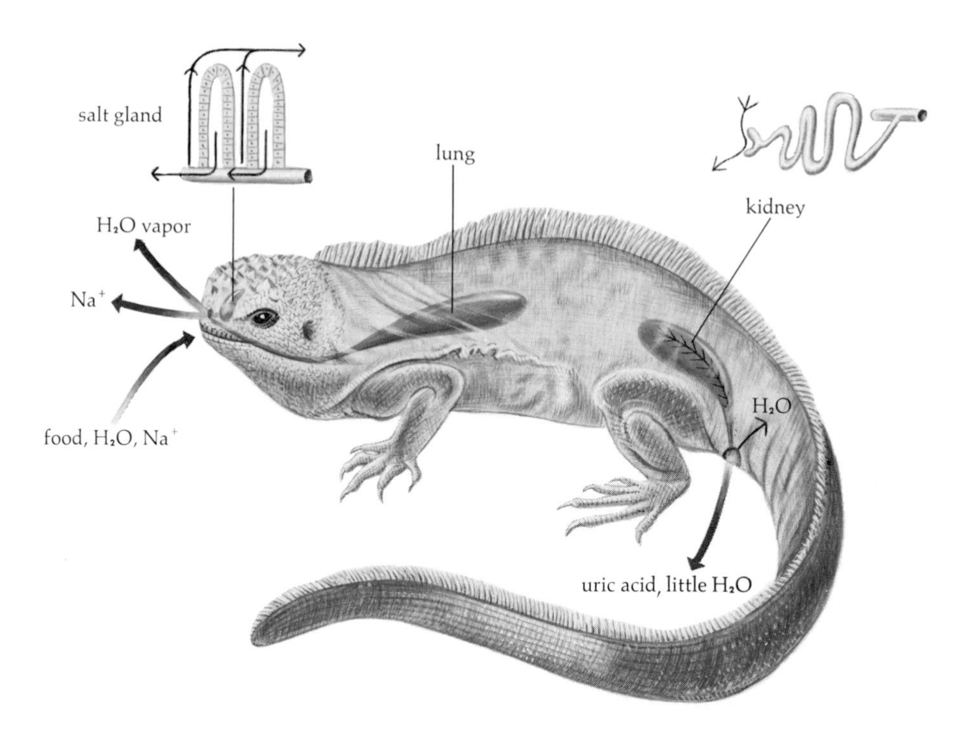

Fig. 17-17. Marine reptiles maintain an internal osmotic pressure which is less than that of seawater. Therefore water tends to diffuse out of them, and sodium tends to accumulate in their tissues. Small glomeruli and long tubules produce a viscous urine of uric acid, and water is resorbed from the cloaca. Excess sodium is excreted by the salt glands.

Birds have larger glomeruli, so that more filtration occurs, but also a larger intermediate segment which resorbs much of the water; some water is also resorbed in the distal convoluted tubule. As in reptiles, the remaining water and metabolic wastes, primarily uric acid, are passed down to the cloaca and mixed with fecal material; water is resorbed and the pasty residue is extruded from the cloaca.

Some reptiles (e.g., sea turtles, sea snakes, the marine iguana) and some birds (e.g., sea gulls, albatross, Savannah swamp sparrow) live in areas with essentially no fresh water and therefore take in sea water. Of these forms, only the Savannah swamp sparrow has kidneys which can excrete a sufficiently concentrated salt solution for osmotic balance. The others have evolved special salt-secreting glands which function analogously to the rectal glands of chondrichthyeans and the chloride cells of marine teleosts (Fig. 17-17). These are modified glands of the head region, but in different groups are nasal glands, orbital glands, or salivary glands. Despite the diverse embryology and independent evolution of these glands, their histological structure is similar in all forms in which they are found, and they involve an active transport system facilitated by a countercurrent distribution system which allows greater concentration. In addition, these marine reptiles and birds maintain water balance by resorbing most of the water at the level of the cloaca (and urinary bladder when present).

MAMMALS

Mammals excrete their nitrogenous wastes as urea rather than uric acid. Their kidneys have very many nephrons, each with a large glomerulus, and there is a great amount of filtration at the renal corpuscle: in man, approximately 125 cm^3 of fluid filters into the nephrons each minute. Most of the water is resorbed by the kidney tubule, primarily in the loop of Henle (which facilitates resorption by its countercurrent arrangement) and secondarily in the distal convoluted tubule; resorption is further facilitated by the hypertonicity of the interstitial fluids in the kidney medulla, where the loop of Henle lies. However, the excretion of urea does require a considerable amount of water, and therefore the urine of mammals, although concentrated, is never as concentrated as the uric acid excreted by reptiles and birds.

The kidneys of most mammals are not capable of producing an extremely concentrated sodium chloride excretion, and, unlike some

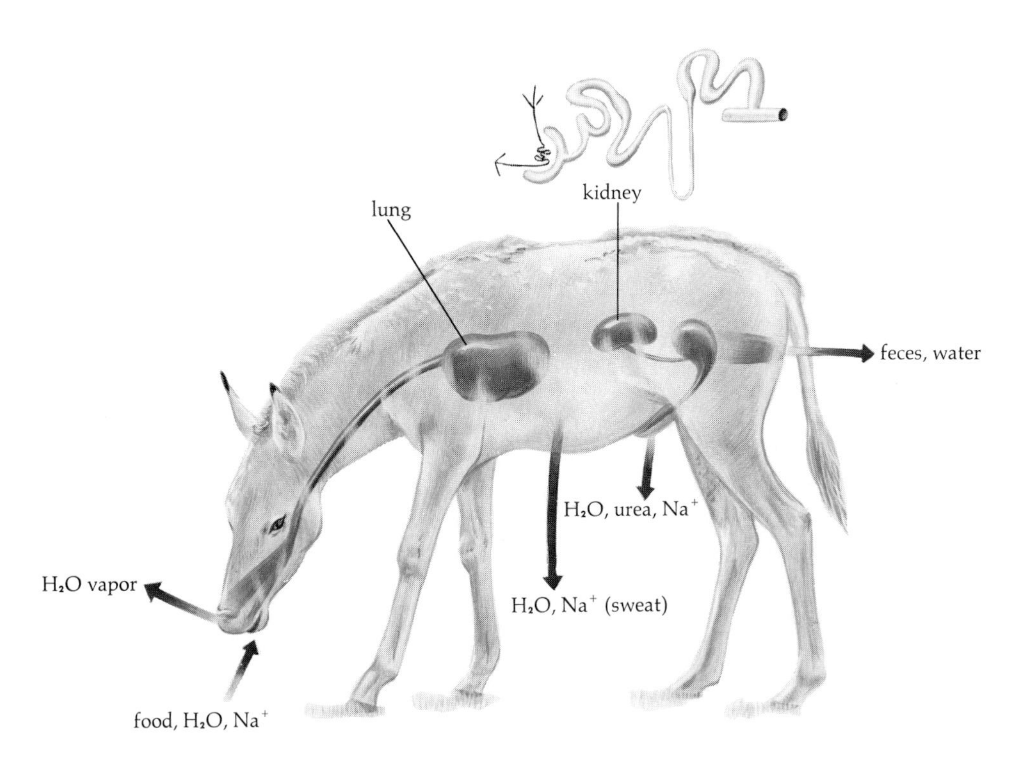

Fig. 17-18. Most mammals have access to only a limited supply of water and lose water through urine, feces, expired air, and sweat (in controlling body temperature). Much filtration occurs with their large glomeruli, but water is resorbed from the forming urine in the loop of Henle and collecting ducts.

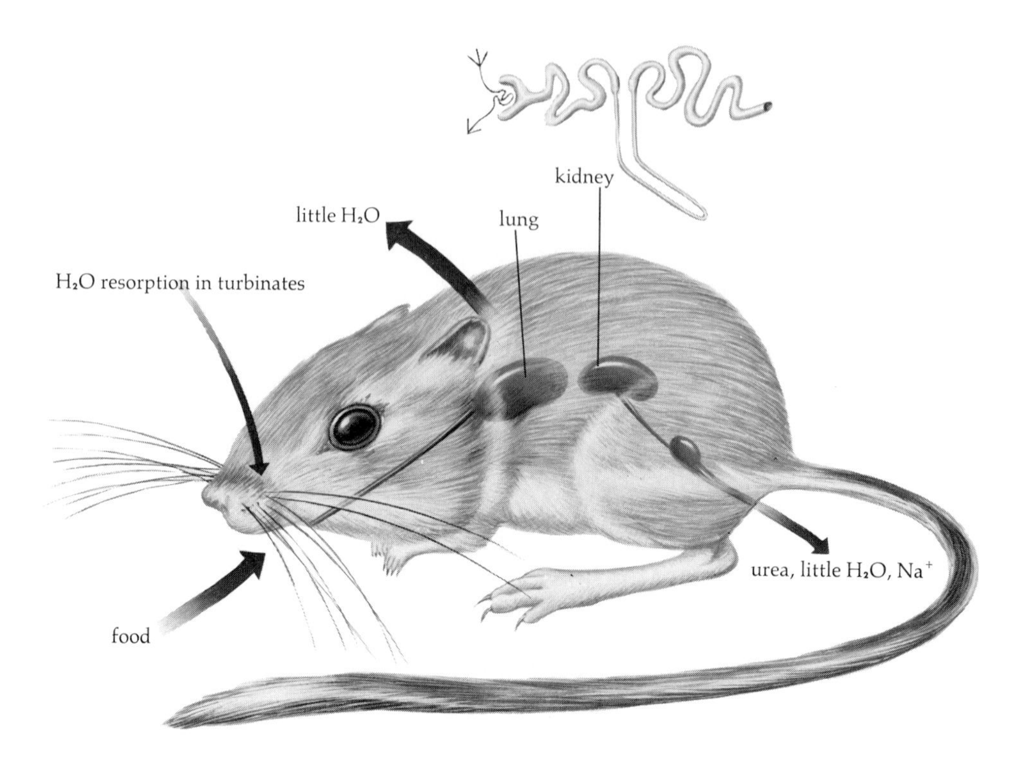

Fig. 17-19. Kangaroo rats are among the desert mammals that can keep a positive water balance solely on metabolic water. Reduction in sweat glands, resorption of respiratory water, extremely long renal papillae, and a high carbohydrate diet all facilitate their water economy.

reptiles and birds, they have evolved no other means for doing so. As a result, few mammals can utilize sea water.

The kidneys are not the only organs through which mammals lose water (Fig. 17-18). Sweat glands cover large parts of the body; they are efficient cooling devices but also permit a large water loss. During the expiration stage of respiration large amounts of water are expelled. There is no cloaca and less water is resorbed from both urine and feces than in reptiles and birds. Finally, some water is lost through the skin (insensible water loss). Therefore, most mammals are quite dependent upon a supply of fresh water.

However, extreme forms are found—particularly in desert environments—which can maintain a positive water balance even on a completely dry diet. The North American kangaroo rat, for instance, normally does not drink. The kidney can concentrate and excrete the sodium chloride which comes from the alkaline desert plants it eats. Even more notably, the kidney has extremely long loops of Henle in

Fig. 17-20. The urinary bladders of a teleost, a frog, and a mammal.

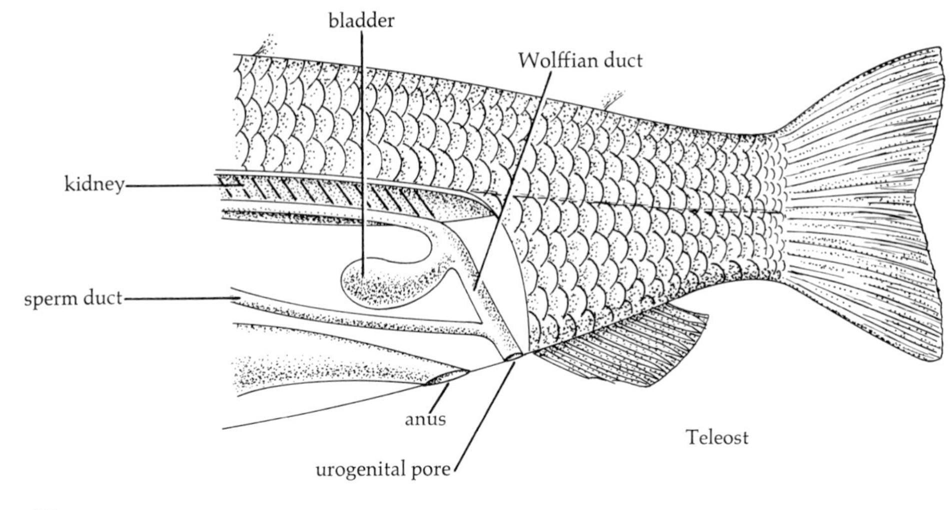
bladder
Wolffian duct
kidney
sperm duct
anus
urogenital pore
Teleost

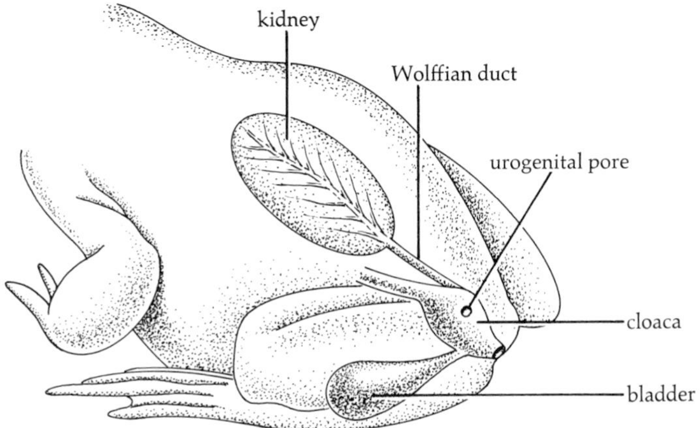
kidney
Wolffian duct
urogenital pore
cloaca
bladder
Frog

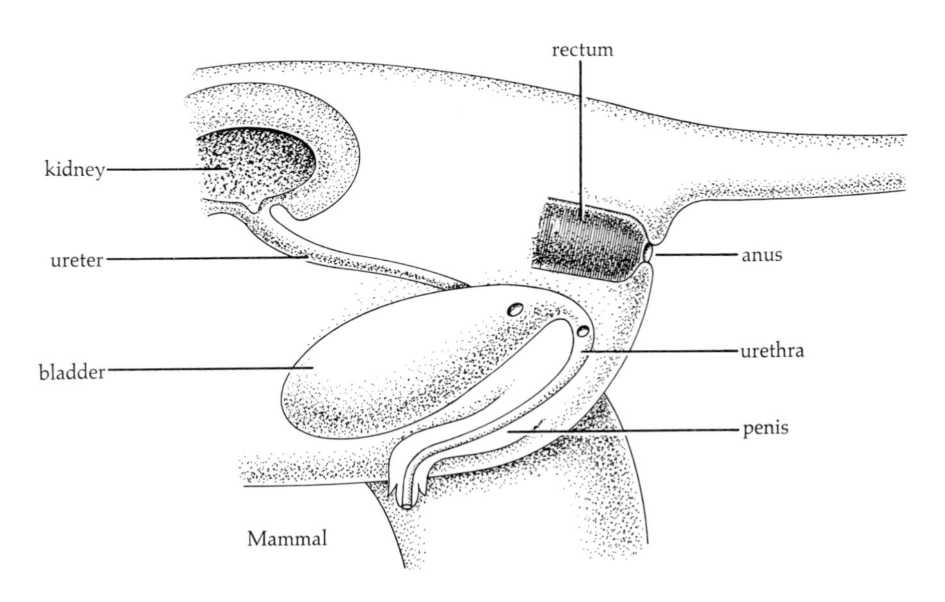
rectum
kidney
ureter
anus
bladder
urethra
penis
Mammal

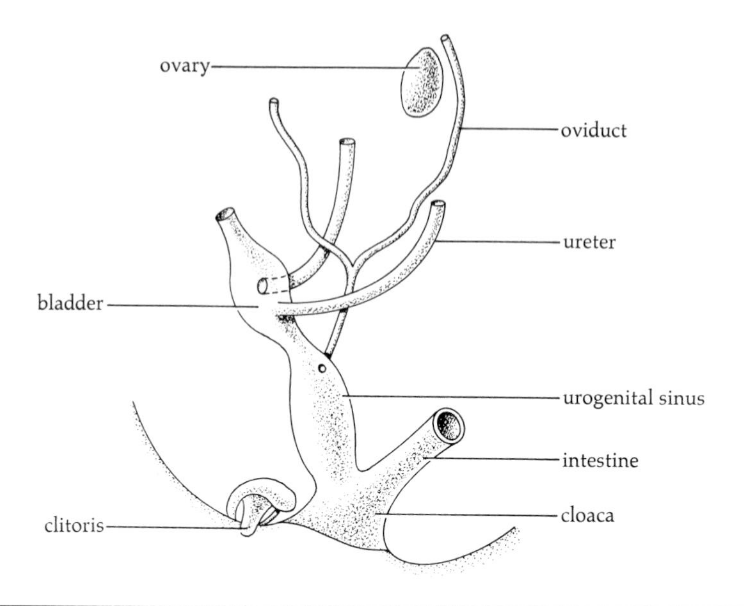

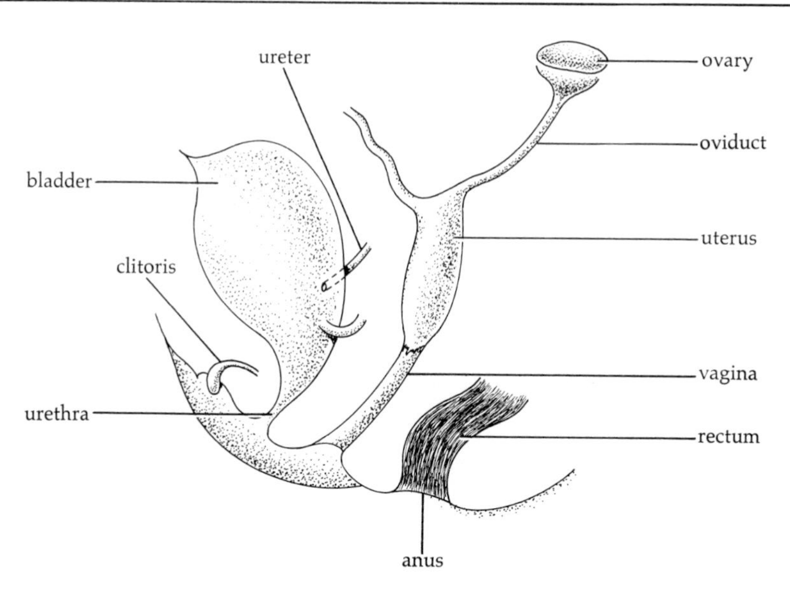

Fig. 17-21. Two stages in the development of the lower urogenital tracts of a human female which show the developmental elimination of the cloaca. After Arey, L. B. (1924). "Developmental Anatomy." Saunders, Philadelphia, Pennsylvania.

which the urine becomes highly concentrated. So efficient is the kidney that metabolic water, formed as a byproduct of carbohydrate and lipid metabolism, normally supplies all necessary water. The kangaroo rat and similar desert forms that live without water are never carnivorous; they eat primarily seeds with a high carbohydrate and lipid content. When under experimental conditions they are deprived of these they will eat protein, but then they must drink in order to get rid of the nitrogenous wastes; in this case they can survive on sea water because of the salt-concentrating ability of the kidney.

Other morphological modifications of desert forms further help to conserve water. They have very few sweat glands. Some, such as the kangaroo rat, have very long nasal bones and complex turbinate bones along the nasal cavity; this provides a large, cool, membrane-lined surface area, which cools exhaled air and condenses some of its moisture (Fig. 17-19).

There has been little work on marine mammals such as whales. Their osmotic equilibrium and salt secretion are probably strongly involved with the kidneys, but this is still open for investigation.

THE URINARY BLADDER

Most fishes have no large storage bladder for urine, and in those that do it is not a special structure but a distal dilation of the Wolffian duct. In amphibians and reptiles a large bladder develops as an outpocketing from the cloaca; urine passes into the cloaca and then backs up into the bladder (Fig. 17-20).

During embryonic life all mammals have a cloaca, from whose walls the urinary bladder and the ureters form; the adult ureters thus open directly into the urinary bladder. By the time of birth a longitudinal septum has formed, separating the dorsal anus, which empties the digestive tract, from the ventral urogenital tube, the urethra. In males and many females the urethra empties both the excretory and genital systems; in female primates the urogenital opening is further divided so that only the urinary tract enters the urethra and the vagina opens separately (Fig. 17-21).

The urinary bladder is lined with an epithelium which can undergo considerable stretching and covered by a complex layering of smooth musculature. A pressure valve prevents the urine from backing up the ureters; a sphincter muscle guards the opening of the bladder into the urethra and is relaxed as the smooth muscle covering contracts and the bladder is emptied.

SUGGESTED READING

Dunson, W. A., Packer, R. K., and Dunson, M. K. (1971). Sea snakes: An unusual salt gland under the tongue. *Science* **173,** 437–441.

Fraser, E. A. (1950). The development of the vertebrate excretory system. *Biol. Rev.* **25,** 159–187.

Hill, L., and Dawbin, W. H. (1969). Nitrogen excretion in the tuatara, *Sphenodon punctatus. Comp. Biochem. Physiol.* **31,** 453–468.

Murrish, D. E., and Schmidt-Nielsen, K. (1970). Water transport in the cloaca of lizards: Active or passive? *Science* **170,** 324–326.

Schmidt-Nielsen, K. (1964). "Desert Animals." Oxford Univ. Press, London and New York.

Schmidt-Nielsen, K. (1959). Salt glands. *Sci. Amer.* **200** (No. 1), 101–116.

Smith, H. W. (1953). "From Fish to Philosopher." Little, Brown, Boston, Massachusetts.

Youson, J. H., and McMillan, D. B. (1970–71). The opisthonephric kidney of the sea lamprey of the Great Lakes, *Petromyzon marinus* L. Parts I through IV. *Am. J. Anat.* **127,** 207–258; **130,** 55–72, 281–304.

18 REPRODUCTIVE SYSTEM

EPILOGUE

SUGGESTED READING

Among vertebrate organ systems the reproductive system is unique. While the adaptations that evolved in other systems help ensure the survival of the individual, those of reproduction ensure the survival of the species but can actually be detrimental to the individual. The individual uses a great deal of energy and often is placed in danger by producing germ cells, participating in reproductive behavior, and anatomically, physiologically, or behaviorally caring for developing young. As if to ensure that reproductive behavior will proceed in the best interests of the species rather than of the individual, it is largely controlled by endocrines and the visceral nervous system; the somatic nervous system influences reproductive behavior much less than it does other types of behavior.

The reproductive system, then, might be characterized as "altruistic," and in this light the evolution of some of its bizarre structures and behavior makes more sense—for instance, the well-known migrations of salmon and birds. Some mammals also migrate: fur seals, for example, travel about 3000 miles from the semitropics to the far northern Pribilof Islands, and on those cold remote islands they mate, deliver, and care for their young in an area of minimal food supply but maximal safety from predators (except, recently, man—the worst of all). Such behavior was selected for because it ensures the continuation of the population without regard for its effect on the individual.

Furthermore, the adaptations of other systems can be viewed as indirect adaptations for reproduction, whose prime adaptive value is to ensure the individual's survival *to reproductive age.* In the context of billions of years of natural selection it is the survival and modification of the species that is important, and it is at the species rather than the individual level that the benefits and consequences of natural laws have been expressed.

The morphological units of reproduction are the sperm and egg cells, collectively called gametes. They require an organ, the gonad, in which they are developed and stored until used. Male and female gametes must be effectively brought together for fertilization. Finally, enough fertilized eggs must develop into adults and reproduce themselves for the species to be continued.

THE GONADS

Gonads are compound glands, which in addition to developing and storing the germ cells also secrete reproductive hormones. They develop from paired masses of mesodermal tissue on either side of the

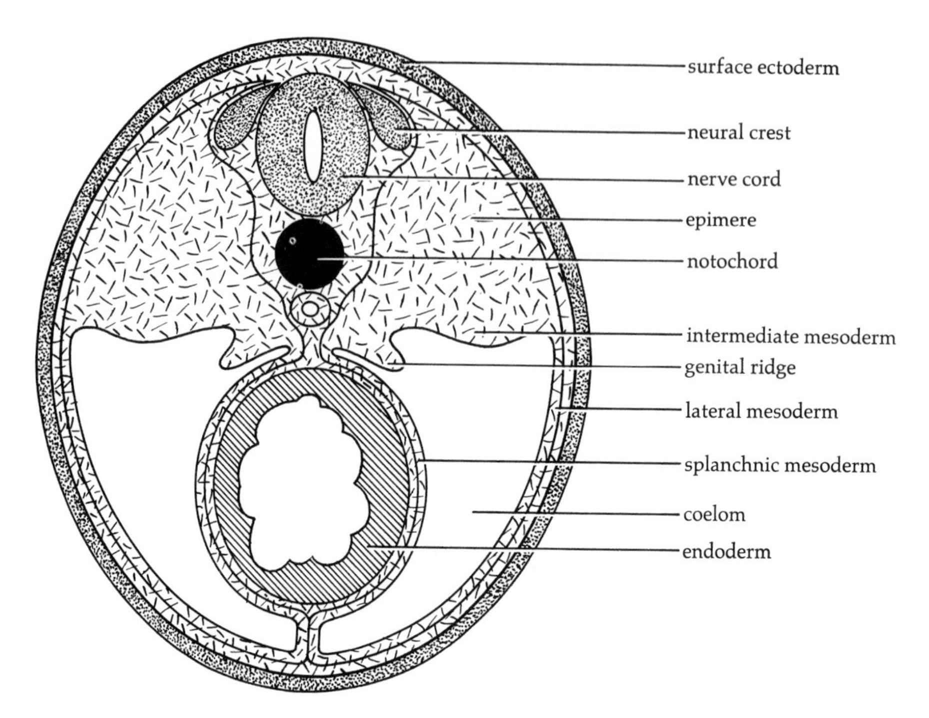

Fig. 18-1. Diagramatic cross section of a pharyngula embryo showing the position of the genital ridge.

developing dorsal mesentery, in the anterior portion of the body cavity; they are contributed to by both the medial portion of the mesomere and the dorsal portion of the adjacent splanchnic mesoderm (Fig. 18-1). However, unlike the developing kidney, the gonad shows no segmentation.

Proliferation of this tissue forms the genital ridge, which bulges into the developing coelom. In the genital ridge is well-defined germinal epithelium which eventually will form the germ cells; at this stage, however, the germinal epithelium proliferates as cordlike buds that grow out into the rest of the genital ridge. These cordlike masses are called the sex cords of the embryonic gonad.

The indifferent (i.e., sexless) gonad gradually takes on a more definitive shape, developing an outer cortex and an inner medulla. From then on the presumptive ovary and presumptive testis develop differently: in the ovary the outer cortical portion differentiates and the inner medullary portion is resorbed; in the testis the inner medullary tissue develops and the outer cortical tissue is resorbed (Fig. 18-2).

In most vertebrates there are symmetrical, paired gonads. However, the lamprey's right and left gonads fuse to form a midline gonad. In the hagfish the left gonad degenerates, and only the right develops. In female birds and female monotremes the right gonad degenerates and the left develops into the adult ovary. The shape of the gonad varies with the vertebrate, and usually mimics body shape; thus urodeles and snakes have long slender gonads, and anurans and turtles have short, oval, compact gonads.

Although the gross shape of the testis varies considerably among vertebrates, its internal structure is remarkably constant. There is an outer connective tissue capsule, the tunica albuginea, which frequently also contains smooth muscles. In most vertebrates there are extensions of this outer layer deep into the testis, dividing it into a number of compartments. Within each compartment are the seminiferous tubules containing the sperm-producing cells, and between the tubules are blood vessels, nerves, loose connective tissue, and the interstitital cells of Leydig which produce the male hormone, testosterone (Fig. 18-3).

A seminiferous tubule is long and coiled and has a narrow lumen; sperm are produced along its entire length. Its outer portion is formed by a basement membrane, upon which rest epithelial cells and the larger nurse cells, or Sertoli cells. Sertoli cells do not themselves produce sperm, but facilitate sperm production, as will be described. The bases of the Sertoli cells lie on the basement membrane, their irregular sides extend toward the lumen, and their apical ends lie at the lumen. Between Sertoli cells are the germinal epithelial cells, which undergo spermatogenesis.

Spermatogenic cells called spermatogonia lie on (or sometimes close to) the basement membrane between the Sertoli cells. During a wave of spermatogenesis the spermatogonia undergo mitotic divisions into

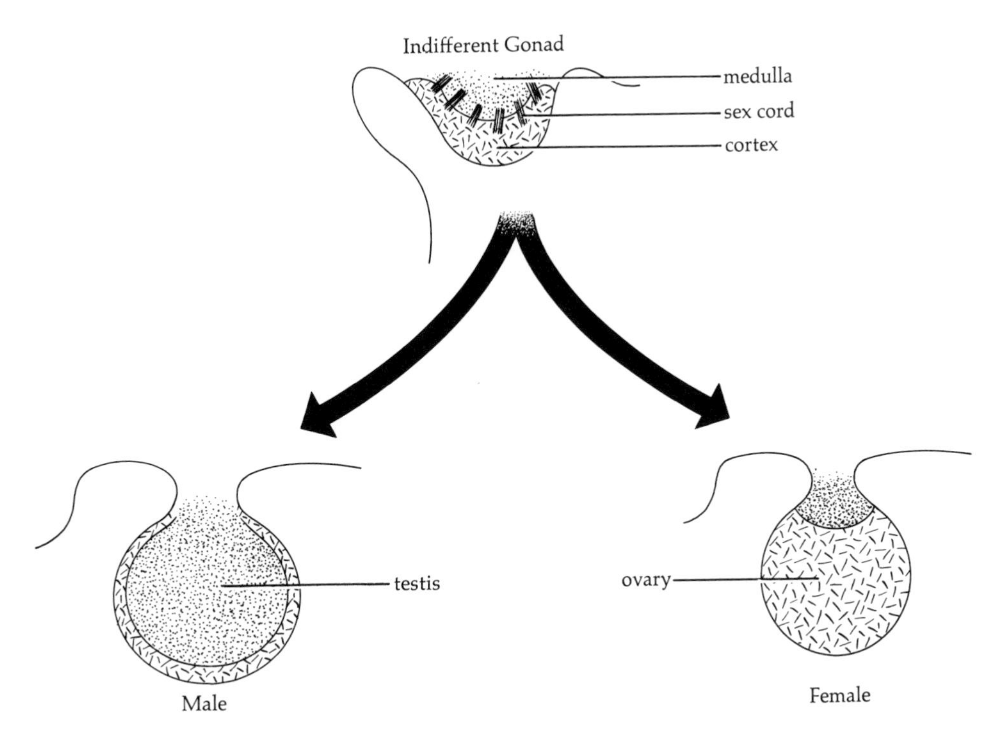

Fig. 18-2. An indifferent, sexless gonad develops from the genital ridge. In males the inner medulla of the gonad forms the testis; in females the outer cortex of the gonad forms the ovary.

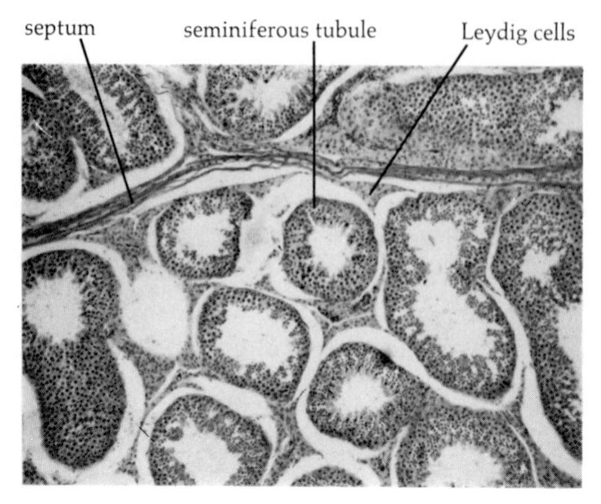

Fig. 18-3. A low-power photomicrograph of a mammalian testis, showing several cross sections of seminiferous tubules, Leydig cells, and a connective tissue septum dividing compartments.

primary spermatocytes. The primary spermatocytes are pushed away from the basement membrane into infoldings of the Sertoli cells (Fig. 18-4). There they undergo meiotic divisions, a two-step process: the first cells produced by meiosis are the secondary spermatocytes; the second cells, which are haploid, are the spermatids. The spermatids mature in a process called spermiogenesis, which involves no further cellular division but their differentiation into spermatozoons (collectively called sperm), which have a head, neck, middle piece, and long tail. Even at this stage the spermatozoons remain in contact with the Sertoli cells.

In many anamniotes the seminiferous tubules are shaped like cysts and are empty except for the period just before and during spawning. At that time spermatogonia just outside these cysts become mitotically active and form many primary spermatocytes which migrate into the

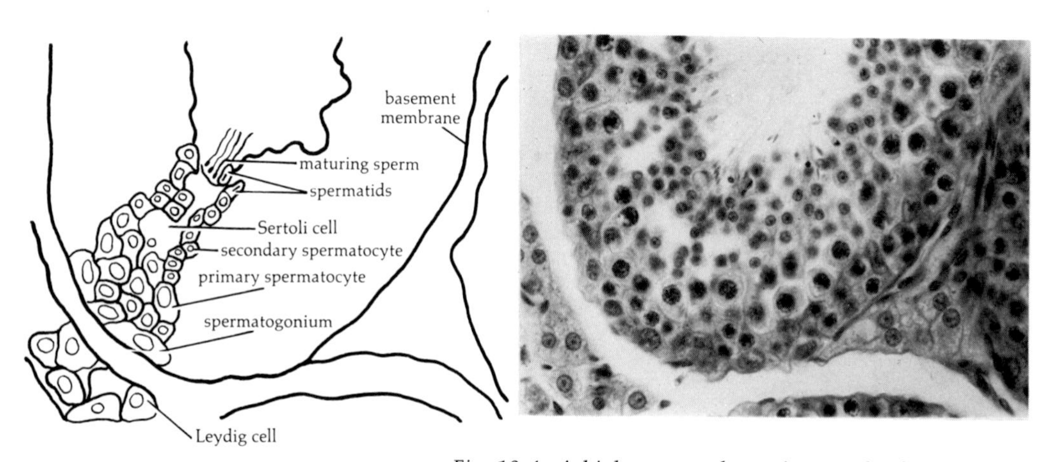

Fig. 18-4. A high-power photomicrograph of part of a single seminiferous tubule.

cysts. There they undergo meiotic division into secondary spermatocytes and spermatids, which differentiate into spermatozoons. Sertoli cells in the walls of the cysts function similarly to those of amniotes.

In most vertebrate groups the seminiferous tubules terminate in modified mesonephric tubules called efferent ductules, with one ductule draining many seminiferous tubules. These efferent ductules carry the sperm out of the testis.

There are two other methods of removing sperm from the testis. In cyclostomes and the Salmonidae family of teleosts, sperm are released from the testis directly into the body cavity. They travel in the coelomic fluid to the posterior part of the cavity and exit the body by a small genital pore leading to the urogenital sinus. In teleosts except Salmonidae the sperm travel from the testis to the urogenital sinus in a sperm duct which develops from the testis and its mesentery, the mesorchium.

In many amniotes, and, to varying degrees in some anamniotes, the testes do not remain in their original embryonic position in the anterior dorsal wall of the body cavity, but migrate caudally. In monotremes, whales, and Sirenia (e.g., dugong) they are retained within the body cavity throughout life, but in other mammals the testes lie at least some of the time in an extra-abdominal extension of the body cavity, the scrotum. They may descend into the scrotum only during periods of sexual activity (as in some marsupials, most rodents, insectivores, bats, some ungulates such as camels, and some primates); or they may descend into the scrotum during late fetal life and remain there permanently (as in some marsupials and most carnivores, ungulates, and primates).

Regardless of when it occurs the migration and descent of the testes are complex matters, controlled primarily by hormones. Part of the process is accounted for by a ligament, the gubernaculum, which forms just behind the early testis and holds it to the peritoneum. As development continues the gubernaculum does not grow, which in effect pulls the testis caudally. However, this is not the whole story, for cutting the gubernaculum does not prevent the descent of the testes.

It seems strange to find these vital organs outside the body cavity where they are vulnerable to many accidents. However, there is a lower temperature in the scrotum than in the body cavity, and it has been found that in mammals whose testes are normally descended spermatogenesis will not occur if the testes do not descend or if the scrotum is artificially kept warm. On the other hand, the testes are kept warm in cetaceans, sirenians, monotremes, and all birds.

The cooling mechanism itself is clearly understood. The scrotum is relatively thin and has a large surface area, so heat is readily dissipated. Moreover, the arteries going to the scrotum are entwined with the veins returning from it, through a plexus of vessels called the pampiniform plexus. This provides a countercurrent heat distribution system which further cools the scrotum (and also helps rewarm the blood returning to the body cavity).

What is not understood is why the testes should have to descend in some animals for spermatogenesis to occur. It is clearly not necessary

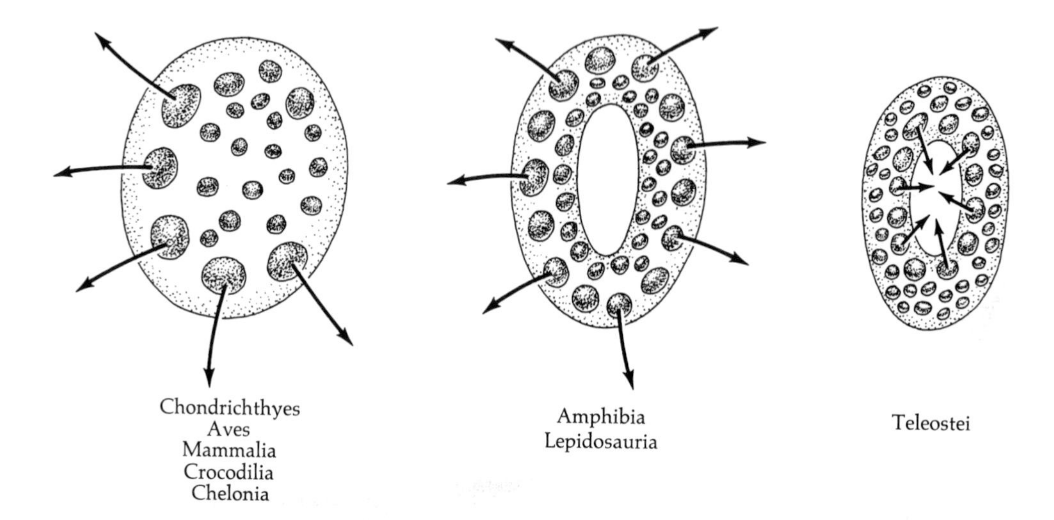

Fig. 18-5. Diagramatic representation of the three types of ovaries found in vertebrates. Arrows indicate direction of ovulation.

in most animals, and one wonders how and for what adaptive reason such a mechanism could have evolved in the first place.

THE OVARY

Ovaries differentiate from the cortical portion of the sexually indifferent gonad; they produce germ cells (eggs, or ova) and female hormones.

There are three ovary types (Fig. 18-5). In most vertebrates the ovary is a solid structure in which the developing eggs are embedded. When mature, the eggs are expelled from the ovary into the body cavity, in a process called ovulation. This is the situation in Chondrichthyes, Aves, Mammalia, and the crocodilians and chelonians of Reptilia. In Amphibia and the lepidosaurians of Reptilia the ovary is hollow, or saclike; the eggs develop within its walls and when mature rupture through the walls into the coelom. In teleosts there is also a saclike ovary, but the mature eggs rupture not into the coelom but into the cavity of the ovary.

Microscopically all ovaries are composed of a loose connective tissue stroma, in which are embedded numbers of ovarian follicles in various developmental stages (Fig. 18-6). Each follicle is made up of a large number of follicle cells surrounding a single diploid oogonium. Only the oogonia will produce the eggs, but the follicular cells must be present for their development and maturation. In most vertebrates, much of the maturation of the egg involves the formation of the nutrient substance, yolk. The proteins of the yolk are synthesized primarily in the liver, carried by the blood vascular system to the ovaries, and there picked up by the follicle cells which deposit yolk in the eggs.

There is a considerable amount of yolk in the eggs of most vertebrates, but the greatest amount is in the so-called large-yolked, or telolecithal, eggs of reptiles, birds, monotremes, and chondrichthyeans. Smaller amounts of yolk are deposited in the mesolecithal eggs of amphibians and teleosts. In the isolecithal eggs of therian mammals the yolk is scanty; a follicular fluid develops within the sphere of the follicle, and the oogonium, now called a primary oocyte, lies within this fluid.

As more yolk is added to the egg its overall size (but not its cytoplasm) increases—in some animals by as much as a million times. As this happens the ovary itself becomes larger; the maturing follicles bulge out, and the outline of developing eggs can be seen on the surface of the ovary.

Cell divisions, particularly meiotic divisions, are also important in the egg's maturation. These do not begin until the egg is almost fully formed in the follicle and are not complete until after the egg has been fertilized (Fig. 18-7). The divisions are asymmetrical, rather than symmetrical as they are in the testes, and only one mature egg is formed from each two-step meiotic division. The other cells which result from the divisions are called polar bodies. They contain the same amount of chromatin but are very much smaller; they do not have the capacity to be fertilized and are eventually resorbed.

Before ovulation occurs the follicle moves to the surface of the ovary, greatly diminishing the amount of ovarian tissue between the egg and the surface. Then the follicle ruptures, casting the egg out into the coelom (in most vertebrates) or into the ovary's cavity (in teleosts). The cells of the ruptured follicle remain in the ovary and undergo rapid cytological changes, becoming the corpus luteum. Although this structure forms in all ovaries, we know it best from studies on mammals where it produces very important hormones for the maintenance of pregnancy.

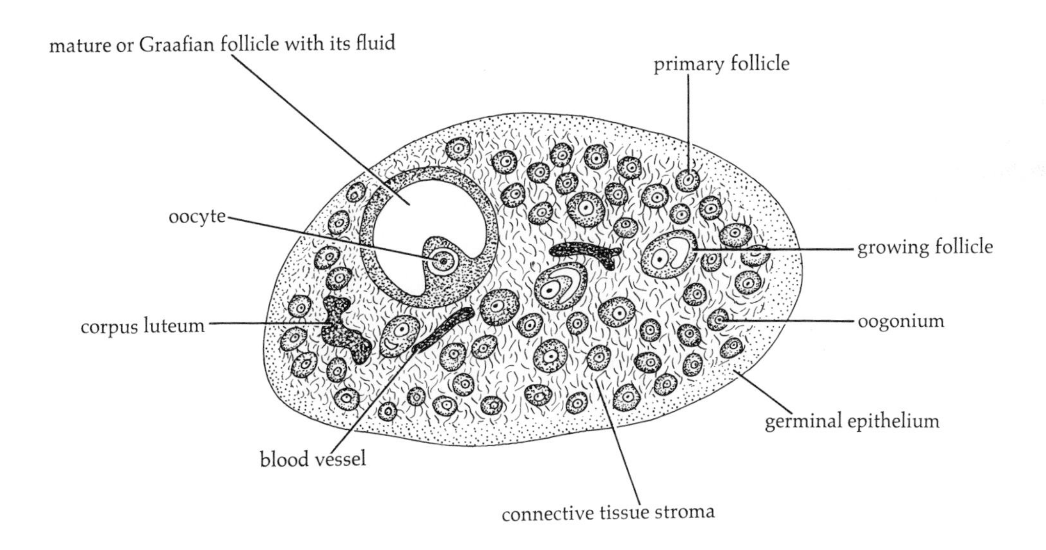

Fig. 18-6. A low-power microscopic section of a mammalian ovary.

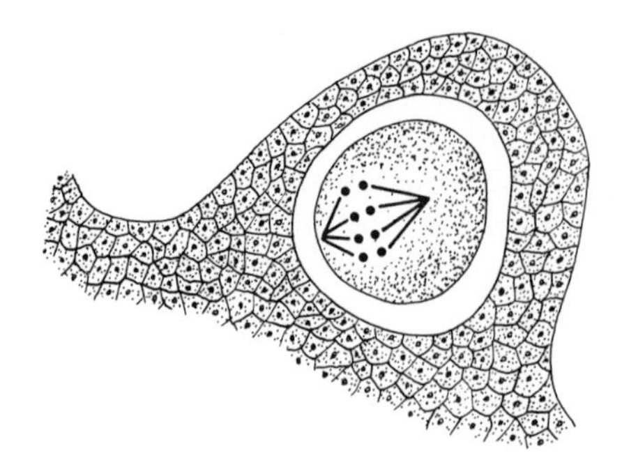

Primary oocyte in follicle with spindle for first meiotic division.

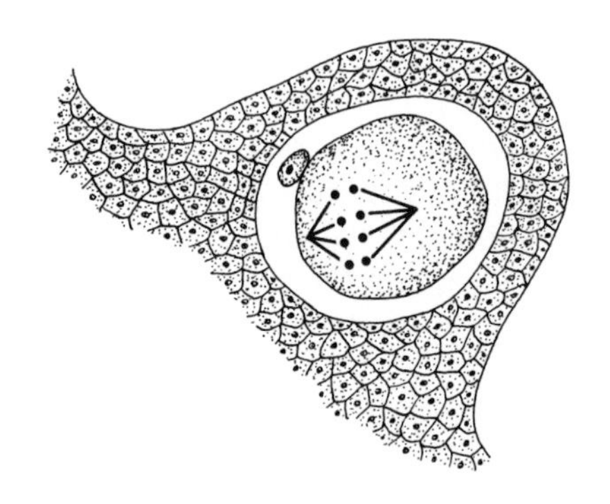

First polar body and secondary oocyte in follicle. Spindle is present for second meiotic division.

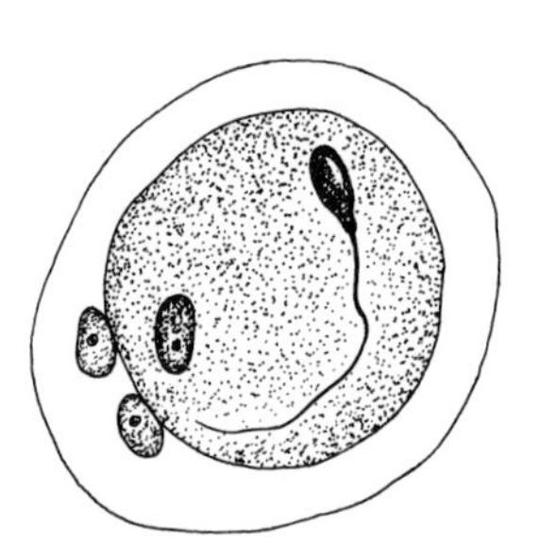

After ovulation, fertilization has occurred with sperm in egg's cytoplasm. Only then does second meiotic division complete, forming second polar body and mature egg still surrounded by egg membrane.

Fig. 18-7. Mammalian oogenesis.

THE REPRODUCTIVE DUCTS

MALE GENITAL DUCTS

As explained above, male cyclostomes and male Salmonidae of teleosts have no genital ducts; the males of other teleosts have a duct formed from the testis itself and its mesentery.

In all other vertebrates the sperm are transported by modified portions of the embryonic kidney (Fig. 18-8). The modified posterior pronephric tubules and anterior mesonephric tubules become the efferent ductules, carrying sperm from the seminiferous tubules out of the testes. The sperm continue down the epididymal duct, or, in mammals, the body and tail of the epididymis; these coiled structures

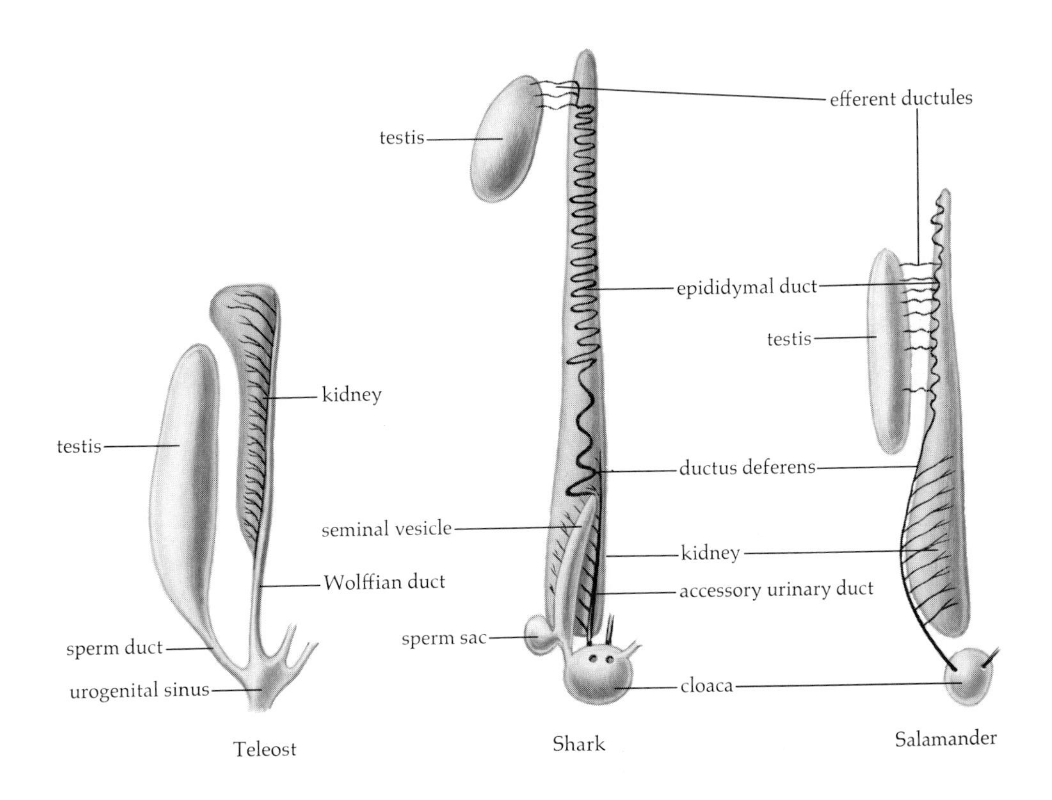

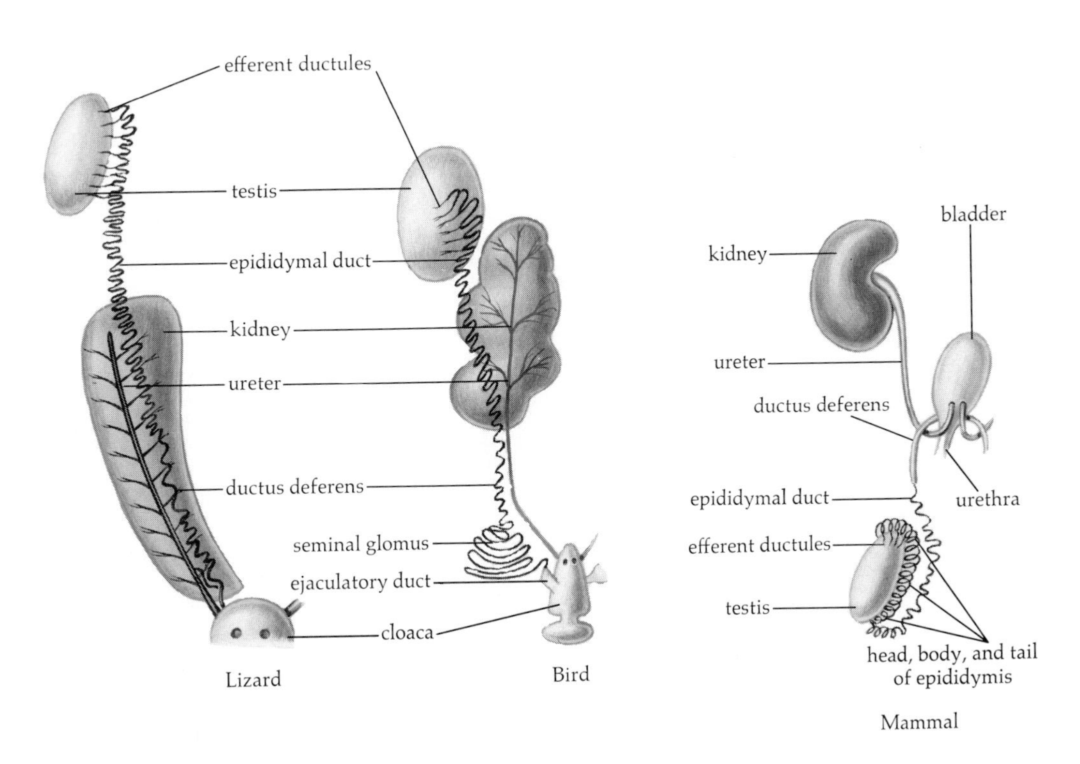

Fig. 18-8. Ventral views of male gonads and genital ducts in diverse vertebrates.

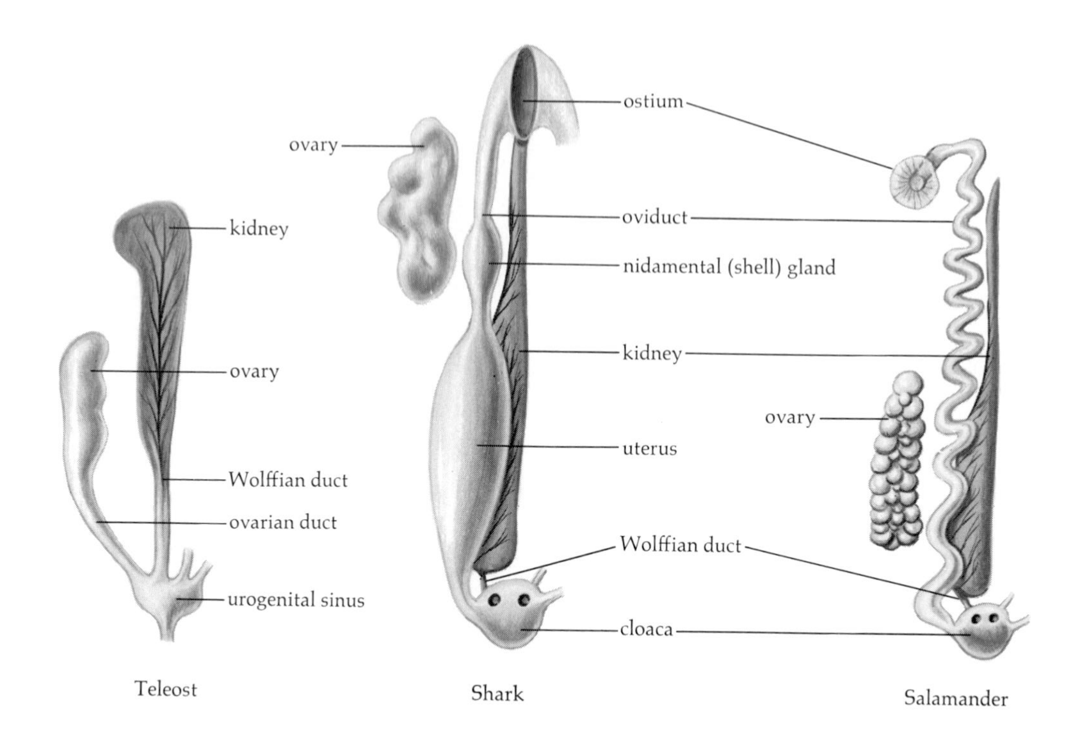

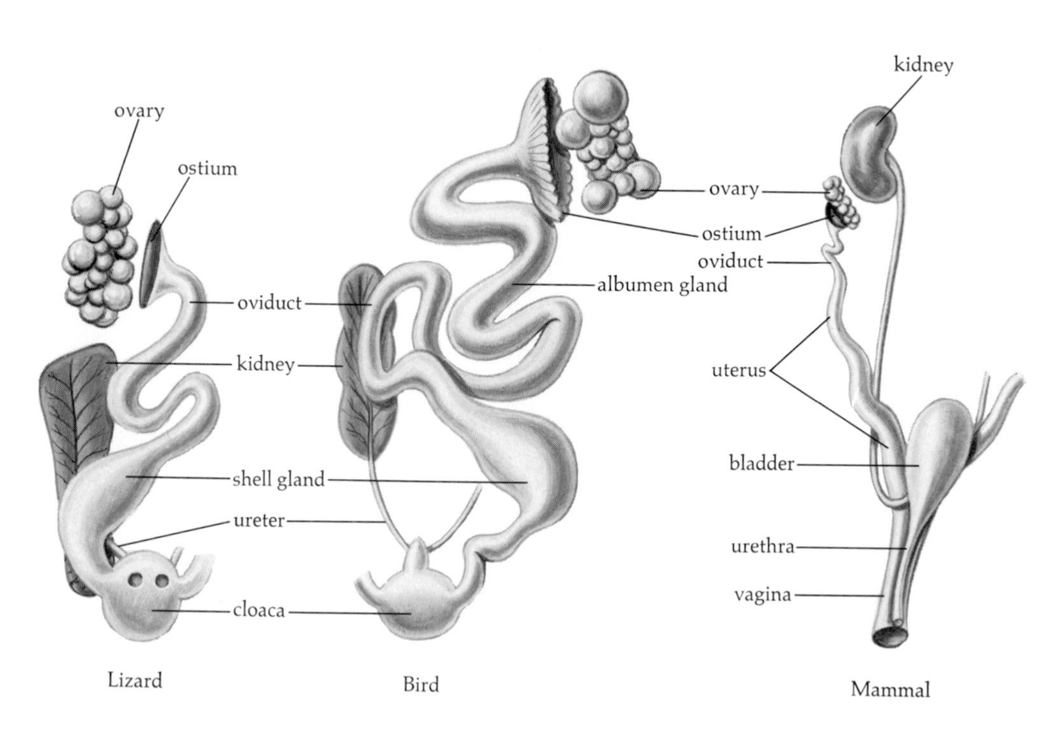

Fig. 18-9. Ventral views of the female gonads and genital tracts in diverse vertebrates. The right ovary and oviduct are shown except in birds, where the right ovary is degenerate.

are the anterior part of the Wolffian duct. The sperm then enter the straight, more caudal portions of the Wolffian duct, called the ductus deferens. Although in different vertebrate groups portions of the Wolffian duct undergo further modifications, this structure is still a modified kidney duct "borrowed" for the purposes of reproduction by most vertebrates.

FEMALE GENITAL DUCTS

There is a parallel situation in the female. Cyclostomes and the Salmonidae of teleosts lack female genital ducts; their eggs pass into the body cavity and out the abdominal pores; in other teleosts an ovarian duct (not oviduct) develops from the ovary and its mesentery, the mesovarium, and leads directly from the cavity of the ovary to the urogenital sinus.

All other female vertebrates have a genital duct, the Müllerian duct. It opens into the body (instead of the ovary), with the opening surrounded by a funnel-shaped, thin membrane called the ostium (Fig. 18-9). The ostium may be quite close to the ovary (as in amniotes), or a considerable distance away (as in amphibians). In Chondrichthyes the right and left Müllerian ducts join anteriorly and have a single ostium.

There is considerable evidence that in elasmobranchs and amphibians the ostium is formed by an enlargement of a pronephric nephrostome, but not, apparently, in other vertebrates. The remainder of the Mullerian duct is formed in either of two ways. In Chondrichthyes and some amphibians the Wolffian duct splits longitudinally, one portion remaining as the Wolffian duct and the other becoming the Müllerian duct. In other vertebrates the Müllerian duct forms from nephrogenic tissue adjacent to the Wolffian duct (Fig. 18-10). A Müllerian duct forms in both male and female vertebrates at about the time that the gonad is in its indifferent stage; functionless bits of the Müllerian duct remain in the adult males of most species.

Since the duct is not continuous with the ovary the eggs must travel through the coelomic fluid to the ostium, where cilia help draw them into the Müllerian duct. Portions of the Müllerian duct may be modified for various needs: to secrete nutrient substances such as the albumen (egg white), to produce shells over the eggs, to form a reservoir where the developing young may be housed. The portion of the duct which functions mainly to convey eggs is called the oviduct; parts of the duct that have other specialized functions are given other names, such as uterus and nidamental gland.

MECHANISMS FACILITATING UNION OF SPERM AND EGG

For species survival, a sufficient number of eggs must be fertilized. There is a close inverse relationship between the mechanisms ensuring fertilization in any taxon and the numbers of germ cells that must be produced (and hence the amount of metabolic energy required).

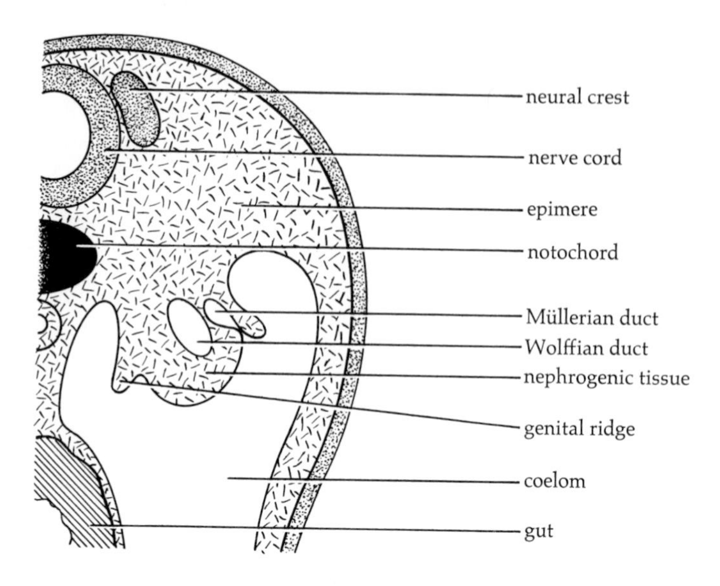

Fig. 18-10. In all vertebrates, except Chondrichthyes and some amphibians, the Müllerian duct develops from the lining of the coelom adjacent to the Wolffian duct.

EXTERNAL FERTILIZATION

Fertilization can take place if both eggs and sperm are released into the water, and this system, called external fertilization, is common among aquatic vertebrates. In many Osteichthyes, fertilization is ensured simply by the enormous numbers of germ cells released (up to a million eggs per female per annum, and even more sperm). Although this system is wasteful, enough eggs will be fertilized purely by chance to continue the species.

Other groups, including some other Osteichthyes and Cyclostomes, engage in more elaborate sexual behavior which decreases the odds against fertilization occurring; in these fish fewer eggs and sperm need be produced. The stickleback, a teleost, goes through a precise courtship ritual which includes building a nest; the female discharges her eggs into the nest, and the male releases his sperm while swimming just above the nest. The female lamprey also builds a nest and then the male and female, wrapped around each other, release their germ cells simultaneously into the nest, where the young develop.

In anuran amphibians the male mounts the female, clasping her trunk in a behavior known as amplexus which aids the release of eggs through the cloaca. As the eggs are released the male releases his sperm, and many eggs are fertilized. In many urodele amphibians there are courtship rituals culminating in a walk; the male goes first, laying down spermatophores (mucous packets with sperm at their tips); the female follows, dragging her cloaca over the spermatophores and thus picking up the sperm. When the eggs later enter the cloaca on their way to the outside they are fertilized by the sperm.

INTERNAL FERTILIZATION

A higher percentage of eggs will be fertilized if the male inserts the sperm directly into the female. This is done by an intromittent organ, several types of which have evolved in different vertebrate lines (Fig. 18-11). In Chondrichthyes, the male's pelvic fin is modified as the clasper organ, which includes a skeletal support, a siphon gland, and a siphon sac; a groove runs from the cloaca through the clasper organ to its tip. The siphon gland produces a mucopolysaccharide secretion, which is stored in the siphon sac and then added to the sperm as it passes along the groove. During copulation the male elasmobranch

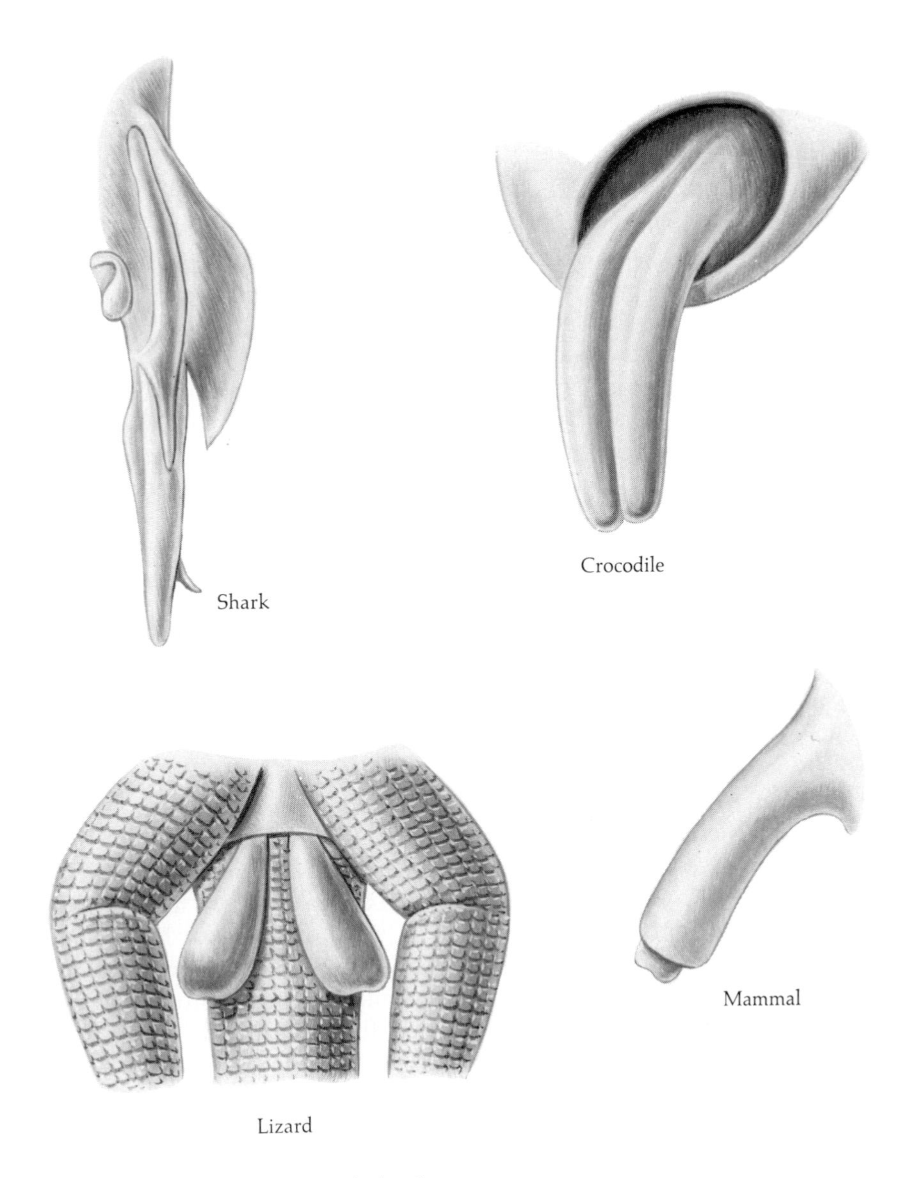

Fig. 18-11. Intromittent organs of vertebrates include the clasper organ of sharks, hemipenes of lizards, and penes of crocodiles and mammals.

wraps himself part way around the female and inserts a single clasper into her cloaca and up into a Müllerian duct. The muscular walls of his urogenital tract contract, forcing the sperm out through the clasper groove and up into her Müllerian duct. The sperm, by their own mobility, swim up the Müllerian duct to the ostium where fertilization occurs.

Although many teleosts have external fertilization, there are also many with internal fertilization. The male's sperm ducts terminate at the base of the anal fin, which is modified as an intromittent organ called the gonopodium. This is inserted into the female's ovarian duct; the sperm, passed directly into the duct, swim up into the saclike ovary where fertilization occurs.

There is at least one amphibian with a functional intromittent organ. *Ascaphus,* of the northwestern United States, can evert its cloaca and place it into the female's cloaca. It is this everted cloaca which gives the animal its common name, the tailed frog.

Two types of intromittent organs have evolved in reptiles. In snakes and lizards there are paired, vascularized, eversible sacs, the hemipenes, opening into the posterior part of the cloaca. When everted, each hemipene has a spiral groove along its medial surface through which the sperm pass when the two hemipenes are brought together and inserted into the female's cloaca. The sperm swim up the Müllerian duct; fertilization occurs at or near the ostium.

In turtles and crocodilians there is a single penis, formed in the ventral wall of the cloaca by large, paired ridges of vascularized erectile tissue called the corpora cavernosa. With sexual excitation the erectile tissue becomes turgid, closing a groove which lies between the two corpora and converting these structures to a tube. When this penis is inserted into the female's cloaca the sperm pass along the tube into the female and swim up the Müllerian duct, and fertilization occurs at or near the ostium.

Few birds have a true copulatory organ, and for those that do (ducks, swans, geese, and flightless birds such as the ostrich) it is morphologically and functionally almost identical with that of turtles and crocodilians. Internal fertilization is necessary for all birds, however, so that shell can be laid down around the fertilized egg before it is passed to the outside. Copulation in birds without a penis involves what is called the "cloacal kiss." In some birds this is a graceful and beautiful maneuveur performed in flight in which the male's cloaca is for an instant placed against the female's cloaca. In order to pass sperm successfully from male to female, this brief event must be precisely timed. To ensure this, it is preceded by a ritualized "courtship dance" in which the male and female fly parallel to one another, the male below and the female above, until one dips down and the other tips up for the climactic event.

The mammalian penis has two masses of erectile tissue called the corpora cavernosa, homologous to the corpora cavernosa of reptiles, and in addition a third, midline mass, the corpus spongiosum. The corpus spongiosum surrounds the urethra, a closed tube (instead of a groove) through which the sperm pass. Both the corpora cavernosa and the corpus spongiosum contain erectile tissue. In many mammals (but not humans) there is an ossification in the connective tissue between the two corpora cavernosa; this is called the os penis, or baculum. The corpus spongiosum is expanded at its distal end to form the glans of the penis, over which is a fold of skin, the prepuce.

Erection occurs by dilation of the arteries supplying the erectile tissues and a simultaneous constriction of their veins, causing a build up of pressure. Almost any stimulation of the penis will cause a reflex erection if the appropriate hormonal balance is present. Ejaculation (release of seminal fluid containing sperm) occurs following erection and further stimulation. It involves violent contractions of the muscles of the ductus deferens and accessory male glands, which force the sperm and seminal fluid out of the urethra.

An intromittent organ, to be successful, requires complementary structures in the female. In most nonmammalian vertebrates the intromittent organ is received by the cloaca only; however, the chondrichthyean clasper organ is pushed up into the distal portion of the Müllerian duct, and in teleosts, where there is no cloaca, the gonopodium is inserted directly into the ovarian duct. Among mammals, only monotremes have a cloaca to receive the penis.

In all other mammals the distal portions of the Müllerian duct are modified to form the genital orifice, which receives the penis (Fig. 18-12). The outer part of the female's genital opening, the urogenital sinus, is formed by an infolded portion of surface skin. Within the wall of the urogenital sinus lies the clitoris, a small homologue of the male's penis containing erectile tissue homologous to the corpora cavernosa (but not the corpus spongiosum). Upon sexual excitation the clitoris is engorged with blood and becomes erect.

Internal to the urogenital sinus is the most distal portion of the Müllerian duct, the vagina. Between urogenital sinus and vagina is a fold of tissue, the hymen, which ruptures and bleeds at the first intercourse. In marsupials the distal ends of the right and left Müllerian ducts have not joined, and there are paired vaginas. It is not surprising, therefore, that the distal end of the marsupial penis is also

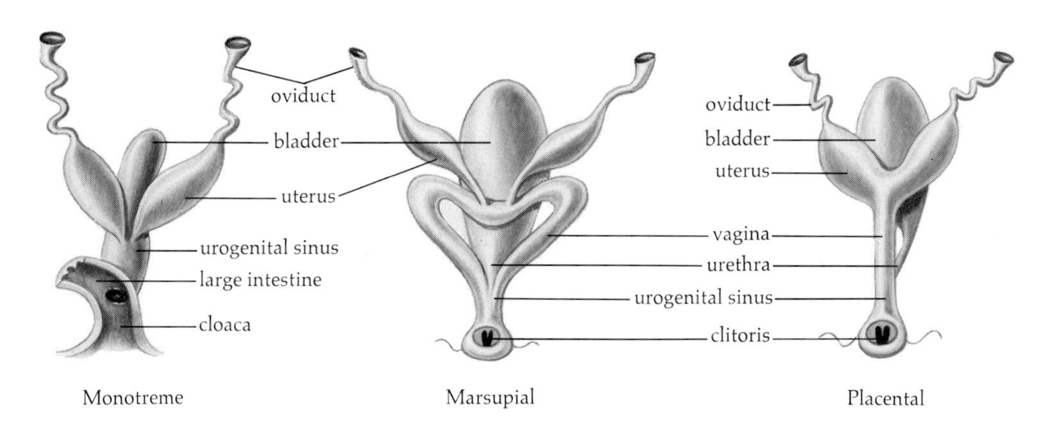

Fig. 18-12. Dorsal views of female genital tracts in each of the three major groups of mammals.

paired, having what is called a bifid glans. In placental mammals the distal ends of the two Müllerian ducts fuse and form a single, midline vagina, which receives the penis during intercourse.

ACCESSORY REPRODUCTIVE GLANDS

The ovaries and testes are not the only glands in the genital system. Accessory genital glands along both the male and female reproductive tracts play important roles in the reproductive process.

Male

In male mammals, only a small portion of the seminal fluid is sperm, the rest being secretions of the accessory male glands which enter the ductus deferens distal to the testes (Fig. 18-13). In fact, if the two ductus deferens are cut and tied off just as they leave the epididymis the male is rendered sterile but his sexuality and potency are unaffected: only the sperm are missing from the ejaculate, and there is no noticeable decrease in its volume or strength. This procedure, incidentally, is called vasectomy. It is simple, safe surgery, the most effective method of birth control, and is sometimes surgically reversible.

In the human male the seminal fluid is produced by a single pair of glands, the seminal vesicles, and a midline gland, the prostate. The seminal vesicles are sacculated appendages from the distal portion of the ductus deferens before it joins the urethra. They produce a nutritious mucous secretion, which, when ejaculation occurs, is forced down the urethra by the muscles covering the seminal vesicles. There are seminal vesicles in most eutherians except cetaceans and many carnivores; there are none in monotremes or marsupials.

The prostate is a large midline gland lying at the point where the two ductus deferens join the urethra. It is a compound tubular alveolar gland with several ducts opening into the urethra. Its slightly alkaline secretion forms the bulk of the seminal fluid. During ejaculation the smooth muscles in its connective tissue capsule contract strongly, forcing the secretion into the urethra.

A pair of smaller glands, the bulbourethral or Cowper's glands, lies distal to the prostate gland and opens into the urethra. During erection but before ejaculation these glands produce an alkaline mucous secretion, which neutralizes the slight acidity of both the male's urethra and the female's vagina and thus prepares the genital tracts for the sperm to come.

Accessory male glands producing seminal fluid are not peculiar to mammals, but in one form or another are found in most male vertebrates. For instance, a mucopolysaccharide is produced by the male elasmobranch's siphon gland. Chondrichthyeans also have a "sexual kidney," or Leydig's gland (no relationship to the interstitial cells of Leydig). This is the modified anterior portion of the opisthonephric kidney, which produces much of the fluid of the ejaculate and passes it through tubules directly into the ductus deferens. Sharks have two distal modifications of the Wolffian duct; one is a simple in-

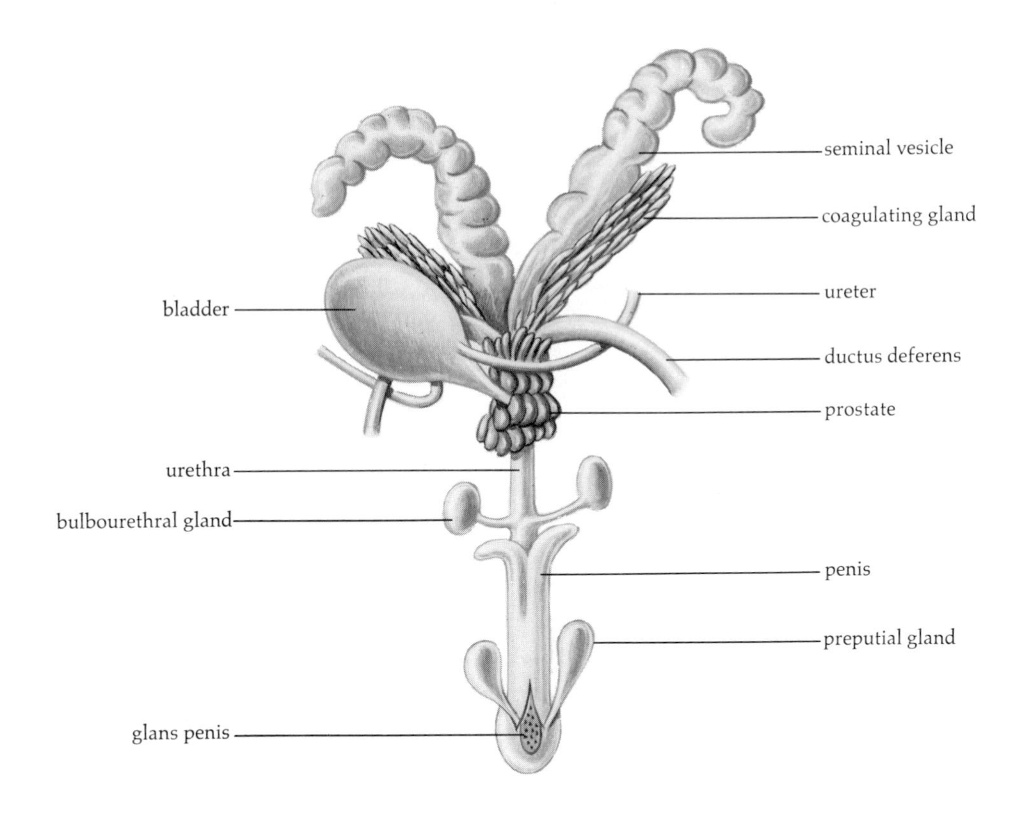

Fig. 18-13. Ventral view of a male rodent genital tract, showing the accessory glands.

crease in the duct's diameter, forming a seminal vesicle, and the other is a saclike dilation off the seminal vesicle, called the sperm sac. Both these structures may store sperm prior to ejaculation. Such modifications of both the Wolffian duct and anterior opisthonephros have evolved in similar but not always identical ways in different groups of vertebrates.

Female

There are also accessory glands in the female's genital tract, which develop as modifications of the Müllerian duct. These are most prominent in nonmammalian vertebrates whose eggs are shelled or have a heavy albumen layer around the yolk (Fig. 18-9). Even mammals, however, have vaginal and cervical glands which help to lubricate the genital tracts. They also have important uterine glands whose albuminous secretion helps repair the epithelial uterine wall following menstruation in primates, or estrous in nonprimates. In addition, there are some accessory sex glands which are not directly connected to the genital tracts. Most notable are scent glands—often located just deep to the anus—which are attractive to the opposite sex (Fig. 18-14).

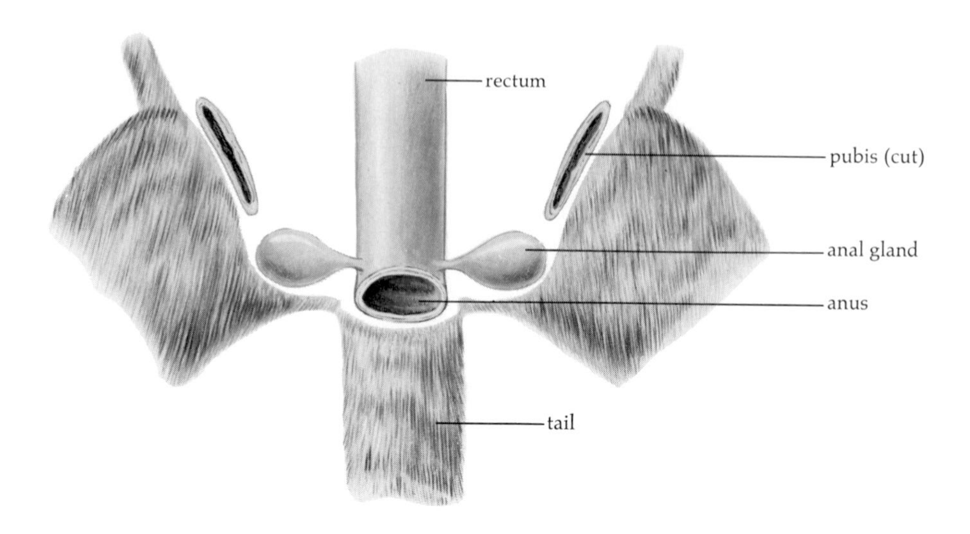

Fig. 18-14. A deep dissection of the cat's pelvis shows the subcutaneous anal glands opening into the anus. Their secretion is under hormonal control and functions as a sexual attractant.

CARE OF EGGS AND FETAL YOUNG

If the eggs, once fertilized, are abandoned in a hostile environment all these modifications to ensure fertilization will be pointless. The problem of protection is less severe for aquatic than for terrestrial vertebrates, since the aquatic environment itself supplies food more readily, gives support, maintains rather even temperatures, and prevents desiccation. But the eggs and young are defenseless throughout their early development and are a potentially good food source for other animals; even by the time development is fairly complete the young animal still does not have its full defensive mechanisms and is vulnerable to predation. To help ensure survival to adulthood a larval stage has developed in some animals. The eggs develop quickly to a point where the animal, although anatomically incomplete, can cope for itself in an aquatic environment while growth and differentiation proceed until it is ready to metamorphose into the adult stage.

It is not surprising that many other devices—morphological, physiological, and behavioral—have also evolved to ensure that a higher percentage of the offspring develop at least to the point of sexual maturity.

One such group of devices involves the degree of anatomical protection and assistance the eggs and developing young receive from the mother. Vertebrates can be divided on this basis into three major groups. In oviparous vertebrates the egg is ejected from the mother before or shortly after fertilization. In ovoviviparous forms the fertilized egg is retained either in a portion of the Müllerian duct or in the ovary itself (in some teleosts), but the developing young is nourished by the egg's yolk. In viviparous vertebrates the fertilized eggs and developing young are similarly retained in a portion of the Müllerian duct or ovary and in addition to physical protection are also given nourishment by the mother through the placenta.

Fig. 18-15. Many oviparous vertebrates have evolved methods to protect eggs and young during development. The male sea horse has a pouch in which the eggs develop. The tropical frog, Pipa, *has pits on its back in which the eggs and tadpoles develop. The spiny anteater develops a pouch during the breeding season in which the eggs and then the hatchlings are kept. The male stickleback builds a nest for the eggs and then fans the nest with its fins to keep fresh water flowing through it. In the oyster catchers, as in most birds, both male and female incubate and protect the eggs and young.*

OVIPAROUS FORMS

There are some oviparous vertebrates in every class. Among anamniotes, the egg is usually laid in a protective gelatin coat of mucus and develops outside the mother. However, in several groups additional protective devices have evolved (Fig. 18-15). The nests built by the lamprey and many teleosts keep the eggs in a relatively secluded area during development. In the oviparous chondrichthyeans the nidamental (shell) gland, located in the anterior portion of the Müllerian duct, secretes a keratinous shell around the eggs as they pass down the duct; the young develop within this shell and then hatch out of it. Other anamniotes have pockets in the skin, called brood pouches, where the eggs are retained during development. The male sea horse has a brood pouch along his ventral trunk, into which the fertilized eggs are placed to develop. In several teleosts and a few amphibians the eggs develop within the oral cavity. The female Surinam toad of South America has many small pockets on her back; the male presses the fertilized eggs into these pockets and there they develop. Brood pouches also occur in some amniotes which are not oviparous—most familiarly in the marsupials whose young are born at an exceptionally early stage and undergo their later developmental stages in the female's ventral brood pouch, called the marsupium.

There are many oviparous amniotes—all birds, most reptiles, and the monotreme mammals. In these, the fertilized egg is protected by layers of albumen, shell membranes, and a calcareous shell, all secreted by specialized portions of the Müllerian duct as the egg passes to the outside. The calcareous shell prevents injury and desiccation and yet permits respiration because it is permeable to air.

OVOVIVIPAROUS FORMS

In the ovoviviparous forms, which include some chondrichthyeans, some teleosts, and many reptiles, the egg is retained in the Müllerian duct but gets all of its nourishment from the yolk. In ovoviviparous teleosts the brood chamber is the ovarian sac; in chondrichthyeans and reptiles it is a dilation of the Müllerian duct, called the uterus. Young in these forms do not emerge from the mother's genital tract until they are well developed and can feed themselves.

VIVIPAROUS FORMS

Viviparity seems a logical extension of ovoviviparity, giving the young both protection and nourishment from the mother. It is dependent upon a placenta—and intimate relationship between the maternal circulatory system and that of the developing young, allowing exchange of gases, nutrients, and waste products. The precise form of this relationship between blood streams varies considerably, particularly in regard to how large and thick the placenta is, and of how many membranes (Fig. 18-16). However, the maternal and fetal blood

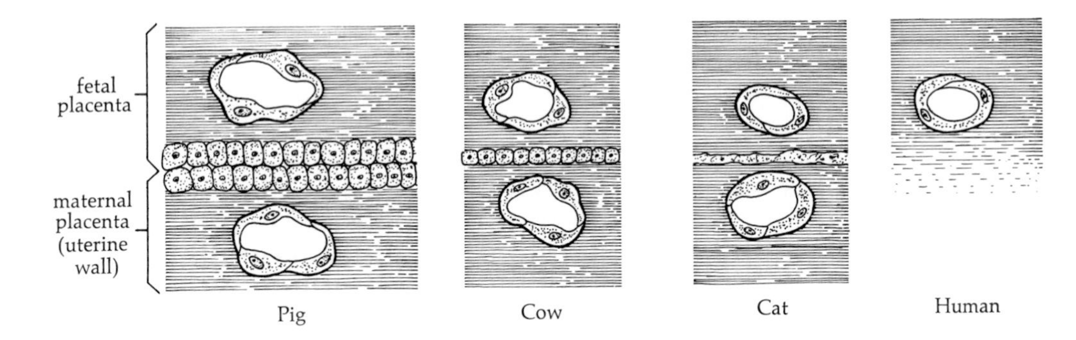

Fig. 18-16. In the therian placenta, diffusion of nutrients and waste products occurs between the maternal and fetal blood streams. However the number of layers of tissue that separates the blood streams varies from six in mammals like the pig to three in mammals like humans.

streams always remain separate except in abnormal situations. Metabolic products pass by diffusion from one to the other, through the placenta.

Viviparity occurs in many taxa, including teleosts whose "placenta" is formed in the ovarian sac, or, in some, in the ovarian follicle. In some sharks and reptiles (e.g., rattlesnakes) and in all mammals except monotremes there is a uterus. This modification of the Müllerian ducts acts as the receptacle for the developing young and the placenta is located there.

ENDOCRINES AND SEX IN MAMMALS

The gonadotropic protein hormones produced by the pituitary gland and the steroid hormones produced by the genital organs themselves profoundly influence the sexual morphology, development, and behavior of all vertebrates. However, they have been most thoroughly studied in mammals, to which the following discussion is limited.

HORMONES OF THE GONADS

Male

Male hormones are called androgens, the most common of which is testosterone. They are produced by the interstitial cells of Leydig in the testes. In the absence of androgens the sperm ducts and accessory glands fail to reach adult structure, and secondary sex characteristics are never formed. If testosterone is given to a castrated adult male, however, it will restore his secondary sex characteristics and even the ducts and glands of the genital tract if these have degenerated; testosterone will also stimulate sexual behavior. Therefore, testosterone and the less common androgens, although not directly involved in the production of sperm, are necessary for full male development and behavior.

Female

Two groups of female hormones are produced by the ovary: estrogens and progesterones. Estrogens act in the female somewhat analogously to the way androgens act in the male, but with less profound effects. They stimulate the formation of primary follicles, but not their later development in the ovary. They are also important in the development of the Müllerian ducts to maturity, although not in their early development. They are necessary for the development of female secondary sexual characteristics. The female's generally smaller size compared to the male's is due to a slight growth-inhibiting effect estrogens have on the skeleton, which in some species is exaggerated by a growth-promoting effect exerted by androgens.

The progesterones maintain pregnancy and, later, lactation. Their effects largely depend upon prior action by estrogens or gonadotropic hormones. For instance, progesterone stimulates ovulation, but only after the gonadotropic hormones have prepared the follicle for ovulation.

HORMONES OF THE PITUITARY

The two primary gonadotropic hormones of the pituitary are the follicle stimulating hormone (FSH) and the luteinizing hormone (LH). These names incorrectly suggest that their action is limited to the female, but in reality they act profoundly on the reproductive tracts of both male and female. In the female, FSH causes the differentiation of secondary and mature follicles, as well as the maturation of the ovum. In the male, FSH is necessary for spermatogenesis, including spermiogenesis. Both sexes also require LH. In the female, it stimulates the secretion of estrogens and progesterone; acting together with progesterone it stimulates ovulation. In males, LH stimulates the secretion of androgens and the release of sperm.

Two other pituitary hormones, namely, prolactin from the anterior pituitary and oxytocin from the posterior pituitary, play significant roles in reproductive behavior and morphology. Prolactin promotes the secretion of milk by the mammary glands if they have already been stimulated to development by other hormones. Prolactin is also important in maintaining parental-care behavior. Oxytocin causes the myoepithelial cells around the alveoli of the mammary glands to contract and thus eject the milk. It is also important in initiating uterine smooth muscle contractions at the time of labor.

SEXUAL CYCLES

In most vertebrates, including all mammals except humans, sexual activity is limited to a relatively short breeding season. At other times the pituitary produces minimal amounts of gonadotropic hormones; this inhibits the gonads from producing sex hormones, keeping most animals in asexual states during most of the year. In humans alone the "breeding season" lasts all year; the male is sexually active and the female undergoes continuous estrous cycles and is almost constantly receptive. Whether this bizarre situation at one time had an adaptive value, or was one of evolution's "mistakes," it has become a maladaptive feature of the uniqueness of this species.

Male

The sexual cycle of nonhuman mammals is relatively simple in males. Environmental factors, most frequently the increasing photoperiod in the spring, stimulate the pituitary to start producing gonadotropic hormones; these cause the initiation of spermatogenesis in the testes and the secretion of testosterone by the testes' interstitial cells bringing about full sexual morphology and behavior. This sexually active state lasts throughout the breeding season, after which the anterior pituitary stops producing gonadotropic hormone; that "turns off" the sexual cycle, and the animal passes through an asexual period for the remainder of the year.

Female

Throughout the breeding season female mammals undergo shorter cyclic changes, known as estrous cycles, which involve follicle differentiation, egg maturation, ovulation, and postovulatory changes in the genital tract, particularly the ovary (Fig. 18-17).

During the proestrous phase, FSH is the primary stimulus to the ovary, causing development of the follicles. The follicular cells themselves secrete both estrogens and progesterones, but much more of the former; the effects of estrogens, therefore, are dominant. They promote the growth of the entire reproductive tract, but most particularly of the endometrium, the epithelial lining of the uterus. By the time the follicles are mature and ready to ovulate the endometrium is extremely

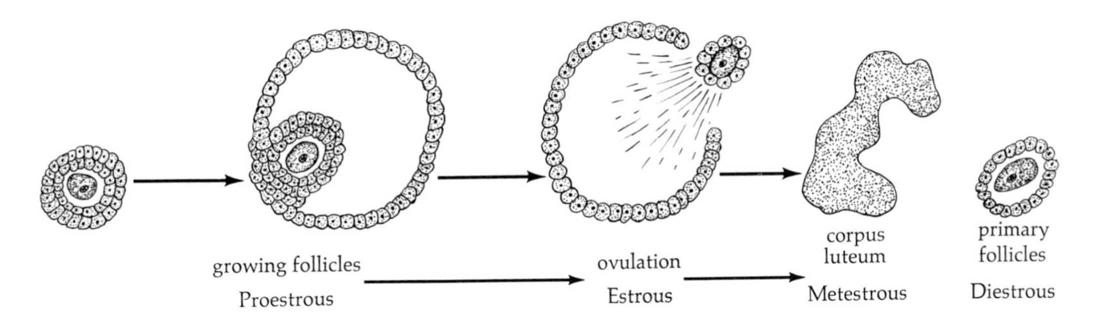

Fig. 18-17. During estrous and menstrual cycles hormonal changes control growth of ovarian follicles, followed by ovulation and formation of the corpus luteum.

thick and ready for the implantation of any eggs which may become fertilized; these fertilized eggs are in the blastula stage by the time they reach the uterus.

"Heat," or the estrous period of the estrous cycle, is next. More progesterone is secreted by the ovary. A hypothalamic center (sometimes called the ovulatory center) releases a releasing factor, which stimulates the anterior pituitary to increase greatly its secretion of LH. This sudden increase in LH, plus the increased progesterone, causes ovulation. In some mammals, such as the rabbit, the hypothalamic ovulatory center is excited only by copulation, so that copulation itself triggers ovulation and pregnancy is almost guaranteed.

After ovulation the follicle contracts, alters, and becomes the corpus luteum; this is the metestrous phase of the estrous cycle. The corpus luteum produces large amounts of progesterone and some estrogen. If fertilization and implantation do not take place, the following events occur.

For a period of time the endometrial glands in the uterine walls continue to grow and secrete a uterine fluid. Progesterone inhibits the contraction of the smooth muscles of the uterine wall and blocks the secretion of FSH and LH by the pituitary. Therefore there is no further follicular development and, of course, no further ovulation. The decrease in LH secretion also causes the corpus luteum to diminish in size and, therefore, to decrease its production of progesterone.

With this withdrawal of progesterone, the thickened endometrium of the uterus atrophies and is gradually resorbed in most mammals, or sloughed off with considerable and abrupt bleeding in the special case of humans; menstruation is thus an active flushing out of the endometrial walls (which is metabolically much less efficient than resorption).

This is the diestrous period. It brings the estrous cycle fully around to where it can either be extended in a long asexual period (called anestrous), or, if environmental factors are correct, yield to a new proestrous period immediately (as is the case in humans.).

PREGNANCY AND BIRTH

The foregoing description of the estrous cycle assumed that fertilization and implantation did not occur. However, if a fertilized egg reaches the uterus, gestation ensues and the estrous cycle is suspended. When the blastocyst (as the blastula stage is called in amniotes) reaches the uterine wall, the endometrial cells around it undergo rapid mitotic division and overgrow it, and the tissue just deep to it is eroded away so that it sinks in and becomes embedded in the walls of the endometrium in very close contact with the uterine blood vessels. This is the process of implantation, and apparently it is the stimulus which extends the period during which progesterone dominates the estrous cycle.

The developing placenta itself produces hormones: progesterone, estrogen, and gonadotropin; the last of these acts very similarly to LH and thus maintains the corpus luteum intact throughout pregnancy. By the end of pregnancy more estrogen is being produced than progesterone, and most of both these hormones are being produced by the placenta rather than the ovary. This high level of estrogen facilitates contraction of the uterine walls and the breakdown of the endometrium. The contractions of the uterus are greatly augmented by the secretion of oxytocin from the posterior pituitary. These powerful contractions of the uterine muscles, accurately described as "labor," discharge first the newborn mammal and then the placenta ("afterbirth").

During the latter portion of pregnancy, as estrogen secretion from the placenta increases, there is also a profound development of the ducts and secretory cells of the mammary glands. Progesterone plays but a small role in this development; estrogen is the primary stimulus. Once the glands are fully formed, prolactin, an anterior pituitary hormone, causes the synthesis of milk and its secretion into the alveoli of the mammary glands. However, the action of prolactin is blocked by progesterone, and thus it is not until the progesterone level falls off, which happens immediately after birth, that the active synthesis of milk begins.

The milk is ejected by contractions of the myoepithelial cells around the alveoli, and these contractions are stimulated by oxytocin. The production and release of oxytocin is reflexive and, after birth, caused by manipulation of the nipples and surrounding skin. Thus the reflexive sucking, stimulated by the need of the newborn and young mammal, in turn stimulates the reflex that causes oxytocin secretion, which in turn stimulates the myoepithelial cells to contract and eject milk—one of the neatest reflex and feedback arrangements.

CARE OF YOUNG

In general, as the size of an animal and the complexity of its nervous system increase, the time of its prenatal and postnatal development is prolonged. In mammals, for instance, the gestation period of small rodents is about 20 days and of large rodents about 30 days; the time from birth to sexual maturity in small rodents is 25 to 50 days, and in large rodents, 8 to 12 months. The beaver has a gestation period of 6 months, and reaches sexual maturity in 2 years. In humans, gestation takes 9 months, but the child requires much care for several years and reaches sexual maturity only at the age of 12 or 15 or even later.

In many taxa the hormonal control of sexual behavior extends to the maintenance of parental care behavior; the reflexes involved in the production and lactation of milk, just described, are a first example, and an interesting one because they ensure that the mother will care for the young at the time when the young has just become a separate individual. As the time after birth lengthens, the hormonal effects produced by the birth process wear off. The sucking reflex may maintain milk production, if the mother will tolerate being suckled, but if care of the young is to go on for several months or even years there needs to be either a constantly renewed hormonal control or some form of external control.

In species like our own, therefore, many forms of social organization have evolved which can be regarded as supplementing and extending the influence of hormones, ensuring that the individual will continue to do what is best for the species. The institution of the human family, for instance, has as its primary adaptive value the nurture and education of the young, and it is buttressed by religion, ethics, government, and cultural myths. Unfortunately these problems are beyond the scope of this book. We can mention, in passing only, that the evolution of these social forms—found not only in humans—is an important extension of the evolutionary process which brought into being anatomical, physiological, and behavioral adaptations for reproduction, and that the evolution of these social forms was necessitated by the prolonged period of care required for the development of a more elaborate nervous system.

EPILOGUE

We noted earlier that the species in which conception and development of young is left largely to chance produce many more gametes than those in which mechanisms have evolved to ensure these processes. However, one might wonder if some species would not be at a greater advantage if they could combine a "high yield" with the capacity to produce large numbers of gametes. On the basis of numbers alone such a species would be in little danger of being wiped out by predators or pestilence.

Such a science fiction proposition tacitly assumes an unlimited supply of the necessities of the environment, but this is never the case. Instead, as Malthus pointed out and Darwin repeated, the numbers of a species, if unchecked, increase exponentially while food supply increases only arithmetically; as we are now learning other resources of the environment, such as oxygen, pure water, and open space, do not increase at all while pollution of the environment, like population, increases exponentially.

The hypothetical species with such phenomenal reproductive capacity would be similar to a species from which the natural checks on population growth have been removed. Not many years ago all major predators, particularly mountain lions, were exterminated in parts of the American southwest, and the deer were "protected" by law. The deer increased enormously in numbers but their food supply did not. As a result the hard-pressed deer gnawed on parts of trees and shrubs which they did not normally use for food. Many trees and shrubs were so badly stripped that they died, and large areas of forest became decimated. The deer died in great numbers from starvation and disease, and the species suffered more from its own "success" than it had when predators kept it under control.

It is not surprising, then, to find that the number of gametes produced dramatically decreases as physiological and morphological mechanisms evolve to ensure development of young. When the process of natural selection is left to itself, this balance is so fine that the number of surviving young generally approximates the number required to maintain a stable population. When environmental factors change more rapidly than natural selection can accommodate them, however—as they did for the southwestern deer—this balance is lost.

This is the situation in which the human species finds itself today. Our biology has not kept pace with our rapidly changing environment. Technological know-how throughout much of the world has greatly lengthened the lifespan of the individual, reduced disease, and greatly increased the percentage of newborn young that survive to reproductive age as well as the ability of individuals to bear children. The social institutions which protect the immature young also reward their production, with everything from social approval to government support. Because the curbs on population growth have been removed, many of the anatomical, physiological, and behavioral adaptations which helped this species become successful are now working to its detriment. The worldwide human population is increasing at an accelerating rate; we have followed the Genesis edict to "be fruitful and multiply, and replenish the earth, and subdue it; and have dominion over . . . every living thing that moveth upon the earth." So great is our success and domination that we now endanger all other species of animals, and we must contemplate the exhaustion of plant life, of oxygen and water supplies, and of space. On the other hand, even before that happens we may have choked and drowned in our own garbage.

It was not a biologist but a poet, T. S. Eliot, who said: "This is the way the world ends / Not with a bang but a whimper." He may turn out to be right, and soon, unless we use our tremendous potential intelligence and our social institutions to impose new and even more stringent curbs on our population growth, and to develop new ways, more productive than wars, for coping with our current human population. If our species is to survive we do not have time to wait for natural selection to solve these problems.

SUGGESTED READING

REPRODUCTION

Barr, W. A. (1968). Patterns of ovarian activity. *In* "Perspectives in Endocrinology." (E. J. W. Barrington, ed.). Academic Press, New York.

Breder, C. M., Jr., and Rosen, D. E. (1966). "Modes of Reproduction in Fishes." The Natural History Press, New York.

Goldberg, S. R. (1970). Seasonal ovarian histology of the ovoviviparous iguanid lizard *Sceloporus jarrovi* Cope. *J. Morph.* **132,** 265–276.

Perry, J. S., and Rowlands, I. W. (eds.) (1969). "Biology of Reproduction in Mammals." Blackwell, Oxford, England.

Sharman, G. B. (1970). Reproductive physiology of marsupials. *Science* **167,** 1221–1228.

Valentine, G. L., and Kirkpatrick, R. L. (1970). Seasonal changes in reproductive and related organs in the pine vole, *Microtus pinetorum,* in southwestern Virginia. *J. Mammal.* **51,** 553–560.

Van Tienhoven, A. (1968). "Reproductive Physiology of Vertebrates." Saunders, Philadelphia, Pennsylvania.

Ehrenfeld, D. W. (1970). "Biological Conservation." Holt, New York.
Ehrlich, P. R. (1968). "The Population Bomb." Ballantine Books, New York.
Etkin, W. (1967). "Social Behavior from Fish to Man." Univ. Chicago Press, Chicago, Illinois.
Hardin, G. (1968). The tragedy of the commons. *Science* **162,** 1243–1248.

GLOSSARY

A

abduct. [L. *ab,* away + *ducere,* to lead] To move an appendage, or part of an appendage, away from the body's axis. Opposite of adduct.

abomasum. [L., *ab,* away from + *omasum,* bullock's tripe] Glandular portion of a ruminant's stomach.

Acanthodii. [Gr. *akantha,* thorn] The subclass of the class Placodermi which is most similar to living bony and cartilaginous fishes.

accommodation. [L. *ac,* toward + *com,* with + *modus,* measure or manner + *-atus,* adapted] Adjustment of the eye's cornea and/or lens to focus images onto the retina.

acelous. [L. *a-,* without + *celom,* hollow] An adjective describing a structure whose surfaces lack concavities; usually referring to the centrum of a vertebra.

acetylcholine. [L. *acetum,* vinegar + *chole,* bile] Acetic acid ester of choline; neurotransmitter of somatic motor nerves to skeletal muscles.

acidophil. [L. *acid(us),* sour + Gr. *philos,* loving] A structure, usually a cell, which stains with acid stains.

acousticolateralis. [Gr. *akouazesthai,* to listen + L. *lateralis,* of the side] Adjective describing vertebrate sense organs with hair cells, that is, the inner ear and the lateral line system.

acrodont. [Gr. *acros,* topmost, highest + *odontos,* tooth] Adjective describing sharp-pointed teeth having no roots.

Actinopterygii. [Gr. *aktinos,* ray + *pterygi,* wing or fin] Subclass of Osteichthyes including all the ray-finned fishes.

adduct. [L. *adducere,* to lead toward or bring into] To move an appendage, or part of an appendage, toward the body's axis. Opposite of abduct.

adenohypophysis. [Gr. *adenos,* a gland + *hypophysis,* outgrowth (from below)] Anterior portion of the pituitary gland; that portion derived from the stomodeum.

adrenal. [L. *ad,* toward + *renalis,* the kidneys] Compound endocrine gland adjacent to the kidneys.

adrenaline. A hormone of the adrenal medulla which functions as a neurotransmitter of the sympathetic nervous system. Syn., epinephrine.

aestivate. [L. *aestivare,* to reside during the summer] To greatly reduce metabolic activity for extended periods of time in response to external stress such as warm weather or lack of food or water. Similar to hibernate, which however is a response to cold.

Agnatha. [Gr. *a-* without + *gnathos,* jaw] The class of vertebrates lacking true jaws.

alar plate. [L. *ala,* wing] Portion of the central nervous system dorsal to the sulcus limitans from which develop the sensory nuclei; in the aggregate these nuclei form a wing shape in the spinal cord.

aldosterone. [NL. *al(cohol) dehyd(rogenatum),* dehydrogenated alcohol + Gr. *stereos,* hard or solid] A powerful steroid hormone of the adrenal cortex.

alveolus. [L. dimin. of *alveus,* hollow sac or cavity] Any small "pocket" such as the smallest chamber in the lung or the socket for a tooth.

ameloblast. [OF. *amel,* enamel + Gr. *blastos,* germ] A cell which produces the enamel or enamel-like substance forming the outer part of a tooth or scale.

amniote. [Gr. *amnos,* lamb + *-ion,* diminutive suffix] Any vertebrate with extraembryonic membranes that include the amnion: reptiles, birds, and mammals.

amphiarthrosis. [Gr. *amphi,* both + *arthrosis,* joint] A skeletal joint with limited movement.

Amphibia. [Gr. *amphibia,* living a double life] The class of vertebrates which includes frogs, toads, salamanders, caecilians, and the extinct labyrinthodonts.

amphicelous. [Gr. *amphi,* both + *celom,* hollow] An adjective describing a structure both of whose ends are concave; usually pertaining to the centrum of a vertebra.

amphiplatyan. [Gr. *amphi,* both + *platys,* flat] Adjective describing a structure flat on both sides. Usually refers to the centrum of a vertebra, in which case it is interchangeable with acelous.

amplexus. [L. *amplexus,* an embrace] The clasping behavior of male and female resulting in simultaneous discharge of sperm and eggs for external fertilization; the term usually is used in reference to anurans.

ampulla. [L. equiv. to *amphora,* a two-handled bottle or storage jar] A dilation in a canal or duct; particularly a dilation containing sensory epithelium in a semicircular canal of the ear.

analogues. [Gr. *ana,* according to + *logos,* ratio or relation] Structures in different animals having the same function irregardless of form or embryonic origin.

anamniote. [Gr. *an,* without + *amniote* (which see)] Any vertebrate which lacks an amnion, although other extraembryonic membranes may be present: all fishes and amphibians.

Anapsida. [Gr. *an,* without + *apsis,* a loop or mesh] Subclass of reptiles which lacks temporal fenestra: turtles.

anastomosis. [Gr. *ana,* again + *stoma,* mouth] A tubular structure, usually a blood vessel, which interconnects two other similar structures.

anatomy. [Gr. *ana,* up + *tome,* a cutting] The study of the structure of organisms, usually as revealed by dissection.

androgen. [Gr. *andro,* male + *-gen,* production (of)] A general term for male hormones.

anestrous. [Gr. *an,* without + *oistros,* mad desire] An extended period of time between estrous cycles.

angiotensin. [Gr. *angeion,* a vessel + L. *tensus,* to stretch] A vasoconstrictive protein produced by the action of renin on an α-globulin.

anlage (pl., anlagen). [Ger. set-up, layout; *an,* on + *lage,* position] The embryonic primordium of a structure.

antidiuretic hormone. [Gr. *anti,* against or opposite + *diouretikos,* from *oureen,* to urinate] Hormone increasing water resorption in the kidney, and maintaining constant osmotic level in tissues; also stimulates contraction of smooth muscles in blood vessels. Syn., vasopressin.

Anura. [Gr. *an,* without + *oura,* tail] Order of amphibians including frogs and toads.

apocrine. [Gr. *apokrino,* to separate] A secretory process in which a small part of the cell forms the secretion; also an adjectival form designating a specific class of large sweat glands.

Apoda. [Gr. *a,* without + *pod,* foot] An order of amphibians which lacks paired appendages: the caecilians.

aponeurosis. [Gr. *apo-,* away + *neuron,* sinew or tendon] A fibrous sheet functioning as a tendon.

arboreal. [L. *arbor,* tree] Adjective designating animals adapted for living in trees.

arborization. [L., *arbor,* tree] A spreading out, as the branches of a tree; used in describing the dendritic pattern of a neuron.

archetype. [Gr. *arche,* first + *typos,* type] The original pattern or model after which a thing is made. In biology, this term means an organism, real or hypothetical, whose morphology is so general that all other members of its group could be derived from it.

archinephros. [Gr. *arche,* first + *nephros,* kidneys] A type of vertebrate kidney running the entire length of the body with but one tubule per segment. Syn., holonephros.

archistriatum. [Gr. *archi,* foremost, old + *striatus,* furrow or channel] The ventrolateral portion of the avian basal ganglia; also used to designate parts of the basal ganglia of other vertebrates.

Archosauria. [*archi,* foremost, old + *sauros,* lizard] Subclass of Reptilia including crocodilians, dinosaurs, pterosaurs, etc.

articulation. [L. *articularis,* pertaining to the joints] A joint; the connection of two skeletal elements.

autonomic nervous system. [Gr. *autos,* self + *nomos,* law] The system of all neurons, both central and peripheral, that have general visceral motor functions.

Aves. [L., *aves,* birds] The class of vertebrates all of whose members are birds.

axial. [L. *axis,* an axis or axle] Relating to the central axis of the body morphologically represented by the spinal cord and notochord.

axon. [Gr. *axon,* an axis or axle] The process of a neuron which transmits nerve impulses away from the perikaryon.

B

baculum. [L. *baculum,* walking stick or staff] A sesamoid bone found in the penis of many mammals.

baleen. [L. *balaena,* a whale] A hard keratinous substance hanging from the roof of the mouth in toothless whales; functionally it acts as a plankton filter.

basal ganglia. [L. *basis,* basis + Gr. *ganglion,* a swelling or knot] A prominent group of nuclei deep to the base of the telencephalon. Syn., corpus striatum.

basal plate. [L. *basis,* basis + Gr. *platys,* broad or flat] (1) The fused parachordal cartilages of the neurocranium which form the skeletal floor of the hindbrain. (2) The portion of the central nervous system which develops ventral to the sulcus limitans; this includes the motor nuclei in the spinal cord and brain.

basement membrane. [L. *basis,* basis + *membrana,* fine skin] Extremely thin connective tissue network supporting an epithelium.

basilar membrane. [NL. *basilaris,* pertaining to the base, especially of the skull + L. *membrana,* fine skin] Connective tissue sheet supporting the organ of Corti.

basilar papilla. [NL. *basilaris,* pertaining to the base, esp. of the skull + L. *papilla,* nipple] A sensory epithelium for hearing in amphibians, reptiles, and birds.

basophil. [Gr. *basis,* base + *philos,* loving] A cell readily stained by basic dyes.

Batoidea. [Gr. *batis,* a ray or skate] The order of Chondrichthyes comprised of skates and rays.

Betz cell. [Named for Vladimir Betz, a nineteenth century Russian anatomist] Large motor cell of the neocortex.

biped. [L. *bi-,* two + *ped,* foot] An animal which locomotes on two legs.

blastula. [Gr. *blastos,* bud or sprout] The embryonic stage between cleavage and gastrulation.

boid. [L. *boa,* water snake. Pronounced bo-id] A family of snakes which includes the boa constrictor.

Botallus, duct of. [L. *ducere,* to lead. Named for Leonardus Botallus, an Italian physician in Paris about 1530] A fetal anastomosis between the aortic arch and the pulmonary arch. Syn., ductus arteriosus.

branchiomeric. [Gr. *branchio,* gills + *meris,* portion] Adjective describing the gill region or its derivatives.

bronchiole. [Gr. *bronchos,* windpipe] Branch off the main bronchus within the lung.

bulbourethral gland. [L. *bulbus,* a bulb (from Gr. *bolbos,* onion) + Gr. *ourethra,* the urethra] A mammalian male secondary genital gland emptying directly into the urethra. Syn., Cowper's gland.

bunodont. [Gr. *bounos,* hill + *odontos,* tooth] Low-crowned, squared-off molar teeth common to omnivorous mammals.

C

calamus. [Gr. *kalamos,* reed or stalk] The proximal end of the shaft of a quill feather.

calcitonin. [L. *calx,* lime] Hormone secreted by the parafollicular cells of the thyroid and parathyroid glands.

calyx. [Gr. *kalyx,* husk or covering] Dilation of the ureter within the metanephric kidney.

carapace. [Sp. *carapacho,* carapace] Dorsal half of a turtle or armadillo shell.

Carboniferous. [L., *carbon,* charcoal + *ferous,* bearing, producing, yielding] Geologic period in the latter Paleozoic era characterized by large deposits of organic carbon from plants.

carnassial teeth. [F. *carnassier,* flesh-eating] Last upper premolar and first lower molar of carnivores; these teeth are specialized for shearing flesh.

cavum arteriosum. [L. *cavum,* hole + Gr. *arteria,* windpipe, artery] Portion of the ventricle in the chelonian and lepidosaurian heart which receives blood from the left atrium.

cavum pulmonale. [L. *cavum,* hole + *pulmo,* lung] Portion of the chelonian and lepidosaurian heart ventricle from which the pulmonary artery exits.

cavum venosum. [L. *cavum,* hole + *venosus,* vein] Portion of the chelonian and lepidosaurian heart ventricle from which the right and left aortae exit.

cementum. [L. *caementum,* rough quarry stones] Layer of dense material holding tooth root into socket.

Cenozoic. [Gr. *kainos,* new or recent + *zoe,* life] The most recent (and current) geologic era, which began about 70 million years ago.

cephalization. [Gr. *kephale,* the head] The concentration of important organs in the head.

cerebellum. [dim. of L. *cerebrum,* brain] Portion of the brain forming the roof of the metencephalon and functioning as the coordinating motor center.

cerebrum. [L. *cerebrum,* brain] The two cerebral hemispheres; the entire telencephalon except olfactory bulbs and tracts.

ceruminous gland. [L. *cera,* wax] Gland of the external auditory meatus secreting ear wax.

Chelonia. [Gr. *chelone,* tortoise] The order of Reptilia comprised of turtles and tortoises.

cholecystokinin. [Gr. *chole,* bile + *kystis,* bladder + *kineo,* to move] A hormone from the upper intestines stimulating contraction of the gallbladder.

Chondrichthyes. [Gr. *chondros,* cartilage + *ichthyes,* fishes] The vertebrate class comprised of cartilaginous fishes.

chondrocyte. [Gr. *chondros,* cartilage + *kytos,* a hollow vessel] A cartilage cell.

Chondrostei. [Gr. *chondros,* cartilage + *osteon,* bone] The superorder of Actinopterygii which includes the sturgeon, the bichir, and the paddlefish.

Chordata. [Gr. *chorde,* gut or string] The phylum to which the subphylum Vertebrata belongs, characterized by a dorsal hollow nerve cord, pharyngeal pouches, and a notochord.

choroid. [Gr. *chorion,* leather] The vascular layer of the eye between sclera and retina.

chromaffin tissue. [Gr. *chroma,* color + L. *affinis,* affinity] Tissue particularly in the adrenal medulla which can be stained deep brown by chromic salts.

chromatophore. [Gr. *chroma,* color + *-phoros,* bearing] Pigment cell.

chromophobe. [Gr. *chroma,* color +

phobos, fright] Cell resistant to stains.

chyme. [Gr. *chymos,* juice] Partially digested food passed from the stomach to the small intestine.

cingulum. [L. *cingere,* to surround] An association tract of the mammalian cerebrum running from the medial cortex to the hippocampus.

cisterna. [Gr. *kiste,* chest] Any cavity or enclosed space; particularly, small cavities made by double-walled endoplasmic reticulum.

cleidoic egg. [Gr. *kleido,* to lock up] An egg covered by a leathery or brittle calcareous shell, characteristic of reptiles, birds, and prototherians.

clitoris. [Gr. *kleitoris,* shut] The female homologue of the penis.

cloaca. [L. *cloaca,* sewer or drain] Common chamber draining genital ducts, excretory ducts, and the alimentary canal; present in most vertebrates but absent in teleosts and therians.

cochlea. [Gr. *kochlias,* snail, with spiral shell] The auditory portion of the inner ear of mammals.

Coelacanthini. [Gr. *koilos,* hollow + *acantho,* spiny] The group of sarcopterygian fishes of which *Latimeria* is the only living form.

colic caecum. [Gr. *kolikos,* colon + L. *caecum,* blind gut] A blind diverticulum occurring at the junction of small and large intestines.

collagen. [Gr. *kolla,* glue + *-gen,* production of] Tough structural protein; the most common and the strongest fibers of connective tissue.

columella. [Dim. of L. *columna,* pillar] Middle ear ossicle of amphibians, reptiles, and birds.

commissure. [L. *committo,* to bring together] A tract of the central nervous system which connects identical structures on right and left sides.

concha. [L. *concha,* shell] A bone, usually scroll-shaped, covered by mucous membranes, which gives added surface area to the nasal chamber.

cone. [Gr. *konos,* pine cone] Visual receptor cell of the retina which responds to specific wavelengths; thus used in color vision.

corium. [L. *corium,* skin, hide, leather] The mesodermally derived, dense connective tissue layer of the skin. Syn., dermis.

cornea. [L. *corneus,* horny] Transparent anterior superficial portion of the eyeball.

corona radiata. [L. *corona,* garland or crown + *radiata,* radiate] The widely radiating fibers of the internal capsule, as they spread out to the entire neocortex.

corpora cavernosa. [L. pl. of *corpus,* body + *cavernosus,* full of hollows] Two dorsal columns of erectile tissue in the amniote penis.

corpora quadrigemina. [L., pl. of *corpus,* body + *quadri-,* four + *geminus,* a twin] Roof of the mammalian midbrain, composed of two superior colliculi and two inferior colliculi.

corpus callosum. [L. *corpus,* body + *callosus,* hard] Neocortical commissure of eutherians; the largest tract in most eutherian brains.

corpus luteum. [L. *corpus,* body + *luteum,* yellow] Remains of the ovarian follicle after ovulation.

corpus spongiosum. [L. *corpus,* body + Gr. *spongos,* a sponge] Ventral medial column of erectile tissue surrounding the urethra in the mammalian penis.

corpus striatum. [L. *corpus,* body + Gr. *striatus,* furrow or channel] Prominent group of ventral medial telencephalic nuclei. Syn., basal ganglia considered collectively.

cortex. [L. *cortex,* bark] Outer portion of an organ, such as the kidney; specifically used to designate the outer portion of the central nervous system when it is gray matter rather than white matter.

cosmoid scale. [Gr. *kosmos,* ornament or form] Heavy bony scales of ancient sarcopterygians which had ornately patterned outer surfaces.

Cotylosauria. [Gr. *kotyle,* cup-shaped + *sauros,* lizard] Stem order of Reptilia.

Cowper's gland. [Named for William Cowper, late seventeenth century British anatomist] A mammalian male secondary genital gland emptying directly into the urethra. Syn., bulbourethral gland.

Cretaceous. [L. *creta,* chalk] Most recent period of the Mesozoic era.

crista. [L. *crista,* crest] The sensory epithelium in the ampulla of the semicircular canal.

Crocodilia. [L. *kroke,* pebble or gravel + *drilos,* worm] The order of Reptilia including alligators, crocodiles, etc.

Crossopterygii. [Gr. *krosso,* tassels or fringe + *pterygion,* little wing or fin] The group of sarcopterygian fishes including the coelacanths and rhipidistians.

ctenoid scale. [Gr. *ktenoeides,* like a comb] Very thin, bonelike scales with small, toothlike projections; present in many teleosts.

cupula. [L. *cupula,* small tub or dome] Gelatinous material in which the hairs of the hair cells of the ampullary cristae are embedded.

Cuvier, duct of. [Named for Georges Cuvier, French anatomist] The vein which collects from the anterior and posterior cardinal veins and drains into the sinus venosus. Syn., common cardinal vein.

cycloid scale. [Gr. *kyklos,* circle] Very thin, bonelike scales, circular in shape, without projections; found in diverse living actinopterygians.

Cyclostomata. [Gr. *kyklos,* circle + *stoma,* mouth] Living order of Agnatha including the lamprey and hagfish.

cytoarchitecture. [Gr. *kytos,* a hollow vessel (now usually meaning a cell) + L. *architectura,* architecture] The structure of the cytoplasm, i.e., the organization of the organelles within a cell.

D

decussate. [L. *decussis,* the number ten] To cross in the form of an X (like the Roman numeral ten).

delamination. [L. *de-,* separation or removal + *lamina,* a thin plate or leaf] Separation into layers.

demibranch. [F. *demi,* half or lesser + Gr. *branch,* gill] A gill arch with lamellae on only one of its two surfaces.

dendrite. [Gr. *dendrites,* pertaining to a tree] Any cytoplasmic process of a neuron which either does not transmit the action potential, or transmits it toward the cell body.

dentine. [L. *dens,* tooth] The ivory portion of a tooth.

dermatocranium. [Gr. *derma,* skin + *kranion,* skull] The portion of the skull formed by dermal bones.

dermatome. [Gr. *derma,* skin + *tome,* cutting] (1) Portion of the pharyngula embryo forming the dermis in the epimeric region. (2) Portion of the skin with sensory innervation from a single spinal nerve.

dermis. [Gr. *derma,* skin] The mesodermally derived, dense connective tissue layer of the skin. Syn., corium.

desiccate. [L. *de-,* prefix indicating intensity + *siccare,* to dry] To dry thoroughly.

desmosome. [Gr. *desmos,* a band + *soma,* body] Tight junctions, usually between epithelial cells.

Devonian. Geologic period of the Paleozoic era characterized by abundance of fish and the first amphibians. So named because strata were first studied in Devonshire, England.

diaphragmaticus. [Gr. *diaphragma,* a partition wall, midriff] Specialized muscle in crocodilians, going from ilium to liver; no structural relationship to the mammalian diaphragm.

diaphysis. [Gr. *dia,* between + *physis,* growth] The shaft of an endochondral bone.

diapsid. [Gr. *di-,* two + *apsis,* a loop or mesh] Referring to a reptile with temporal fenestrae above and below the squamosal–postorbital joint.

diarthrosis. [Gr. *di-,* two + *arthron,* a joint] A freely movable joint.

diencephalon. [Gr. *dia-,* between +

enkephalos, brain] Posterior portion of the prosencephalon, lying between the cerebral hemispheres and the midbrain.

diestrous. [Gr. *dia-,* between + *oistros,* mad desire] Portion of the estrous cycle characterized by the deterioration of the endometrium after metestrous and before proestrous.

digastricus. [Gr. *di-,* two + *gaster,* belly] The mammalian muscle which lowers the jaw to open the mouth.

digitiform gland. [L. *digitus,* finger + *forma,* form + *glans,* acorn] Salt-secreting gland opening into the large intestine of chondrichthyeans. Syn., rectal gland.

diplospondylous. [Gr. *diplo-,* double + *spondylos,* vertebra] Condition of having two centra per vertebra.

Dipnoi. [Gr. double-breathing, from *di-,* two + *pneo,* breath, air] The order of sarcopterygians comprised of lungfish.

diverticulum. [L. *diverticulum,* a by-road, from *de* + *verto,* to turn aside] A pouch extending from a tubular or saccular organ.

ductus arteriosus. [L. *ducere,* to lead + *arteria,* artery] A fetal anastomosis between the aortic arch and the pulmonary arch. Syn., duct of Botallus.

duodenum. [L. *duodeni,* twelve] The initial portion of the small intestine; both bile and pancreatic ducts open into it. So named because of its length (12 fingerbreadths) in the human.

E

eccrine glands. [Gr. *ekkrinein,* to secrete (from *ex,* out of + *krino,* separate) + L. *glans,* acorn] Small, unbranched, merocrine-secreting sweat glands.

ectoderm. [Gr. *ektos,* outside + *derma,* skin] The outer epithelial cells of the early gastrula, and their derivatives.

ectostriatum. [Gr. *ektos,* outside + *striatus,* furrow or channel] A portion of the avian cerebral hemisphere receiving visual information.

egestion. [L. *ex,* out of + *gestus,* carry] The elimination of unabsorbed food remains, usually as feces.

Elasmobranchii. [Gr. *elasmos,* plate metal + *branchia,* gills] Subclass of Chondrichthyes including sharks, skates, and rays.

endocrine. [Gr. *endon,* within + *krino,* separate] Referring to glands that lack ducts and release their secretions into the bloodstream.

endoderm. [Gr. *endon,* within + *derma,* skin] The inner epithelial cells of the early gastrula and their derivatives.

endolymph. [Gr. *endon,* within + L. *lympha,* a clear fluid] Fluid contained in the membranous labyrinth of the inner ear.

endometrium. [Gr. *endon,* within + *metra,* uterus] The epithelial lining of the uterus.

endoplasmic reticulum. [Gr. *endon,* within + *plasma,* something molded or formed + L. *reticulum,* little net] The complex of cytoplasmic membranes within a cell.

endostyle. [Gr. *endon,* within + *stylos,* pillar] The iodine-concentrating structure of *Amphioxus* and the ammocoete larva.

entorhinal cortex. [Gr. *entos,* within + *rhis* (*rhin*), nose + L. *cortex,* bark] Portion of the mammalian cortex intermediate between piriform cortex and hippocampus.

epaxial. [Gr. *epi,* upon + L. *axis,* axis] Above or beyond any axis; in anatomy, the portion of the body which develops dorsal to an axis formed by the developing notochord.

epicritic. [Gr. *epi,* upon + *krino,* separate] Referring to the system of sensory nerves carrying the finest, most discriminative sensations.

epidermis. [Gr. *epi,* upon + *derma,* skin] The epithelial layer of the skin; i.e., that portion of the skin that is ectodermally derived.

epididymis. [Gr. *epi,* on + *didymos,* twin] The extremely convoluted portion of the Wolffian duct; that portion lying nearest the testes.

epimere. [Gr. *epi,* upon + *meros,* part] The embryonic epaxial mesoderm.

epimysium. [Gr. *epi,* upon + *mys,* muscle] The dense connective tissue sheath around a skeletal muscle.

epinephrine. [Gr. *epi,* on + *nephros,* kidney] Neurotransmitter of the sympathetic nervous system, and hormone secreted by the adrenal medulla. Syn., adrenaline.

epiphysis. [Gr. *epi,* upon + *physis,* growth] The end of an endochondral bone when it has formed from a different ossification than the shaft.

epithalamus. [Gr. *epi,* upon + *thalamos,* a chamber] Dorsalmost portion of the diencephalon, including the habenular nuclei.

erythrocyte. [Gr. *erythros,* red + *kytos,* a hollow vessel] Red blood cell.

estrogen. [Gr. *oistros,* mad desire + *-gen,* production of] Any of several female hormones responsible for secondary sex characteristics and the initiation of estrous.

estrous. [Gr. *oistros,* mad desire] 1. Portion of the female reproductive cycle, also known as heat, during which ovulation occurs. 2. Estrous cycle is another term for female reproductive cycle.

euryapsid. [Gr. *eury,* broad + *apsis,* a loop or mesh] Referring to reptiles with a single temporal fenestra dorsal to the postorbital–squamosal suture.

Eutheria. [Gr. *eu,* good + *therion,* wild beast] Infraclass of mammals commonly known as placental mammals.

extracolumella. [L. *extra,* without, outside + *columella,* small column] The outer, usually cartilaginous portion of the auditory ossicle of amphibians, reptiles, and birds.

F

fasciculus. [L. *fasciculus,* little bundle] A bundle of muscle or nerve fibers running together.

fenestration. [L. *fenestra,* window] The evolution of gaps in the dermatocranium usually covered by connective tissue membrane.

fibroblast. [L. *fibra,* fiber + Gr. *blastos,* bud, sprout] Connective tissue cells which produce connective tissue fibers.

follicle. [L. *folliculus,* a small sac]. A general term for a spherical group of cells surrounding a cavity.

fornix. [L. *fornix,* arch or vault] A tract of the brain running from the hippocampus to the mammillary bodies and septal nuclei.

fundic. [L. *fundus,* bottom] Referring to that portion of the stomach which contains gastric glands.

G

gamete. [Gr. *gamete,* wife or *gametes,* husband] Any germ cell, either male or female.

ganglion. [Gr. *ganglion,* a swelling or knot] An aggregate of nerve cell bodies outside the central nervous system.

ganoid scale. [Gr. *ganos,* brightness + *-oide,* resembling, like] Heavy scales consisting of bone, cosmine, and many layers of ganoine, giving them a lustrous, metallic color.

ganoine. [Gr. *ganos,* brightness] A specific hard, enamel-like protein.

gastrin. [Gr. *gaster,* paunch or belly] A hormone secreted by the stomach, causing the gastric glands to secrete HCl.

gastrula. [Gr. *gaster,* paunch or belly + *-ule,* a Latin diminutive suffix] Embryonic stage of development following the blastula and preceding the pharyngula.

genu. [L. *genu,* knee] A general term used to describe any structure shaped like a bent knee.

globus pallidus. [L. *globus,* ball + *pallidus,* pale] Portion of the mammalian basal ganglia just deep to the putamen.

glomerulus. [L., dim. of *glomus,* a ball of yarn] Knot of capillaries within Bowman's capsule of the nephron.

glottis. [Gr. *glottis,* aperture of the larynx, a variation of the Athenian form of *glossa,* tongue] Opening from the pharynx to the laryngeal chamber.

glucagon. [Gr. *glykys,* sweet] Hormone secreted by the alpha cells of the islets of Langerhans; causes mobilization of liver glycogen.

glycogen. [Gr. *glykys,* sweet + *-gen,* production of] A branched polysaccharide of glucose; the principal storage carbohydrate of vertebrates.

gnathostome. [Gr. *gnathos,* jaw + *stoma,* mouth] Vertebrate having jaws.

gonad. [Gr. *gonos,* seed] Primary sexual organ, i.e., the organ which produces germ cells.

gonopodium. [Gr. *gono,* reproductive + *podion,* little foot] Modified anal fin of some fishes, used as an intromittent organ.

guanine. [Sp. *guano,* dung] A purine occurring in crystalline form in iridophores.

gubernaculum. [L. *gubernaculum,* a helm] A fibrous cord attaching testis to abdominal wall; significant in testicular descent.

gustatory. [L. *gusto,* taste] Having to do with taste.

gyrencephalic cortex. [Gr. *gyros,* ring + *enkephalos,* brain + L. *cortex,* bark] Convoluted cerebral cortex. Opposite of lissencephalic cortex.

gyrus. [Gr. *gyros,* circle] A raised area or convolution of the cerebral hemisphere, delimited by grooves (sulci).

H

hair cell. A mechanoreceptor cell in the acousticolateralis system, characterized by many stereocilia and usually a single kinocilium on its free surface.

Haversian canal. [Named for Clopton Havers, a seventeenth century English anatomist] A canal within a bone, containing blood vessels and surrounded by concentric layers of compact bone.

helicotrema. [Gr. *helix,* spiral + *trema,* a hole] The continuity at the apex of the cochlea between scala tympani and scala vestibuli.

hemipenes. [Gr. *hemi,* half + L. *penis,* a tail or a penis] Eversible intromittent organs of snakes and lizards.

hemocytoblast. [Gr. *hemo,* blood + *kytos,* hollow vessel + *blastos,* bud or sprout] The cell type from which all blood cells differentiate.

hemoglobin. [Gr. *hemo,* blood + *globulin,* from L. *globulus,* round body] The iron-containing protein of red blood cells which loosely binds oxygen.

hemopoiesis. [Gr. *hemo,* blood + *poiesis,* a making] The process of blood formation.

Henle, loop of. [Named for Friedrich G. J. Henle, a nineteenth century German anatomist] Long, thin intermediate segment of the mammalian nephron, characterized by its hairpin turn.

heterocelous. [Gr. *hetero,* other, different + *celom,* hollow] Centrum of a vertebra with saddle-shaped ends, characteristic of Aves.

heterocercal. [Gr. *hetero,* other, different + *kerkos,* tail] Describing a caudal fin where the axis extends dorsally and the fin itself is larger dorsally than ventrally.

heterodont dentition. [Gr. *hetero,* different + *odontos,* tooth] A condition in which tooth morphology varies with the region of the jaw; cf. homodont dentition.

heteroplastic. [Gr. *hetero,* different + *plastikos,* formed, molded] Describing a type of bone which forms within other, already differentiated tissues, such as tendons, ligaments, or heart muscle.

hippocampus. [Gr. *hippocampos,* seahorse] Portion of the telencephalon; in mammals it becomes folded deep to the lateral ventricle.

holobranch. [Gr. *holos,* whole, entire + *branch,* gill] A gill arch with lamellae on both sides.

Holocephali. [Gr. *holo,* entire + *kephalikos,* of the head] Subclass of

Chondrichthyes comprising the ratfish.

holocrine. [Gr. *holo,* entire + *krinein,* to separate] Referring to a gland or its secretion, characterized by the destruction of the gland cell during the release of the secretion.

holonephros. [Gr. *holos,* entire + *nephros,* kidney] A type of vertebrate kidney running the entire length of the body with but one tubule per segment. Syn., archinephros.

Holostei. [Gr. *holos,* entire + *osteon,* bone] Superorder of Actinopterygii with abbreviated heterocercal tail; e.g., *Amia.*

homeostasis. [Gr. *homoios,* like + *stasis,* a standing] The ability to maintain internal integrity unaffected by external environment.

homeotherm. [Gr. *homoios,* like + *therme,* heat] An animal capable of maintaining a constant internal body temperature.

homocercal. [Gr. *homos,* one and the same + *kerkos,* tail] Describing a caudal fin which is externally symmetrical although internally asymmetrical, characteristic of Teleostei.

homodont dentition. [Gr. *homo,* one and the same + *odontos,* tooth] A condition in which all the teeth have similar shape; cf. heterodont dentition.

homology. [Gr. *homo,* one and the same + *logos,* proportional] "The same structure in different animals under every variety of form and function." (Richard Owen)

hyaline. [Gr. *hyalinos,* of glass] Describing a homogeneous, translucent appearance.

hymen. [Gr. *hymen,* membrane] A thin fold of tissue partly occluding the vaginal orifice.

hypaxial. [Gr. *hypo,* beneath + L. *axis,* axis] Below or under any axis: in anatomy, the portion of the body which develops ventral to an axis formed by the developing notochord.

hyperstriatum. [Gr. *hyper,* above + *striatus,* furrow or channel] A portion of the avian cerebral hemisphere lying dorsally and superficially.

hypodermis. [Gr. *hypo,* beneath + *dermis,* skin] The deep, loose, fatty connective tissue of the skin. Syn., subcorium; subcutaneous tissue.

hypomere. [Gr. *hypo,* beneath + *meros,* part] The mesoderm of the hypaxial region.

hypophysis. [Gr. *hypo,* beneath + *phys,* grow; together, an outgrowth from below] Compound endocrine gland lying at the base of the hypothalamus. Syn., pituitary.

hypothalamus. [Gr. *hypo,* beneath + *thalamos,* a chamber] The most ventral portion of the diencephalon.

hypsodont. [Gr. *hypsos,* height + *odontos,* tooth] High-crowned teeth.

I

Ichthyopterygia. [Gr. *ichthys,* fish + *pterygion,* little wing or fin] Subclass of Reptilia comprised of extinct marine reptiles which externally resemble sharks.

ileum. [L. adaptation of Gr. *eileo,* to roll up, twist] The most posterior portion of the mammalian small intestine.

infundibulum. [L. *infundibulum,* funnel] The stalk of nervous tissue connecting the hypothalamus with the posterior pituitary.

ingestion. [L. *ingestus,* poured or thrown into] Process of bringing food into the digestive system.

innervation. [L. *in,* in + *nervus,* nerve + *-ate,* provided with] The distribution of nerves to an organ of the body.

insulin. [L. *insula,* island] The hormone of the beta cells of the islets of Langerhans; concerned with sugar metabolism.

integument. [L. *integumentum,* a covering] Compound organ including epidermis, dermis, and hypodermis. Syn., skin.

intercalary. [L. *intercalarius,* concerning an insertion] Occurring between two structures or events, such as intercalary plates between neural arches of sharks.

internal capsule. [L. *internus,* interior, away from the surface + *capsula,* diminutive of *capsa,* a chest or box] The mammalian fiber tract between the dorsal thalamus and the neocortex.

interrenal. [L. *inter,* between + *renes,* kidneys] The steroidogenic tissue of the adrenal gland or its homologue.

intromittent. [L. *intro,* inwardly + *mittere,* to send] Carrying into a body; used to describe the penis and analogous organs.

iridophore. [L. *irido-,* rainbow + Gr. *-phore,* bearer] A "pigment" cell filled with light-reflecting purine crystals.

iris. [L. *iris,* crystal] Portion of the eye usually pigmented, surrounding the pupil.

isolecithal. [Gr. *iso,* equal + *lekithos,* egg yolk] Describing an egg with scant, evenly distributed yolk.

J

jejunum. [L. *jejunus,* empty] Portion of the mammalian small intestine between the duodenum and ileum.

Jurassic. Middle of the three periods in the Mesozoic era, characterized by great diversity of large reptiles.

juxtaglomerular apparatus. [L. *juxta,* near + *glomerulus* (which see)] A group of myoepithelial cells adjacent to the kidney's glomerulus.

K

keratin. [Gr. *keras,* horn] A structural protein usually formed in the epidermis or its derivatives.

Kerckring, valves of. [L. *valva,* leaf of a folding door; named for Theodor Kerckring, seventeenth century Dutch anatomist] Large folds of the submucosa and mucosa in the small intestine of birds and mammals.

kinocilium. [Gr. *kineo,* move + L. *cilium,* an eyelid] Long, narrow cytoplasmic process with internal structure suggesting capability of movement.

L

Labyrinthodontia. [Gr. *labyrinthos,* labyrinth + *odontos,* tooth] An extinct subclass of Amphibia comprising the earliest known fossil amphibians.

Lacertilia. [L. *lacerta,* lizard] Suborder of Squamata comprised of lizards.

lacteal. [L. *lacteus,* milky] The blind lymphatic capillary of a villus.

lacuna. [L. *lacuna,* pit, dim. of *lacus,* a hollow, lake] A small space or cavity, such as the spaces in bone matrix containing osteocytes.

lagena. [L. *lagena,* flask] A portion of the nonmammalian inner ear, of still unknown function.

lamella. [L. dim. of *lamina,* leaf or plate] A thin sheet or layer.

Langerhans, islets of. [Named for Paul Langerhans, nineteenth century German anatomist] Endocrine portion of the pancreas.

lemniscus. [Gr. *lemniskos,* ribbon or fillet] A flattened tract of nerve fibers in the central nervous system.

Lepidosauria. [Gr. *lepidos,* scale + *sauros,* lizard] Subclass of reptiles comprised of Rhynchocephalia and Squamata.

leukocyte. [Gr. *leukos,* white + *kytos,* a hollow vessel] White blood cell.

Leydig's gland. [L. *glans,* acorn; named for Franz von Leydig, nineteenth century German anatomist] Intermediate portion of chondrichthyean kidney.

Lieberkühn, crypt of. [Gr. *kryptos,* hidden; named for Johann Lieberkühn, eighteenth century German anatomist] Deep grooves in the mucosa of the small intestine of birds and mammals.

ligamentum arteriosum. [L. *ligamentum,* band, tie + Gr. *arteria,* windpipe or artery] Ligament that remains after the closing of the ductus arteriosus, which see.

limbic lobe. [L. *limbus,* a border + Gr. *lobos,* lobe as of the ear or liver] Medial telencephalic structures of mammals, including hippocampus, septal nuclei, mammillary bodies, and cingulate cortex.

linea alba. [L. *linea,* line + *alba,* white] Midventral line of white connective tissue.

Lissamphibia. [Gr. *lissos,* smooth + *amphibia,* living a double life] The subclass of Amphibia which includes all the living forms.

lissencephalic cortex. [Gr. *lissos,* smooth + *enkephalos,* brain + L. *cortex,* bark] Smooth cerebral cortex. Opposite of gyrencephalic.

lophodont. [Gr. *lophos,* a crest + *odontos,* tooth] High-crowned teeth with transverse ridges, e.g., the molars and premolars of horses.

M

macula. [L. *macula,* spot] Sensory epithelium of either the sacculus or utriculus of the inner ear.

macula densa. [L. *macula,* spot + *densus,* thick] Specialized portion of the distal convoluted tubule of the kidney, concerned with controlling osmolarity.

Mammalia. [L. *mamma,* breast or teat] Class of vertebrates characterized by hair, mammary glands, and a squamosal–dentary jaw joint.

mastication. [L. *masticatus,* chewed; from *masticare,* to chew] The process of chewing.

median eminence. [L. *medius,* middle + *eminentia,* prominence] Ventral projection from the hypothalamus.

Meibomian gland. [L., *glans,* acorn; named for Hendrik Meibom, a seventeenth century German anatomist] Oil-secreting gland of the mammalian eyelid.

melanin. [Gr. *melas,* black] Brown, black, or rust-colored pigment.

melanophore. [Gr. *melas,* black + *-phore,* bearer] Pigment cell containing melanin.

merocrine. [Gr. *meros,* a part + *krino,* to separate] Describing a gland or its secretion in which the secretory product is released without damage to the cell.

mesencephalon. [Gr. *mesos,* middle + *enkephalos,* brain] Midbrain, composed of the dorsal tectum and ventral tegmentum.

mesentery. [Gr. *mesos,* middle + *enteron,* intestine] Double-walled serous membrane through which nerves and blood vessels pass to and from the alimentary canal or other viscera.

mesobronchus. [Gr. *mesos,* middle + *bronchos,* windpipe] The primary branch of the trachea in birds.

mesoderm. [Gr. *mesos,* middle + *derma,* skin] The middle of the three germ layers of the gastrula.

mesomere. [Gr. *mesos,* middle + *meros,* a part] The intermediate mesoderm between epimere and hypomere.

mesonephros. [Gr. *mesos,* middle + *nephros,* kidney] The intermediate nephrogenic tissue between pronephros and metanephros; also used to describe the functional kidney formed from this tissue.

mesorchium. [Gr. *mesos,* middle + *orchis,* testis] The mesentery of the testis.

Mesozoic. [Gr. *mesos,* middle + *zoion,* animal] Geologic era after the Paleozoic and before the Cenozoic, characterized by an abundance of reptiles, and the origin of mammals and birds.

metanephros. [Gr. *meta-,* after, between, among, or denoting change + *nephros,* kidney] Kidney formed from posterior nephrogenic tissue; the adult kidney of amniotes.

Metatheria. [Gr. *meta-,* after, between, among, or denoting change + *therion,* wild beast] The infraclass of Theria comprised of marsupials.

metencephalon. [Gr. *meta-,* after + *enkephalos,* brain] Anterior portion of the hindbrain, including cerebellum and pons.

metestrous. [Gr. *meta-,* after + *oistros,* mad desire] Portion of the estrous cycle following ovulation, characterized by the formation of the corpus luteum.

microfilament. [Gr. *mikros,* small + L. *filum,* thread] Submicroscopic strand within the cytoplasm, found in nerve fibers, muscle fibers, etc.

microtubule. [Gr. *mikros,* small + L. *tubus,* pipe] Submicroscopic cytoplasmic ductule, such as in nerve fibers.

microvillus. [Gr. *mikros,* small + L. *villus,* shaggy hair] Submicroscopic, very narrow cytoplasmic projection from an epithelial cell.

modality. [L. *modus,* manner] A term to designate any specific form of sensation, such as hearing, taste, etc.

modiolus. [L. *nave* of a wheel; literally, small measure; derived from *modus,* manner] The bony core of the cochlea.

morphology. [Gr. *morphe,* form + *-logy,* a combining form used in names of sciences, etc.] The branch of biology dealing with form or structure of organisms.

mucosa. [L. *mucosus,* slimy] The epithelium and lamina propria of the alimentary canal.

mucous (adj.) or **mucus** (noun). [L., *mucosus,* slimy] A clear viscid secretion, containing globular proteins and inorganic salts.

Müllerian duct. [L. *ductus,* a leading; named for Johannes Müller, a nineteenth century German anatomist] The oviduct, including its derivatives.

myelencephalon. [Gr. *myel-,* indicating relationship to spinal cord (or bone marrow) + *enkephalos,* brain] Posterior portion of the hindbrain, continuous with the spinal cord.

myelin. [Gr. *myel-,* indicating relationship to spinal cord (or bone marrow)] A mixed group of lipids formed in glial or Schwann cells surrounding nerve fibers.

myeloid tissue. [Gr. *myel-,* indicating relationship to bone marrow (or spinal cord)] Hemopoietic tissue producing red blood cells and granular white blood cells.

myocardium. [Gr. *mys,* muscle + *kardia,* heart] Musculature of the heart.

myoepithelium. [Gr. *mys,* muscle + *epi-,* upon + *thele,* nipple] Groups of cells derived from epithelium which surround glands and apparently possess contractile properties.

myofilament. [Gr. *mys,* muscle + L. *filum,* thread] Submicroscopic strand of contractile protein in a muscle cell.

myoglobin. [Gr. *mys,* muscle + L. *globus,* round body] A globular protein structurally and functionally similar to hemoglobin but found in muscle cells.

myotome. [Gr. *mys,* muscle + *tomos,* a cut] Portion of the epaxial mesoderm which gives rise to the epaxial musculature.

N

neocortex. [Gr. *neos,* new + L. *cortex,* bark] Six-layered mammalian cerebral cortex.

neostriatum. [Gr. *neos,* new + *striatus,* furrow or channel] A caudal portion of the avian telencephalon.

nephrostome. [Gr. *nephros,* kidney + *stoma,* mouth] The opening of a kidney tubule to the coelom, found in certain fishes.

nerve. [L. *nervus,* sinew, tendon] A group of nerve fibers running together in the peripheral nervous system.

neuroblast. [Gr. *neuron,* nerve + *blastos,* germ] Embryonic cell giving rise to neurons.

neurocranium. [Gr. *neuron,* nerve + *kranion,* skull] Endochondral portion of the skull which forms structures of the braincase and sense organ capsules.

neurohypophysis. [Gr. *neuron,* nerve + *hypophysis,* an undergrowth (*hypo* + *physis*)] Portion of the pituitary gland formed from the hypothalamus. Syn., pars nervosa.

neuron. [Gr. *neuron,* nerve] Nerve cell.

neuropil. [Gr. *neuron,* nerve + *pilos,* felt] Complex interweaving of axons and dendrites within the central nervous system.

niche. [Fr. an ornamental recess as in a wall, for a statue or similar object] The totality of environmental factors necessary for a species' continuity.

nidamental. [L. *nidamentum,* the materials of which a nest is made] The shell gland, formed from a specialized part of the Müllerian duct in Chondrichthyes, Reptilia, Aves, and Prototheria.

Nissl substance. [Named for Franz Nissl, early twentieth century German neurologist] Large aggregates of ribonucleoprotein in the cytoplasm of neurons.

noradrenaline. Hormone and neurotransmitter, similar in structure and function to adrenaline, which see. Syn., norepinephrine.

norepinephrine. See noradrenaline.

notochord. [Gr. *notos,* back + *chorde,* cord or string] The embryonic axial skeleton of all chordates. In most forms it is reduced or lost in the adult.

nucleus. [L. dim. of *nux,* nut, and hence meaning a little nut, the kernel of a nut, or the inside of something] 1. The portion of a cell which contains the chromosomes and nucleoplasm. 2. An aggregate of nerve cell bodies within the central nervous system.

O

octapeptide. [L. *octo,* eight + Gr. *pepsis,* digestion] Eight amino acid polypeptide.

odontoblast. [Gr. *odontos,* tooth + *blastos,* sprout, germ] Cell which produces dentine.

olfaction. [L. *olfactus,* smell] The sense of smell.

omasum. [L. *omasum,* bullock's tripe] Heavily muscular portion of a ruminant's stomach.

ontogeny. [Gr. *on,* being + *-gen,* production of] An individual's development from egg to adult.

oogonium. [Gr. *oon,* egg + *gone,* generation] Primary germ cell of the ovary.

operculum. [L. *operculum,* a lid or cover, from *operio,* to cover] Bony flap covering gill slits in actinopterygians.

Ophidia. [Gr. *ophis,* serpent + *idion,* diminutive suffix] Suborder of Squamata comprised of snakes.

optic tectum. [Gr. *optikos,* from *ops,* eye + L. *tectum,* roof (from *tego,* to cover)] The visual portion of the roof of the midbrain; called optic lobes in nonmammalian vertebrates and superior colliculi in mammals.

opisthocelous. [Gr. *opisthen,* behind or in back + L. *celom,* hollow] Referring to a vertebra whose centrum is concave posteriorly and convex anteriorly.

opisthonephros. [Gr. *opisthen,* behind or in back + *nephros,* kidney] The adult kidney in most anamniotes, derived from the intermediate and posterior nephrogenic tissue.

Ordovician. [Named for the Ordovices, Latin name of an ancient Celtic tribe in north Wales] Period of the Paleozoic era in which the earliest vertebrate fossils are found.

osmoregulation. [Gr. *osmos,* a pushing or thrusting + L. *regula,* rule or pattern] Control of the internal concentration of ions.

ossification. [L. *os,* bone] The formation of bone.

Ostariophysi. [Gr. *ostarion,* a little bone + *physis,* growth] The group of physostomous teleosts having Weberian ossicles; e.g., goldfish, catfish.

Osteichthyes. [Gr. *osteon,* bone + *ichthyes,* fishes] Class of bony fish including all actinopterygians and sarcopterygians.

osteoblast. [Gr. *osteon,* bone + *blastos,* bud, sprout] A cell which gives rise to osteocytes.

osteoclast. [Gr. *osteon,* bone + *klastos,* broken] A large, multinucleated cell capable of resorbing bone matrix.

osteocyte. [Gr. *osteon,* bone + *kytos,* a hollow vessel] A mature bone cell.

osteon. [Gr. *osteon,* bone] A single Haversian canal system within a bone.

ostium. [L. *ostium,* door, entrance, mouth] Any small opening to a cavity or canal; specifically referring to the opening of the Müllerian duct to the body cavity.

Ostracoderm. [Gr. *ostrakon,* shell + *derma,* skin] Order of agnathan fishes comprising the earliest known fossil vertebrates.

otolithic membrane. [Gr. *ous,* ear + *-lith,* noun ending meaning stone] The gelatinous membrane containing inorganic or organic crystals, overlying maculae of the inner ear.

oviparous. [L. *ovum,* egg + *pario,* to bear] Referring to animals whose fertilized eggs develop outside the mother.

ovoviviparous. [L. *ovum,* egg + *vivus,* alive + *pario,* to bear] Referring to animals whose fertilized eggs develop within the mother but gain no nourishment from the mother.

ovulation. [L. *ovulum,* little egg] Release of eggs from the ovary.

oxyhemoglobin. [oxy-, a combining form of oxygen + Gr. *hemoglobin* (*hemo,* blood + L. *globus,* ball or sphere)] Hemoglobin with loosely bound oxygen.

oxytocin. [Gr., *oxys,* sharp, keen + *tokos,* childbirth] Hormone released from the posterior pituitary causing the initiation of labor and the ejection of milk.

P

Pacinian corpuscle. [L. *corpusculum,* dim. of *corpus,* body; named for Filippo Pacini, Italian anatomist of the nineteenth century] A pressure-sensitive sense organ located in deep layers of the skin, alimentary canal, and mesenteries.

palaeoniscoid. [Gr. *palaios,* old] Generalized fish of the superorder Chondrostei, particularly those of the Paleozoic era.

paleontology. [Gr. *palaios,* old + *onto-,* being + *-logy,* science or body of knowledge] The study of fossils.

paleostriatum. [Gr. *palaios,* old + *striatus,* furrow or channel] (1) A deep portion of the avian telencephalon. (2) In mammals, the globus pallidus.

Paleozoic. [Gr. *palaios,* old + *zoion,* animal] The geologic era before the Mesozoic.

papilla. [L. *papilla,* nipple, dim. of *papula,* a pimple] Any small protuberance, usually with a canal opening at its tip. A dermal papilla is a vascularized projection of the dermis, approaching the surface of the epidermis.

parabronchus. [Gr. *para,* beside or near + *bronchos,* windpipe] Tertiary branch or bronchiole of the avian lung.

parachordal. [Gr. *para,* beside or near + *chorde,* cord] Paired cartilaginous plates of the neurocranium forming under the hindbrain.

parafollicular cell. [Gr. *para,* beside or near + L. *folliculus,* a small sac] A cell adjoining follicle cells of the thyroid or parathyroid and secreting calcitonin.

parasympathetic nervous system. [Gr. *para,* beside or near + *syn,* with + *pathetikos,* sensitive] Portion of the autonomic nervous system originating from cranial and sacral regions, having acetylcholine as its neurotransmitter. It increases vegetative functions and inhibits somatic functions.

parathyroid. [Gr. *para,* beside or near (the thyroid)] Endocrine glands concerned with calcium and phosphate metabolism.

parenchyma. [Gr. *parencheo,* to pour in beside] The distinguishing cells of an organ; e.g., the parenchymal cells of bone are osteocytes, those of liver, hepatic cells.

pelage. A French word meaning the hair or fur of a mammal.

penis. [L. *penis,* a tail, or the penis] The male intromittent organ of mammals, chelonians, crocodilians, and ratite birds.

Permian. [Named for Perm, former province of Russia] Most recent period of the Paleozoic Era.

pericardium. [Gr. *perikardion,* from *peri,* around + *kardia,* heart] The serous membranes lining the pericardial (heart) cavity.

perichondrium. [Gr. *peri,* around + *khondrion,* dim. of *khondros,* cartilage] The fine connective tissue coating of a cartilage.

perikaryon. [Gr. *peri,* around + *karyon,* nut, kernel] The central portion, or cell body, of a neuron.

perimysium. [Gr. *peri,* around + *mys,* muscle] Fine, loose connective tissue separating bundles of muscle fibers within a muscle.

periosteum. [Gr. *peri,* around + *osteon,* bone] Fine connective tissue coating of a bone.

peristalsis. [Gr. *peri,* around + *stalsis,* constriction] The rhythmic contractions of smooth muscles lining the gut.

peritoneum. [Gr. *peritonaion,* from *periteino,* to stretch over] Serous membrane lining the peritoneal and pleuroperitoneal cavities.

pessulus. [L. *pessulus,* a bolt] Small bone of the avian syrinx, supporting one of the vibrating membranes.

pharyngula. [Gr. *pharynx,* throat] Developmental stage of vertebrates following gastrulation, when the basic body plan is morphologically evident.

pharynx. [Gr. *pharynx,* throat] The most anterior portion of the endodermally lined gut, characterized by the laterally extending pharyngeal pouches.

photopigment. [Gr. *phos,* light + L. *pigmentum,* paint] Molecules reactive to wavelengths of visible light.

phylogeny. [Gr. *phylon,* race, tribe + *-geneia,* origin] The evolutionary history of an organism.

physoclistous. [Gr. *physa,* bladder + *kleistos,* shut] Describing those teleost fish which lack a pneumatic duct as adults.

physostomous. [Gr. *physa,* bladder + *stoma,* mouth] Describing those teleost fish which retain a pneumatic duct as adults.

pilaster cell. [L. *pilastrum,* dim. of *pila,* pillar] Specialized cells lining the blood channels in the secondary lamellae of teleost gills.

pineal. [L. *pinus,* pine] A gland of questionable function in the epithalamus.

piriform cortex. [L. *pirum,* pear + *forma,* form + *cortex,* bark] Surface gray matter on the ventrolateral surface of the telencephalon; it receives olfactory information.

pituicyte. [L. *pituitarius,* pertaining to or secreting phlegm + Gr. *kytos,* a hollow vessel] The parenchymal cells of the pars nervosa of the pituitary.

Placoderm. [Gr. *plax,* anything flat or broad + *derma,* skin] Class of extinct fish; the first vertebrates to evolve jaws and paired appendages.

placoid denticles. [Gr. *plax,* flat plate + L. *denti-,* having to do with tooth + *-culus,* dim. suffix] Small, hard appendages of the chondrichthyean integument; similar in structure to teeth.

plasticity. [Gr. *plastikos,* formed, molded] In anatomy, the capability of being molded or changing form.

pleurodont dentition. [Gr. *pleuron,* side + *odontos,* tooth] Condition of having teeth attached to the jaws by one side of the root only, as seen in many lizards.

plumage. [L. *pluma,* soft feathers; plural, down] All the feathers a bird has at any one time.

pneumatic duct. [Gr. *pneuma,* air or gas + L. *ductus,* a leading] Duct connecting the alimentary canal and the gas bladder in physostomous fishes.

podocyte. [Gr. *pous,* foot + *kytos,* a hollow vessel] Specialized cell of the inner wall of Bowman's capsule, facilitating filtration.

poikilotherm. [Gr. *poikilos,* varied + *therme,* heat] Animals not capable of maintaining a constant internal body temperature independent of external temperature.

polyphyodont dentition. [Gr. *polys,* many + *phyo,* to produce + *odontos,* tooth] Describing the ability to replace lost teeth an indefinite number of times.

pons. [L. *pons,* bridge] Structure in the base of the metencephalon containing cells and fibers which send information to the cerebellum.

porphyropsin. [Gr. *porphyra,* purple + *opsis,* sight] The photosensitive pigment of rods in fresh-water vertebrates.

procelous. [Gr. *pro,* before or in front of + *celom,* hollow] Describing vertebrae whose centra have concave anterior surfaces and convex posterior surfaces.

proctodeum. [Gr. *proktos,* anus + *hodaios,* on the way] The portion of cloaca or posterior alimentary canal which is formed by an invagination of surface ectoderm.

proestrous. [Gr. *pro,* before or in front of + *oistros,* mad desire] Portion of the estrous cycle preceding ovulation, characterized by maturation of the eggs and growth of the entire reproductive tract.

progesterone. [Gr. *pro,* in front of or before + L. *gestatus,* to bear] Female hormones produced by ovary and placenta which maintain pregnancy and, later, lactation.

prolactin. [Gr. *pro,* before + L. *lac,* milk] Hormone of the anterior pituitary stimulating sexual behavior, synthesis of milk, and maintenance of corpus luteum; found in both males and females. Syn., lactogenic hormone.

pronephros. [Gr. *pro,* before or in front of + *nephron,* kidney] The functional kidney developed from the most anterior nephrogenic tissue.

proprioception. [L. *proprius,* one's own + *capio,* to take] The receiving of sensory information by the nervous system, from end organs located in muscles, ligaments, and joints.

prosencephalon. [Gr. *pros,* before + *enkephalos,* brain] Forebrain, made up of olfactory bulbs, cerebral hemispheres, and diencephalon.

prostate. [Gr. *prostates,* one standing before] Accessory male sex gland in mammals.

protopathic. [Gr. *protos,* first + *pathos,* suffering] Describing the nerve fibers and cell bodies concerned with the reception and transmission of pain and temperature.

Prototheria. [Gr. *protos,* first + *therion,* wild beast] Subclass of Mammalia comprised of duck-billed platypus and spiny anteater; i.e., the monotremes, egg-laying mammals.

proventriculus. [L. *pro,* before + *ventriculus,* dim. of *venter,* belly] Glandular portion of the avian stomach.

Purkinje fiber. [Named for Johannes E. von Purkinje, Bohemian anatomist and physiologist of the early nineteenth century] (1) Axons of the large, multipolar Purkinje cells of cerebellar cortex. (2) Specialized neuromuscular fibers in the heart which conduct and coordinate contractions.

putamen. [L. *putamen,* that which falls off in pruning] The most lateral portion of the basal ganglia of mammals.

pygostyle. [Gr. *pyge,* rump + *stylos,* pillar] The abbreviated caudal vertebrae of birds, which forms a skeletal support of the tail feathers.

pyloric caecum (pl., caeca). [Gr. *pyloros,* gatekeeper (from *pyle,* gate + *ouros,* a warder) + L. *caecus,* blind] Blind diverticulum of the alimentary canal of teleosts, leaving from the small intestine just behind the pyloric sphincter and providing extra surface area for absorption.

Q

quadruped. [L. *quattuor,* four + *pes,* foot] Four-footed animal.

Quarternary. [L. *quartus,* fourth] The most recent (and current) period of the Cenozoic Era, only about one million years old.

R

rachis. [Gr. *rhachis,* spine, ridge] The portion of the shaft of a contour feather which supports the vane.

ramus (pl., rami). [L. *ramus,* branch] A primary division of a nerve or blood vessel.

Rathke's pouch. [Named for Martin H. Rathke, nineteenth century German anatomist] Evagination of the stomodeum that forms the anterior pituitary.

reflex. [L. *reflexus,* bent back] A fixed motor pattern resulting from a specific sensory stimulus.

renin. [L. *renes,* kidneys] A proteolytic enzyme released by the kidney into the bloodstream.

Reptilia. [L. *reptilis,* creeping] Class of vertebrates; first animals to have evolved a cleidoic egg.

rete mirabile. [L. *rete,* net + *mirabilis,* marvelous] An extremely tortuous vascular network, usually involving hairpin turns and the interruption of an arterial continuity.

reticulum. [L. *reticulum,* little net (from *rete,* net) + *-culum,* dim. suffix] A fine network of connective tissue fibers and cells.

retroperitoneal. [L. *retro,* backward, behind + Gr. *peritones,* stretched around] Describing the position of an organ in which one surface abuts a serous membrane and the other surface, somatic muscles of the body.

Rhipidistii. [Gr. *rhipidos,* a fan] Extinct group of sarcopterygian fish which gave rise to the first amphibians.

rhodopsin. [Gr. *rhodon,* rose + *opsis,* sight] Visual pigment in the rods of salt-water and terrestrial vertebrates.

rhombencephalon. [Gr. *rhombos,* a diamond-shaped figure + *enkephalos,* brain] The hindbrain, composed of metencephalon and myelencephalon; includes medulla oblongata, pons, cerebellum.

Rhynchocephalia. [Gr. *rhynchos,* snout + *kephalikos,* of the head] An order of Reptilia; the only living member is *Sphenodon.*

rod. Photoreceptor of the eye sensitive to black and white.

ruga. [L. *ruga,* a wrinkle] A fold, particularly of stomach mucosa.

rumen. [L. *rumen,* throat, gullet] Largest chamber of ruminant's stomach, harboring massive numbers of symbiotic bacteria.

S

sacculus. [L. dim. of *saccus,* sack] Chamber of the vestibular portion of the membranous labyrinth of the inner ear.

sacral. [L. *os sacrum,* holy bone; so called because of its use in sacrifices] Portion of the vertebral column attached to the pelvic girdle.

saltatorial. [L. *saltare,* to jump about] Referring to a leaping mode of locomotion.

sarcolemma. [Gr. *sarx,* flesh + *lemma,* shell or husk] Cell membrane of a muscle cell.

sarcoplasm. [Gr. *sarx,* flesh + *plasma,* something molded or formed] The cytoplasm of a muscle cell.

Sarcopterygii. [Gr. *sarx,* flesh + *pterygi,* wing or fin] Subclass of Osteichthyes comprised of the lobe-finned fishes; i.e., dipnoans, coelacanths, and rhipidistians.

scala naturae. [L. *scala,* ladder + *natura,* nature] The arrangement of living beings into a single linear scale, from simple to complex; now discredited.

scale. [Fr. *écaille,* fish-scale] Overlapping dermal and/or epidermal hardenings, formed in skin folds.

sclera. [Gr. *skleros,* hard] The outer tough connective tissue layer of the eye.

sclerotome. [Gr. *skleros,* hard + *tomos,* a cut] Portion of the epimere giving rise to the vertebral column.

sebaceous gland. [L. *sebum,* grease; Low Latin, *sebaceous,* like lumps of tallow] Holocrine lipid-secreting gland.

secretin. [L. *secretus,* past participle of *secernare,* to separate] A "hormone" produced by the duodenum which stimulates the flow of pancreatic enzymes.

Selachii. [Gr. *selachos,* shark] The order of Elasmobranchii comprised of sharks.

selenodont. [Gr. *selene,* moon + *odontos,* tooth] Teeth usually high-crowned with longitudinal crescent-shaped ridges.

septal nuclei. [L. *septum,* a partition] Anterior ventral medial portion of the telencephalon.

serous. [L. *serum,* whey] Relating to or containing serum or a thin, watery, proteinaceous fluid.

Silurian. [Named after territory presumably held by the Silures in southeast Wales] Period of the Paleozoic Era following the Ordovician Period.

sinoatrial node. [L. *sinus,* sinus + *atrium,* entrance court] The "pacemaker"; the neuromuscular tissue initiating waves of contractions of the heart.

sinusoid. [L. *sinus,* sinus] Capillary with greatly expanded lumen.

sinus venosus. [L. *sinus,* sinus + *venosus,* vein] The chamber of the heart in fishes, amphibians, and reptiles which receives the systemic venous blood.

somatotropin. [Gr. *somato,* body + *trophe,* nourishment] Hormone secreted by the anterior pituitary, significant in stimulating bodily growth. Also called the growth hormone.

spermatid. [Gr. *sperma,* seed + *id,* young] A haploid male germ cell before it has matured into a spermatozoon.

spermatocyte. [Gr. *spermo,* seed + *kytos,* hollow vessel] Cells arising from a spermatogonium that will give rise to spermatozoa.

spermatogonium. [Gr. *spermo,* seed + *gone,* generation] The diploid stem germ cells of the testis.

spermatophore. [Gr. *spermo,* seed + *phoros,* bearing] The mucous secretion of male salamanders, containing a packet of spermatozoa.

spermatozoon (plural, -zoa). [Gr. *spermo,* seed + *zoon,* animal] Mature male germ cells.

Sphenodon. [Gr. *sphen,* wedge + *odontos,* tooth] Genus of the only living Rhynchocephalian.

splanchnocranium. [Gr. *splanchna,* entrails or viscera + *kranion,* skull] Portion of the skull derived from neural crest and forming in relationship to the pharynx.

Squamata. [L. *squama,* scale] Order of reptiles containing lizards and snakes.

stereocilia. [Gr. *stereos,* solid + L. *cilium,* eyelid] Cytoplasmic processes having the shape of kinocilia but lacking their internal structure.

steroidogenic tissue. [steroid + Gr. *genesis,* production] Endocrine tissue of the adrenal cortex or its homologues, that secretes steroid hormones.

stomodeum. [Gr. *stoma,* mouth + *hodaios,* on the road] Ectodermal invagination at the presumptive mouth, whose epithelium lines the oral cavity.

subcorium. [L. *sub,* under + *corium,* skin] See subcutaneous tissue.

subcutaneous tissue. [L. *sub,* under + *cutis,* skin] The deep, loose, fatty connective tissue of the skin. Syn., hypodermis, subcorium.

submucosa. [L. *sub,* under + *mucosus,* mucus] Dense connective tissue layer of the alimentary canal.

substantia nigra. [L. *substantia,* substance + *niger,* black] Nuclear area of the mammalian brain forming the upper part of the cerebral peduncle.

subunguis. [L. *sub,* under + *unguiculus,* nail or claw] Soft keratin underportion of a claw, nail, or hoof.

sudoriferous gland. [L. *sudor,* sweat + *ferare,* to bring forth] Sweat gland.

sulcus (pl., sulci). [L. *sulcus,* furrow or ditch] A groove on the surface of the cerebral hemisphere.

sulcus limitans. [L. *sulcus,* furrow or ditch + *limes,* limit] Groove in the lateral aspect of the developing central canal and ventricular system of the central nervous system, separating dorsal, sensory portions from ventral, motor portions.

sustentacular. [L. *sustento,* to hold upright] Supporting, as in a sustentacular cell.

sympathetic nervous system. [Gr. *syn,* with + *pathetikos,* sensitive] Portion of the autonomic nervous system originating from thoracic and lumbar regions, having adrenaline or noradrenaline as its neurotransmitter. It increases somatic functions and inhibits vegetative functions.

symphysis. [Gr. *syn,* with + *physis,* growth] A joint between two bones, formed by fibrocartilage.

synapse. [Gr. *synapsis,* junction] The approximation of nerve to nerve or nerve to muscle, across which information is passed.

synapsid. [Gr. *syn,* with + *apsis,* a loop or mesh] Referring to reptiles with a single temporal fenestra ventral to the postorbital–squamosal suture.

synarthrosis. [Gr. *syn,* with + *arthrosis,* articulation] An immovable joint.

synergy. [Gr. *syn,* with + *ergon,* work] Coordinated activity of different parts.

syrinx. [Gr. *syrinx,* a pipe] Vocal organ of birds, located at the bifurcation of the trachea.

T

tapetum lucidum. [Gr. *tapes,* carpet + L. *lucidus,* from *lux,* light] A portion of the choroid layer of the eye which reflects light, found in various vertebrates.

taxon (pl., taxa). [Derived from taxonomy] Any of the formal categories used in classifying organisms.

taxonomy. [Fr. *taxonomie,* from Gr. *taxis,* from *tassein,* to arrange or put in order] The branch of biological sciences dealing with the classification of organisms.

telencephalon. [Gr. *telos,* end + *enkephalos,* brain] Portion of the brain which develops from the anterior end of the central nervous system; cerebral hemispheres and olfactory bulbs.

tectorial membrane. [L. *tectorium,* a covering, from *tego,* to cover + *membrane,* fine skin] Gelatinous membrane of the cochlea, onto which the tips of the stereocilia adhere.

tectum. [L. *tectum,* roof] The roof of the midbrain, derived from its alar plate.

tegmentum. [L. *tegmentum,* cover] The floor of the midbrain, derived from its basal plate.

teleology. [Gr. *telos,* end + *-logy,* suffix for branch of knowledge] The study of events or processes which assumes that each has its ultimate goal or purpose.

Teleostei. [Gr. *telos,* end + *osteon,* bone] The superorder of Actinopterygii characterized by having a terminal mouth, a homocercal tail, and thin, stiff scales.

telodendria. Gr. *telos,* end + *dendron,* tree] The branching of nerve endings at the end of an axon.

telolecithal. [Gr. *telos,* end + *lekithos,* yolk] Eggs with a very large amount of yolk which is separated from the cytoplasm.

Tertiary. [L. *tertiarius,* of third part or rank] The first geologic period of the Cenozoic era, characterized by the great diversification of mammals, birds, and teleosts.

testosterone. Male hormone produced by the interstitial cells of the testis.

tetraiodothyronine. The primary hormone produced by the thyroid gland, which causes an increase in an animal's metabolic rate. Syn., thyroxine.

thalamus. [Gr. *thalamos,* a bed or bedroom] That portion of the diencephalon receiving information from the brainstem and sending massive projections to the cerebral hemispheres. Syn., dorsal thalamus.

thecodont dentition. [Gr. *theke,* box + *odontos,* tooth] The condition of having teeth whose roots extend into sockets in the bone.

Theria. [Gr. *therion,* wild beast] Subclass of the class Mammalia, which includes both placentals and marsupials.

thyrocalcitonin. The hormone produced by parafollicular cells of the thyroid gland. It causes a lowering of blood calcium levels.

thyroglobulin. Major chemical component of the thyroid colloid; it is a large globular protein involved in the synthesis of thyroxine.

thyroxine. [Gr. *thyreos,* large shield + *-eides,* form] The primary hormone produced by the thyroid gland, which causes an increase in an animal's metabolic rate. Syn., tetraiodothyronine.

tone. [Gr. *tonos,* tone, strain; literally a stretching, as *teinein,* to stretch] Term designating the state of partial contraction of a muscle.

torus semicircularis. [L. *torus,* bulge or

swelling] Portion of the nonmammalian midbrain just deep to the optic ventricle.

tract. [L. *tractus,* a drawing out] A group of nerve fibers running together in the central nervous system.

transduce. [L. *trans,* across + *ducere,* to lead] To change form of energy, e.g., light to nerve impulses.

transformer ratio. [L. *trans,* across + *forma,* form or shape] Term used in quantifying the pressure amplification of the middle ear apparatus.

Triassic. [L. *trias,* triad; so named for its three divisions] First period of the Mesozoic era, characterized by the beginning of the great adaptive radiation of reptiles.

triiodothyronine. An active form of the thyroid hormone, which contains three iodine atoms whereas thyroxine has four.

tuber cinereum. [L. *tuber,* bump, swelling + *cinereus,* ashy gray] The region of the hypothalamus surrounding the infundibulum.

typhosole. [Gr. *typhos,* snake] A fold along the lumen of the small intestine of the lamprey.

typology. [L. *typus,* type; from Gr. *typos,* blow or impression] An approach to taxonomy which assumes that the structure of all members of the species is similar and individual variation is an abnormality.

U

ultimobranchial body. [L. *ultimus,* last + Gr. *branch,* gill] Endocrine glands of fish producing calcitonin. Also designating the embryonic precursors of the parafollicular cells of the thyroid gland of tetrapods.

undulation. [L. *undula,* dim. of *unda,* wave] Movement in a curved, wavy pattern.

unguis. [L. *unguiculus,* nail or claw] The hard keratin upper part of a claw, hoof, or nail.

ureter. [Gr. *oureter,* from *ourein,* to urinate] The duct draining the metanephric kidney.

urethra. [same derivation as *ureter*] The duct draining the mammalian urinary bladder.

Urodela. [Gr. *oura,* tail] Order of Amphibia comprised of salamanders.

urophysis. [Gr. *oura,* tail + *phys,* grow] The enlarged caudal tip of the spinal cord of teleosts, chondrosteans, and elasmobranchs, which is apparently an endocrine gland.

uropygeal gland. [Gr. *oura,* tail + *pyge,* rump or buttocks] Large, lipid-secreting, holocrine glands found at the base of the pygostyle in birds.

urostyle. [Gr. *oura,* tail + *stylos,* pillar] The bone formed from fused caudal vertebrae in Anura.

utriculus. [L. *utriculus,* dim. of *uter,* bag] The largest sac-like portion of the membranous labyrinth of the inner ear.

V

vagina. [L. *vagina,* sheath] Distal end of the mammalian Müllerian duct; that portion which receives the penis during intercourse.

valvula. [Mod. L. dim. of *valva,* leaf of a folding door] The portion of the teleost cerebellum which extends into the optic ventricle.

vasopressin. [L. *vas,* a vessel] Hormone released by posterior pituitary, functioning to increase water resorption and stimulate contraction of smooth muscle. Syn., antidiuretic hormone.

ventilation. [L. *ventilare,* to fan] The process of bringing fresh air or water to the respiratory membrane.

vibrissa. [L. *vibrare,* to shake] The stiff tactile hairs of the facial region of mammals; e.g., cat's "whiskers."

villus. [L. *villus,* shaggy hair] A projection of epithelium and lamina propria toward the lumen of the small intestine in birds and mammals.

viviparous. [L. *vivus,* alive + *pario,* to

bear] Describing the condition of having young retained in the uterus or analogous organ of the female and receiving nourishment from the maternal bloodstream during development.

vomeronasal organ. [Named for the vomer and nasal bones] A chemical-sensitive sense organ adjacent to the olfactory organ, found in most tetrapods.

W

Wolffian duct. [L. *ductus,* a leading; named for Kaspar F. Wolff, German embryologist in Russia in the eighteenth century] The duct which drains the embryonic kidney in all vertebrates; it has various functions in the adult.

wulst. [Ger. *wulst,* bulge, prominence] A dorsomedial elevated portion of the avian telencephalon.

X

xanthophore. [Gr. *xanthos,* yellow + *phoros,* bearing] Pigment cell containing carotenoid or pteridine pigments.

Z

zygapophysis. [Gr. *zygon,* yoke + *apophysis,* offshoot] The articulating process of a tetrapod vertebra.

Index

B

C

D

E

H

I

N

O

P

Q

R

S

T

U

CAMROSE LUTHERAN COLLEGE
LIBRARY

B C D E F G H I J 4 5 6 7 8 9 0 1 2 3